THE LANDS OF THE EASTERN CALIPHATE

THE LANDS OF THE EASTERN CALIPHATE

Mesopotamia, Persia, and Central Asia
from the Moslem conquest to the time of Timur

by

G. LE STRANGE
Author of *Baghdad during the Abbasid Caliphate*,
Palestine under the Moslems etc.

WIPF & STOCK · Eugene, Oregon

Wipf and Stock Publishers
199 W 8th Ave, Suite 3
Eugene, OR 97401

The Lands of the Eastern Caliphate
Mesopotamia, Persia, and Central Asia
from the Moslem conquest to the time of Timur
By Le Strange, G.
Softcover ISBN-13: 978-1-6667-7307-1
Hardcover ISBN-13: 978-1-6667-7308-8
eBook ISBN-13: 978-1-6667-7309-5
Publication date 3/6/2023
Previously published by Cambridge, 1930

This edition is a scanned facsimile of the original edition published in 1930.

PREFACE.

IN the following pages an attempt is made to gather within a convenient compass the information scattered through the works of the medieval Arab, Persian, and Turkish geographers, who have described Mesopotamia and Persia, with the nearer parts of Central Asia. The authorities quoted begin with the earlier Moslem writers, and conclude with those who described the settlement of these lands which followed after the death of Tîmûr,—the last great Central Asian wars of conquest,—for with the fifteenth century the medieval period in Asia may be said to come to an end.

The present work is also the complement of *Baghdâd under the Abbasid Caliphate* published in 1900, and carries forward the geographical record which I began in *Palestine under the Moslems*, a work that appeared in 1890.

To keep the volume within moderate compass, the geography of Arabia, with the description of the two Holy Cities of Mecca and Medina, though these for the most part were under the dominion of the Abbasids, has been omitted. Perhaps some other scholar may take up the subject, with fuller knowledge than I have, and write the historical geography of Arabia with Egypt across the Red Sea under the Fatimid Caliphs; completing the circuit of Moslem lands by describing the various provinces of North Africa, with the outlying and shortlived, though most splendid, western Caliphate of Spain.

If Moslem history is ever to be made interesting, and indeed to be rightly understood, the historical geography of the nearer East during the middle-ages must be thoroughly worked out. I have made a first attempt, but how much more needs to be done, and better done than in the present volume, I am the first to recognise. The ground, however, for future work is now cleared; the authorities for each statement are given in the footnotes; some mistakes are corrected of previous writers, and a beginning made of a complete survey for this period of the provinces of the Abbasid Caliphate. But my book is only a summary, and does not pretend to be exhaustive; also to keep down the size, I have been obliged to omit translating in full the Itineraries, which our Moslem authorities give us. In this matter a new edition, duly corrected from recently published texts, is indeed much needed of Sprenger's *Post und Reise Routen des Orients*, though the translation of the Itineraries which Professor De Goeje has appended to his edition of Ibn Khurdâdbih and Ḳudâmah, goes far to supply the lack.

With each province I have given such information as our authorities afford of the trade and manufactures; the record, however, is very fragmentary, and for a general survey of the products of the Moslem east, during the middle-ages, the chapter on the subject (*Handel und Gewerbe*) in A. von Kremer's *Culturgeschichte des Orients* is still the best that I know.

A chronological list of the Moslem geographers referred to in the notes by initial letters is given at the end of the Table of Contents. The fuller titles of other works quoted in the notes are given on the first reference to each author, and the names of their works will easily be recovered, for subsequent references, by consulting the index for the first mention made of the book.

In the introductory chapter a summary description will be found of the works of the Arab geographers;

but this matter has already been more fully discussed in *Palestine under the Moslems.*

The dates are given according to the years of the Hijrah, with the corresponding year A.D. (in brackets). The method of transcription adopted needs no comment, being that commonly in use; it may be noted that the Arab *w* is usually pronounced *v* in Persian; and that besides the emphatic *ẓ* the Arab *dh* and *ḍ* are both indifferently pronounced *z* in modern Persian, while the *th* has the sound of *s*.

In a work like the present, almost entirely composed from eastern sources, many errors will doubtless be found; also, with the great number of references, mistakes are unavoidable, and I shall feel most grateful for any corrections, or notice of omissions.

My hope is that others may be induced to set to work in this field of historical geography, and if this essay be soon superseded by a more complete survey of the ground, it will have served its purpose in having prepared the way for better things.

G. LE STRANGE.

3, Via San Francesco Poverino,
Florence, Italy.
May, 1905.

CONTENTS.

CHAPTER I.

INTRODUCTORY.

CHAPTER II.

'IRÂḲ.

CHAPTER III.

'IRÂḲ (*continued*).

CHAPTER IV.

'IRÂḲ (*continued*).

CHAPTER V.

'IRÂḲ (*continued*).

CHAPTER VI.

JAZÎRAH.

CHAPTER VII.

JAZÎRAH (*continued*).

CHAPTER VIII.

THE UPPER EUPHRATES.

CHAPTER IX.

RÛM OR ASIA MINOR.

CHAPTER X.

RÛM (*continued*).

CHAPTER XI.

ADHARBÂYJÂN.

CHAPTER XII.

GÎLÂN AND THE NORTH-WEST PROVINCES.

CHAPTER XIII.

JIBÂL.

CHAPTER XIV.

JIBÂL (*continued*).

CHAPTER XV.

JIBÂL (*continued*).

CHAPTER XVI.

KHÛZISTÂN.

CHAPTER XVII.

FÂRS.

CHAPTER XVIII.

FÂRS (*continued*).

CHAPTER XIX.

FÂRS (*continued*).

CHAPTER XX.

FÂRS (*continued*).

CHAPTER XXI.

KIRMÂN.

CHAPTER XXII.

KIRMÂN (*continued*).

CHAPTER XXIII.

THE GREAT DESERT AND MAKRÂN.

CHAPTER XXIV.

SIJISTÂN.

CHAPTER XXV.

ḲÛHISTÂN.

CHAPTER XXVI.

ḲÛMIS, ṬABARISTÂN AND JURJÂN.

CHAPTER XXVII.

KHURÂSÂN.

CHAPTER XXVIII.

KHURÂSÂN (*continued*).

CHAPTER XXIX.

KHURÂSÂN (*continued*).

CHAPTER XXX.

KHURÂSÂN (*continued*).

CHAPTER XXXI.

THE OXUS.

CHAPTER XXXII.

KHWÂRIZM.

CHAPTER XXXIII.

SUGHD.

CHAPTER XXXIV.

THE PROVINCES OF THE JAXARTES.

LIST OF MAPS.

ABBREVIATIONS AND CHRONOLOGICAL LIST OF MOSLEM GEOGRAPHERS.

				A. H.	A. D.
I. K.	...	Ibn Khurdâdbih	...	250	(864)
Kud.	...	Ḳudâmah	...	266	(880)
Ykb.	...	Ya'ḳûbî	...	278	(891)
I. S.	...	Ibn Serapion	...	290	(903)
I. R.	...	Ibn Rustah	...	290	(903)
I. F.	...	Ibn Faḳîh	...	290	(903)
Mas.	...	Mas'ûdî	...	332	(943)
Ist.	...	Iṣṭakhrî	...	340	(951)
I. H.	...	Ibn Ḥawḳal	...	367	(978)
Muk.	...	Muḳaddasî	...	375	(985)
N. K.	...	Nâṣir-i-Khusraw	...	438	(1047)
F. N.	...	*Fârs Nâmah*	...	500	(1107)
Idr.	...	Idrîsî	...	548	(1154)
I. J.	...	Ibn Jubayr	...	580	(1184)
Yak.	...	Yâḳût	...	623	(1225)
Kaz.	...	Ḳazvînî	...	674	(1275)
Mar.	...	*Marâṣid*	...	700	(1300)
A. F.	...	Abu-l-Fidâ	...	721	(1321)
Mst.	...	Mustawfî	...	740	(1340)
I. B.	...	Ibn Baṭûṭah	...	756	(1355)
Hfz.	...	Ḥâfiẓ Abrû	...	820	(1417)
A. Y.	...	'Alî of Yazd	...	828	(1425)
J. N.	...	*Jahân Numâ*	...	1010	(1600)
A. G.	...	Abu-l-Ghâzî	...	1014	(1604)

EMENDATIONS (1930)

p. 63, line 2

The naphtha spring of Khâniḳîn is at the modern Nafṭ Khânah. (R. Levy.)

line 11 from bottom

Bandanîjîn is the modern Mendelîj or Mendelî; and Bâdarâyâ is modern Badrâi. Both these names are found on the Survey of India Map, 1914. (R. Levy.)

p. 65, line 23

Shafâthâ is an oasis lying 40 miles west of Karbalâ: 'Ayn-at-Tamr has not been identified.

p. 74, line 13

The site of Pombedita is probably to be sought for at Kifl on the Shaṭṭ-al-Hindiyah where the tomb of the Prophet Ezekiel is shown.

p. 86, line 7

The settlement of the tribes of Rabî'ah and Bakr in Upper Mesopotamia, according to Balâdhurî (p. 178), took place after Sassanian days at the time of the Moslem Conquest. (A. A. Bevan.)

p. 111, line 24

For "Maypharkath" read "Mayperkat." See W. Wright, *Catalogue of Syriac MSS*, p. 1304. (A. A. Bevan.)

p. 141, line 4 from bottom

Read "drowned in the river Calycadnus near Salûḳîyah (Seleucia of Cilicia)," and omit next three lines to foot of page.

p. 144, line 8 from bottom

For "Bythia" read "Bithynia."

p. 155, line 5

After "*Agiou Theologou*" add as a note: "Hence not Loukas as often stated wrongly."

p. 167, line 6

Delete "cochineal" and insert "see below p. 184."

p. 168, line 17

The mountain called Sarâhand lay on the Ḳiblah (south-west) side of Abhar: see below p. 222. Also p. 169, line 3, for "north" read "south."

p. 175, line 20

For "east" read "west" coast of Caspian.

p. 182, line 11 add as note

Erivan (at present the chief town of the province) was founded as a frontier fortress by Rivan (or Erivan) Khân in the days of Shâh Isma'îl I (1502 to 1524). See Hammer Purgstall, *Geschichte des osmanischen Reiches*, IV, 85.

p. 184, line 12 add as note

The Ḳirmiz insect (Kermes ilicis of Linnæus) feeds on the leaves of the oak, and is allied to the American cochineal insect which lives on the cactus.

p. 187, line 9 from bottom

For "Here" read "Three leagues from here."

p. 195, line 14

For "Juhastah" read "Jûhastah," and line 21, for "Ẓabîyah" read "Zabyah."

p. 222, line 7

After "Abhar river" add "(which took its rise in the Sarâhand mountain to the south-west)."

p. 329

"Mukrân" is the more exact spelling.

p. 393. Add to footnote 2

The ruins of Shahr-i-Bilḳîs have been described and planned by Sir Percy Sykes: see 'A sixth Journey in Persia' in the *Geographical Journal*, 1911, p. 6. Clavijo visited them in 1404, and writes the name of the town Zabrain.

p. 513, first column line 17

Read "328" for "327."

p. 520, first column, as line 3

Insert "Mûḳân, 175."

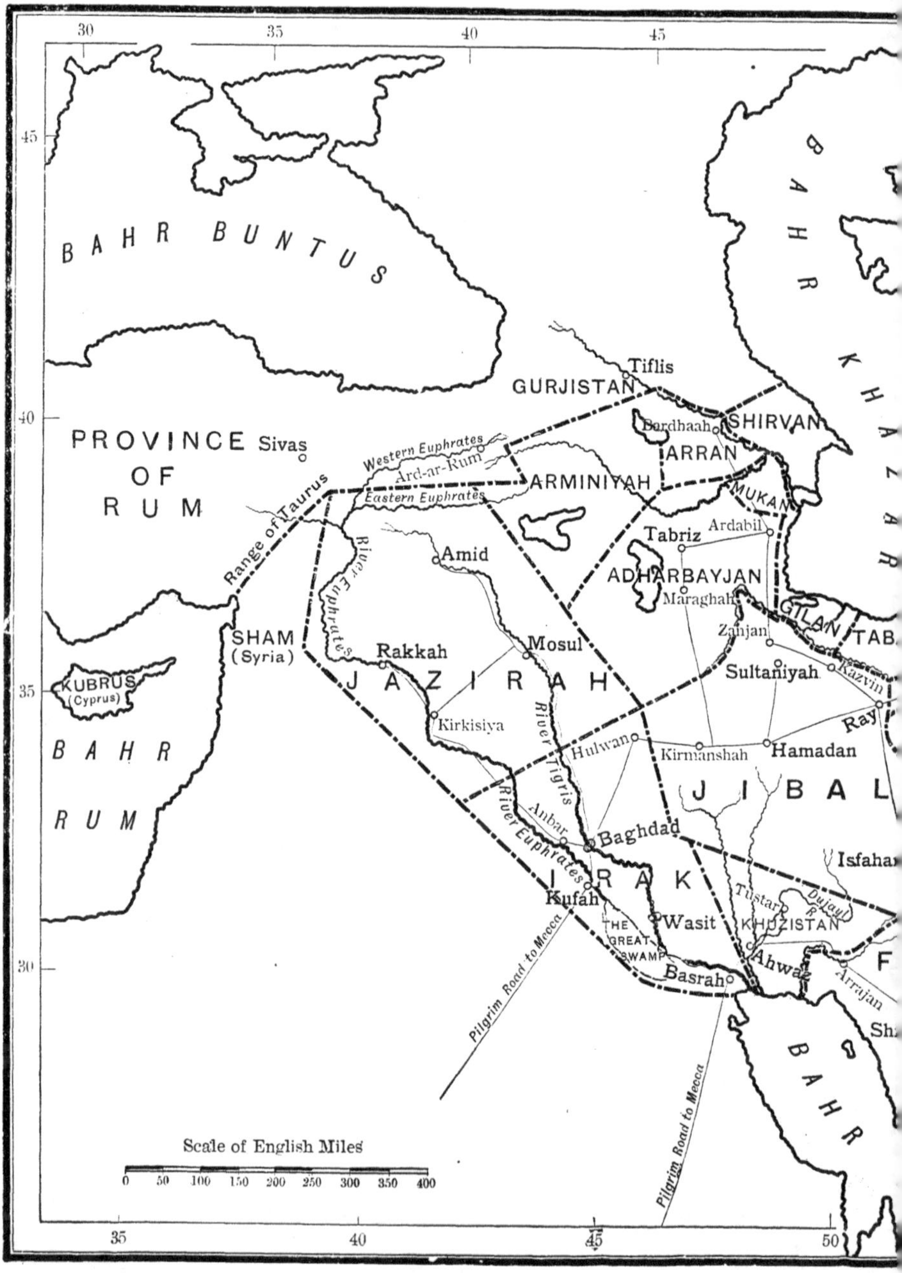

THE PROVINCES OF THE ABBASID CALI

MAP I

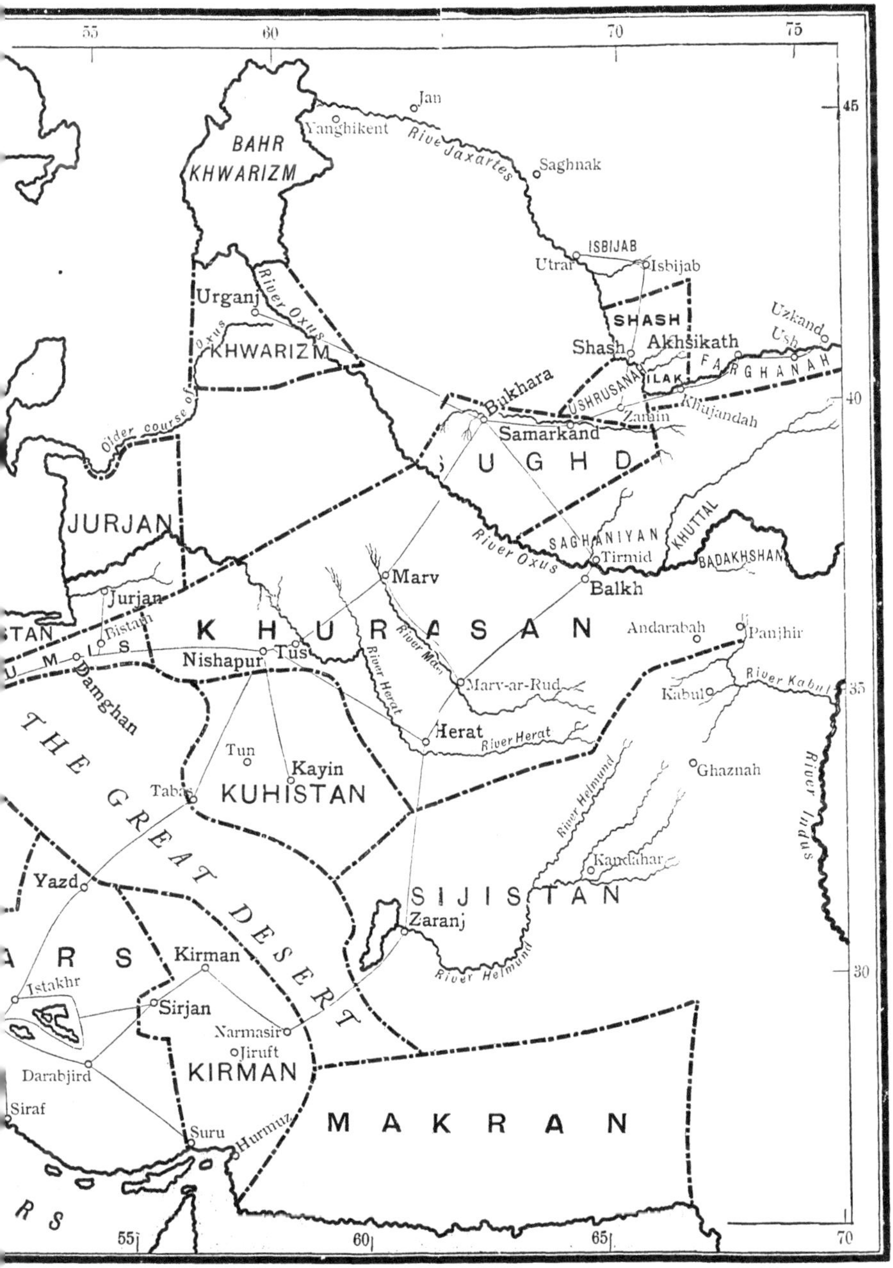

TE, SHOWING THE CHIEF HIGH ROADS

CHAPTER I.

INTRODUCTORY.

Mesopotamia and Persia, their provinces under the Abbasid Caliphs. The outlying provinces to the north-west and the north-east. The high roads from Baghdâd to the Moslem frontier. The Moslem geographers, and their works. Other authorities. Place-names in the Arabic, Turkish, and Persian provinces.

Mesopotamia and Persia had formed the kingdom of the Sassanian Chosroes, which the Arabs utterly overthrew when, after the death of Muḥammad, they set forth to convert the world to Islam. Against the Byzantines, the other great power which the Moslems attacked, they achieved only a partial victory, taking possession, here and there, of rich provinces, notably of the coast lands to the south and east of the Mediterranean; but elsewhere the Emperors successfully withstood the Caliphs, and for many centuries continued to do so, the Roman empire in the end surviving the Caliphate by over two hundred years.

The kingdom of the Sassanians, on the other hand, the Arabs completely overran and conquered; Yazdajird, the last of the Chosroes, was hunted down and slain, and the whole land of Îrân passed under the rule of Islam. Then further, and to no inconsiderable extent, the empire of the Caliphs, which had taken over bodily the administration of the older Persian kingdom, came itself to be modelled on the pattern in government which the Chosroes had established; this more especially under the Abbasids, who, rather more than a century after the death of the Prophet, overthrew their rivals the Omayyads, and changing the seat of the Caliphate from Syria to Mesopotamia, founded Baghdâd on the Tigris, a few miles above Ctesiphon, the older winter capital of the Sassanians.

Baghdâd forthwith became, for the East, the centre of the Moslem empire, but from the time of the first Abbasid Caliph this empire no longer remained, even nominally, undivided. Spain fell off, and before long an Omayyad Caliph at Cordova was the rival of the Abbasid Caliph at Baghdâd. In rather more than a century after their establishment in power, the Abbasids also lost Egypt, which, at about the date when the Omayyad prince at Cordova had recently proclaimed himself Commander of the Faithful, passed into the power of the Fatimids, who likewise took the style of Caliph, and renounced allegiance to Baghdâd. Syria had for the most part followed the fortunes of Egypt; Arabia was the debateable land between the two; in the Further East many provinces became independent of the Abbasid Caliph, but there no permanent rival Caliphate was established; so that in general terms all those broad provinces, which had formed the Sassanian kingdom before the days of Islam, remained to the last nominally, if not really, subject to the Abbasids. This vast stretch of country, bounded to the eastward by the deserts of Central Asia, with the mountains of Afghanistân, and westward by the Byzantine empire, was divided among the many provinces which will be described in detail in the succeeding chapters of the present work. The names of the provinces, and their boundaries, for the most part (and as far as is known), were under the Arabs identical with those that had existed under the Chosroes; indeed the East alters so little that in the majority of cases both names and boundaries have remained almost unchanged to the present day, though, as was to be foreseen, the political state, and especially the economical or material conditions of the country, have varied considerably during the last thirteen hundred years.

It will be convenient, before proceeding further, to give a brief summary of these various provinces, taking them in the order in which they are described in the succeeding chapters.

The great lowland province, which the Greeks called Mesopotamia, is the gift of its two rivers the Euphrates and the Tigris; and the latter in its lower course (as will be more fully explained in Chapter II) did not, in Abbasid times, run in the channel which its waters follow at the present day. A glance at the map shows that the sterile Arabian desert comes close up to the western border of the Euphrates, and this river, therefore, has no right bank affluents. With the Tigris, on the other hand, it is different; the highlands of Persia follow a line standing back at a considerable distance from the eastern side of this river, and many streams flow down from the Persian mountains, these forming numerous left bank affluents of the Tigris. The Moslems inherited from the Sassanians a system of irrigation for Mesopotamia which made this province one of the richest in the known world. The system will be more fully explained later; but briefly it may be said that the Arabs effectually watered the country lying between the two rivers by draining the surplus of the Euphrates through a number of transverse canals flowing to the Tigris; while the districts to the eastward of the Tigris, extending up to the foot-hills of the Persian highlands, were watered in part by the streams which flowed down from these mountains, in part by a series of loop canals, taken from the left bank of the Tigris, and returning to it again, which in turn absorbed the flood-waters of the many small rivers rising in the eastern hills.

The Arabs divided Mesopotamia into two provinces, Lower and Upper, of which the Lower comprised the rich alluvial lands known anciently as Babylonia. Lower Mesopotamia was called Al-'Irâḳ, and its northern limit (which, however, varied at different times) was a line going east and west, from points on the Euphrates and Tigris, respectively, where these two rivers first began to flow near each other through the Mesopotamian plain. The largest city of 'Irâḳ, under the Abbasids, was of course Baghdâd; but already a century before that dynasty had come to power, the first Moslems, on conquering this part of Mesopotamia, had founded three great towns, Wâsiṭ, Kûfah, and Baṣrah, which continued to flourish for many centuries; and these, with Anbâr (already a city in Sassanian days) lying on the Euphrates in the latitude of Baghdâd, were the great centres of population in the 'Irâḳ province under the Abbasid Caliphs.

North of the limit of the alluvial lands stretched the hard and somewhat stony plains of Upper Mesopotamia, where had been the kingdom of Nineveh in ancient times. Upper Mesopotamia the Arabs called Al-Jazîrah, 'the island,' or rather 'the peninsula,' or partial island, for these great plains were almost enclosed by a ring of waters, formed by the upper courses of the Euphrates and Tigris, and by streams or canals joining the two to the southward of the stony plains. The province of Jazîrah extended north to the mountains in which the two great rivers had their sources; it was divided into three districts, named after the Arab tribes which had settled here in the times of the Chosroes, and its chief towns were Mosul near the ruins of Nineveh, Âmid on the Upper Tigris, and Raḳḳah at the great bend of the Euphrates, near the desert border on the further side of which is Damascus.

The chapter following deals with the mountainous countries in which the twin rivers, which are the head streams of the Euphrates, take their rise. This country formed the debateable land between the Caliphate and the empire. Time and again its towns and fortresses were taken and retaken, by Moslems and Christians, as the tide of war ebbed and flowed. The country was never permanently settled by the Arabs, and detailed description of it is for the most part lacking in our earlier authorities. The same remark, and in a higher degree, applies to the province called Rûm (the Roman Territory) which, till the latter part of the 5th (11th) century, remained an integral part of the Byzantine empire; for between this province and the Caliphate the great rampart of the Taurus chain formed the line of demarcation. Almost yearly the Moslems made incursions through the Taurus passes into Anatolia; more than once they laid ineffectual siege to Constantinople; and at times they garrisoned and occupied divers fortress towns up on the great plateau of Asia Minor. But beyond such temporary occupation the Abbasid Caliphs did not succeed in conquering the upland country; they made many raids through Asia Minor, but they held no land, and Moslem rule was not established there, until in the decline of the Caliphate, the Saljûḳ Turks settled in these highlands which they wrested from the Byzantines, and then finally Asia Minor, or Rûm, came to be counted as Moslem land, in which condition it still remains.

To the east of Jazîrah, or Upper Mesopotamia, came the province of Adharbâyjân, the ancient Atropatene, bounded above and below, respectively, by the Araxes and the White River, the Safîd-

Rûd, both of which streams flowed into the Caspian. The most notable natural feature of this province was the great salt lake, now known as the lake of Urmîyah, near which stood Tabrîz and Marâghah, the provincial capitals, while Ardabîl, another great town, lay to the eastward nearer the shore of the Caspian. The chapter following describes a number of smaller provinces of the north-western border. First Gîlân, or Jîlân, on the Caspian, where the Safîd-Rûd, breaking through the Alburz range, the mountain barrier of the Persian highlands, flows through an alluvial plain of its own making, pushing out a small delta into the Caspian. Next, the province of Mûghân at the mouth of the combined Araxes and Cyrus rivers; then Arrân lying to the westward between the courses of these two rivers; with Shirvân to the north of the Cyrus, and Gurjistân (Georgia) at its head waters. Lastly we have Moslem Armenia lying at the head waters of the Araxes, which is the mountainous province surrounding the lake of Vân.

South-east of Adharbâyjân spreads the rich province of Media, which the Arabs very appropriately called Al-Jibâl, 'the mountains,' for its mountains overhang the lowlands of Lower Mesopotamia, and, range behind range, stretch across eastward to the border of the Great Desert of Central Persia. The western part of the Jibâl province, in later times, when the Kurds attained fame and power, came to be known as Kurdistân; and in the later middle-ages, but by a misnomer, as will be explained in due course, the province of Al-Jibâl was often called 'Irâḳ 'Ajamî, or *Persian* 'Irâḳ, in contrast to *Arabian* 'Irâḳ, which was Lower Mesopotamia. The Jibâl province included many great cities; in the west Kirmânshâh and Hamadân (the latter the ancient Ecbatana); in the north-east Ray (Rhages), and to the south-east Ispahân. At a later period the Mongols of Persia founded Sulṭânîyah in its northern plains, which for a time taking the place of Baghdâd, became the capital of this portion of their empire, which included both Mesopotamia and Persia under the rule of the Îl-Khân. In the mountains of the Jibâl province many rivers take their rise, among the rest the Kârûn, which the Arabs called Dujayl or Little Tigris, and which after a long and tortuous course flows out at the head of the Persian Gulf, a little to the east of the combined mouth of the Euphrates and Tigris.

The province of Khûzistân, lying south of Media and east of Lower Mesopotamia, occupies the lower course of the Kârûn river, or Dujayl, with its numerous affluents. This country was extremely rich; Tustar and Ahwâz were its chief towns; and its lands being plentifully irrigated were most productive. East of Khûzistân, and bordering the Gulf, lay the great province of Fârs, the ancient Persis and the cradle of the Persian monarchy. Under the Abbasids it still kept the division into the five Kûrahs, or districts, which had been organized under the Sassanians, and Fârs was closely studded with towns, great and small, the most important of which were Shîrâz the capital, Iṣṭakhr (Persepolis), Yazd, Arrajân, and Dârâbjird. The islands of the Gulf were counted as of Fârs, and Ḳays island was an important commercial centre before the rise of Hurmuz. The chief physical feature of Fârs was the great salt lake of Bakhtigân, which with other smaller sheets of water stood in the broad highland valleys, whose mountains were offsets of the ranges in the Jibâl province, already referred to. In Fârs, the Dârâbjird district under the Mongols came to be counted as a separate province, and was in the 7th (13th) century called Shabânkârah; the Yazd district also, in the later middle-ages, was given to the Jibâl province.

To the east of Fârs lay the province of Kirmân, far less fertile, almost lacking in rivers, and bordering on the Great Desert. Of this province there were two capitals in Abbasid times, Sîrjân and Kirmân city; and the two other most important towns of the province were Hurmuz, on the coast; and Jîruft, inland, a centre of much commerce. The Great Desert of Central Persia is the most remarkable physical feature of the high tableland of Îrân. This immense salt waste stretches south-east diagonally across Persia, from Ray, at the base of the mountains which on their northern side overlook the Caspian, spreading in a broad band —or rather, in a dumb-bell-shaped depression—the lower end of which merges into the hills of Makrân, the province bordering on the Indian Ocean. In the Great Desert there are few oases; a salt efflorescence covers much of the barren levels, but the desert in winter time is not difficult to pass, and many well marked tracks connect the towns on either side. But on the other hand the Great Desert is a real barrier to any continuous intercourse between the provinces of Fârs and Kirmân, which lie on its south-western side, and the eastern provinces which are beyond its other limit, namely Khurâsân with Sîstân to the south-east, and this desert barrier has played an important part all through the history of Persia. After describing what the Moslem geographers have to say of the Great Desert, the same chapter deals with the Makrân province, which on the east touched India, running up to the highlands overlooking the Indus valley, part of which is now known as Balûchistân. On these regions, however, our authorities are not very fully informed.

North of Makrân, and across the narrow part of the desert opposite Kirmân, lay the province of Sijistân or Sîstân, to the east of the extensive, but very shallow lake of Zarah. Into this lake drained the waters of the Helmund, and numerous other rivers flowing south-west from the high mountains of Afghanistân lying above Kâbul and Ghaznah. Here Ḳandahâr stood in a plain between two of the affluents of the Helmund, and where this great river flowed into the Zarah lake lay Zaranj, the capital of Sijistân. North-west of the Zarah lake, and on the border of the Great Desert, was the very hilly province aptly called Ḳûhistân (Land of Mountains), the chief towns of which were Tûn and Ḳâyin, well known as the Tunocain of Marco Polo; Sijistân and Ḳûhistân thus forming the southern border of Khurâsân, the great eastern province of Persia.

Before describing this last, however, the three small provinces of Ḳûmis, Ṭabaristân and Jurjân, which form the subject of the succeeding chapter, require notice. Ḳûmis, of which the capital was Dâmghân, lay in length along the north border of the Great Desert eastward of Ray, comprising the southern foot-hills of the mountain chain of Alburz which shuts off the high plateau of Persia from the Caspian Sea. These mountains, and more particularly their northern flank descending to the Caspian, formed the province of Ṭabaristân, otherwise called Mâzandarân, which extended from Gîlân and the delta of the White River (Safîd-Rûd), on the west, to the south-eastern corner of the Caspian. Here Ṭabaristân joined Jurjân, or Gurgân, the ancient

Hircania, which included the valleys watered by the rivers Atrak and Jurjân, on which last stood Jurjân city. The Jurjân province extended eastward from the Caspian Sea to the desert which separated Khurâsân from the cultivated lands of the Oxus delta, namely the province of Khwârizm.

The modern province of Khurâsân is but a moiety of the great tract of country which, from Abbasid times down to the later middle-ages, was known under this name; for Khurâsân of those days included what is now become the north-western part of Afghanistân. On the east, medieval Khurâsân bordered on Badakhshân, its northern frontier was the Oxus and the desert of Khwârizm. The Moslem geographers divided Khurâsân into four quarters, named after its four capital cities; *viz.* Nîshâpûr, Marv, Herat, and Balkh. From a physical point of view the remarkable feature of Khurâsân consisted in the two great rivers of Herat and of Marv, which rising in the mountains of what is now Afghanistân, turned north and flowed out to waste in the sands of the desert towards Khwârizm, reaching no sea or lake.

The chapter following deals with the upper waters of the Oxus, and a number of small provinces, stretching from Badakhshân westwards, which lie to the north, on the right bank affluents of the great river. Its delta, forming the province of Khwârizm to the south of the Aral Sea, is next described, of which Urganj was the older capital, and in this chapter some pages are devoted to clearing up the much debated subject of the older course of the Oxus to the Caspian. Beyond the great river, and between the Oxus and the Jaxartes, lay the province of Sughd, the ancient Sogdiana, with its two noble cities, Samarḳand and Bukhârâ, both on the Sughd river. This is the penultimate chapter of the present work; and the last chapter deals with the provinces along the Jaxartes, from Farghânah near the borders of the Chinese deserts, of which the capital was Akhsîkath, to Shâsh, modern Tâshkand, with the Isbîjâb province to the north-west, beyond which the Jaxartes flowed out, through the bleak wilderness, into the upper part of the Aral Sea. Of these northern countries of the Further East, however, lying beyond Central Asia, the earlier Arab geographers give but a succinct account. They were the Turk lands, and it was only after the Mongol invasion that they rose to importance; of this period unfortunately there is a lack of precise information, the Arab geographers failing us for the most part, and their place being but ill-supplied by the later Persian and Turkish authorities.

The Moslems, by the injunction of their Prophet, were bound each, once in a lifetime, to make the pilgrimage to Mecca. Under the Abbasids, when the Moslem empire reached its fullest extent, the pilgrimage was facilitated by the elaborate system of high roads, all made to radiate from Baghdâd, where the Tigris was crossed by those coming from the further east and bound for the Ḥijâz. Of this road system (which the Arabs had inherited from the earlier Persian kingdom) we possess detailed contemporary descriptions; and the chief lines, running through the provinces named in the foregoing paragraphs, may here be summarily described.

The most famous of the trunk roads was the great Khurâsân road, which, going east, united the capital with the frontier towns of the Jaxartes on the borders of China. This, too, is perhaps that which of all the roads is best described. Leaving East Baghdâd by the Khurâsân gate, it went across the plain, passing over numerous streams by well-built bridges, to Ḥulwân at the foot of the pass leading up to the highlands of Persia. Here it entered the Jibâl province and after a steep ascent reached Kirmânshâh, the capital of Kurdistân. Crossing the Jibâl province diagonally, north-east, the road passed through Hamadân to Ray. From Ray onwards it went almost due east through Ḳûmis, having the Ṭabaristân mountains on the left, and the Great Desert on the south, till it entered the province of Khurâsân near the town of Bisṭâm. Continuing onwards it came to Nîshâpûr, then to Ṭûs, and on to Marv, beyond which it crossed the desert to the Oxus bank at Âmul, thence reaching successively Bukhârâ and Samarḳand in the province of Sughd. At Zâmîn a short distance east of Samarḳand, the road bifurcated: on the left hand one road proceeded to Shâsh (Tâshkand) and ultimately to the ford at Utrâr on the lower course of the Jaxartes; the other road, leaving Zâmîn, turned off to the right, towards Farghânah and the Upper Jaxartes, coming to Akhsîkath the capital, and finally to Ûzkand on the borders of the Chinese desert.

This in its full extent was the great Khurâsân road; and to the present day the post-roads crossing Persia, but centring in Ṭihrân, near the older Ray, follow the same long track which the earlier Arab geographers have described. After the fall of the Abbasid Caliphate, the road system was in part altered by the building of Sulṭânîyah, which became the capital of the Mongols. But all that this entailed was a branch rcad north from Hamadân direct to Sulṭânîyah, which, for a time, took the place of Ray as the centre point of the roads in this quarter.

In earlier days, under the Abbasids, cross-roads had branched off, right and left, to various parts of Persia from the chief towns along the Khurâsân high road. Thus from near Kirmânshâh a road went north to Tabrîz and other towns on the Urmîyah lake, with prolongations to Ardabîl and to places on the Araxes. From Hamadân, going south-east, there was a high road to Isfahân; and from Ray, going north-west, the distances to Zanjân are given, whence a highway led up to Ardabîl. Nîshâpûr in Khurâsân was a centre for many branch roads; southwards one went to Ṭabas on the borders of the Great Desert in Ḳûhistân; another road went to Ḳâyin; while south-east was the highway to Herat, whence Zaranj in Sijistân was reached. From Marv a high road followed up the Marv river to Lesser Marv (Marv-ar-Rûd), where, joining a road coming from Herat, it went on to Balkh and the eastern frontier lands beyond the Oxus. Finally from Bukhârâ there was direct communication, north-west, with Urganj in Khwârizm; and, south-west, with Tirmid on the Oxus opposite Balkh.

This completes the system of the Khurâsân road; and now returning to Baghdâd, the central point, the highways going in other directions must be sketched. Down the Tigris, the distances and stations being given both by land and by water, was the highway through Wâsiṭ to Baṣrah, the great port for the trade of the Persian Gulf. From both Wâsiṭ and Baṣrah, Ahwâz in Khûzistán was reached, and thence the high road went due east to Shîrâz in Fârs. This was a centre of many roads. North was the road to Isfahân and on to Ray; north-east, through Yazd and across the Great Desert Ṭabas was reached, which communicated with Nîshâpûr; eastward by more than one route Sîrjân and

Kirmân were in communication, and thence eastward across the Great Desert was the way to Zaranj in Sijistân; while south-east and south from Shirâz two roads branched towards the Persian Gulf ports, one passing through Dârâbjird to Sûrû near Hurmuz, the other to Sîrâf, at one time the chief harbour of Fârs.

Returning once again to Baghdâd, the central point, we find that the great Pilgrim road to Mecca and Medina left West Baghdâd, going south to Kûfah on the border of the Arabian desert, which it crossed almost in a direct line to the Ḥijâz. A second Pilgrim road started from Baṣrah, running at first nearly parallel with the other, which it finally joined two stages north of Mecca. Then from Baghdâd, north-west, a road went to the Euphrates at Anbâr, and thence up that river to Raḳḳah, a centre point for roads across the Syrian desert to Damascus, and for many other highways going north to the Greek frontier towns. Finally from Baghdâd, north, there were high roads up both banks of the Tigris to Mosul, whence Âmid was reached on the one hand, and Ḳirḳîsiyâ on the Euphrates to the south-west. From Âmid there were roads communicating with most of the frontier fortresses towards the Greek country.

This in brief was the road system under the Abbasids, which, centring in Baghdâd, connected the capital by a system of post-stages with the outlying provinces of the empire. The system is very carefully described by the Arab geographers, and for purposes of reference it may be well now to give in chronological order a short account of our contemporary authorities, on whose works we rely for the facts set down in the following chapters[1].

The earlier of our authorities date from the middle of the 3rd (9th) century, and the first geographical treatises of the Arabs take the form of Road Books. These set forth in detail the various itineraries, are interspersed with short accounts of the towns passed through, and give the revenues and products, in turn, of each province. Of these Road Books we possess four, in particular, which are of primary importance, and they complement

[1] For further particulars of the Arab geographers see *Palestine under the Moslems* (London, 1890), the Introductory chapter; also for more detail, the Introduction to the French translation of Abu-l-Fidâ, by M. Reinaud (Paris, 1848).

each other, for their texts have in many passages come down to us in a mutilated condition. The authors of these Road Books of the 3rd (9th) century are Ibn Khurdâdbih, Ḳudâmah, Ya‘ḳûbî and Ibn Rustah.

The first two are almost identical in substance. Ibn Khurdâdbih was post-master of the Jibâl province, Kuḍâmah was a revenue accountant; their itineraries give stage by stage the distances along the great Khurâsân road and the other trunk roads, as sketched in the preceding paragraphs, which radiated from Baghdâd. The work of Ya‘ḳûbî has unfortunately not reached us in its entirety; to it we owe the account of Baghdâd which, with the description written by Ibn Serapion, has made it possible to work out in detail the topography of the Abbasid capital. Ya‘ḳûbî gives further a number of valuable notes on many other cities, and the details of the high roads traversing the ‘Irâḳ province are found fully set forth only in his work. Of Ibn Serapion, his contemporary, only a fragment has reached us; but this, in addition to the account given of Baghdâd, is of capital importance for the river and canal system of Mesopotamia; he gives also shorter descriptions of the rivers in other provinces. Ibn Rustah has written a similar work to Ya‘ḳûbî, adding many notices of towns; but above all he has given us a most minute account of the great Khurâsân road as far as Ṭûs, near Mashhad, with some of its branch roads, notably those going to Isfahân, and to Herat; also the road from Baghdâd south to Kûfah, and to Baṣrah, with the continuation eastward to Shîrâz. On all these trunk lines, not only are the distances and stages given, but an exact description is added of the nature of the country passed through; whether the way be hilly, ascending or descending, or whether the road lies in the plain; and this description of Ibn Rustah is naturally of first-rate importance for the exact identification of the line traversed, and for fixing the position of many lost sites. Another authority is Ibn-al-Faḳîh, a contemporary of Ibn Rustah, who wrote a very curious geographical miscellany, of which unfortunately only an abridgment has come down to us. Some of his notices of places, however, are of use in completing or correcting the earlier accounts[1].

[1] The texts of Ibn Khurdâdbih, Ḳudâmah, Ya‘ḳûbî, Ibn Rustah and

The systematic geographers begin with the 4th (10th) century. They describe fully and in turn each province of the Moslem empire, only incidentally giving the high roads, and generally piecemeal for each province. Their works are of course a great advance on the Road Books; to them we owe such fulness of geographical detail as will be found in the following chapters, and the three first names on the list, Iṣṭakhrî, Ibn Ḥawḳal, and Muḳaddasî, are those to whose labours we are most materially indebted. The work of Ibn Ḥawḳal is but a new edition, partly enlarged and emended, of Iṣṭakhrî; on the other hand Iṣṭakhrî, a native of Persepolis, gives the description of his native province, Fârs, in far greater detail than is to be found in Ibn Ḥawḳal, who reduced his chapter on Fârs to the due proportion of the remainder of the book. Muḳaddasî, their contemporary, wrote his geography entirely on independent lines, and chiefly from his personal observations of the divers provinces. His work is probably the greatest, it is certainly the most original, of all those which the Arab geographers composed; his descriptions of places, of manners and customs, of products and manufactures, and his careful summaries of the characteristics of each province in turn, are indeed some of the best written pages to be found in all the range of medieval Arab literature.

It is further to be remarked that to these last three systematic geographers we owe the exact identification of most of the names displayed on the accompanying maps. At the close of each chapter they give a table of 'the distances,' namely the stages or sections of the great high roads, already described, which crossed the province in question, and in addition to the high roads an immense number of cross-distances are added, going between

Ibn-al-Faḳîh are edited by Professor De Goeje in volumes V, VI, and VII of his series *Bibliotheca Geographorum Arabicorum* (Leyden, 1885—1892); further in vol. VI he has added a French translation, with many important notes, of the first two authorities. Of Ibn Serapion the text, describing Mesopotamia, will be found in the *Jour. R. Asiat. Soc.* for 1895, p. 9; and the MS. referred to is that in the British Museum, numbered Add. 23,379. Ya'ḳûbî, in addition to his work on geography, also wrote a history, the text of which has been edited by Professor M. T. Houtsma (*Ibn-Wâdhih, qui dicitur Al-Ja'qubî, Historiae*, Leyden, 1883), and this often contains valuable information in matters of geography.

neighbouring towns. These distances, plotted out and starting from known points, enable us to cover the map with a system of triangulation, by means of which the positions of some towns, long ruined, and the very vestiges of which have in many cases disappeared, can be approximately laid down; as, for instance, in the case of Tawwaj in Fârs, the ruins of which have not yet been identified, though their situation can now be fixed within narrow limits. Another writer of the 4th (10th) century is Mas'ûdî, who has left two works; the first for the most part historical, and well known under the title of *The Golden Meadows*; the second, a sort of commonplace book, full of curious details and notes, which is called *At-Tanbîh*, '*The Admonishment*[1].'

Coming to the 5th and 6th (11th and 12th) centuries, we have the works of two famous travellers, pilgrims, whose descriptions of the places they passed through are of considerable importance. Nâṣir, son of Khusraw, the Persian, in the middle of the 5th (11th) century went from Khurâsân to Mecca and back, visiting Egypt and Syria on his way out, and crossing Arabia on the homeward journey, and his diary, written in Persian, is one of the earliest works we possess in that language. Ibn Jubayr, the Spanish Arab, a century later made the pilgrimage starting from Granada; and his account of Mesopotamia, particularly of Baghdâd, is one of the most interesting that has come down to us. Dating from the beginning of the 6th (12th) century is another Persian work, called the *Fârs Nâmah* (Book of Fârs), describing most minutely that province, and invaluable as far as it goes. Also dating from the middle of this century we have the systematic geography of Idrîsî, who lived at the court of the Norman king, Roger II of Sicily. He wrote in Arabic, and very inconveniently has composed

[1] The texts of Iṣṭakhrî, Ibn Ḥawḳal, and Muḳaddasî form volumes I, II, and III, respectively, of the already-mentioned series of the *Bibl. Geogr. Arab.* (Leyden, 1870—1877). Of Mas'ûdî the text of the *Tanbîh* has been edited by Professor De Goeje in vol. VIII of the same series (Leyden, 1894); and a translation in French of this has been published (Paris, 1896) by Baron Carra de Vaux under the title of *Le Livre de l'Avertissement.* The history, called *The Golden Meadows* (Murûj-adh-Dhahab), was published (Paris, 1861), the Arabic text being given with a French translation, by Messrs Barbier de Meynard and Pavet de Courteille; the two last works under the auspices of the French Société Asiatique.

his description of the known world in 'Climates,' that is according to zones of latitude, whereby the various provinces are often divided up arbitrarily, Mesopotamia, for instance, being partly described in the 3rd Climate, partly in the 4th. He had, unfortunately for our purpose, no personal knowledge of Persia or the regions east of the Mediterranean, but had visited Asia Minor, then still a province of the Roman empire, and his description of this region would be invaluable, but for the fact that the place-names (by reason of incorrect MSS.) are in many cases illegible, or so corrupt as to be at present mostly beyond recognition[1].

Coming to the 7th (13th) century, the period of the Mongol invasion and the fall of the Abbasid Caliphate, we have the voluminous *Geographical Dictionary* of Yâḳût, a compilation it is true from earlier writers, but illustrated by the author's own far extended travels, which, when it is used with due criticism, is perfectly invaluable. The articles are arranged in alphabetical order, and Yâḳût quotes freely from almost all his predecessors in Arab geographical literature, some of whose works, as for instance those of the traveller Ibn-al-Muhalhal, who wrote in 330 (942), are only known to us by his excerpts. This great dictionary was epitomised, three-quarters of a century after its appearance, in a work called *Al-Marâṣid*, 'the Observatories,' and the author of this epitome, a native of Mesopotamia, often gives valuable corrections, of first-hand authority, for places in the regions round Baghdâd. Of about the same date is Ḳazvînî, who wrote a work in two parts on cosmography, which gives interesting notes on the products and the commerce of divers towns and provinces; and in the earlier part of the 8th (14th) century we have the systematic geography of Abu-l-Fidâ, a Syrian prince, who, though he compiled largely from the works of his predecessors, in addition gives

[1] The Persian text of Nâṣir-i-Khusraw, with an annotated French translation, has been brought out by C. Schéfer, in the series of the École des Langues Orientales Vivantes (Paris, 1881). The Arabic text of Ibn Jubayr was well edited by W. Wright (Leyden, 1852). The *Fârs Nâmah* exists only in manuscript: that quoted is in the British Museum, numbered Or. 5983. Idrîsî has been translated into French (indifferently well) by A. Jaubert (Paris, 1836); passages quoted I have verified with the Arabic text, preserved in the Bibliothèque Nationale, *Manuscrits Arabes*, Nos. 2221 and 2222.

facts from his own observation of the countries which he had visited[1].

Of the same date, namely the first half of the 8th (14th) century, are the travels of Ibn Baṭûṭah the Berber, who rivalled the Venetian Marco Polo in the extent of his voyages. His book is written in Arabic; his contemporary, Mustawfî, wrote in Persian a description of the Mongol kingdom of Îrân (Mesopotamia with Persia), which shows the condition of the country after the Mongol settlement, when this region was governed by the Îl-Khâns. Mustawfî also wrote an historical work called the *Târîkh-i-Guzîdah*, 'the Select History,' which, besides being of considerable value for Mongol times, often contains geographical notes of great importance[2].

For the time of Tîmûr we have primarily the notices in the historical work of 'Alî of Yazd, then the *Geography* written by Ḥâfiẓ Abrû; both are in Persian, and date from the first half of the 9th (15th) century. Lastly for the settlement after the conquests of Tîmûr, the works of two Turkish authors, one writing in Eastern Turkish, the other in 'Othmanlî, have to be mentioned, both being of the earlier half of the 11th (17th) century. These are the *History of the Turks and Mongols* by the Khwârizm prince Abu-l-Ghâzî, and the *Universal Geography* called the *Jahân Numâ*

[1] The *Mu'jam-al-Buldân*, the great dictionary of Yâḳût, has been edited in Arabic by F. Wüstenfeld (Leipzig, 1866—1873); the articles relating to places in Persia will be found translated into French, with additions from Mustawfî and later authorities, in the *Dictionnaire de la Perse* (Paris, 1861) of M. Barbier de Meynard. The *Marâṣid-al-Iṭṭilâ*, which is the epitome of Yâḳût, has been edited by Juynboll (Leyden, 1852). The two volumes of the *Cosmography* of Ḳazvînî have been edited by Wüstenfeld (Göttingen, 1848). The text of the *Geography* of Abu-l-Fidâ was edited by Reinaud and De Slane (Paris, 1840), and Reinaud also began (Paris, 1848) a translation of this work in French, prefixing to it a valuable Introduction on the Arab Geographers, which translation S. Guyard afterwards (Paris, 1883) completed.

[2] The Travels of Ibn Baṭûṭah, the Arabic text with a French translation, have been published (Paris, 1874—1879) by Defrémery and Sanguinetti. The Persian Geography of Ḥamd Allah Mustawfî (the text of the *Nuzhat-al-Ḳulûb*) was lithographed at Bombay in 1311 (1894), and the *Guzîdah* is quoted from the British Museum MS. numbered Add. 22,693, MSS. Add. 7630 and Egerton 690 having been collated. Part of the *Guzîdah* has now been printed, with a French translation, by M. J. Gantin (Paris, 1903).

(World Displayer) by the celebrated bibliographer Ḥâjj Khalfah[1].

For elucidating points of detail the works of many of the Arab historians are of primary importance. By earlier writers history and geography were often treated of in one and the same work. An instance of this is the *Book of the Conquests*, written by Balâdhurî, and dating from the middle of the 3rd (9th) century. It describes in turn, east and west, all the conquests of the Moslems, and is of great interest as showing the state of the country when Islam first became the dominant creed. Of the chronicles, besides the History written by Ya'ḳûbî, already mentioned, there is, dating from the 3rd (9th) century, the work of Ibn Mashkuwayh, of which the Sixth Section only has been printed. The annals of Ḥamzah of Isfahân, written in the middle of the 4th (10th) century, likewise give useful information, and though of course composed in Arabic. the work was evidently based on many Persian books, now lost, and it relates facts of which we should otherwise be ignorant.

The most complete, however, of the Arabic chronicles, down to the beginning of the 4th (10th) century, at which date he flourished, is that of Ṭabarî, and his work is for geography a primary authority. For later Abbasid history Ibn-al-Athîr has to be relied upon; also the entertaining summary of Moslem history generally known by the name of Fakhrî. The *Universal History* of Ibn Khaldûn is often of use to supplement the meagre chronicle of Ibn-al-Athîr; and the great *Biographical Dictionary* of Ibn Khallikân occasionally adds details. These authors all wrote in Arabic. In Persian the two histories called the *Rawḍat-aṣ-Ṣafâ* and the *Ḥabîb-as-Siyâr*, respectively by Mîrkhwând and by Khwândamîr

[1] The Persian text of the history of Tîmûr by 'Alî of Yazd, known as the *Ẓafar Nâmah*, is published in the Bibliotheca Indica (Calcutta, 1887). A French translation called *Histoire de Timour Bec* was published (Paris, 1722), by Petis de la Croix. Ḥâfiẓ Abrû exists only in manuscript; the one quoted is that of the British Museum, numbered Or. 1577. The Turkish text of the *Jahân Numâ* was printed in Constantinople in 1145 (1732) by Ibrâhîm Efendî, and a Latin translation of part of this work was published by M. Norberg (Lund, 1818). The Turkî text, with a French translation, of the History of the Mongols, by Abu-l-Ghâzî, has been published by Baron Desmaisons (St Petersburg, 1871).

his grandson, must be mentioned, for especially in the Persian provinces both these works give valuable geographical information. Two other Persian chronicles, relating to the Saljûḳ dynasties in Asia Minor and in Kirmân, are likewise of importance, and are more than once quoted in the following pages, being referred to under the names of the chroniclers Ibn Bîbî, and Ibn Ibrâhîm[1].

To complete our survey, a few pages in conclusion of this preliminary chapter may be devoted to some general remarks on the place-names which occur in the following chapters, and are set down on the maps. In the two provinces of Mesopotamia the great majority of the place-names are notably either Arabic or Aramaic, this last having been the common language of the people here, prior to the Moslem conquest. The Arabic names of towns generally have, or had, a meaning, as for instance Al-Kûfah, Al-Baṣrah, and Wâsiṭ. The Aramaic names, as a rule, are easily recognisable by their form, and by the termination in long â, for example Jabultâ; and the meaning of these too is generally not far to seek: e.g. 'Abartâ, 'the passage, or crossing place,' marking a bridge of boats; and Bâjisrâ, which is equivalent

[1] The text of Balâdhurî has been edited by Professor De Goeje (Leyden, 1866). He has also given us Ibn Mashkuwayh, forming the latter part of his *Fragmenta Historicorum Arabicorum* (Leyden, 1871). The History by Ḥamzah of Isfahân has been edited (with a Latin translation) by I. M. E. Gottwaldt (Leipzig, 1844). The numerous volumes of the great Chronicle of Ṭabarî have been published, in three series, under the editorship of Professor De Goeje (Leyden, 1879—1901). The Chronicle of Ibn-al-Athîr is edited by Tornberg (Leyden, 1867—1876). Fakhrî, more correctly named Ibn-aṭ-Ṭiḳṭaḳâ, has been edited by Ahlwardt (Gotha, 1860). Of Ibn Khaldûn, the text quoted is that printed at Bulâḳ in 1284 (1867): the text of Ibn Khallikân has been edited by Wüstenfeld (Göttingen, 1837), and an English translation was made by De Slane, for the Oriental Translation Fund (London, 1843). The references to the Persian texts of the histories by Mîrkhwând (or Amîrkhwând) and by Khwândamîr are to the lithographed editions, published in Bombay, of the *Rawḍat-aṣ-Ṣafâ* in 1266 (1850), and of the *Ḥabîb-as-Siyâr* in 1273 (1857). The two Saljûḳ chronicles are edited by Professor Houtsma in vols. I and IV of his *Recueil de Textes relatifs à l'Histoire des Seljoucides* (Leyden, 1886—1902). The first of these is by Ibn Ibrâhîm (otherwise called Muḥammad Ibrâhîm, or Muḥammad ibn Ibrâhîm), who flourished about the year 1025 (1616); and the second chronicle is by Ibn Bîbî, who wrote about 680 (1281). See also an article by Professor Houtsma in the *Zeit. Deutsch. Morg. Gesell.* 1885, p. 362.

to the Arabic Bayt-al-Jisr, meaning 'bridge-house.' Older Persian names like Baghdâd, 'the god-given place,' are rare; and here and there a Greek name survives, as for instance Al-Ubullah, representing Apologos.

The Greek province of Asia Minor, as already said, only became Moslem land after the Saljûḳ conquest, in the latter half of the 5th (11th) century; and hence the Greek names are often known to us in two forms, an earlier (Arabic) and a later (Turkish); as, for example, Seleucia given first as Salûḳîyah, later as Selefkeh; and Heraclia which we find at first as Hiraḳlah, and in more modern times as Arâkliyah. After the Saljûḳ occupation of the country and the subsequent Ottoman supremacy, Turkish names naturally come to supplant the earlier Greek nomenclature; but in the matter of orthography it must be remembered that the Arabic alphabet is quite as foreign to Turkish as it is to Greek, hence Turkish words (as every Turkish dictionary shows) often have alternative spellings, and the place-names are in like case. Thus we find both Ḳarâ Ḥiṣâr and Ḳarah Ḥiṣâr; Ḳarah-sî and Ḳarâsî; Ḳaramân and Ḳarâmân, with many other examples.

Looking over the maps of the Persian provinces, it is striking how few names there are of Arabic origin. With the exception of Marâghah in Adharbâyjân, and the hamlet of Bayzâ (Al-Bayḍâ, 'the white town') in Fârs, there is hardly an Arabic town name to be met with. The Moslems indeed changed little or nothing when they took over the Sassanian kingdom[1]. Very often villages and post-stations had names taken from some natural and notable object; as for example Myrtle village, Camel village, and Salt village; which in Persian were called Dih Murd, Dih Ushturân, and Dih Namak. These names the Arab geographers constantly

[1] It has been remarked that in all Moslem Spain, where rich cities abounded, there is only one that bears an Arabic name, to wit the port of Almeria, for Al-Marîyah, 'the Watch Tower.' A place-name like Calatayud, which might be taken for another instance, is not primarily the name given to the *town*, but was only the fortress—*Ḳal'at Ayyûb*, Job's Castle—below which a town afterwards sprang up. In many cases the original Iberian, Roman, or Visigothic name is for lack of documents unknown; as for instance in the case of Granada. *Mutatis mutandis*, the same remarks apply to Persia.

translate, and in their pages we find the above, for instance, given as Ḳaryat-al-Âs, Ḳaryat-al-Jamâl, and Ḳaryat-al-Milḥ, but there is every reason to believe that in Persia, at all times, the Persian name was in use; in other words it is here, as with us, when we speak of the Black Forest (Schwarz-Wald) or the Cape of Good Hope, such names likewise commonly varying on the maps, and in books, according to the language of the speaker.

It will be observed that we have sometimes in the Arabic lists the name of a post-stage, in Arabic, of which the Persian equivalent has not come down to us; e.g. in the case of Râs-al-Kalb, 'Dog's Head,' possibly the place later called Samnân. Also occasionally the Arabs gave a nickname to a Persian town, and both names continued simultaneously in use; as for instance Kanguvâr, which from the stealing of their mules here the early Moslems had called Ḳaṣr-al-Luṣûṣ, 'Robber Castle'; but Persian Kanguvâr has in the end survived the Arab nickname. Even when the Moslem conquerors founded a new provincial capital, as was the case with Shîrâz, which soon came to eclipse the older Iṣṭakhr (Persepolis), they seem to have taken and perpetuated in the new town the name of the original Persian village. The origin and etymology of the name Shîrâz, like many others, appears to be unattainable, for unfortunately the geography of the old Sassanian kingdom is almost entirely unknown to us.

The pronunciation of names, as is natural, varied with the lapse of time; Ṭuraythîth becomes Turshîz: Hamadhân is in later books spelt Hamadân[1]; further there was evidently an Arabic and a Persian pronunciation (or spelling) of the same name contemporaneously current, thus Arabic Ḳâshân is written Kâshân in Persian, Ṣâhik appears later as Châhik, and Ṣaghâniyân is Chaghâniyân. Then again, as the Arabic grammar demanded tri-consonantal roots, the Persian Bam had to be written in Arabic *Bamm*, and Ḳum *Ḳumm*; but this was merely to suit the rules of Arabic orthography, and the doubled final

[1] It is to be remarked that the *dh*, which the modern Persians pronounce *z* (e.g. Azarbâyjân, written Adharbâyjân), was apparently sometimes not given the *z* sound; thus Hamadhân is now called Hamadân, and never pronounced Hamazân. In Persian the Arabic *w* is generally, but not always, pronounced *v*, e.g. Ḳazwîn or Ḳazvîn.

consonant was never in use in the Persian. In some cases a name would fall into disuse for some unknown reason, to be replaced by another name, but Persian like the first; an instance occurs in Ḳirmâsîn or Ḳirmîsîn, later known as Kirmânshâhân, shortened to Kirmânshâh at the present day. But we are alike ignorant of the true import of these names, and the cause of the change.

In the matter of the prefixing of the Arabic article *Al* to place-names, the usage appears to be extremely arbitrary. The strict grammatical rule appears to be that the article is only prefixed to Arabic, not to foreign names. This rule, however, never was kept; for instance in Mesopotamia, where most of the names were of course of Semitic origin, the Tigris is always named Dijlah (without the article), but the Euphrates is Al-Furât, though this last is like the first a foreign word[1]. In the Persian provinces, the tendency was, with the lapse of time, to drop the Arabic article, e.g. (Arabic) As-Sîrajân becomes (Persian) Sîrjân. The usage however is quite arbitrary, for no explanation can be given why the ancient Rhages should be invariably called by the Arabs Ar-Ray, while Jay, the old name for one part of Isfahân, is always given without the article[2].

The Arabs were somewhat poverty-stricken in the matter of their nomenclature, and the lack is cause of much confusion. With them the capital of a province, as a rule, may be called by the name of the province, even when it has a name of its own; thus Damascus still is commonly known as Ash-Shâm, '(the capital of) Syria'; and Zaranj, the chief town of Sijistân, was

[1] Thus we have Al-Ubullah (an original Greek name) with the article, and a number of other instances occur. Purely Arab towns sometimes took the article, sometimes not; e.g. Al-Kûfah, said to mean 'the (city of the) Reed-huts'; but on the other hand, Wâsiṭ, 'the Middle-town,' is always written without the article, though here too it would have seemed equally appropriate.

[2] How little any rule holds is shown by the case of Jiddah, the port of Mecca, given both as Juddah, and as Al-Juddah by all the earlier writers. In the following pages where a place-name commonly occurs in the Arabic authors preceded by the article, this is, on first mention, so given. Subsequently, however, when the name is repeated, for the sake of brevity, and in the maps for distinctness, the article as a general rule is omitted. The use or disuse of the article varies with the different Arab geographers, and like their spelling of foreign names is the reverse of consistent.

more often known simply as Sijistân, for Madînat-Sijistân, 'the City' of that province. From this usage much confusion naturally arises when the province had two capitals. This for example is the case with the Kirmân province, where the name Kirmân (*scilicet* city) in the earlier books stands for the first capital Sîrjân, and in later times for the present city of Kirmân, a totally different town, which only became the capital when Sîrjân had gone to ruin. Also, on comparing together the maps, as deduced from the statements of the medieval geographers, with the map of the present day, it will often be found that the name of a lost city has been preserved in the modern district; thus of the lost Sîrjân city, for example, the name is still met with in the modern Sîrjân district; the same is the case with both Bardasîr and Jîruft, formerly each the name of an important town, now only preserved in the district. In short the district and its chief city being always, possibly, known by the same name, either one or the other with the lapse of time might become obsolete. Hence, and conversely to the foregoing examples, the name of the older Aradûn district is now given to the little town known as Aradûn, which of old was called Khuvâr (of Ray).

In physical geography the Arab nomenclature was not rich. Single and notable mountain peaks generally had proper names (e.g. Damâvand, Alvand), but as a rule no chain of mountains had any particular designation. The great Taurus range shutting off the Byzantine lands was often (and incorrectly) referred to as the Jabal Lukkâm, but this is properly only one mountain group of the Anti-Taurus; and the very notable range of the Alburz, dividing off the high Persian plateau from the Caspian, has, with the Arab geographers, no common term for its long chain of peaks. The great lakes generally had each its special name (e.g. Mâhâlû, Zarah, and Chîchast), but more commonly the lake was known by the name of the principal town on its shores; as for example the Urmîyah lake, and the lake of Vân also called after Arjîsh. Seas were even less distinctively named, being referred to by a variety of appellations, taken from the provinces or chief towns on their coasts. Thus the Caspian was indifferently termed the Sea of Ṭabaristân, or of Gîlân, or of Jurjân, also of Bâkû, and it was latterly known as the Khazar

Sea, from the kingdom of the Khazars which in the earlier middle-ages lay to the northward of it. In a similar way the Aral was known as the Sea of Khwârizm, and the Persian Gulf as the Sea of Fârs.

In conclusion it is to be understood that only a selection from our authorities is given in the following chapters; the number of towns and villages, the names of which are reported as being situated in this or that province, is very great, certainly more than double the sum catalogued in the index of the present work. But where the site could not even approximately be fixed, the mere name, one in a list, has been omitted. In regard to the maps, these, it will be noted, are simply diagrams to illustrate the text, and they do not show the country as it was at any one particular epoch. Thus towns, which in fact succeeded one another, are often marked as though existing at one and the same time, but the text will duly explain whether this was, or was not the case[1].

[1] Perhaps some apology is due for the inordinate number of references which crowd the footnotes of the following pages; though doubtless by the student, wishing to verify a fact, this will not be counted as a fault. All, or none, seemed the only course. The Moslem writers, Arabs, Persians and Turks, as is well known, are the greatest plagiarists in all literature, and seldom acknowledge their indebtedness. On the other hand, each geographer or historian generally adds something of his own to what he copies (unacknowledged) from a predecessor, and often by combining many authorities sufficient scraps of information are obtained definitely to substantiate a fact or fix a position. As an instance I may quote the case of the not very important town of Khurḳân, in the Ḳûmis province. Nothing much is known of it, but it seemed not unimportant to mark that this Khurḳân of Ḳûmis, though now disappeared from the map, was to be kept separate from the like-written name (in Arabic) of Kharraḳân in the Jibâl province. All that is known of the Ḳûmis town is its position; but to fix this, (1) Ḳazvînî has to be cited, who says the town stood four leagues from Bisṭâm; to which information (2) Yâḳût adds the fact that it stood on the road going to Astarâbâd; while (3) Mustawfî further tells us that in his day Khurḳân was an important village with a saint's tomb, and plentiful water supply, hence it was not a mere post-station. Yet to record all this, which amounts to so little, three authors have to be quoted, with references to their works, in the footnote.

CHAPTER II.

'IRÂḲ.

The division of Mesopotamia, Northern and Southern. 'Irâḳ or Babylonia. Change in the courses of the Euphrates and Tigris. The great irrigation canals. Baghdâd. Madâin and the cities on the Tigris thence down to Fam-aṣ-Ṣilḥ.

The great plain of Mesopotamia, through which the Euphrates and the Tigris take their course, is divided by nature into two parts. The northern half (the ancient kingdom of Assyria) consists mostly of pasture lands covering a stony plain; the southern half (the ancient Babylonia) is a rich alluvial country, where the date palm flourishes and the land is watered artificially by irrigation channels, and this for its exceeding fertility was accounted, throughout the East, as one of the four earthly paradises. The Arabs called the northern half of Mesopotamia Al-Jazîrah, 'the Island,' the southern half was known as Al-'Irâḳ, meaning 'the Cliff' or 'Shore,' but it is doubtful how this term came originally to be applied; possibly it represents an older name, now lost, or it was used originally in a different sense. The alluvial plain was also commonly known to the Arabs under the name of As-Sawâd, 'the Black Ground,' and by extension As-Sawâd is frequently used as synonymous with Al-'Irâḳ, thus coming to mean the whole province of Babylonia[1].

The frontier between 'Irâḳ and Jazîrah varied at different epochs. By the earlier Arab geographers the limit generally

[1] In its secondary sense *Sawâd* means 'the District' round a city, hence we have the Sawâd of Baghdâd, of Kûfah, and of Baṣrah frequently employed to designate respectively the environs of these cities.

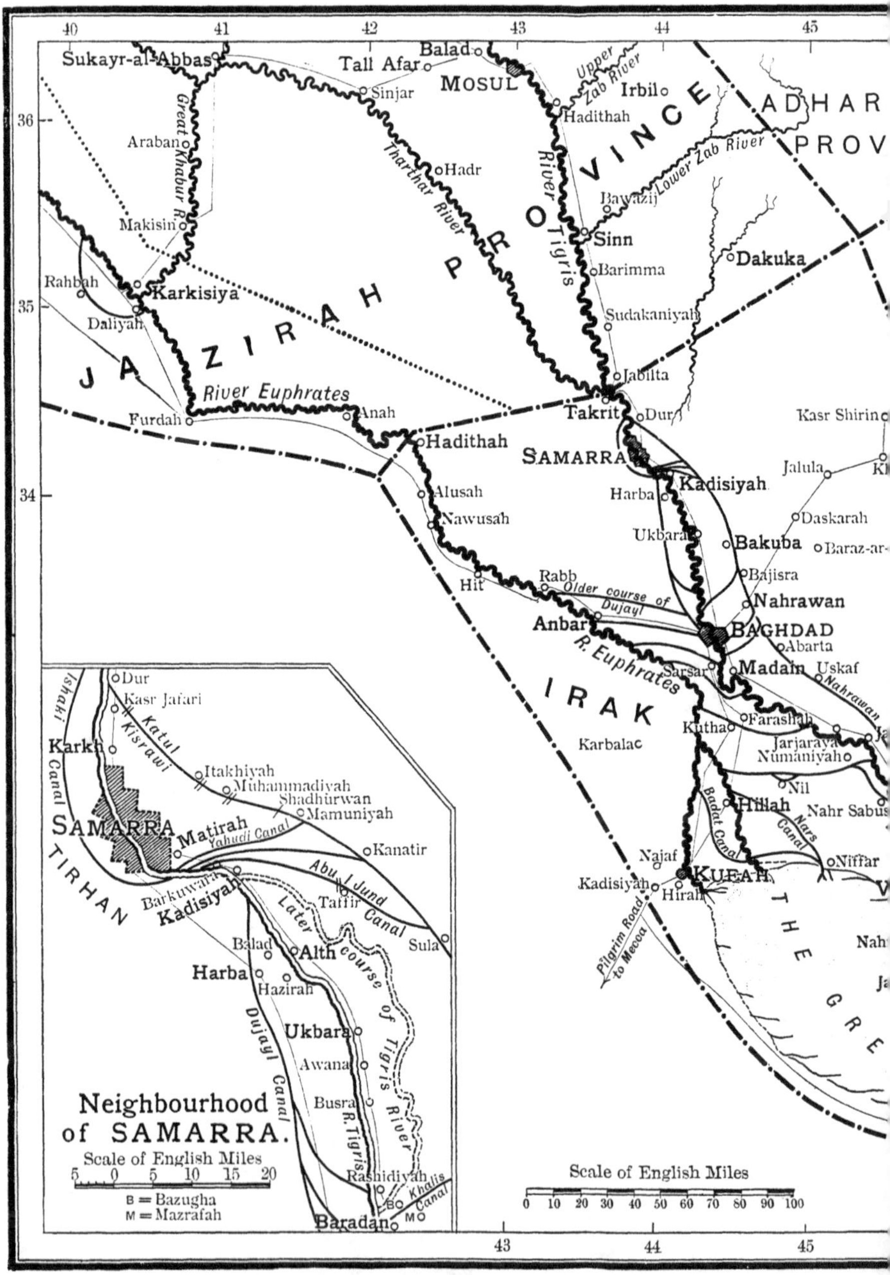

PROVINCES OF IRAK, KHU

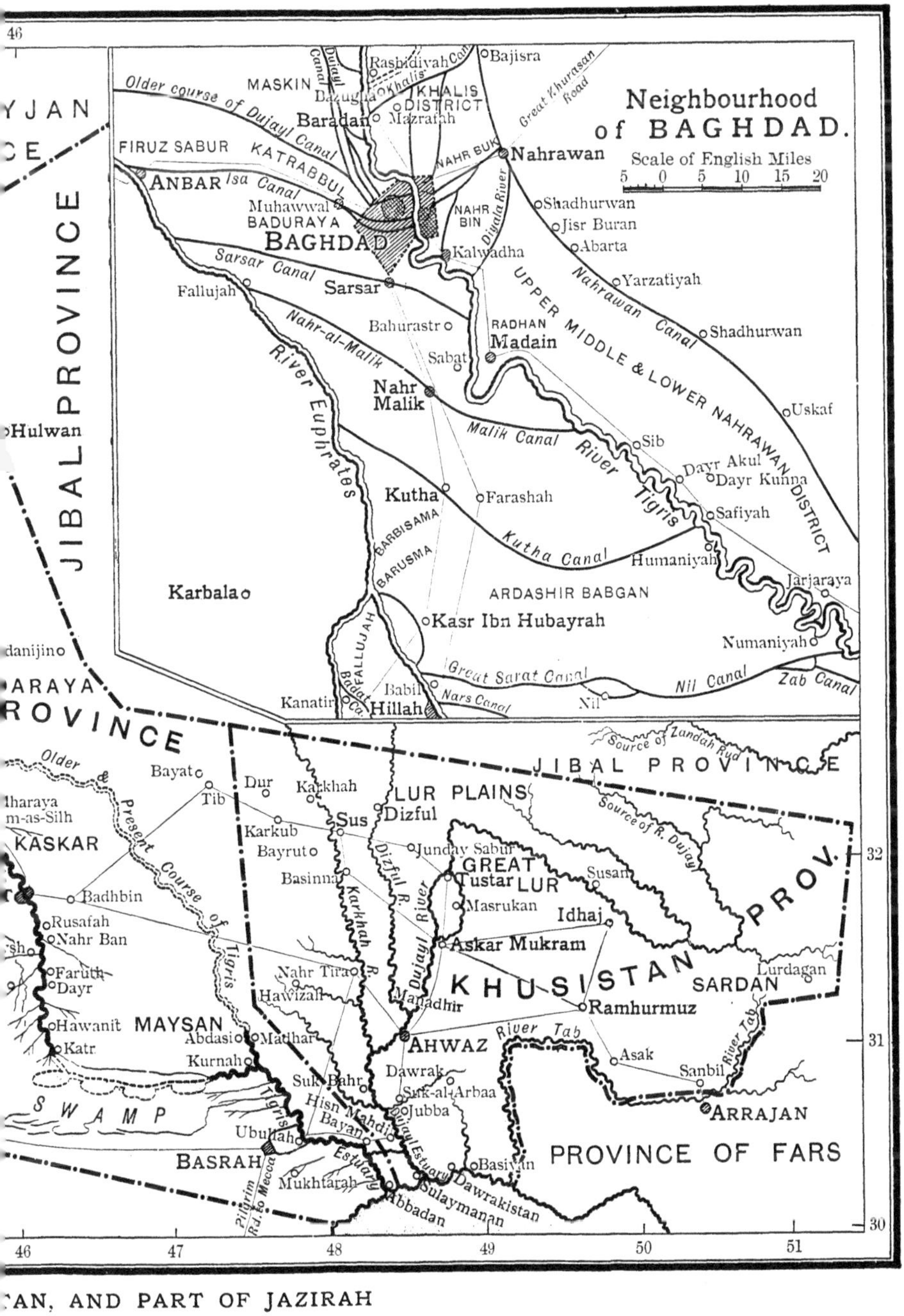

:AN, AND PART OF JAZIRAH

coincided with a line going north from Anbâr on the Euphrates to Takrît on the Tigris, both cities being reckoned as of 'Irâḳ. Later authorities make the line go almost due west from Takrît, so as to include in 'Irâḳ many of the towns on the Euphrates to the north of Anbâr; this, physically, is the more natural division between the two provinces, and it crosses the Euphrates below 'Ânah, where the river makes a great bend to the southward. The Euphrates was known to the Arabs as Al-Furât; the Tigris they called Dijlah (without the article), a name which occurs in the Targums as *Diglath*, corresponding to the latter part of *Hiddekel*, the form under which the Tigris is mentioned in the book of Genesis. When the Moslems conquered 'Irâḳ in the middle of the 1st (7th) century Ctesiphon, which they called Madâin, on the Tigris, was the chief city of the province, and the winter capital of the Sassanian kings. The Arabs, however, required cities for their own people, also to serve as standing camps, and three were before long founded, namely, Kûfah, Baṣrah, and Wâsiṭ, which rapidly grew to be the chief towns of the new Moslem province, Kûfah and Baṣrah more particularly being the twin capitals of 'Irâḳ during the Omayyad Caliphate[1].

With the change of dynasty from the Omayyads to the Abbasids a new capital of the empire was required, and the second Abbasid Caliph founded Baghdâd on the Tigris some miles above Ctesiphon (Madâin). Baghdâd soon eclipsed all the recent glories of Damascus under the Omayyads, becoming the metropolis of the Abbasid Caliphate, and naturally also the capital city of 'Irâḳ, which province now rose to be the heart and centre of the Moslem empire in the east.

During the middle-ages the physical conditions in 'Irâḳ were entirely different from what they are now, by reason of the great changes which have come to pass in the courses of the

[1] As such Kûfah and Baṣrah were known as Al-'Irâḳân (vulgarly Al-'Irâḳayn), meaning 'the two capitals of Al-'Irâḳ.' At a later date, however, when Kûfah and Baṣrah had lost their pre-eminence, the name Al-'Irâḳayn or 'the two 'Irâḳs' came to be used incorrectly, as though meaning the two *provinces* of 'Irâḳ, namely Arabian and Persian 'Irâḳ, the latter standing for the province of Al-Jibâl, but this will be more particularly explained in Chapter XIII.

Euphrates and Tigris, and the consequent ruin of the numerous irrigation canals which, under the earlier Caliphs, made 'Irâḳ a very Garden of Eden for fertility. At the present day, the Tigris, following a winding course in a direction mainly south-east, is joined at a point about 250 miles (as the crow flies) below Baghdâd by the waters of the Euphrates at Ḳurnah. The combined rivers, now known as the Shaṭṭ-al-'Arab (the Arab Stream), thence flow out to the Persian Gulf by a broad channel or tidal estuary measuring in length about a hundred miles in a direct line. This is what the modern map shows; but in early Moslem times, and, as will be demonstrated, in all probability as late as the middle of the 10th (16th) century, the Tigris, when it came about a hundred miles below Baghdâd, turned off south, from what is its present bed, flowing down by the channel now known as the Shaṭṭ-al-Ḥayy (the Snake Stream) to Wâsiṭ. This city occupied both banks of the river, and the Tigris some 60 miles below Wâsiṭ, after expending most of its waters by irrigation channels, finally spread out and became lost in the Great Swamp.

Throughout the middle-ages the Great Swamp, which covered an area 50 miles across, and very nearly 200 miles in length, came down to the immediate neighbourhood of Baṣrah. At its north-western end the swamp received the waters of the Euphrates a few miles to the south of Kûfah; for the main channel of the Euphrates was in those days the Kûfah arm of the river, that which flows by Ḥillah (now the main stream) being then only a great irrigation canal, called the Nahr Sûrâ. Along the northern edge of the lower part of the Great Swamp a line of lagoons, connected by open channels, made navigation possible; boats passing where the Tigris entered the swamp at Al-Ḳaṭr, to where (near modern Ḳurnah) the swamp surcharged by the waters of both Euphrates and Tigris drained out by the Abu-l-Asad canal into the head of the estuary of the Shaṭṭ-al-'Arab. By this waterway cargo-boats went down without difficulty from Baghdâd to Baṣrah, which last, the seaport of Baghdâd, lay at the end of a short canal, leading west out of the tidal estuary—the Blind Tigris as the Shaṭṭ-al-'Arab was then more commonly called.

The present course of the Tigris, as shown on the modern map, keeps to the eastward of the Shaṭṭ-al-Ḥayy channel, turning off at the village now known as Ḳûṭ-al-'Amârah, which stands for the medieval Mâdharâyâ; and this, the present channel down to Ḳurnah, was also apparently that occupied by the river during the period of the Sassanian monarchy, when the Great Swamp, described by the Arab geographers, did not as yet exist. The historian Balâdhurî dates the origin of the swamp as far back as the reign of Ḳubâdh I, the Sassanian king who reigned near the end of the 5th century A.D. In his day the dykes existing along the Tigris channel, as it then ran, having been for many years neglected, the waters suddenly rose, and pouring through a number of breaches, flooded all the low-lying lands to the south and south-west. During the reign of Anûshirwân the Just, son and successor of Ḳubâdh, the dykes were partially repaired and the lands brought back under cultivation; but under Khusraw Parwîz, the contemporary of the prophet Muḥammad, and in about the year 7 or 8 after the Flight (A.D. 629) the Euphrates and the Tigris again rose, and in such flood as had never before been seen. Both rivers burst their dykes in innumerable places, and finally laid all the surrounding country under water. According to Balâdhurî King Parwîz himself, when too late, superintended the re-setting of the dykes, sparing neither treasure nor men's lives, 'indeed he crucified in one day forty dyke-men, at a certain breach (Balâdhurî reports), and yet was unable to master the flood.' The waters could in no wise be got back, and the swamps thus formed became permanent; for during the succeeding years of anarchy and when the Moslem armies began to overrun Mesopotamia and the Sassanian monarchy perished, the dykes, such as still existed, naturally remained uncared for, 'and breaches came in all the embankments, for none gave heed, and the Dihḳâns (namely the Persian nobles, who were the landlords) were powerless to repair the dykes, so that the swamps every way lengthened and widened.'

The above well accounts for the formation of the Great Swamp, and Ibn Rustah refers to this epoch, under the last Sassanians, the first great shifting of the Tigris from the eastern channel, beyond Mâdharâyâ, to the western channel (the Shaṭṭ-al-Ḥayy) which passed down through the site subsequently occupied by the Moslem city of Wâsiṭ. This change, says Ibn Rustah, had turned all the country bordering the older eastern course into a desert, and so it remained in the 3rd (9th) century when he wrote. He then describes the back-water, six leagues long (above Ḳurnah), which ran up north to 'Abdasî and Madhâr, where the channel was stopped by a dam; this being evidently the last reach of the former, and present, eastern course of the Tigris. Ibn Rustah states that the dam, which in his time stopped all navigation above this point, had not existed in Sassanian days, when the channel was still open north of 'Abdasî and Madhâr right up to where this rejoined the Tigris course (of his day) in the district north of Wâsiṭ (at Mâdharâyâ), whence up stream the river was clear to Madâin. He continues:—'and of old, sea-going ships sailing in from India came up the Tigris (estuary, of the later) Baṣrah, and thence could attain to Madâin (Ctesiphon), for sailing on they came out above (the present) Fam-aṣ-Ṣilḥ into the Tigris reach of (the river below where, in later times, was) Baghdâd.'

The lower Tigris at the present day, therefore, flows in the bed which, in the main, it had followed during Sassanian times. But during all the centuries of the Abbasid Caliphate it poured into the swamps down the western channel past Wâsiṭ, and the question arises—when did the change back to the present eastern channel take place? The answer is that doubtless the change was brought about gradually, and from the silting up of the western arm; in any case, all our Moslem authorities, down to the age of Tîmûr and the beginning of the 9th (15th) century, describe the lower Tigris as still passing through Wâsiṭ, this fact being confirmed by Ḥâfiẓ Abrû writing in 820 (1417). One of the first travellers to speak of the eastern arm as the navigable channel, was John Newberie, who in 1581, after visiting Baghdâd, went down by boat in six days to Baṣrah, passing on the fifth day Ḳurnah, 'a castle which standeth upon the point where the river Furro (Euphrates) and the river of Bagdet (the Tigris) doe meet.' In the following century the Frenchman Tavernier made the same journey down the Tigris. He left Baghdâd in February 1652, and he states that at some considerable distance below this city the Tigris divided into two branches. The western channel (that by Wâsiṭ) was in his time no more navigable, but it ran—as he expresses it—'vers la pointe de la Mésopotamie.' The French traveller followed in his boat the present eastern channel, which took its course 'le long de l'ancienne Chaldée,' after leaving (Ḳût-al-)'Amârah; and just before coming to Baṣrah he passed Ḳurnah where, he says, the Tigris and Euphrates joined their streams[1].

The existence of the Great Swamp, and the consequent change in the courses of both Euphrates and Tigris, is the chief matter of note in the physical condition of Lower Mesopotamia during the Caliphate; but of almost equal importance was the system of canalisation inherited by the Arabs when, after the conquest, they took over the country from the Persians. Briefly, as already stated, we find that all 'Irâḳ north of the swamp, and between the two rivers, was then traversed, like the bars of a gridiron, by a succession of canals which drained eastward into the Tigris; while east of the Tigris a canal, 200 miles in length, called the Nahrawân, starting from below Takrît and re-entering the river fifty miles north of Wâsiṭ, effected the irrigation of the lands on the further or Persian side of the Tigris. The details of this great system of waterways will be explained more fully in due course, but a glance at the accompanying map, drawn

[1] Baladhuri, 292. I. R. 94. Yak. i. 669. In 1583 John Eldred went down from Baghdâd to Baṣrah, and also describes how one day's journey before the latter place 'the two rivers of Tigris and Euphrates meet, and there standeth a castle called Curna': see his voyage in Hakluyt, *Principal Navigations* (Glasgow, 1904), vi. 6; also v. 371, for in 1563 Cæsar Frederick had made the same journey and speaks of 'the castle of Corna' in similar terms. For the voyage of John Newberie, see *Purchas, His Pilgrimes* (folio, 1625—26), v. 1411, 1412; *Six Voyages en Turquie de J. B. Tavernier* (Utrecht, 1712), i. 240. Other travellers do not afford any detailed information. The earliest mention of the western (present) Tigris arm as navigable appears to be the anonymous Portuguese traveller, a copy of whose manuscript is in the possession of Major M. Hume (see *The Athenæum* for March 23rd, 1901, p. 373), who speaks of the castle (of Ḳurnah) six leagues above Baṣrah where the Euphrates and Tigris flowed together. His voyage from internal evidence must have been made in about the year 1555. The conclusion therefore appears to be that, from the time of Muḥammad, and during the nine following centuries, the Tigris took the western arm down to the swamps; afterwards, in the early part of the 16th century A.D., changing back into the eastern channel, which it had followed in Sassanian times before the rise of Islam, and which its main stream now follows at the present day.

up from the accounts of contemporary authorities, shows how the marvellous fertility of 'Irâḳ during Abbasid times was due to a strict economy of the water supply; and that while nearly all the land between the Euphrates and the Tigris was irrigated by the waters of the Euphrates led off through canals flowing eastward, the lands along the left bank of the Tigris, and towards the foot-hills of the Persian highlands, were made fertile by the canals of the Nahrawân, which economically distributed the surplus waters of the Tigris to the eastward, and caught the flood of the numerous streams flowing down from the mountains of Kurdistân.

The topography of Baghdâd has been dealt with in a previous volume[1], and all that is necessary in this place is to summarise the most important facts, in order to make clear the position of the Abbasid capital among the other cities of 'Irâḳ, and explain the details of the road system (already referred to in Chapter I) of which Baghdâd was the central point.

The first of the great canals which ran from the Euphrates to the Tigris was the Nahr 'Îsâ[2], and just above where its waters flowed out into the latter river, the Caliph Manṣûr about the year 145 (762) built the Round city, which became the nucleus of Baghdâd. The Round city had four equidistant gates lying one Arab mile apart each from the other, and from every gate went a high road. Great suburbs were in time built on these four roads, and these before long came to be incorporated in the circuit of the great metropolis. The four gates of the Round city were (1) the Baṣrah Gate to the S.E. opening on the suburbs along the Tigris bank where the various branches of the 'Îsâ canal flowed out; (2) the Kûfah Gate to the S.W. opening on the high road

[1] *Baghdâd during the Abbasid Caliphate* (Oxford, 1900). It is to be noted that the number of districts, towns, and villages in 'Irâḳ of which information has come down is very great, and a volume would be needed to report all that is known of this, the capital province of the Abbasids. The map constructed for the paper on Ibn Serapion (*Jour. Roy. Asiat. Soc.* 1895, p. 32) gives all the places lying on the rivers and canals, but this does not exhaust the list, and the reader may be referred to the work of Professor M. Streck, *Die alte Landschaft Babylonien* (Leyden, 1901), for fuller details, which it is impossible to find place for in the present chapter.

[2] *Nahr* means both 'canal' and 'river' in Arabic; 'Îsâ was the name of the Abbasid prince who dug the canal.

going south, which was the Pilgrim road to Mecca; (3) the Syrian Gate to the N.W. where the high road branched left to Anbâr on the Euphrates, and right to the towns on the western Tigris bank north of Baghdâd; and (4) the Khurâsân Gate leading to the main bridge of boats for crossing the river. By this bridge East Baghdâd was reached, at first known as the Camp of Mahdî, son and successor of the Caliph Manṣûr, and Mahdî built his palace here, also founding the great Friday Mosque of East Baghdâd. The settlement on the east side was divided into three quarters, that near the bridge head was known as the Ruṣâfah quarter, the Shammâsîyah quarter lay above it along the river bank, and the Mukharrim quarter below it. These three quarters of East Baghdâd were surrounded by a semicircular wall, going from the river bank above the Shammâsîyah to the river again below the Mukharrim; and across the middle and narrow part of East Baghdâd went the beginning of the great Khurâsân road, starting from the Khurâsân Gate of the Round city, and crossing the main bridge to the (second) Khurâsân Gate of East Baghdâd, whence, as explained in the previous chapter, the trunk road went east to the limits of the Moslem empire.

From the Kûfah Gate of the Round city, as already stated, led the Kûfah or Pilgrim road, going south, and the great suburb which here stretched to a point nearly a league distant from the walls of the Round city was known as Karkh. The suburb of the Muḥawwal Gate lay to the westward of the Round city, being reached from both the Kûfah Gate and the Syrian Gate, where the roads converging fell into the great western high road going through the town of Muḥawwal to Anbâr. North of the Syrian Gate was the Ḥarbîyah quarter (balancing Karkh on the south of the Round city), and beyond the Ḥarbîyah and surrounded on two sides by a bend in the river were the northern cemeteries of West Baghdâd, at a later time famous as the Kâẓimayn, and so named from the tombs of two of the Shî'ah Imâms.

The city of Baghdâd occupied the central point of four districts, two being on either bank of the Tigris. On the western side the Ḳaṭrabbul district was north of the 'Îsâ canal, and Bâdûrayâ lay to the south of the same; while on the eastern bank the Nahr Bûḳ district was to the north of the line of the Khurâsân road, and Kalwâdhâ district to the southward; the town of Kalwâdhâ standing on the river bank a short distance below the southernmost gate of East Baghdâd. From Baghdâd, as the central point of the road system of the empire, two roads (as already said), going south and west, bifurcated at the Kûfah Gate of the Round city; and two, going north and east, passed through East Baghdâd, having their starting-point at the further end of the main bridge of boats. The southern road, to Kûfah (and Mecca), after leaving the suburb of Karkh, came before long to the town of Ṣarṣar, on the Nahr Ṣarṣar, the second of the great canals from the Euphrates to the Tigris, which flowed parallel with the Nahr 'Îsâ on the south. The western or Anbâr road turning off at the Kûfah Gate, and passing through the suburb of Barâthâ, came after about a league to the town of Muḥawwal which stood on the 'Îsâ canal. The eastern or Khurâsân road left East Baghdâd (as already said) at the Khurâsân Gate, north of the Mukharrim quarter, and the first town reached was Nahrawân Bridge at the crossing of the great canal of this name. Finally, the northern road passed through the Shammâsîyah quarter to the Baradân Gate of East Baghdâd, and shortly came to the town of Baradân lying on the east bank of the Tigris; whence, keeping along the left bank of the river, the high road reached Sâmarrâ and the towns of northern Mesopotamia.

During the five centuries of the Abbasid Caliphate the plan of Baghdâd with its suburbs changed considerably as the city grew and in parts fell to ruin. What has been sketched in the foregoing paragraphs was the city as it existed in the time of Hârûn-ar-Rashîd. The civil war which broke out after his death brought about the ruin of the Round city. In 221 (836) the seat of the Caliphate was removed to Sâmarrâ, and during the reigns of seven Caliphs Baghdâd was reduced to the condition of a provincial town. When finally in 279 (892) Sâmarrâ was abandoned and the Caliph re-established his court in the old capital, it was East Baghdâd, where many new palaces came to be built, which succeeded to the glories of the Round city, now falling more and more to ruin; and for the next four centuries, down to the invasion of the Mongols, the Caliphs permanently established their residence on the east bank.

These palaces of the later Caliphs were built on the land to the south of Mukharrim, the lowest of the three quarters included within the wall of East Baghdâd as it had existed in the time of Hârûn-ar-Rashîd. These three quarters, at the date in question, had fallen to ruin, but the new palaces quickly came to be surrounded by new suburbs, which in their turn were before long enclosed by a great semicircular wall. The new wall of East Baghdâd, including in its circuit a part of the older Mukharrim, went from the river bank above the palaces to the river bank below (adjacent to Kalwâdhâ), and it was built by the Caliph Mustaẓhir in 488 (1095). This was the wall, more than once repaired, which finally in 656 (1258) proved impotent to withstand the Mongol attack, and the Abbasid Caliphate fell. At the present day this ruined wall remains, enclosing within its wide circuit the few relics that time has left of the city of the Caliphs, and still protecting modern Baghdâd, which is as heretofore the capital of ‘Irâḳ, and the residence of its Turkish Governor.

Seven leagues below Baghdâd, and occupying both banks of the Tigris, lay Al-Madâin, ‘the Cities,’ as the Arabs called the ruins of the twin capitals, Ctesiphon and Seleucia, which had been founded under the earlier Seleucids three centuries before Christ. Seleucia of the west bank had received its name from Seleucus Nicator. The name of Ctesiphon, which the Arabs give under the shortened form of Ṭaysafûn, is of uncertain etymology; though in appearance it is Greek, it probably is a corruption of the old Persian name of the city, for it is not known to us how the Sassanians called this capital of their empire[1]. In 540 A.D. Anûshirwân the Just had taken Antioch of Syria, with Seleucia on the Orontes, and after the fashion of Persian monarchs had transported the inhabitants of this Seleucia to his capital at Ctesiphon. Here he settled them in a new suburb on the east side of the Tigris, opposite therefore to the site of Seleucia of

[1] It has been plausibly suggested that Ctesiphon is to be identified with Casiphia of the book of Ezra (viii. 17), which lay between Babylon and Jerusalem, and which in the Septuagint version is named ‘the Silver City.’ *Madâin* is merely the Arabic plural of *Madînah*, ‘a city’; and *Casiphia* would be the Chaldee form of the Persian name, now lost, of the capital of the Chosroes.

Mesopotamia; and this suburb existed when the Arabs conquered the country in the following century, being still known as Rûmîyah, the Roman (or Greek) town, which some report to have been built on the plan of Antioch.

Al-Madâin, according to the Moslem authors, consisted of seven cities, whose names, with divers readings, are duly chronicled; but five cities only appear to have been in existence and inhabited when Ya'ḳûbî wrote in the 3rd (9th) century. These were, on the east bank, Al-Madînah-al-'Atîḳah, 'the Old Town,' corresponding with Ctesiphon, and one mile south of it Asbânbur, adjacent to which lay Rûmîyah. On the opposite bank of the Tigris was Bahurasîr, a corruption of Bih-Ardashîr—'the good town of King Ardashîr'—and one league below it was Sâbâṭ, which according to Yâḳût was called by the Persians Balâsâbâd.

The great Sassanian palace, of which the ruins still exist, on the eastern bank of the Tigris, was known to the Arabs under the name of the Aywân-Kisrâ, 'the Hall of the Chosroes,' and this, according to Ya'ḳûbî, stood in Asbânbur; while another great building known as Al-Ḳaṣr-al-Abyaḍ, 'the White Palace,' was to be seen in the Old Town a mile distant to the north. This last, however, must have disappeared by the beginning of the 4th (10th) century, for all later authorities give the names of 'the White Palace,' and 'the Hall of the Chosroes' indifferently to the great arched building which to the present day exists here as the sole relic of the Sassanian kings. This building had a narrow escape from complete destruction in the middle of the 2nd (8th) century, when Manṣûr was founding Baghdâd; for the Caliph expressed his intention of demolishing the Sassanian palace, and using the materials for his new city. His Persian Wazîr, Khâlid the Barmecide, in vain attempted to dissuade him from this act of barbarity, but the Caliph was obstinate; the Wazîr, however, gained his point for, when the order came to be carried into effect the demolition was found to be more costly than the materials were worth for the new buildings, and the Arch of the Chosroes, as Yâḳût calls it, was left to stand. At a later period much of its stone work was carried off for the battlements of the new palace of the Tâj in East Baghdâd, which the Caliph 'Alî Muktafî finished building in the year 290 (903).

In the 4th (10th) century Madâin, which is at the present day a complete ruin, was a small and populous town, with a fine Friday Mosque dating from the days of the Moslem conquest; near which stood the tomb of Salmân the Persian, one of the best known Companions of the prophet Muḥammad. The markets of Madâin were built of burnt brick and were well provided. In the neighbouring Rûmîyah, the Caliph Manṣûr had for a time held his court, while at Sâbâṭ on the opposite bank Mamûn had also resided. The grandeur of the ancient palace of the Chosroes is a theme on which the Arab geographers relate many details. Ya'ḳûbî says that the summit of the great arch is 80 ells in height; Yâḳût refers to the magnificent kiln-burnt bricks, each near an ell in length by somewhat less than a span in width. Mustawfî, who gives the legendary account of Madâin and its palace, reports that in the 8th (14th) century both Madâin and Rûmîyah had come to be uninhabited ruins, though the villages opposite, on the western bank, still retained their inhabitants. Of these, he adds, the most important was Bahurasîr, already mentioned, which Yâḳût, who had been there, calls Ar-Rûmaḳân. To the south of it lay Zarîrân, a stage on the Pilgrim road, and to the west Ṣarṣar, already mentioned, on the Ṣarṣar canal, which last fell into the Tigris a short distance above Madâin. The district round Madâin, which stretched eastward from the Tigris to the Nahrawân canal, was known as Râdhân (Upper and Lower), of which Yâḳût names numerous villages, and Mustawfî praises the magnificent crops harvested here[1].

Dayr-al-'Âḳûl, 'the Convent of the (river) Loop,' is still marked on the map, situate on the east bank 10 leagues below Madâin, and the name is descriptive of the Tigris course at this point. It was a Christian monastery, surrounded by a town of considerable size, the latter being counted as the chief city of the district of Middle Nahrawân. In the town was a Friday Mosque[2], standing

[1] Ykb. 320, 321. I. S. 9. I. H. 167. Muk. 122. Yak. i. 425, 426, 768, 809; ii. 729, 929; iii. 3. Mst. 139, 140.

[2] This convenient, but of course incorrect term translates the Arabic *Masjid-al-Jâmi'*, otherwise rendered a Great Mosque. The Moslems have two categories of mosques. Small mosques (*Masjid*) where any one could pray at any time, often equivalent to a *Maḳâm* or *Mashhad*, the 'shrine' or 'place of

at some distance from the market place. Ibn Rustah at the close of the 3rd (9th) century describes the toll-barrier which was set across the Tigris here, and kept closed by the officer of the customs. He writes:—'the toll-bar (*Al-Maasir*) is the name given to the places on the Tigris where two boats have been moored on the one bank of the river, opposite two other boats on the further bank, which two likewise are firmly moored. Then across the stream they have carried cables, the ends being fastened on either bank to these boats, and thus ships are prevented from passing at night without paying toll.' Muḳaddasî in the 4th (10th) century refers to Dayr-al-'Âḳûl as one of the finest cities of this region of the river bank, but afterwards the bed of the Tigris changed and Yâḳût in the 7th (13th) century says that the great convent then lay a mile distant from the Tigris, standing solitary in the midst of the plain. Mustawfî, however, in the following century still counts Dayr-al-'Âḳûl as a large town, having, he adds, a damp climate on account of its surrounding palm-groves.

Also on the east bank, but lying three leagues above Dayr-al-'Âḳûl, was the small town of As-Sîb, for distinction called Sîb of the Banî Kûmâ, which was noted for its olive-groves, and famous in history for the battle which took place here in 262 (876), when Ya'ḳûb the Saffarid was defeated by the troops of the Caliph Mu'tamid. A short distance below Dayr-al-'Âḳûl stood the monastery of Marmârî, surnamed the Disciple, otherwise called Dayr Ḳunnâ (or Ḳunnah), which lay a mile to the east of the Tigris, and 16 leagues from Baghdâd. The historian Shâbustî in

martyrdom' of a saint. The *Muṣallâ* or 'praying-place' was more especially that used at the services of the great festivals. The Great Mosque, on the other hand, was where weekly the Friday prayers were said, and the sermon (*Khuṭbah*) preached; and it was called Masjid-al-Jâmi', 'the Mosque of the Congregation'—terms often translated by 'the Cathedral,' or 'Congregational Mosque.' The possession of a *Jâmi'* or *Mimbar* (pulpit, for the Friday Sermon) generally is a criterion of the size of a town, or village; and the fact is often mentioned as such by the Arab geographers; Iṣṭakhrî for instance gives a long list of places in Fârs which had, or had not a Mimbar; and this comes to much the same as if it were said that in such and such a village, in a Christian land, stood the parish church. At a later date the term Masjid-al-Jâmi' became changed to Masjid-al-Jum'ah, meaning 'the Friday Mosque,' but this is not the classical usage.

the 4th (10th) century (quoted by Yâḳût) describes it as a great monastery surrounded by so high and strong a wall as to be like a fortress and impregnable. Within the wall were a hundred cells for the monks, and the right to a cell was only to be bought for a price ranging from two hundred to a thousand dînârs (£100 to £500). Each cell stood in its own garden, watered by a small canal and planted with fruit trees which produced a crop that yearly might be sold for from 50 to 200 dînârs (£25 to £100).

Over against Dayr Ḳunnâ, but on the Tigris bank, was the small town of Aṣ-Ṣâfîyah, which Yâḳût writes was in his day already a ruin; and opposite this on the western side lay Humânîyah (or Humaynîyah) which is still found on the map, two leagues S.E. of Dayr-al-'Âḳûl. In the beginning of the 3rd (9th) century Humânîyah was a place of some importance, for after the death of the Caliph Amîn, his two sons and his mother, Zubaydah, widow of Hârûn-ar-Rashîd, were for a time sent to be kept in prison here by Mamûn; and Yâḳût in the 7th (13th) century describes Humânîyah as a large village surrounded by well cultivated lands[1].

Jarjarâyâ, or Jarjarây, which still exists, lay four leagues S.E. of Dayr-al-'Âḳûl. It is described by Muḳaddasî in the 4th (10th) century as having been a large town, and its Friday Mosque stood close to the Tigris, which surrounded the town on two sides. Ya'ḳûbî writing in the previous century states that its population chiefly consisted of Persian nobles, and it was the capital of the district of Lower Nahrawân. In the 7th (13th) century, according to Yâḳût, it was, like most of the towns of the Nahrawân districts, in a state of complete ruin. On the western bank of the Tigris, four leagues below Jarjarâyâ, at the ruins now called Tall-Nu'mân stood the town of An-Nu'mânîyah, which Yâḳût counts as the half-way stage between Baghdâd and Wâsiṭ. An-Nu'mânîyah was the capital of the Upper Zâb district, its Friday Mosque standing in the market place, and Ya'ḳûbî adds that near by stood the monastery called Dayr Hizḳil, where mad people were looked after by the monks. Nu'mânîyah was celebrated according to Ibn Rustah for its looms, where carpets like those of Ḥîrah were

[1] I. R. 185, 186. Ykb. 321. Kud. 193. Muk. 122. Mas. *Tanbih* 149. Yak. ii. 676, 687; iii. 362; iv. 980. Mst. 139. Ibn-al-Athir, vi. 207.

manufactured. In the 8th (14th) century Mustawfî still speaks of Nu'mânîyah as a flourishing town surrounded by date-groves. The small town of Jabbul lay on the eastern bank, nine leagues below Jarjarâyâ, where Ibn Rustah in the 3rd (9th) century says that there were government bake-houses. It was then a large hamlet, having a Friday Mosque standing in the market place, and Muḳaddasî describes it as of the size of Dayr-al-'Âḳûl; but when Yâḳût wrote, Jabbul had sunk to the size of a big village[1].

The town of Mâdharâyâ occupied the position where at the present day Ḳûṭ-al-'Amârah stands, namely at the bifurcation of the Shaṭṭ-al-Ḥayy from the eastern, and modern bed of the Tigris, which now goes thence south-eastward down to Ḳurnah. Mâdharâyâ was on the east bank, and in the 3rd (9th) century it was inhabited by Persian nobles. Here the great Nahrawân canal flowed back into the Tigris; and immediately below Mâdharâyâ came Al-Mubârak, a town which lay opposite Nahr Sâbus on the western bank of the Tigris. The town of Nahr Sâbus was at the mouth of the canal of this name, which will be spoken of later, and this was the chief town of the Lower Zâb district; it was counted as five leagues distant from Jabbul. On the opposite bank, and five leagues down stream, was the Ṣilḥ canal with the town called Fam-aṣ-Ṣilḥ at its 'mouth' (*Fam*), or point of origin, which latter lay seven leagues above Wâsiṭ. Fam-aṣ-Ṣilḥ town stood on the Tigris bank, it had fine markets and a Friday Mosque, according to Ibn Rustah. This place was famous in Moslem history for the magnificent palace built here by Ḥasan ibn Sahl, the Wazîr of Mamûn, in which he celebrated the marriage of his daughter Bûrân with the Caliph, spending fabulous sums in banquets and gifts, as will be found chronicled in the pages of Mas'ûdî. Fam-aṣ-Ṣilḥ afterwards fell to ruin, and Yâḳût who visited it in the 7th (13th) century, found the town and neighbouring villages along the canal for the most part uninhabited[2]. From the town of Fam-aṣ-Ṣilḥ the buildings of the Great Mosque in Wâsiṭ were visible on the southern horizon.

[1] Kud. 193. Ykb. 321. I. R. 186, 187. Muk. 122. Yak. ii. 23, 54; iv. 796. A. F. 305. Mst. 141.

[2] Ykb. 321. Kud. 194. I. R. 187. Yak. ii. 903; iii. 917; iv. 381. Mas. vii. 65.

CHAPTER III.

‘IRÂḲ (*continued*).

Wâsiṭ. The Great Swamps. Madhâr and Ḳurnah. The Blind Tigris. Baṣrah and its canals. Ubullah and ‘Abbâdân. The Tigris above Baghdâd. Baradân. The Dujayl district. ‘Ukbârâ, Ḥarbâ, and Ḳâdisîyah.

Wâsiṭ, the ‘middle city,’ was so called because it lay equidistant (about 50 leagues) from Kûfah, Baṣrah, and Ahwâz. It was the chief town of the Kaskar district, and before the foundation of Baghdâd, as already said, was one of the three chief Moslem cities of ‘Irâḳ.

Wâsiṭ was founded about the year 84 (703) by Ḥajjâj, the famous viceroy of Mesopotamia in the reign of the Omayyad Caliph ‘Abd-al-Malik. The city occupied both banks of the Tigris, the two halves being connected by a bridge of boats, and there were two Friday Mosques, one for each half of the city. Ya‘ḳûbî states that eastern Wâsiṭ had been a town before the days of Ḥajjâj, and here in the 3rd (9th) century the population was still for the most part Persian. In the western half of the city stood the Green Palace, built by Ḥajjâj, and called Al-Ḳubbat Al-Khaḍrâ, celebrated for its great dome, from the summit of which Fam-aṣ-Ṣilḥ seven leagues distant to the north could be seen. The lands round Wâsiṭ were extremely fertile, and their crops provisioned Baghdâd in time of scarcity; also paying yearly into the treasury a million of dirhams (£40,000) from taxes, as reported by Ibn Ḥawḳal, who was at Wâsiṭ in 358 (969). Muḳaddasî states that the mosque in the eastern half of Wâsiṭ likewise was built by Ḥajjâj. The town markets were magnificent and well stocked, also at either end of the bridge of boats were two small harbours where boats moored for convenience of discharging cargo.

During the whole period of the Caliphate Wâsiṭ continued to be one of the most important cities of 'Irâķ, and apparently the eastern quarter was the first to fall to ruin, for Ķazwînî, who was Judge at Wâsiṭ in the latter half of the 7th (13th) century, speaks of the town as lying solely on the western Tigris bank. Ibn Baṭûṭah, who was here in the early part of the following century, praises the fine buildings of the city, especially a great Madrasah, or college, with 300 rooms for students, and Mustawfî his contemporary speaks of the immense palm-groves lying round the town which made its climate very damp. At the close of the 8th (14th) century Wâsiṭ is frequently mentioned as a place of importance during the various campaigns of Tîmûr, who kept a strong garrison here; but about a century after this, as already described in the beginning of the last chapter, the Tigris ceased to flow past Wâsiṭ, taking the eastern course down by Ķurnah, and the city fell to complete ruin. Ḥâjjî Khalfah, writing in the beginning of the 11th (17th) century, speaks of it as then standing in the desert, but the canal was famous for its reeds from which pens were made[1].

Below Wâsiṭ, according to Yâķût, the Tigris flowed out into the Great Swamp by five navigable waterways, the names of which he gives, and this statement is corroborated by the accounts of earlier writers. Ibn Serapion mentions a number of towns lying on the main arm of the river below Wâsiṭ, and above Al-Ķaṭr, where in the 4th (10th) century the swamp began. The first of these towns was Ar-Ruṣâfah, 'the Causeway,' lying on the left bank, ten leagues from Wâsiṭ, and near it flowing eastward into the swamp was the canal called Nahr Bân, with the town of the same name,

[1] Ykb. 322. I. R. 187. Ist. 82. I. H. 162. Muk. 118. Kaz. ii. 320. I. B. ii. 2. Mst. 141. A. Y. i. 640, 657; ii. 517. J. N. 463. The ruins of Wâsiṭ do not appear to have been examined by any recent explorer. Their position on the Shaṭṭ-al-Ḥayy is fixed within narrow limits by the Arab itineraries. Chesney (*Report of the Euphrates and Tigris Expedition*, i. 37) states that these ruins were visited by Ormsby and Elliott in 1831—2, but he does not mark their position.

also spelt Nahr Abân, at its exit. Below this came Al-Fârûth and then Dayr-al-'Ummâl, 'the Convent of the Governors.' These were on the eastern bank, opposite to which and flowing west into the swamp were three canals, first the Nahr Ḳuraysh with a great village on it of the same name; then Nahr-as-Sîb, on which stood the towns of Al-Jawâmid, 'the Dried-lands,' and Al-'Uḳr; finally, the Nahr Bardûdâ on which lay the town of Ash-Shadîdîyah. All these were important towns lying in the swamp, round and about Al-Jâmidah, otherwise called (in the plural) Al-Jawâmid; further, Muḳaddasî describes a large town in this region called Aṣ-Ṣalîḳ, standing on an open lagoon which was surrounded by farmsteads and well cultivated lands. Over against these places and on the eastern bank of the main arm of the Tigris was Al-Ḥawânît, 'the Taverns,' where there was a toll-bar moored across the river, like the one already described at Dayr-al-'Âḳûl (p. 36), and this was close to Al-Ḳaṭr, 12 leagues below Ruṣâfah, where, according to Ibn Rustah, the Tigris in the 3rd (9th) century dividing into three arms finally entered the swamp[1].

The Swamps were called Al-Baṭâiḥ (the plural form of *Al-Baṭîḥah*, signifying a 'lagoon') and their history has been already described (p. 26). The whole area covered by them was dotted with towns and villages, each standing on its canal, and though the climate was very feverish the soil, when drained, was most fertile. Ibn Rustah writing at the close of the 3rd (9th) century describes the Great Swamp as everywhere covered by reed-beds, intersected by water channels, where immense quantities of fish were caught, which, after being salted, were despatched to all the neighbouring provinces. In regard to the Tigris waters, it appears that from Ḳaṭr eastward—and probably following, approximately, the line of the present channel of the Euphrates—the waterway led through a succession of open lagoons to the Abu-l-Asad canal, by which the waters of the swamp drained out to the Baṣrah estuary. These lagoons of open water, clear of reeds, were called *Hawr* or *Hawl* by the Arabs, and the lagoons were connected by channels navigable for small boats. The great river barges,

[1] I. S. 9, 20. Kud. 194. I. R. 184, 185. Muk. 119. Yak. ii. 10, 553; iii. 209, 415, 840; iv. 217, 758.

according to Ibn Rustah, did not pass below Ḳaṭr, but here transferred their cargoes to wherries, so light of draught as to pass through the channels threading the lagoons. All along these channels, stations on platforms had been made, where in huts built of reeds, and thus raised above the plague of gnats, guards were posted to keep the course clear and to protect wayfarers, for the recesses of the Great Swamp were the natural hiding-place of outlaws.

Ibn Serapion gives the names of four of the great lagoons (Hawr, or Hawl) through which the waterway went towards Baṣrah. The first was called Baḥaṣṣâ, the second was the Bakamṣî lagoon, then the Baṣrayâthâ, and the fourth was the Hawr-al-Muḥammadîyah, the largest of all, on which stood the tower called Minârah Ḥassân, after Ḥassân the Nabathæan who had been employed by the Omayyad viceroy Ḥajjâj to drain and reclaim lands in the Great Swamp. Beyond this last lagoon came the channel passing the villages of Al-Ḥâlah and Al-Kawânîn, and ending in the canal of Abu-l-Asad, which finally carried the waters of the swamp to the head of the Tigris estuary. This Abu-l-Asad, whose canal roughly corresponds with the last reach of the present course of the Euphrates above Ḳurnah, had been a freedman of the Caliph Manṣûr, and when in command of troops at Baṣrah he dug, or more probably widened, the boat channel which, as Yâḳût remarks, had doubtless existed here from Sassanian times. Ḳurnah, at the present point of junction of the Euphrates and Tigris, is not mentioned by any of the Arab geographers, and the first notice of this castle appears in the Turkish *Jahân Numâ* at the beginning of the 11th (17th) century.

The last reach of the eastern course of the Tigris—that of Sassanian times, as also of the present day—existed, as already said, in the middle-ages as a back-water, stopped at its northern end by a dam. This back-water, called the Nahr-al-Madhâr, was six leagues in length, and led to the two cities of 'Abdasî (or 'Abdâsî) and Al-Madhâr; the exact sites of which are unknown. The surrounding district—along the then desiccated eastern bed of the Tigris—was called Jûkhâ, and it stretched north-westward to Kaskar, the district of Wâsiṭ. Madhâr had been a city of much importance at the time of the Moslem conquest, and was then the capital of the district of Maysân, otherwise called Dasti-Maysân. Madhâr is described as lying four days' journey from Baṣrah, and was celebrated for its beautiful mosque and the much venerated tomb of 'Abd-Allah, son of the Caliph 'Alî. The neighbouring town of 'Abdasî, according to Yâḳût, was of Persian origin, that name being the Arabic form of the older Afdâsahî, which had been a hamlet of the Kaskar district before the conquest. Kaskar and Maysân were the two districts of the eastern part of the Great Swamp, and Kaskar, according to Ḳazwînî, produced much excellent rice which was exported. On its pastures buffaloes, oxen, and goats were fattened; the reed-beds sheltered ducks and water-fowl that were snared and sent in to the markets of the surrounding towns, while in its canals the shad-fish (called *Shabbût*) was caught in great numbers, salted and exported. Further, in Maysân might be seen the tomb of the prophet 'Uzayr, otherwise Ezra, which Ḳazwînî says was at a place settled entirely by Jews, who served the shrine. This was renowned throughout the countryside as a spot where prayers were answered, and in consequence the shrine was made rich by votive offerings[1].

The broad estuary formed by the combined Tigris and Euphrates waters, nearly a hundred miles in length, began at the exit of the Abu-l-Asad canal, and flowed out to the Persian Gulf at 'Abbâdân. This estuary was variously known as the Blind Tigris (Dijlah-al-'Awrâ), or the Fayḍ (the estuary) of Baṣrah, and the Persians named it Bahmanshîr; at the present day it is generally known as the Shaṭṭ-al-'Arab, 'the Arab River.' The tide from the Persian Gulf came up it, reaching as far north as the head of the channel at Madhâr and 'Abdasî, also filling and emptying the numerous canals of Baṣrah, and those irrigating the lands east and west of the estuary. Baṣrah, the great commercial port of 'Irâḳ, lay close to the border of the desert, at some distance to the west of the estuary, with which it was in water communication by means of two canals. Both north and south of Baṣrah numerous canals drained the lower waters of the Great Swamp

[1] I. R. 94, 185. I. S. 28. Kud. 240. Baladhuri, 293, 342. Kaz. ii. 297, 310. Yak. i. 669; iii. 603; iv. 468, 830. J. N. 455.

into the Blind Tigris, and on the east side of the estuary several other canals came in, while a broad artificial channel called the Nahr Bayân, at a point about 30 miles above 'Abbâdân, joined the estuary of the Tigris with that of the Dujayl (the Kârûn river), which flows down from the Khûzistân province into the Persian Gulf at Sulaymânân[1].

Al-Baṣrah—the name is said to mean 'the Black Pebbles'—was founded in the reign of 'Omar in the year 17 (638), and its lands were divided among the Arab tribes who were then in garrison here after the conquest of the Sassanian empire. The city grew quickly to be, with Kûfah, one of the new capitals of 'Irâḳ; and in the year 36 (656) near Baṣrah 'Alî gained the barren victory, the famous Battle of the Camel, over those who were responsible for the death of the Caliph 'Othmân; in which battle Ṭalḥah and Zubayr, two well-known Companions of the Prophet, were slain. Baṣrah lay about 12 miles in a direct line from the Tigris estuary, being reached by two great canals, the Nahr Ma'ḳil from the N.E. down which ships came from Baghdâd, and the Nahr-al-Ubullah by which the traffic passed from Baṣrah going S.E. to the Persian Gulf at 'Abbâdân. These two canals, with the waters of the estuary to the east for the third side, formed the Great Island as it was called; and the city of Ubullah stood at its S.E. angle, above where the Ubullah canal joined the estuary.

Baṣrah city had its greatest length along the junction canal, of the two arms just named, and its houses extending westward in a semicircle reached the border of the desert, where a single gate called Bâb-al-Bâdiyah (the Desert Gate) gave egress. The width of the city, from the canal bank to this gate, was in the 4th (10th) century three miles, but its length greatly exceeded this measurement. The houses of the town were for the most part of kilnburnt bricks, the walls were surrounded by rich pasture lands,

[1] I. S. 28. The word *'Awrâ*, meaning 'blind of an eye,' is applied to rivers that have silted up, and to roads along which there is no thoroughfare. At first the name of the Blind Tigris appears to have been given to the 'Abdasî channel; and only at a later date to the lower estuary. Mas. *Tanbîh* 52. Yak. i. 770. J. N. 454. This last gives the Tigris estuary under the name of the Shaṭṭ-al-'Arab.

watered by numerous minor canals, and beyond these lay extensive palm-groves. Mukaddasî states that Basrah had three Friday Mosques, one at the western gate, close on the desert, and this was the oldest; a second mosque, the finest, built with beautiful columns, stood in the chief market place, and it was 'unequalled among the mosques of all 'Irâk'; the third was situated among the houses of the town. There were also three great market streets, full of shops and warehouses, and these equalled the Baghdâd markets in extent. The Mirbad (the Kneeling-place for Camels) was the famous quarter at the western gate, where the desert caravans halted, and this was one of the busiest parts of the city. Near here were the shrines at the tombs of Talhah and Zubayr, but even when Mukaddasî wrote many quarters of the city had already gone to ruin[1]. Among other institutions, Mukaddasî mentioned a public library, which existed in Basrah during the 4th (10th) century, having been founded and endowed by a certain Ibn Sawwâr, who had also provided the town of Râmhurmuz in Khûzistân with a similar institution. In both a stipend provided for the entertainment of students, and for the copying of books; and the number of these stored in the Basrah library was considerable.

During the many wars and insurrections recorded in the history of the Abbasids Basrah suffered much. In 257 (871) when the great rebellion of the Zanj was at its height, their leader—who gave himself out as a descendant of the Caliph 'Alî—stormed Basrah, burnt the greater part of the town including the Great Mosque, and for three days his troops plundered the city. Then in 311 (923) Basrah was again sacked, and this time during 17 days, by the chief of the Carmathians. But the place in time partly regained its former opulence. In 443 (1052) it was visited by the Persian traveller Nâsir-i-Khusraw, who describes it as most populous, the city wall being in good repair though many quarters of the town were still in ruin. The palace of the Caliph 'Alî near the Great Mosque still existed, and there were thirteen shrines recalling divers events of the days when 'Alî was in

[1] The tomb of Zubayr is still marked by the ruins of that name which stand on the site of medieval Basrah. Modern Basrah, lying on the Tigris estuary, occupies the position of Ubullah at the exit of the canal.

residence here. Nâṣir also carefully enumerates the twenty districts surrounding the city.

In 517 (1123) the city wall, running half a league within the old line, was rebuilt by the Ḳâḍî 'Abd-as-Salam, and in the 8th (14th) century, after the Mongol invasion, when Ibn Baṭûṭah was here, Baṣrah was still a very populous city. He speaks of the mosque of 'Alî, a fine tall edifice with seven minarets, which however was only opened for the Friday prayers and already stood two miles distant from the inhabited quarters of the town, being surrounded by ruins. The older city wall, lying two miles beyond this mosque, could still be traced, near which were the shrines of Ṭalḥah and Zubayr; but the town proper then consisted of only three inhabited quarters. Mustawfî, writing in the same century, gives a long account of Baṣrah. Its mosque, which he reports had only been rebuilt by the Caliph 'Alî, was the largest in Islam—and any mosque planned larger it was impossible ever to complete—and of this mosque 'Alî had set the Ḳiblah (or Mecca point) quite exactly in its right direction. Here, too, there was a minaret which shook or remained still according as an oath sworn to before it was true or false: a perpetual miracle established by the Caliph 'Alî who had built it. Mustawfî gives some further account of the Baṣrah shrines, and then speaks in high praise of the beautiful gardens and palm-groves surrounding the city, 'so thickly planted that you cannot see a hundred paces distant,' and the dates of so fine a quality that they were profitably exported to India and to China.

Baṣrah had at all times been famous for its canals, which according to Ibn Ḥawḳal, in the 4th (10th) century, exceeded 100,000 in number, and of these 20,000 were navigable for boats. The Nahr Ma'ḳil, already mentioned as the main channel from the direction of Baghdâd, had been dug during the reign of 'Omar by Ma'ḳil ibn Yasar, a Companion of the Prophet. This and the Ubullah canal, going from Baṣrah towards the south-east, were each four leagues in length, and the gardens of the Ubullah canal along the south side of the Great Island were held to be one of the four earthly paradises[1].

[1] As generally reported (but different authorities give different lists) the other three were, the Ghawṭah, or Garden Lands, of Damascus; the Sha'b

Al-Ubullah, the Arab form of the Greek Apologos, dated from Sassanian or even earlier times, but it lay on the estuary and was feverish, and the Moslems when they founded their new city, Baṣrah, built this further inland near the desert border. Ubullah, as already said, was to the north at the mouth of its canal, and on the Great Island. Opposite, on the south side of the canal, was the town called Shiḳḳ 'Othmân, 'Othman's breach' in the dyke (he is said to have been a grandson of his namesake the third Caliph); and over against the canal mouth, but on the east side of the estuary, was the station whence those who had crossed the Tigris took the road for Khûzistân. This was called 'Askar Abu Ja'far—'the Camp of Abu Ja'far,' in other words, of the Caliph Manṣûr. Ubullah was in the 4th (10th) century a town of considerable size, having its own Friday Mosque, and the like was the case with Shiḳḳ 'Othmân, both according to Muḳaddasî being fine buildings. Nâṣir-i-Khusraw, who was here half a century later, speaks of the palaces, markets, and mosques of both towns as then in excellent state, but the Mongol inroad a couple of centuries later affected all this countryside, and Ḳazwînî writing in the 7th (13th) century describes these places as gone to ruin, though Shiḳḳ 'Othmân was held famous for its great Sidr or lotus trees. In the next century Ibn Baṭûṭah describes Ubullah as a mere village, from which condition it has arisen in modern times by the building, on the older site, of New Baṣrah.

Where the Nahr-al-Ubullah flowed into the Tigris estuary there had been a dangerous whirlpool, ships being often wrecked here in earlier times. According to Ibn Ḥawḳal this peril to all mariners was done away with by a certain Abbasid princess—some say Zubaydah—who, loading many ships with stones, sunk them at this spot, and thus blocked the whirlpool. Ibn Serapion carefully enumerates the nine canals which came into the Tigris estuary on the western side; namely, three above the Nahr Ma'ḳil, and four south of Baṣrah, between the Ubullah canal and the

Bavvân, or Vale of Bavvân, in Fârs, which will be described in Chapter XVIII; and lastly the Wâdî-aṣ-Ṣughd, or Valley of Soghdiana, lying between Samarḳand and Bukhârâ, which will be mentioned in Chapter XXXIII. Ist. 80. I. H. 159, 160, note *c*. Muk. 117, 130, 413. N. K. 85—89. Yak. i. 636; iv. 845. I. B. ii. 8, 13, 14. Mst. 137.

mouth of the estuary. The only one of these canals which is of importance is the Nahr Abu-l-Khaṣîb—so called after a certain freedman of the Caliph Manṣûr—on which in the middle years of the 3rd (9th) century the great stronghold of the Zanj rebels was built. This city, which they named Al-Mukhtârah, was so strongly fortified as to resist for a considerable time the armies sent against it by the Abbasid Caliph, and it was only after fifteen years of continuous warfare that the rebellion of the Zanj was finally crushed[1].

The chief canals on the eastern side of the Tigris estuary, according to Ibn Serapion, were the following. First the Rayyân, on or near which lay the two towns of Al-Maftaḥ and Ad-Daskarah (the Flat-land); the exact position of these is unknown, though the first-named town was of sufficient importance for the estuary to be often named the Tigris of Al-Maftaḥ. Below this was the Nahr Bayân, with the town of Bayân lying at its mouth five leagues distant from Ubullah on the opposite side of the estuary. The port of Muḥammarah on the Ḥaffâr channel occupies its site at the present day, this channel connecting the upper reach of the Tigris estuary with that of the Dujayl (Kârûn). Muḳaddasî, writing three-quarters of a century later than Ibn Serapion, says that this channel, four leagues in length, was widened and dug out by the order of 'Aḍud-ad-Dawlah the Buyid. Already in the previous century it is spoken of by Ḳudâmah under the name of the New Canal (An-Nahr-al-Jadîd), and it was navigable for cargo-boats coming to Baṣrah from Ahwâz, which before the widening of the 'Aḍudî channel (as Muḳaddasî calls it) had had to pass down the Dujayl estuary, out to sea, and then up the Tigris estuary past Bayân to Ubullah[2].

The great island between the two estuaries which Yâḳût names (in Persian) Miyân Rûdân (Betwixt the Rivers) is described by Muḳaddasî as a *Sabkhah* or salt-marsh, with the town of 'Abbâdân on the seaboard at one angle, and Sulaymânân at the other angle on the Dujayl estuary. 'Abbâdân still exists, but now lies up the

[1] Ist. 81. Baladhuri, 362. I. H. 160, 161. Muk. 118, 135. I. S. 29, 30. N. K. 89. Kaz. ii. 190. Yak. ii. 675. I. B. ii. 17. Tabari, iii. 1982.

[2] I. S. 30. I. K. 12. Kud. 194. Ist. 95. I. H. 171. Muk. 419. Mas. *Tanbih* 52. Yak. iv. 586.

estuary more than twenty miles from the present coast-line of the Persian Gulf, for the sea has been pushed back thus far by the delta of the great river. Muḳaddasî in the 4th (10th) century, however, describes 'Abbâdân as having only the open sea beyond it. It was inhabited by mat-weavers, who used the Ḥalfâ grass of the island for their trade; and there were great guard-houses round the town for the protection of the mouth of the estuary. Nâṣir-i-Khusraw, who was here in 438 (1047), says that in his day the low tide left a couple of leagues dry between 'Abbâdân and the sea, and to serve as a lighthouse to warn mariners they had built a scaffolding with great beams of teak-wood, very broad below and narrowing above, 40 yards in height, which was known as the Khashâb (Wood-works). On its summit was the watchman's cabin, and the platform being stone-flagged and supported on arches was used at night for a brasier where a beacon-fire was lighted. 'Abbâdân was still a flourishing town in the 7th (13th) century with many mosques and Rubâṭs (guard-houses), but in the next century when Ibn Baṭûṭah passed through, it had sunk to the size of a village and already was three miles distant from the coast-line. Mustawfî, however, the contemporary of Ibn Baṭûṭah, speaks of 'Abbâdân as a considerable port, and states that its revenues, which amounted to 441,000 dînârs in the currency of his day, were paid in to the Baṣrah treasury. The harbour of Sulaymânân, a few leagues east of 'Abbâdân, was often counted as of the Khûzistân province, and all that is recorded of it appears to be the fact that it was founded by a certain Sulaymân ibn Jâbir, surnamed 'the Ascetic[1].'

Returning to the latitude of Baghdâd the towns lying along the Tigris to the north of the capital as far as the limits of 'Irâḳ have now to be described, with those which stood near the bank of the great Nahrawân canal. As already said (see p. 32) the

[1] Baladhuri, 364. Ist. 90. I. H. 173. Muk. 118. Kaz. ii. 280. N. K. 89, 90. Yak. iv. 708. I. B. ii. 18. Mst. 137. Mas. i. 230. Yâḳût (i. 645) notes that the people of Baṣrah had the habit of turning proper-names into place-names by the terminal syllable *ân*: e.g. *Ṭalḥatân*, 'the Ṭalḥah canal.' This explains the forms Sulaymânân and 'Abbâdân, the latter being called after a certain 'Abbâd. The shore line at the mouth of the Tigris estuary advances at the rate of about 72 feet in the year, or a mile and a half in the century; hence the present inland position of 'Abbâdân.

chief high road from Baghdâd to Mosul and the northern towns went along the left or eastern bank of the Tigris. It left East Baghdâd by the Baradân Gate of the Shammâsîyah quarter, and in about four leagues reached the small town of Al-Baradân, which still exists under the slightly altered form of Badrân. Close to Baradân were two other important villages, Bazûghâ and Al-Mazrafah, the latter lying three leagues above Baghdâd. At Ar-Râshidîyah near Baradân the Khâliṣ canal joined the Tigris, as will be explained presently; and immediately above this, at the present day, ends a great bend of the Tigris to the eastward, which bend begins at Ḳâdisîyah 60 miles north of Baghdâd. The river bed, however, during the middle-ages took an almost straight line from Ḳâdisîyah to Baradân, and the ruins still exist on the eastern side of the dry channel, the names being marked on the map, of towns mentioned by Ibn Serapion and other early authorities.

The bed of the Tigris would indeed appear to have changed here more than once. What is the present (eastern) channel of the river the author of the Marâṣid, writing about the year 700 (1300), speaks of as the Shuṭayṭah or 'Lesser Stream'; and one of the great alterations must have taken place during the reign of the Caliph Mustanṣir, namely between the years 623 and 640 (1226 to 1242), for it is chronicled that he dug many canals to irrigate the lands left dry by the shifting of the main stream. As early as the 4th (10th) century also, Mas'ûdî speaks of law-suits, to which this changing of the Tigris bed had given rise, between the landowners on the eastern and western banks above Baghdâd. Of these towns then lying on the east bank of the Tigris (their ruins being now found on the dry channel far to the westward of the present river) one of the best known was 'Ukbarâ, close to which lay Awânâ, and then Buṣrâ further down stream, the three places standing some 10 leagues from Baghdâd. They lay surrounded by gardens, to which pleasure-seekers from the capital resorted, and Muḳaddasî especially praises the grapes of 'Ukbarâ, which he says was a large and populous town. A short distance above 'Ukbarâ was 'Alth or Al-'Alth, which is still marked on our maps, but now of the western bank, and Muḳaddasî describes this as a large and very populous city, lying on a branch canal from the Tigris.

North-west of 'Alth, where the river at the present day turns off eastward for the great bend, stands Ḳâdisîyah of the Tigris—not to be confused with the place of the same name to the west of the Euphrates. It was famous for its glass-works, and opposite to it the Dujayl canal branched from the Tigris going south[1].

The Dujayl canal (this also not to be confounded with the Dujayl river, the Kârûn), as will be explained in the next chapter, had originally been a channel from the Euphrates to the Tigris, but by the beginning of the 4th (10th) century its western part had become silted up, and its eastern and lower course was then kept clear by a new channel, taken from the Tigris immediately below Ḳâdisîyah. The Dujayl—meaning 'the Little Tigris'—watered all the rich district of Maskin lying to the north of West Baghdâd beyond Ḳaṭrabbul. The later Dujayl was therefore a loop-canal of the Tigris, which it rejoined opposite 'Ukbarâ after throwing off a number of branches, some of which ran so far south as to bring water to the Ḥarbîyah, the great northern suburb of West Baghdâd (see above, p. 31). The district of the Dujayl, otherwise called Maskin, included a great number of villages and towns, lying westward of 'Ukbarâ and the Tigris channel, the chief of which was Ḥarbâ, which was visited by Ibn Jubayr in 580 (1184) and still exists. Here may be seen at the present day the ruins of a great stone bridge across the canal which, as the historian Fakhrî records and the extant inscription still testifies, was built by the Caliph Mustanṣir in 629 (1232). Near Ḥarbâ was Al-Ḥaẓîrah (the Enclosure), where the cotton stuffs called Kirbâs were manufactured, being largely exported. Yâḳût further names a considerable number of villages—there were over a hundred in all—which were of this district, and many of these, as for example Al-Balad (the Hamlet) near Ḥaẓîrah, are still to be found on the map. As late as the 8th (14th) century the Dujayl district, with Ḥarbâ for its chief town, is described by Mustawfî as of amazing fertility, and its pomegranates were the best to be found in the markets of Baghdâd.

[1] Kud. 214. Muk. 122, 123. Mas. i. 223. Yak. i. 395, 552, 606, 654; iii. 705; iv. 9, 520. Mar. ii. 270, 429.

Many other towns were of this district. About ten miles above Ḳâdisîyah is Sâmarrâ, which will be described in the next chapter, and Maṭîrah lay half-way between the two, immediately above where three small canals branched from the left (east) bank of the Tigris. Midway between Maṭîrah and Ḳâdisîyah, below the exit of these canals, stood Barkuwârâ, otherwise Balkuwârâ, or Bazkuwâr. The village of Al-Maṭîrah, according to Yâḳût, had derived its name from a certain Maṭar of the Shaybân tribe, who was a notable man of the Khârijite sect, and it had been originally called Al-Maṭarîyah, this in time becoming corrupted to Al-Maṭîrah[1]. Ten miles north again of Sâmarrâ was Karkh Fîrûz, also called Karkh of Sâmarrâ, to distinguish it from Karkh the southern quarter of West Baghdâd, and further to the north lay Dûr, where the great Nahrawân canal branched from the left bank of the Tigris. At this point, but from the right or western bank of the Tigris, began the Isḥâḳî canal which making a short loop rejoined the river again opposite Maṭîrah. The positions of all these places are fixed by the canals, some of them, in ruin, also still exist, but nothing is known of them beyond their names.

[1] Ykb. 265. I. S. 14. I. J. 233. Yak. i. 178, 605; ii. 235, 292, 555; iv. 529, 568. Mst. 138. Fakhri, 380. Commander J. F. Jones in the *Records of the Bombay Government* (new series, number XLIII, 1857, p. 252) gives a drawing of the Ḥarbâ bridge. He gives (p. 47) Barkuwârâ under the form Bez-guara.

CHAPTER IV.

'IRÂḲ (*continued*).

Sâmarrâ. Takrît. The Nahrawân canal. Ba'ḳûbâ and other towns. Nahrawân town, and the Khurâsân road. Jâlûlâ and Khâniḳîn. Bandanîjân and Bayât. Towns on the Euphrates from Ḥadîthah to Anbâr. The 'Îsâ canal. Muḥawwal, Ṣarṣar and the Nahr-al-Malik. The Kûthâ canal.

Sâmarrâ, which for more than half a century and during the reigns of seven Caliphs, from 221 to 279 (836 to 892), became the Abbasid capital, had existed as a town before the Arab conquest, and long after it had fallen from its temporary pre-eminence continued to be an important city. The name in Aramæan is written Sâmarrâ, which the Caliph Mu'taṣim when he took up his residence here changed, officially, to Surra-man-raa, 'for good augury,' these words in Arabic signifying 'Who sees it, rejoices.' Under this form it is a mint city on Abbasid coins; but the name was pronounced in many different ways, six forms are cited by Ibn Khallikân, Sâmarrâ being that most commonly used, and the one selected by Yâḳût as the heading to his article on this city.

Ya'ḳûbî writing at the close of the 3rd (9th) century has left us a long and detailed account of Sâmarrâ and its palaces, for the seven Caliphs who lived here, mostly as the prisoners of their Turk bodyguard, occupied their enforced leisure in building, and in laying out pleasure-grounds. The city proper stood on the eastern bank of the Tigris and extended with its palaces for a distance of seven leagues along the river. On the western bank also many palaces were built, each Caliph in succession spending fabulous sums on new pleasure-grounds. The land where the

Caliph Mu'taṣim (a younger son of Hârûn-ar-Rashîd) built his first palace when he came to settle at Sâmarrâ in 221 (836) belonged to a Christian monastery (*Dayr*) which was bought for 4000 dînârs (£2000) and it was known as Aṭ-Ṭîrhân. His Turk body-guard were granted fiefs at Karkh, and further up stream to Dûr, some also lay south of Sâmarrâ towards Maṭîrah; and the Caliph proceeded to build the first Friday Mosque near the east bank of the Tigris, and lay the foundations of his palace. Artificers were brought together from all parts of the empire, and immense quantities of teak-wood (*Sâj*) were imported, also palm beams from Baṣrah and divers marbles from Antioch and Laodicea. A thoroughfare called the Great Road (Ash-Shâri'-al-A'ẓam) was laid out along the Tigris bank, being bordered by the new palaces and the fiefs, and this road went from Maṭîrah right up to Karkh, many by-roads and market streets branching from it. The new Treasury and Government Offices also were built, and the Great Hall called Dâr-al-'Âmmah (the Public Audience Chamber) where the Caliph sat in state on Mondays and Thursdays.

Besides his palace in Sâmarrâ, Mu'taṣim laid out a pleasance on the west side of the Tigris opposite the new capital, with which it was connected by a bridge of boats, and the gardens were planted with palms brought up from Baṣrah, and with exotics sent for from provinces as far distant as Syria and Khurâsân. These lands on the western side were irrigated by branch canals from the Nahr-al-Isḥâķî, already mentioned, which was dug by Isḥâķ ibn Ibrâhîm, Chief of Police to Mu'taṣim, and this was more especially the district called Ṭîrhân, which Ya'ķûbî speaks of as 'the plain' of Sâmarrâ. When the Caliph Mu'taṣim died in 227 (842) Sâmarrâ was in a fair way to rival Baghdâd in the grandeur of its palaces and public buildings. His two sons Wâthiķ and Mutawakkil, who became Caliphs in turn, completed the work of their father. Hârûn-al-Wâthiķ built the palace, called after his name the Ķaṣr-al-Hârûnî, on the Tigris bank, and at either end of this, east and west, was a great platform. Wâthiķ also dug a harbour from the river, where cargo-boats coming up from Baghdâd might conveniently unload. He was succeeded by his brother Ja'far-al-Mutawakkil in 232 (847) who at first lived in the Hârûnî palace, but in 245 (859) he began to build himself a new palace three leagues north of Karkh, to which he extended the Great Road, and this with the new town which sprang up round it was called Al-Mutawakkilîyah or the Ḳaṣr-al-Ja'farî. The ruins of the Ja'farî palace still exist in the angle formed by the branching of the Nahrawân canal, and the older town of Al-Mâḥûzah came to be incorporated with it.

Mutawakkil also built a new and more magnificent Friday Mosque to replace that of his father, which had become too small for the population of the new capital, for the houses now extended in a continuous line with palaces and gardens from Maṭîrah to Dûr. In his palace of the Mutawakkilîyah, otherwise called the Ja'farîyah, Mutawakkil was murdered by his son Muntaṣir in 247 (861), and, during the troublous times that followed, the four next Caliphs had their abode at the Ḳaṣr-al-Jawsaḳ (the Palace of the Kiosque) on the western side of the Tigris opposite Sâmarrâ, this being one of those built by Mu'taṣim. Mu'tamid, son of Mutawakkil, and the last of the Caliphs to reside at Sâmarrâ, lived first at the Jawsaḳ, but afterwards built himself a new palace on the eastern bank, known as Ḳaṣr-al-Ma'shûḳ (the Palace of the Beloved), from whence he finally removed the seat of Government to Baghdâd a short time before his death in 279 (892). The names of many other palaces are given by our authorities. Ibn Serapion for instance mentions the celebrated Ḳaṣr-al-Jiṣṣ (the Gypsum Palace) built by Mu'taṣim on the Isḥâḳî canal; and Yâḳût, who names a great number of palaces, adds a long account of the almost fabulous prices which each had cost its builder, and the total he makes to be 204 million dirhams, equivalent to about eight million sterling.

The glory of Sâmarrâ, however, naturally came to an end with the return of the Caliphs to Baghdâd, and its many palaces rapidly fell to ruin. In the 4th (10th) century Ibn Ḥawḳal praises its magnificent gardens, especially those on the western side of the Tigris, but Muḳaddasî says that Karkh on the north was, in his day, become the more populous quarter of the town. The great Friday Mosque of Sâmarrâ, however, still remained, which Muḳaddasî says was the equal of that of Damascus in magnificence. Its walls were covered with enamelled tiles (*mînâ*), it was paved with marble, and its roof was supported on marble columns. The minaret was remarkable for its great height, and, Yâḳût asserts, it had been the minaret of the first mosque, having been built by Mu‘taṣim, who wished the Call to Prayer to be audible over all the city. It was visible from a league distance all round. It is apparently this ancient minaret which still exists as the well-known Malwîyah tower, having a spiral outside stairway going to the top, which stands about half a mile to the north of modern Sâmarrâ; such was in any case the belief of Mustawfî who, in the early part of the 8th (14th) century, says that the minaret then existing of the Friday Mosque was 170 ells (*Gez*) in height, ‘with the gangway going up outside, the like of which was to be seen nowhere else,’ and he adds that the Caliph Mu‘taṣim had been its builder.

Later authorities add little to our knowledge of Sâmarrâ, and in after years it came chiefly to be inhabited by Shî‘ahs; for here were the tombs of the tenth and eleventh Imâms, ‘Alî-al-‘Askarî and his son Al-Ḥasan, and here above all, said they, was the mosque with the underground chamber where the twelfth Imâm had disappeared in 264 (878), he being Al-Ḳâim, the promised Mahdî, who was to reappear in the fulness of time. The shrines where these Alids were buried stood in that part of Sâmarrâ called ‘Askar Mu‘taṣim, ‘the Camp of Mu‘taṣim,’ and it is from this that the tenth Imâm had his title of Al-‘Askarî. Writing in the early part of the 8th (14th) century Mustawfî, the Shî‘ah, especially mentions these shrines, and adds that the Friday Mosque near by these tombs, besides its great minaret already referred to, was possessed of a famous stone basin called Kâs-i-Fir‘awn (Pharaoh's Cup) measuring 23 paces in circumference by 7 ells high, and half an ell in thickness, which stood in the mosque court for the Ablution, and which the Caliph Mu‘taṣim had caused to be made. Mustawfî, however, adds that, in his day, Sâmarrâ was for the most part a ruin, only in part inhabited, and this statement is confirmed by the description left us by his contemporary Ibn Baṭûṭah, who was here in the year 730 (1330)[1].

[1] Baladhurî, 297, 298. Ykb. 255—268. I. K. 94. I. S. 18. Ist. 85. I. H. 166. Muk. 122, 123. A. F. 289. Yak. iii. 14—22, 82, 675; iv. 110. Ibn Khallikan, No. 8, p. 15. Mst. 139. I. B. ii. 132.

Takrît, lying thirty miles north of Sâmarrâ on the west bank of the Tigris, was commonly counted as the last town of 'Irâḳ, and was famous for its strong castle which overlooked the river. Ibn Ḥawḳal in the 4th (10th) century states that the majority of its population were Christians, and that they possessed a great monastery here. Muḳaddasî says the wool-workers of this town were famous, and in its neighbourhood much sesame was grown; Mustawfî adds, also water melons, of which three crops a year were produced in spite of the somewhat raw climate of Takrît. Ibn Jubayr states that the city wall was 6000 paces in circuit, with towers in good repair, when he passed through Takrît in 580 (1184), and Ibn Baṭûṭah gives praise to both its markets and its numerous mosques[1].

The great Nahrawân canal left the Tigris a short distance below Dûr, as already said, and in its upper course was known as Al-Ḳâṭûl-al-Kisrawî, 'the Cut of the Chosroes,' for it owed its origin to the Sassanian kings. It served to irrigate all the lands along the east bank of the Tigris from above Sâmarrâ to about a hundred miles south of Baghdâd, and Ibn Serapion mentions a great number of towns along its banks with bridges and weirs, but most of these have now disappeared, though the line of the canal is still marked on the map. Leaving Dûr[2], which, for distinction among the many towns of this name, was called Dûr-al-'Arabâyâ or of Al-Ḥârith, the canal passed to the back of the Mutawakkilîyah and other outlying quarters north of Sâmarrâ, and here it was crossed by a stone bridge. It next came to Îtâkhîyah, a village and fief called after Îtâkh the Turk, sometime captain of the guard to the Caliph Mu'taṣim; this had originally been a monastery called Dayr Abu-Ṣufrah, and here stood the bridge of the Chosroes (Ḳanṭarah Kisrawîyah). The monastery took its name from Abu Ṣufrah the Khârijite. Next the Nahrawân came to Al-Muḥammadîyah, a small town, where it was crossed by a bridge of skiffs (Jisr Zawârîḳ)[3], and according to

[1] Ist. 77. I. H. 156. Muk. 123. I. J. 234. Mst. 138. I. B. ii. 133.

[2] *Dûr* means the 'Houses' or 'Habitations,' and is a common place-name, being the plural form of *Dârah*, 'a homestead.'

[3] It is to be noted that in the classical usage *Jisr* stands for 'a bridge of boats,' while *Ḳanṭarah* is 'a masonry bridge of arches.' *Shâdhurwân*, trans-

Yâḳût this Muḥammadîyah was but a later name of Îtâkhîyah, the change having been effected by Mutawakkil in honour of his son Muḥammad-al-Muntaṣir, who afterwards became Caliph by the murder of his father. At some distance below these places the Nahrawân was joined successively by the three lesser Ḳâtûls, namely the Yahûdî, the Mamûnî, and the canal of Abu-l-Jund, which were all three taken from the left bank of the Tigris near Maṭîrah below Sâmarrâ, and which irrigated the fertile districts south of that city. Above their inflow, the Nahrawân was dammed back by the first of its many weirs (Ash-Shâdhurwân), and where the first canal came in stood the large village of Al-Mamûnîyah. This, the Yahûdî (or Jews') canal, was crossed between Maṭîrah and Mamûnîyah by a stone bridge called Ḳanṭarah Waṣîf, after Waṣîf, one of the captains of the Turk bodyguard, in the reign of Mu'taṣim. The second canal, called Al-Mamûnî, fell into the Nahrawân below the village of Al-Ḳanâṭir, 'the Bridges.' The third canal was called Abu-l-Jund—'Father, or Supplier, of the Army'—from the fact that the crops raised on the lands watered by it were used as rations for the troops. It was the largest canal of the three, and had been dug by Hârûn-ar-Rashîd, who built a palace there while superintending its construction. On its banks stood the town of Ṭaffir, and here it was crossed by a bridge of boats. Yâḳût, who had himself visited Ṭaffir, describes it as occupying in the 7th (13th) century a waterless and pastureless plain, where wild animals dwelt, lying between Ba'ḳûbâ and Daḳûḳâ. He passed through this going from Baghdâd to Irbil; no habitations were to be met with, and Yâḳût says that his guide, when the caravan travelled by night over this plain, 'was wont to take his direction by the Pole-star, until, with the day, the plain had been crossed.'

lated by 'weir,' more properly designates a portion of a canal, or river bed, that has been paved and embanked to confine the stream. It should, however, be added that Jisr undoubtedly sometimes also designated a stone bridge of arches, as in the celebrated Jisr-al-Walîd, the name given to the bridge over the river Sarus, between Adana and Mopsuestia, which was built by Justinian. The word Ḳanṭarah also designates any arched structure, as a viaduct or aqueduct, being borrowed from the Byzantines, who used the word *κέντρον* (the Latin *centrum*) to denote the central arch of a bridge, and by extension applied it to mean the whole structure.

Four leagues below where the last of these three canals joined the Nahrawân lay the town of Ṣûlâ or Ṣalwa, otherwise called Bâb Ṣalwâ or Bâṣalwâ. Below this again was the town of Ba'ḳûbâ, some ten leagues north of Baghdâd, and the capital of the Upper Nahrawân district. At Ba'ḳûbâ the Great Ḳâṭûl canal changed its name, and became the Tâmarrâ, under which name it passed on to Bâjisrâ and thence to the city called Jisr Nahrawân, beyond which the main waterway was more especially known as the Nahrawân canal. Near Bâjisrâ (the Aramaic form of Bayt-al-Jisr, 'the bridge-house') which stood in a well cultivated district, surrounded by palm-trees, the Tâmarrâ sent off a branch from its right bank known as the Nahr-al-Khâliṣ, which flowed out into the Tigris at Baradân to the north of Baghdâd, and from the Khâliṣ many of the canals of East Baghdâd derived their water.

Jisr Nahrawân, the Bridge-town, where the Khurâsân road from Baghdâd crossed, will be described presently; and here a canal called the Nahr Bîn branched from the right bank of the Nahrawân, flowing ultimately into the Tigris at Kalwâdhâ. From this the water channels of the lower quarters of East Baghdâd derived their supply. One mile below Jisr Nahrawân the Diyâlâ canal branched south from the main stream, and after irrigating the outer gardens of East Baghdâd, reached the Tigris three miles below the capital.

South of Jisr Nahrawân the great canal took the name of the Nahrawân exclusively, and after passing the Upper Weir (Shâdhurwân) it came to Jisr Bûrân, the bridge named after the wife of the Caliph Mamûn. Below this stood Yarzâṭiyah (or possibly Barzâṭiyâ), and then the town of 'Abartâ, which Yâḳût describes as of Persian origin, having important markets. Beyond 'Abartâ lay the Lower Weir and next Iskâf (or Uskâf) of the Banî Junayd, a city lying on both banks of the canal, and the Banî Junayd, Yâḳût reports, had been chiefs of this district and famous for their hospitality. Yâḳût adds that by the 7th (13th) century, when he wrote, the lands round here had entirely gone out of cultivation, for the Nahrawân had gradually silted up during the previous two centuries, the Saljûḳ Sultans having ever been too much occupied with their wars to attend to the needful dredging, and the mending of dykes: 'further,' he adds, 'their armies had made a roadway of this same canal, whereby both district and canal have now gone to ruin.'

Beyond Uskâf the Nahrawân flowed on for nearly 60 miles, between a continuous line of villages and farmsteads, down to Mâdharâyâ where its waters finally rejoined the Tigris. Mâdharâyâ, as already said, stood to the south of Jabbul and above Al-Mubârak, which lay opposite the town of Nahr Sâbus. When Yâkût wrote it was in ruin, and its name is now no longer marked on the map, but it must have stood just below the present Kût-al-'Amârah where, as already explained, the Tigris now divides off from the Shatt-al-Hayy channel[1].

This triple division of the Nahrawân canal (namely the Kâtûl, the Tâmarrâ, and the Nahrawân proper), with the three branch canals (the Khâliṣ, the Nahr Bîn, and the Diyâlâ) which flowed back to the Tigris after watering the East Baghdâd region, is the explanation which Ibn Serapion has given of a very complicated skein of waterways. In later times the names were not always applied as he gives them. A glance at the present map shows that the Nahrawân, two hundred miles in length, must have taken up all the streams from the Persian highlands which, had it not been dug, would have flowed (at flood time) down to the left bank of the Tigris. The Tâmarrâ section was originally one of these streams, and Yâkût describes how its bed had been artificially paved for a length of seven leagues to prevent the sands absorbing its waters, which were divided up to irrigate the several districts of East Baghdâd. The Khâliṣ and the Diyâlâ were according to his account branches of the Tâmarrâ (in any case the Khâliṣ of the Arab geographers cannot be the river known by this name at the present day, for this now flows at some distance to the north-west of Ba'kûbâ), and Khâliṣ in the time of Yâkût was the name of the district, to the north of the Khurâsân road, which on one side came right up to the walls

[1] Yarzâtiyah is possibly the present Razatiyah or Zatariyah lying *above* 'Abartâ. Ykb. 321. I. S. 19, 20. Baladhuri, 297. I. R. 90. I. K. 175. Mas. *Tanbih* 53. Yak. i. 252, 454; iii. 539, 604; iv. 16, 381, 430.

of East Baghdâd. In the 3rd (9th) century Ibn Rustah and Ibn Khurdâdbih give Nahrawân as the name of the mountain stream, which came into the Great Ḳâṭûl at Ṣalwâ; in the 8th (14th) century Mustawfî writes that the Nahrawân was the name of the Diyâlâ river, which rose in the mountains of Kurdistân, and which was formed by the junction of two streams, one the Shirwân river which lower down was called the Taymarrâ, the other the Ḥulwân river, which flowed down past Ḳaṣr Shîrîn and Khâniḳîn; and these two streams united above Ba'ḳûbâ where they flowed into the Nahrawân canal.

In regard to Nahrawân town, otherwise called Jisr Nahrawân (Nahrawân Bridge), this was the first stage out of Baghdâd along the great Khurâsân road, and it was of old a place of much importance, though now represented by the insignificant hamlet of Sifwah. Ibn Rustah in the 3rd (9th) century describes Nahrawân town as lying on both banks of the canal; in the western half were the chief markets, a Friday Mosque, and many waterwheels for irrigation purposes; while on the eastern side there was a second Friday Mosque, and other markets, with many hostelries round the mosque where the Mecca pilgrims and travellers were wont to put up. Ibn Ḥawḳal in the following century speaks of the fertile lands lying round the town, and Muḳaddasî adds that the eastern part in his day was the most populous, its Friday Mosque being then the only one in use. In the 8th (14th) century, when Mustawfî wrote, Nahrawân town was in ruin, for the Khurâsân road no longer passed through it, but went north by Ba'ḳûbâ. The fertile district about here was still called the Ṭarîḳ-i-Khurâsân (the District of the Khurâsân road) of which Ba'ḳûbâ, Mustawfî states, was the chief town, and it was formed by a continuous line of gardens and palm-groves from which magnificent crops of oranges and shaddocks were harvested[1].

The town of Barâz-ar-Rûz (the Rice Field), now known as Bilâd-ar-Rûz, lay north-east of Nahrawân town, and is frequently mentioned by Yâḳût. The Caliph Mu'taḍid had built a palace here; it was counted as of the Tâmarrâ district, and lay eastward

[1] I. R. 90, 163. I. K. 175. Ist. 86. I. H. 167. Muk. 121. Yak. i. 812; ii. 390, 638. Mst. 139, 141, 216.

off the Khurâsân high road, being also noticed by Mustawfî. Leaving Nahrawân town the next stage of the Khurâsân road was Daskarah-al-Malik, 'of the King,' which Ibn Rustah describes as a considerable city, possessing a great walled castle of Sassanian times, to which a single gateway on the west side gave access. From its position this 'Daskarah of the King' appears to be identical with the celebrated Dastagird, where Khusraw Parwîz had his great palace, which history relates was plundered and burnt to the ground by Heraclius in 628 A.D. This palace, the ruins of which it would seem were in the 4th (10th) century still known as Dastagird Kisrawîyah (of the Chosroes), was seen by the traveller Ibn Muhalhal (quoted by Yâḳût) who says that it then consisted of a wonderful edifice containing many halls and domes, so finely built as to appear carved, each wall in a single block of stone. In regard to the Arab town, Ibn Ḥawḳal in the 4th (10th) century describes Daskarah as possessing a strong castle, doubtless of Moslem foundation, and Muḳaddasî speaks of the place as a small market town, with a Friday Mosque that had a finely vaulted roof. Not far distant from Daskarah was the village of Shahrâbân, mentioned by both Yâḳût and Mustawfî, the latter adding that eighty villages belonged to this town, which had been founded by Princess Gulban, a daughter of one of the Chosroes.

The town of Jalûlâ was the next stage on the Khurâsân road, surrounded by many trees but unfortified. Not far from the town, standing in the village of Hârûnîyah, was an ancient bridge of stone wrought with leaden joints, which had been built by one of the Chosroes, and this crossed the river by which, according to Yâḳût, boats went down to Ba'ḳûbâ and Bâjisrâ. In history Jalûlâ was famous for the great victory gained over the Persians by the Moslems here in the year 16 (637), which resulted in the final overthrow and flight of King Yazdajird. At a later date Mustawfî names the place Rubâṭ Jalûlâ, from the guard-house which had been built here by Malik Shâh the Saljûḳ; and the position of Jalûlâ corresponds with the modern station of Ḳizil Rubâṭ, 'the Red Guard-house.' East of Jalûlâ was the town of Khâniḳîn, which is noticed by Muḳaddasî as a city on the road to Ḥulwân. Here Ibn Rustah says there was a great bridge of many arches over the river, built of well-mortared kiln-bricks. Near Khâniḳîn was a naphtha spring that produced a large revenue, and Yâḳût describes the bridge aforesaid as having 24 arches in his day, the 7th (13th) century, across which passed the Khurâsân road. When Mustawfî wrote in the next century Khâniḳîn had fallen to ruin, and was merely a large village, but its district was still extremely productive.

Six leagues beyond Khâniḳîn, and half-way to Ḥulwân the first town of the Jibâl province, lay Ḳaṣr Shîrîn, 'the Palace of Shîrîn,' the mistress of King Khusraw Parwîz. There was a large walled village here, and the ruins of the Sassanian palace, which Ibn Rustah describes as consisting in the 3rd (9th) century of a mighty arched hall, built of burnt brick, rising in the midst of chambers, the walls of which were of solid masonry. Further there was a great platform before the arched hall, paved with marble slabs. Yâḳût and Mustawfî give long descriptions of Ḳaṣr Shîrîn, the ruins of which still exist; and it is to be noted that the legends of Farhâd the lover of Queen Shîrîn, and of Pahlabâdh the musician, and of Shabdîz the famous horse of King Parwîz, are found localised in many places of the surrounding district[1]. Overhanging Ḳaṣr Shîrîn is the great mountain wall forming the outpost of the Persian plateau, and Ḥulwân, the next stage on the Khurâsân road, though often counted as of 'Irâḳ, being in the mountain pass, will be described in a later chapter.

South of the line of the Khurâsân road, and on the Khûzistân frontier, two important towns remain to be noticed—Bandanîjîn and Bayât. Bandanîjîn, a name no longer found on the map, was the chief town of the districts of Bâdarâyâ and Bâkusâyâ, and the village of Bâkusâyâ still exists near which the town of Bandanîjîn must have been situated. The two districts lay beyond and north-east of the Nahrawân canal, and comprised a great number of fertile villages. Bandanîjîn the capital, according to Yâḳût, was called in Persian Wandanîgân, and Mustawfî says in his day the name was pronounced Bandanîgân, being of the Liḥf district, the 'Foot-hills' of the Kurdistân mountains, and its river came down from Ariwajân. According to Ibn

[1] I. R. 164. Ist. 87. I. H. 168. Muk. 121. Kaz. ii. 295. Yak. i. 534; ii. 107, 393, 573, 575, 813; iv. 112. Mst. 137, 138, 139, 193.

Khurdâdbih Bandanîjîn was counted as of the same district as Barâz-ar-Rûz. Bayât, the ruins of which still exist, is mentioned by Mustawfî; he adds that its river, which rose in the Kurdistân mountains, became lost in the plains before reaching the Tigris, and though its water was brackish, many fertile districts were irrigated by it. Bayât appears to be practically the same place as the town of Aṭ-Ṭîb, mentioned by Ibn Ḥawḳal, where excellent belts, like the Armenian belts, were made. It was a city of some importance under the Abbasids, and its ruins lie close to those of the later town of Bayât. Yâḳût says that in his day the inhabitants of Ṭîb were Nabathæans, and still spoke their Aramaic dialect, tracing their descent direct from Seth, son of Adam[1].

The cities of 'Irâḳ which lay on the Euphrates, and between the two rivers along the transverse canals, must now be described. As already said, a line carried west from the Tigris at Takrît to the Euphrates would cross that river a little below 'Ânah, where its course makes a great bend south, and this is the natural frontier between Jazîrah and 'Irâḳ, as marked by Mustawfî. To the south of this line begins the Sawâd, or alluvial land, of Babylonia; to the north lie the more stony plains of Upper Mesopotamia. The city of Al-Ḥadîthah on the Euphrates, about 35 miles below 'Ânah, is the northernmost town on this side. The name signifies 'the New Town,' and to distinguish it from Al-Ḥadîthah on the Tigris, it was called Ḥadîthah-an-Nûrah, 'of the Chalk' pit. Yâḳût describes it as possessing a strong castle surrounded by the waters of the Euphrates, and it was founded during the Caliphate of 'Omar, not long after the Moslem conquest. Mustawfî describes it as in every way the opposite of Takrît, both in situation and climate. Between Ḥadîthah and Hît, down stream, came the two towns of Alûsah and An-Nawûsah, lying on the Euphrates seven leagues distant one from the other, and Alûsah, which Yâḳût refers to as a small town, still exists. Both are frequently mentioned in the records of the Moslem conquest;

[1] I. K. 6. Ist. 94. I. H. 176. Yak. i. 230, 459, 477, 745; iii. 566; iv. 353. Mst. 137, 138, 220. The Bâdarâyâ district of Bandanîjîn must not be confused with Bâdûrayâ, the name of the southern district of West Baghdâd.

further, An-Nawûsah was counted as a village of Hît, which last was a walled town with a strong castle, celebrated for its palm-groves and lying on the western side of the Euphrates. Ibn Ḥawḳal speaks of Hît as very populous, and Mustawfî in the 8th (14th) century describes more than 30 villages, among the rest Jibbah, as of its dependencies. Immense quantities of fruit, both of the cold and the hot regions, were grown here; nuts, dates, oranges and egg-plants all ripening freely, but the town was unpleasant to live in on account of the overpowering stench of the neighbouring bitumen springs.

At the time of the Moslem conquest the famous Trench of King Sapor II (Khandaḳ Sâbûr) still existed. This had been dug by Sâbûr Dhû-l-Aktâf, as the Arabs called him, in the fourth century A.D. It began at Hît and ran down to Ubullah (near the later Baṣrah) where it reached the Gulf. Originally it carried water, being intended as a line of defence for the rich lands of Lower Mesopotamia against the desert tribes; and its dry bed may still, in part, be traced. 'Ayn-at-Tamr, 'the Spring of the Date Palm,' due south of Hît in the desert, is described by Muḳaddasî as a small fortress, and a stream running from here entered the Euphrates below Hît. Dates and sugar-cane were exported from its district, the latter more especially from a neighbouring town called Shafâthâ; but the exact site of these two places is unknown[1].

Twelve leagues below Hît was the village of Ar-Rabb, where previous to the 4th (10th) century the (earlier) Dujayl canal left the Euphrates; and taking its course due east, after watering the Maskin and Ḳaṭrabbul districts, reached the northern suburbs of West Baghdâd. As already mentioned, this western portion of the Dujayl soon became silted up; and by the time when Iṣṭakhrî wrote in 340 (951) the Dujayl already took its waters from the Tigris opposite Ḳâdisîyah, as described in the paragraphs on the Maskin district. Al-Anbâr, 'the Granaries,' standing on the left bank of the Euphrates, was one of the great cities of 'Irâḳ in Abbasid times. It dated from before the Moslem conquest, and by the Persians was called Fîrûz Sâbûr (or Fayrûz Sâbûr, in

[1] I. S. 10, 13. I. R. 107. Kud. 217. Baladhuri, 179. Ist. 77. I. H. 155. Muk. 117, 123, 135. Yak. i. 352; ii. 223; iii. 759; iv. 734, 997. Mst. 135, 141.

Greek Perisabor) from its founder King Shâpûr[1]; and under the Arabs Fîrûz Sâbûr became the name of the surrounding district. It is said that the town was called 'the Granaries' because of old the Persian kings had stored the wheat, barley, and straw for the rations of their troops in this city. The first Abbasid Caliph, Saffâḥ, had for a time made Anbâr his residence, and he died in the palace which he had built here. His brother Manṣûr also for a time lived at Anbâr, and from here went to Baghdâd, where the new Abbasid capital had begun to be built. Mustawfî gives the tradition that the Jews whom Nebuchadnezzar brought from Jerusalem to Babylonia were interned at Anbâr. In the 8th (14th) century the town walls, he says, were 5000 paces in circuit.

The importance of Anbâr lay in its position at the head of the first great navigable canal which flowed from the Euphrates to the Tigris, which it entered at the harbour (Al-Farḍah) to the south of the Round City of West Baghdâd. This canal, the Nahr 'Îsâ, took its name from an Abbasid prince 'Îsâ who was either 'Îsâ ibn Mûsâ, nephew of Manṣûr, or 'Îsâ ibn 'Alî (the more usual ascription), the uncle of that Caliph. In either case Prince 'Îsâ gave the canal its name, he having re-dug it, making thus a navigable channel from the Euphrates into Baghdâd. Where the canal left the Euphrates, a little below Anbâr, it was crossed by a magnificent bridge, called Ḳanṭarah Dimimmâ, from the village of Dimimmâ which was on the Euphrates bank close to the hamlet of Al-Fallûjah. The Nahr 'Îsâ, passing by many villages and farms of the Fîrûz Sâbûr district, at length came to the town of Al-Muḥawwal, one league distant from the suburbs of West Baghdâd. Just before reaching this town the Ṣarât canal branched from the left bank of the Nahr 'Îsâ, and this canal formed the dividing line between the Ḳaṭrabbul district to the north and Bâdûrayâ to the south of West Baghdâd. The Ṣarât canal, following an almost parallel curve to the Nahr 'Îsâ, poured its waters into the Tigris immediately below the Baṣrah Gate of the Round City, and from these two streams all the watercourses

[1] Sâbûr is the Arab form of the Persian Shâpûr or Shâh-pûr, which the Greeks wrote Sapor.

of West Baghdâd were derived, with the exception of the few coming from the Dujayl canal.

Al-Muḥawwal means 'the place of unloading,' and the town took its name from the fact that the river barges going from the Euphrates towns to Baghdâd, had here to unload into small boats that could pass under the numerous bridges which below Muḥawwal spanned the 'Îsâ canal where this traversed the suburb of Karkh. Muḥawwal was a fine town, famous for its markets and its gardens, and as late as the 8th (14th) century possessed some magnificent buildings, among which Mustawfî counts a palace built by the Caliph Mu'taṣim which stood on the summit of a mound, and which, by the spell of a powerful incantation, had been freed from the plague of mosquitoes. The exact site of Muḥawwal is not now known, but it must lie to the north-east of the ancient Babylonian mound called the Hill of 'Aḳarḳûf, which is frequently mentioned by the Arab geographers, and which Mustawfî connects with the legends of the tyrant Nimrod who threw Abraham into the fiery furnace[1].

Three leagues below the village of Dimimmâ the second of the great transverse canals, the Nahr Ṣarṣar, flowed off towards the Tigris, which it entered four leagues above Madâin. This canal, in its lower reaches, traversed the Bâdûrayâ district, which lay south of West Baghdâd, and Ibn Serapion describes how along its banks numerous waterwheels (*dâliyah*) and levers (*shadûf*) were set up for irrigating the fields. Some way above where, near Zarîrân, the canal flowed into the Tigris, and almost in sight of the White Palace of the Chosroes at Madâin, was the flourishing town of Ṣarṣar, where a great bridge of boats carrying the Kûfah road crossed the canal. Ṣarṣar town lay a couple of leagues only from Karkh, the great southern suburb of West Baghdâd; the Ṣarṣar canal, Ibn Ḥawḳal writes, was navigable for boats, and Ṣarṣar

[1] I. S. 10, 14. I. K. 7, 72, 74. Kud. 217. Ist. 77. I. H. 155, 166. Muk. 123, 134. Yak. i. 367; ii. 600; iii. 697; iv. 432. Mst. 136, 138, 140, 141. The lower courses of the Nahr 'Îsâ and of the Ṣarât canal belong to the topography of Baghdâd, and have been fully described in a former work. The site of Anbâr appears to be that marked by the ruins at Sufayrah, or possibly those to the north of this village of which Mr J. P. Peters has given a plan in *Nippur*, i. 177.

town stood in a forest of date-palms. Muḳaddasî likens it to the towns of Palestine for the manner of its building; and Ṣarṣar continued to be a place of importance down to the close of the 8th (14th) century when Tîmûr took possession of Baghdâd and garrisoned the surrounding districts.

The third transverse canal was the Nahr-al-Malik, which began at the village of Al-Fallûjah[1] five leagues below the head of the Nahr Ṣarṣar, and flowed into the Tigris three leagues below Madâin. This, 'the King's Canal,' dated from ancient times, and is mentioned by the Greeks as the Nahar Malcha. Yâḳût reports that tradition gave it as having been dug either by King Solomon or by Alexander the Great. On its banks was the town called Nahr-al-Malik, with a bridge of boats on the Kûfah road, this lying seven miles south of Ṣarṣar. According to Ibn Ḥawḳal Nahr-al-Malik town was larger by a half than the latter town, being likewise famous for its corn lands and palm-groves; Mustawfî adding that over 300 villages were of its district.

The fourth transverse canal was the Nahr Kûthâ, its point of origin on the Euphrates being three leagues below that of the Nahr-al-Malik, and its outflow 10 leagues below Madâin. The Kûthâ canal watered the district of this name, which was also known as the Ardashîr Bâbgân district (after the first Sassanian king), though part of it was counted as the Nahr Jawbar district on a branch canal. The city of Kûthâ Rabbâ, with its bridge of boats, stood on the banks of the main channel, and is said to be identical with the Biblical Cuthah, mentioned in 2 Kings xvii. 24, an important town of the neighbourhood of Babylon. According to Moslem tradition Kûthâ was the place where Abraham was thrown into the fire by the tyrant Nimrod, and the town took its name from Kûthâ, the grandfather of Abraham, according to the Moslem tradition. In the 4th (10th) century Ibn Ḥawḳal describes the place as a double city, Kûthâ-aṭ-Ṭarîḳ, 'of the Road,' and Kûthâ Rabbâ, which last was a city larger than Bâbil (Babylon), and near here, he says, were great mounds of

[1] This is the Feluchia (Feluge or Felugia) of Cæsar Frederick, and other Elizabethan merchants, where coming down the Euphrates they left their boats and went by land across to Baghdâd: as narrated in Hakluyt, *Principal Navigations* (Glasgow, 1904), v. 367, 455, 466; vi. 4.

ashes said to mark the place of Nimrod's fiery furnace; Muḳaddasî adding that near the high road might be seen an ancient tower, about which many legends were told. The Itineraries state that Kûthâ town, the site of which appears to be that marked on the map as Tall Ibrâhîm, 'the Hill of Abraham,' was four miles south of Nahr Malik town.

Some few miles to the north of the Kûthâ canal stood the large village of Al-Farâshah, the half-way stage between Baghdâd and Ḥillah, on the high road followed at the close of the 6th (12th) century by the Mecca pilgrims going down to Kûfah. Ibn Jubayr, who was here in 580 (1184), describes it as a populous well-watered village, where there was a great caravanserai for travellers, defended by battlemented walls; and Mustawfî also gives Farâshah in his itinerary, placing it seven leagues south of Ṣarṣar[1].

[1] I. S. 15. I. R. 182. Ist. 85, 86. I. H. 166, 168. Muk. 121. I. J. 217. Yak. i. 768; iv. 317, 846. Mar. ii. 363. A. Y. i. 633. Mst. 141, 193. The course of the Nahr 'Îsâ is more or less that of the modern Saklawîyah canal: the Ṣarṣar appears to have followed the line of the Abu Ghurayb canal; the Nahr-al-Malik is the Radhwânîyah, and the Nahr Kûthâ is the Ḥabl Ibrâhîm, 'Abraham's rope,' of the modern maps. These identifications, however, are only approximate, for naturally in over a thousand years the face of the alluvial Sawâd is entirely changed from what it was in Abbasid times.

CHAPTER V.

‘IRÂḲ (*continued*).

The bifurcation of the Euphrates. The Sûrâ channel. Ḳaṣr Ibn Hubayrah. Nîl and its canal. The Nahr Nars. The Badât canal, and Pombedita. The Kûfah channel. Kûfah city. Ḳâdisîyah. Mashhad ‘Alî, and Karbalâ. The twelve Astâns of ‘Irâḳ. Trade. The high roads of ‘Irâḳ.

The river Euphrates in the 4th (10th) century bifurcated at a point some six leagues below where the Kûthâ canal was led off. The western branch, to the right, which was then considered the main stream of the Euphrates, passed down by Kûfah and thence to the Great Swamp; while the eastern branch, to the left, which now is the main stream of the river, is by Ibn Serapion and the other Arab geographers called the Nahr Sûrâ, or As-Sûrân; and this by many channels likewise poured its waters finally into the swamp. Taking the Sûrâ branch first (the present Euphrates channel) we find that Ibn Serapion admits this was greater even in his day than the Kûfah branch and more broad. Where the bifurcation took place, the Upper Sûrâ canal watered the three sub-districts of Sûrâ, Barbîsamâ, and Bârûsmâ, which formed part of the middle Bih Ḳubâdh district; then bearing south the channel passed a couple of miles to the westward of the city called Ḳaṣr Ibn Hubayrah, and here it was crossed by the great bridge of boats known as the Jisr Sûrâ (or Sûrân) by which the Pilgrim road went down from Ḳaṣr Ibn Hubayrah to Kûfah.

The town of Al-Ḳaṣr, as it was called for short, the Castle or Palace of Ibn Hubayrah, took the name from its founder, who had been governor of ‘Irâḳ under Marwân II, the last Omayyad

Caliph. Ibn Hubayrah had not lived to complete his work, but after the fall of the Omayyads, the first Abbasid Caliph, Saffâḥ, took up his residence here, finished the palace, and called it Hâshimîyah in honour of his own ancestor Hâshim. The town which rapidly sprung up round the palace of the Caliph none the less continued to be called after the Omayyad governor, and even though Manṣûr made Hâshimîyah for a time his residence, before the foundation of Baghdâd, Ḳaṣr Ibn Hubayrah, or Madînah (the City of) Ibn Hubayrah, was always the name of the place in common use. In the 4th (10th) century Ḳaṣr Ibn Hubayrah was the largest town between Baghdâd and Kûfah, and it stood on a loop canal from the Sûrâ, called the Nahr Abu Raḥâ, 'the Canal of the Mill.' The city was extremely populous, it had fine markets, many Jews residing here, as Muḳaddâsî writes, and the Friday Mosque was in the market place. By the early part of the 6th (12th) century, however, it appears to have fallen to decay, being eclipsed by the rising importance of Ḥillah; and at the present day even the site of it is unknown, though it is doubtless marked by one of the numerous ruins which lie a few miles north of the great mounds of ancient Babylon, or Bâbil as the Arabs name these.

The city of Ḥillah, lying a few miles below the Bâbil ruins, on the Euphrates, otherwise the Sûrâ canal as it was called in the 4th (10th) century, was at this date known as Al-Jâmi'ân, 'the Two Mosques,' and the town at first stood mostly on the eastern bank. It was a populous place, and its lands were extremely fertile. Then Al-Ḥillah, 'the Settlement,' was built on the opposite right bank, by Sayf-ad-Dawlah, chief of the Banî Mazyad, in about the year 495 (1102); and this quickly grew to importance, for its bridge of boats became the new Euphrates crossing for the Pilgrim road from Baghdâd to Kûfah, the high road no longer passing down by Ḳaṣr Ibn Hubayrah (then a ruin) and the Sûrâ bridge. By the 6th (12th) century, also, the Sûrâ arm comes to be considered the main stream of the Euphrates, as at the present day, and the name Nahr Sûrâ gradually goes out of use. In 580 (1184) Ibn Jubayr crossed the Euphrates by 'a great bridge of boats, bound by iron chains,' at Ḥillah, then already a large town stretching along the western side of the Euphrates. Ibn Baṭûṭah, who followed in his footsteps in the early part of the 8th (14th) century, gives a long account of this famous bridge of boats at Ḥillah, the double iron chains of which were secured at either end to immense wooden piles. He praises the town markets, and his account is fully borne out by Mustawfî, his contemporary, who speaks of Ḥillah as beginning to occupy the east as well as the west bank of the Euphrates. It was surrounded by date-groves and hence had a damp climate. Mustawfî adds that the population of Ḥillah were all bigoted Shî'ahs, and they possessed a shrine (Maḳâm) here, where they believed that, in the fulness of time, the promised Mahdî, who had disappeared at Sâmarrâ in 264 (878), would reappear and convert all mankind to their faith (see above, p. 56)[1].

Returning once more to the account given by Ibn Serapion in the 4th (10th) century of the Sûrâ canal, this, as already said, passed to the west of the great ruins of Babylon, or Bâbil. These ruins Muḳaddasî describes as then occupied by the site of a village near a bridge of boats, and Mustawfî gives a long account of the great magicians who had lived in Bâbil, and of the well at the summit of the hill in which the fallen angels Hârût and Mârût were imprisoned until the day of judgment.

Above Bâbil, the last of the many canals flowing from the Euphrates to the Tigris branched from the Sûrâ. This waterway, now known as the Shaṭṭ-an-Nîl—'the Nile Stream'—Ibn Serapion calls the Great Ṣarât, the name is the same as that of the more famous canal of West Baghdâd (see p. 66) in the upper reach lying to the west of the city of Nîl. From its point of origin the Great Ṣarât flowed eastward past many rich villages, throwing off numerous water channels, and shortly before reaching the city of Nîl a loop canal, the Ṣarât Jâmasp, branched left and rejoined the main stream below the city. This loop canal had been re-dug by Ḥajjâj, the famous governor of 'Irâḳ under the Omayyad Caliphs, but took its name, as was reported, from Jâmasp, the chief Mobed, or Fire-priest, who in ancient days had aided King Gushtâsp to establish the religion of Zoroaster in Persia. The

[1] I. S. 10, 16. Ykb. 309. Ist. 85, 86. I. H. 166, 168. Muk. 121. Yak. ii. 322; iii. 861; iv. 123. I. J. 214. I. B. ii. 97. Mst. 138.

city of An-Nîl likewise was founded by Ḥajjâj; it became the chief town of all this district, its ruins being still marked on the map under the name of Nîlîyah; and the Nîl canal was reported to have taken its name from the Nile of Egypt which it was said to recall. The main canal here, opposite Nîl city, was spanned by a great masonry bridge named the Ḳanṭarah-al-Mâsî. In the time of Abu-l-Fidâ that portion of the canal which lay west of the town, namely the Great Ṣarât of Ibn Serapion, was also known as the Nahr-an-Nîl, but Ibn Serapion gives this name exclusively to the reach beyond, east of Nîl city.

This reach, therefore, passing on, watered the surrounding districts till it came to a place called Al-Hawl—'the Lagoon'—near the Tigris, and opposite Nu'mânîyah (see p. 37), whence a branch, called the Upper Zâb canal, communicated directly with the river. The main channel of the Nîl, here turning off south, flowed for some distance parallel to the Tigris, down to a point one league below the town of Nahr Sâbus which lay one day's march above Wâsiṭ, where the canal finally discharged its waters into the Tigris, probably in part by the Lower Zâb canal. It is to be added that this last reach of the Nîl, below the Lagoon, was known as the Nahr Sâbus, 'the Canal of Sâbus,' and this gave its name to the town on the right bank of the Tigris, already mentioned (see p. 38). The nomenclature of these channels changed at different epochs; in the 7th (13th) century Yâḳût says that all the reach from Nîl city to Nu'mânîyah was called the Upper Zâb canal, while his Lower Zâb canal is apparently identical with the Nahr Sâbus of Ibn Serapion; both canals in the 7th (13th) century had, however, gone much to ruin, though still bordered by fertile lands.

Returning now to the ruins of Babylon on the Euphrates, the Sûrâ below here was crossed by a masonry bridge called the Ḳanṭarah-al-Ḳâmighân, 'through which its waters pour with a mighty rush' as Ibn Serapion reports. Six leagues below this bridge, and near Jâmi'ân, the later Ḥillah, the Sûrâ canal bifurcated, the right arm going south past that city, and the left arm, called the Nahr-an-Nars, turning off to the south-east, and after watering Ḥammâm 'Omar with other villages reached the town of Niffâr. This canal took its name from Nars (or Narses), the

Sassanian king who came to the throne in 292 A.D.; he having caused it to be dug. After running south for some distance both the Nahr Nars and the Sûrâ channel poured their waters finally into the Badât canal, which traversed the northern limit of the Great Swamp; and this Nahr-al-Badât (or Budât) was a long drainage channel taken from the left bank of the Kûfah arm of the Euphrates, at a point a day's journey to the north of Kûfah city, probably near the town of Ḳanṭarah-al-Kûfah, otherwise called Al-Ḳanâṭîr, 'the Bridges,' which doubtless carried the high road across the Badât. This city of 'the Bridges' lay 27 miles south of the great Sûrâ bridge of boats, and 28 miles north of Kûfah; and it probably lay adjacent to, or possibly was identical with, the Hebrew Pombedita (Arabic *Fam-al-Badât*, 'mouth of the Badât canal'), mentioned by Benjamin of Tudela in the 6th (12th) century as a great centre of Jewish learning in Babylonia. The Badât canal after a course of over 50 miles, and after receiving on its left bank the drainage of the Lower Sûrâ and Nars canals, discharged itself finally into the Great Swamp near the town of Niffâr[1].

The districts lying between the bifurcation of the Lower Euphrates, having the Sûrâ canal to the east and the main stream to the west, were known as the Upper and Lower Al-Fallûjah. Below these the main stream passed down by the town of Al-Ḳanṭarah and the outflow of the Badât canal to the city of Kûfah, which lay on the western bank of the Euphrates over against the bridge of boats, and south of this its waters were discharged by various channels into the Great Swamp. This older arm of the river is named by Ḳudâmah and Mas'ûdî the channel of Al-'Alḳamî, and it appears to be identical with the modern Nahr Hindîyah which branches from the present Euphrates stream below Musayyib and, flowing past the ruins of Kûfah, rejoins the present main stream of the Euphrates by a winding course through marshes that are a part of the Great Swamp of Abbasid times.

The city of Al-Kûfah was founded immediately after the

[1] I. S. 16. Baladhuri, 254, 290. I. R. 182. I. H. 167. Muk. 121. A. F. 53. Yak. i. 770; ii. 31, 903; iii. 4, 379; iv. 773, 798, 840, 861. Mst. 136. *The Travels of Benjamin of Tudela* (Asher), i. 112. See also De Goeje in *Zeit. Deutsch. Morg. Gesell.* for 1885, p. 10.

Moslem conquest of Mesopotamia, at the same time as Baṣrah was being built, namely, about the year 17 (638), in the Caliphate of 'Omar. It was intended to serve as a permanent camp on the Arab, or desert, side of the Euphrates, and occupied an extensive plain lying above the river bank, being close to the older Persian city of Al-Ḥîrah. Kûfah rapidly increased in population, and when in the year 36 (657) 'Alî came to reside here the city during four years was the capital of that half of Islam which recognised 'Alî as Caliph. In the mosque at Kûfah 'Alî was assassinated in the year 40 (661). Iṣṭakhrî describes Kûfah as the equal in size of Baṣrah in the 4th (10th) century, but the former had the better climate, and its buildings were more spacious; also its markets were excellent, though in this point it stood second to Baṣrah. The Great Mosque, where 'Alî received his death-wound, was on the eastern side of the city, and had tall columns brought from the neighbouring town of Ḥîrah, which fell to ruin as Kûfah became more populous. One of the chief quarters of Kûfah was Al-Kunâsah—'the place of the Sweepings'—which lay on the desert side of the town, and all round stood palm-groves which produced excellent dates. When Ibn Jubayr passed through Kûfah in 580 (1184) it was an unwalled town mostly in ruins, but its Friday Mosque still existed, and Ibn Baṭûṭah, in the 8th (14th) century, describes its roof as supported by pillars, formed of stone drums joined with lead. A Miḥrâb or niche marked the place where 'Alî had been assassinated. Mustawfî, who gives a long account of Kûfah, says that its walls, 18,000 paces in circuit, had been built by the Caliph Manṣûr. The sugar-cane grew here better than anywhere else in 'Irâḳ, and cotton crops yielded abundantly. In the mosque, on a column, was the mark of 'Alî's hand; and they also preserved here the oven (*tannûr*) from the mouth of which the waters had poured forth at the time of the Deluge of Noah.

Less than a league south of Kûfah are the ruins of Ḥîrah, which had been a great city under the Sassanians. Near by stood the famous palaces of As-Sadîr and Al-Khawarnaḳ, the latter built, according to tradition, by Nu'mân, prince of Ḥîrah, for King Bahrâm Gûr, the great hunter. The palace of Khawarnaḳ with its magnificent halls had mightily astonished the early Moslems when they first took possession of Ḥîrah on the conquest of

Mesopotamia. In later times Khawarnaḳ was sometimes used as a hunting lodge by the Caliphs, and apparently, though nothing now remains of it, some walls and domes were still standing, though in ruin, when Ibn Baṭûṭah passed by here in the beginning of the 8th (14th) century.

On the actual desert border, five leagues west of Kûfah, and the first stage on the road to Mecca, was the large hamlet of Al-Ḳâdisîyah surrounded by palm-groves, near which, in the year 14 (635), the Moslems had won their first great battle against the Persians, which led almost immediately to the subjugation of Mesopotamia. Muḳaddasî describes Ḳâdisîyah—called Ḳâdisîyah of Kûfah to distinguish it from the city of the same name on the Tigris (see p. 51)—as a town much frequented during the season of the Pilgrimage. It was defended by a small fort, and had two gates. Its lands were watered by a canal from the Euphrates which entered the town at the Baghdâd Gate; and at the Desert Gate (Bâb-al-Bâdiyah) was the Friday Mosque, before which, when the Pilgrims came, a great market was held. In the 8th (14th) century when Ibn Baṭûṭah travelled through Ḳâdisîyah it had sunk to the size of a large village, and Mustawfî describes it as for the most part in ruin[1].

Najaf, where the tomb of ‘Alî (Mashhad ‘Alî[2]) is to the Shî‘ahs a most venerated shrine, lies about four miles to the westward of the ruins of Kûfah, and is a populous town to the present day. The Shî‘ah tradition, as given by Mustawfî, is that on receiving the fatal stab in the Kûfah mosque, ‘Alî, knowing his death to be imminent, had immediately given orders that when the breath was out of his body, it was to be put on a camel and the beast turned loose; where the camel knelt, there his corpse was to be buried. All this was forthwith done, but during the time of the Omayyads no

[1] I. S. 10, 16. Kud. 233. Mas. *Tanbih* 52. Ist. 82. I. H. 162, 163. Muk. 116, 117. Yak. ii. 492; iii. 59; iv. 322. I. J. 213. I. B. i. 414; ii. 1, 94. Mst. 133, 138, 140. The broad shallow lake—known as the Baḥr Najaf—which now extends to the westward of the ruins of Kûfah and the Najaf shrine, did not exist in the middle-ages, and the Pilgrim road from Kûfah to Mecca passed across what is now its bed.

[2] *Mashhad* means ‘the place of Martyrdom,’ hence equivalent to Shrine; Al-Maḳâm, ‘the Place,’ is used in the same sense.

tomb was erected at Mashhad 'Alî, for the place was kept hidden for security. Subsequently, however, in the year 175 (791), the holy site was discovered by the Abbasid Caliph Hârûn-ar-Rashîd. For, when hunting one day near Kûfah, he chased his quarry into a thicket, but on attempting to follow the Caliph discovered that no force could prevail on his horse to enter the place. On enquiring of the peasants they informed him that this spot was known as the burial-place of the Caliph 'Alî, an inviolate sanctuary, where even wild beasts were safe from harm. Orders were given by Hârûn to dig, and the body of 'Alî being discovered, a Mashhad or shrine was, according to Mustawfî, forthwith built over the spot, which soon became a holy place of visitation. The early history of the shrine is obscure, the foregoing is the usual Shî'ah account, but though Hârûn-ar-Rashîd at one period of his reign favoured the Alids, the Arab chronicles certainly do not relate that he invented the tomb of 'Alî.

The earliest notice in detail of Mashhad 'Alî is of the middle of the 4th (10th) century by Ibn Ḥawḳal. He says that the Ḥamdânid prince Abu-l-Hayjâ—who was governor of Mosul in 292 (904) and died in 317 (929)—had built a dome on four columns over the tomb at Mashhad 'Alî, which shrine he ornamented with rich carpets and hangings: also he surrounded the adjacent town with a wall. Iṣṭakhrî and Ibn Ḥawḳal, however, add that in their day the burial-place of 'Alî was shown in the corner of the Great Mosque at Kûfah, and this was credited by many persons of note, as is affirmed by other authorities. Mustawfî says, further, that in the year 366 (977) 'Aḍud-ad-Dawlah the Buyid erected the mausoleum which in his (Mustawfî's) day still existed, and the place then became a little town, 2500 paces in circuit. In the chronicle of Ibn-al-Athîr it is recorded that 'Aḍud-ad-Dawlah, at his own wish, was buried here, likewise his sons Sharâf and Bahâ-ad-Dawlah; and in subsequent times various other notable persons followed the example. In the year 443 (1051) the shrine was burnt to the ground by the Baghdâd populace, who were zealous in persecuting the Shî'ahs. It must however have been quickly rebuilt, for Malik Shâh and his Wazîr, the Niẓâm-al-Mulk, made their visitation here in 479 (1086).

Writing in the early part of the 8th (14th) century Mustawfî adds that Ghâzân, the Îl-Khân of his day, had recently erected at Mashhad 'Alî a home for Sayyids (descendants of the Prophet) called the Dâr-as-Siyâdah, also a Khânḳâh or Darvish monastery. Yâḳût in the previous century describes the dyke at Najaf which kept back the waters of the Euphrates from overflowing the town, but he gives no account of the shrine. The traveller Ibn Baṭûṭah was here in the year 726 (1326) and speaks of Mashhad 'Alî as a fine city, which he entered by the Bâb-al-Ḥaḍrat—'the Gate of the Presence'—leading direct to the shrine. He gives a long description of its great markets and colleges, also of the mosque where 'Alî's tomb was shown, the walls of which were covered with enamelled tiles of Ḳâshânî work. He reports that at the tomb cripples were frequently healed of their infirmities, and he gives a long account of the many gold and silver lamps hung up as offerings, as well as the magnificent carpets, and describes the actual tomb as enclosed in a railing of chiselled gold plates, secured by silver nails. Four gates gave access to the shrine, each curtained, and having a silver doorstep, the walls also being hung with silk embroideries; and his account closes with the enumeration of the miracles vouchsafed here to all true believers[1].

Karbalâ, or Mashhad Ḥusayn, lies eight leagues to the north-west of Kûfah, and marks the site of the battlefield where in the year 61 (680) Ḥusayn, son of 'Alî, and grandson of the Prophet, was slain, with nearly all his family. The place of martyrdom of Ḥusayn is to Shî'ahs of the present day a more venerated place than Mashhad 'Alî. By whom the shrine was first built is not mentioned, but in the 3rd (9th) century some monument must have existed here, for in the year 236 (850) the Caliph Mutawakkil earned the lasting hatred of all good Shî'ahs by ordering the shrine of Ḥusayn to be destroyed by flooding the place with water, also he forbade the visitation of the sacred spot under heavy penalties. Mustawfî adds, when describing the palaces at Sâmarrâ, that this iniquity on the part of Mutawakkil was requited to him, in that none of the buildings he began at Sâmarrâ could ever be completed, but soon fell to the same state of ruin in which

[1] Ist. 82. I. H. 163. Muk. 130. Ibn-al-Athir, ix. 13, 42, 169, 394; x. 103. Mst. 134. Yak. iv. 760. I. B. i. 414—416.

the wicked Caliph had left the tomb of Ḥusayn. How long the place remained a ruin is not stated, but 'Aḍud-ad-Dawlah the Buyid in 368 (979) built a magnificent shrine here, doubtless an enlargement of the building noticed incidentally by the geographers Iṣṭakhrî and Ibn Ḥawḳal who wrote a little before this date.

In 407 (1016) the dome at Mashhad Ḥusayn was burnt down, but must have been shortly afterwards restored, for the place was visited by Malik Shâh in 479 (1086), when he went hunting in these districts. Yâḳût unfortunately gives no description of the shrines at Karbalâ, merely mentioning incidentally that the name Al-Ḥâir, meaning 'a garden pool,' was commonly given to the enclosure round the tomb of Ḥusayn. Mustawfî in the 8th (14th) century speaks of the little town that had grown up round the shrine as being some 2400 paces in circuit, and his contemporary Ibn Baṭûṭah describes the fine college (Madrasah) which he visited here. The Holy Theshold of the actual tomb, which the pilgrims kissed on entry, was he says of solid silver; the shrine was lighted by numerous gold and silver lamps, and the doorways were closed by silken curtains. Ibn Baṭûṭah adds that the little town was then mostly a ruin, from the ceaseless fighting of rival factions among its inhabitants, but it stood among many groves of date palms, well watered by canals coming from the Euphrates[1].

When describing 'Irâḳ in the 3rd (9th) century, Ibn Khurdâdbih and Ḳudâmah state that the province was then divided into twelve districts called *Astân*, each containing a varying number of sub-districts, called *Ṭassûj*, and of these latter the total number was sixty. This division, which probably in its origin was made for fiscal purposes, is repeated in part by Muḳaddasî in the following century, and it will be worth while to enumerate the twelve Astâns, giving at the same time the best known of their sub-districts or Ṭassûj. The list is divided into three groups according to the irrigation channels, and whence the water was taken.

The first group of four districts consists of those lying on the east side of the Tigris, and watered from that river and from the Tâmarrâ. These were (1) the Astân of Shâd Fîrûz or Ḥulwân

[1] Ist. 85. I. H. 166. Muk. 130. Yak. ii. 189. Mst. 134, 139. I. B. ii. 99. Ibn-al-Athir, vii. 36; viii. 518; ix. 209; x. 103.

(otherwise Shâdh Fayrûz) comprising the sub-districts of Tâmarrâ, and Khâniḳîn, with three others; five in all: (2) the Shâd Hurmuz district, round Baghdâd, with the sub-districts of Nahr Bûḳ, of Kalwâdhâ and Nahr Bîn, of Al-Madînah-al-'Atîḳah (otherwise Madâin), of Upper and of Lower Râdhân, with two others; seven in all: (3) the Shâdh Ḳubâdh district, with the sub-districts of Jalûlâ, of Bandanîjîn, of Barâz-ar-Rûz, and of Daskarah, with four others, making a total of eight. Of these two last districts this is the nomenclature given by Ibn Khurdâdbih; Ḳudâmah on the contrary transposes the names, making the Astân of Shâdh Ḳubâdh the Baghdâd district, and giving Khusraw Shâdh Hurmuz as the name of the Jalûlâ Ṭassûj with its seven neighbours. The last Astân to the east of the Tigris was (4) the district of Bâzîjân Khusraw, otherwise of Nahrawân, which Ḳudâmah names Arandîn Kird, and this comprised five sub-districts, to wit: Upper, Middle, and Lower Nahrawân (with Iskâf of the Banî Junayd and Jarjarâyâ), next the Bâdarâyâ Ṭassûj, and lastly Bâkusâyâ.

The next group of two districts was of those watered partly from the Tigris, partly from the Euphrates; it consisted of (5) the Astân of Kaskar, otherwise called Shâdh Sâbûr, with four sub-districts lying round Wâsiṭ; and (6) the Astân of Shâdh Bahman, or the Kûrah Dijlah, on the Lower Tigris, with four sub-districts, Maysân and Dasti-Maysân being two of them, the latter lying round Ubullah.

The remaining six districts all lay to the west of the Tigris, and were watered by the old Dujayl canal previously mentioned and by the great canals flowing eastward from the Euphrates to the Tigris. The first of these was: (7) Astân-al-A'lâ, 'the Upper District,' with the four sub-districts lying along the Nahr 'Îsâ, namely Fîrûz Sâbûr or Al-Anbâr, Maskin, Ḳaṭrabbul, and Bâdûrayâ. Next below came: (8) the Astân of Ardashîr Bâbgân, lying along the Kûthâ canal and the Nîl, with the sub-districts of Bahurasîr and Rûmaḳân opposite Madâin, of Kûthâ, and of the two canals called Nahr Jawbar and Nahr Durḳît. To the east of this was: (9) the district of the Zâb canals, called the Astân of Bih Dhîvmâsufân, comprising the sub-districts of the Upper, Middle, and Lower Zâb canals.

The last three districts were those respectively of Upper, Middle, and Lower Bih Ḳubâdh, and of these the first (10) Upper Bih Ḳubâdh comprised six sub-districts, namely, Bâbil (the ruins round Babylon), Upper and Lower Al-Fallûjah, with two others, and the Ṭassûj of 'Ayn-at-Tamr some distance to the west of the Euphrates. The Astân (11) of Middle Bih Ḳubâdh included four sub-districts, to wit, those of the Badât canal, of Sûrâ with Barbîsamâ, of Bârûsamâ, and of Nahr-al-Malik. Finally (12) Lower Bih Ḳubâdh comprised five sub-districts, all of which apparently lay adjacent to the lower course of the Euphrates where it entered the Great Swamp. The names in these lists show clearly that we have here the division of the country which the Arabs took over from the Sassanians; Ardashîr Bâbgân was the founder of the dynasty: Shâd Fîrûz or Shâdh Fayrûz means 'glorious fortune' in Persian. Bih Ḳubâdh is 'the Goodness, or good land, of King Ḳubâdh,' and the 'Glory' (Shâdh) of Hurmuz, of Ḳubâdh, of Shâpûr, and of Bahman recall the names of four of the most famous kings of Persia[1].

The trade of 'Irâḳ consisted of imports rather than of exports, the capital province consuming the products of the outlying regions. Muḳaddasî, however, gives a list of commodities and manufactures for which several cities were famous, and this though not very full is worth examining.

The markets of Baghdâd were noted for all kinds of curious wares brought together here from foreign lands. Its manufactures were coloured silks—the famous 'Attâbî or 'Tabby' silk in particular, named after one of its quarters—fine strong cloth, curtains and veils, stuffs for turbans, napkins of all sorts, and mats woven of reeds. In Baṣrah many stuffs were manufactured of raw silk and its markets were famous for the jewellers, who sold all manner of curiosities; further Baṣrah was the chief emporium for various ores and minerals, antimony, cinnabar, Mars-saffron, litharge and many others being mentioned. There were also exported dates, ḥenna-dye and raw silk, as well as rose-water and essence of violets: while at Ubullah excellent linen was woven. Kûfah was famous for its dates, for its essence of violets, and for raw-silk stuffs of which turbans were made; Wâsiṭ exported lupins and

[1] I. K. 5—8. Kud. 235, 236. Muk. 133.

dried fish called *Shîm*; finally Nu'mânîyah manufactured much cloth, and was famous for all sorts of woollen stuffs[1].

As explained in the introductory chapter, the central point of the system of high roads during the Abbasid Caliphate naturally was Baghdâd; whence five main roads—to Baṣrah, Kûfah, Anbâr, Takrît and Ḥulwân—set forth, communicating ultimately with the outposts of the empire.

The easiest route to Baṣrah from Baghdâd was naturally by boat down the Tigris, and this, noting all the towns passed to right and left on the river bank, is given in much detail by both Ibn Rustah and Ya'ḳûbî. Down as far as Al-Ḳaṭr the Tigris main channel was followed, then came the Great Swamp through which boats passed threading the lagoons (Hawl, see above, p. 42). The Abu-l-Asad canal led out to the head of the Tigris estuary, and from this Baṣrah was reached by the Nahr Ma'ḳil. The Ubullah canal led back to the estuary, and was followed by those bound for 'Abbâdân and the Persian Gulf. The way by land from Baghdâd to Wâsiṭ, which went down the eastern side of the Tigris through Madâin, is also given by Ibn Rustah, and this enables the towns on the river bank to be set down on the map, for the distances are stated in farsakhs (leagues); Ḳudâmah also gives this route in detail, and in one or two cases where lacunæ occur they can generally be filled up from Abu-l-Fidâ. The road from Wâsiṭ to Baṣrah by land, along the northern edge of the Great Swamp, is given by Ḳudâmah, and this too is the way by which Ibn Baṭûṭah travelled in the 8th (14th) century. Ibn Rustah and Ḳudâmah likewise give the road from Wâsiṭ, eastward, to Ahwâz the capital of Khûzistân; and from the stage at Bâdhbîn, one march east of Wâsiṭ on this road, a bifurcation went north-east to Ṭîb, from which Sûs (Susa) in Khûzistân was reached[2].

The Pilgrim road, going south from Baghdâd to Kûfah, left the Round City by the Kûfah Gate and passed through the Karkh quarter to Ṣarṣar, and thence on to Ḳaṣr Ibn Hubayrah. Beyond this it crossed the eastern arm of the Euphrates (the present main

[1] Muk. 128.

[2] I. R. 184, 186—188. Ykb. 320. Kud. 193, 225, 226. Mst. 195. A. F. 305. I. B. ii. 8.

channel) called in the 4th (10th) century the Nahr Sûrâ, at the Sûrâ bridge of boats, and thence came down to Kûfah, opposite to which the western arm of the Euphrates was crossed by the bridge of boats which led to the eastern suburbs of the city. From Kûfah the Pilgrim road went south-west to Ḳâdisîyah, where it entered the Arabian desert. This road is given by all the earlier geographers, and in much detail by Ibn Rustah, who for some parts of the way from Baghdâd to Kûfah gives alternative routes, with the distances in miles and in leagues. After the beginning of the 6th (12th) century Ḳaṣr Ibn Hubayrah, the half-way stage between Baghdâd and Kûfah, fell to ruin; Ḥillah taking its place (see p. 71) to which the high road went down from Ṣarṣar by Farâshah. At Ḥillah the eastern arm of the Euphrates was crossed by a great bridge of boats similar to that which had formerly existed at Sûrâ. This is the route followed by Ibn Jubayr and all later travellers. From Kûfah to Baṣrah along the southern border of the Great Swamp was reckoned as 80 or 85 leagues, and this road, which branches to the left at the second desert stage south of Ḳâdisîyah, is described by Ibn Rustah and Ibn Khurdâdbih[1].

As already said, two Pilgrim roads crossed the deserts of Arabia going from Mesopotamia to the Ḥijâz, one starting from Kûfah, the other from Baṣrah, and they came together at the stage of Dhât 'Irḳ, which was two days march north-east of Mecca. These two famous Pilgrim ways are described stage by stage, and the half-stage is also given, where the caravan halted for supper (*Al-Muta'ashshâ*), with the number of miles between each halt carefully noted, in the Road Books of the 3rd (9th) century and by Muḳaddasî. The road from Kûfah passed through Fayd,

[1] I. R. 174, 175, 180, 182. Ykb. 308. I. K. 125, 145. Kud. 185. A. F. 303. I. J. 214—219. Mst. 193. Muḳaddasî (p. 252) estimates the distance from Baṣrah to Kûfah along the edge of the desert at ten long marches (Marḥalah), and at the shortest reckoning it is over 250 miles. It is famous in history for having been traversed in a night and a day by a certain Bilâl ibn Abi Burdah, riding swift dromedaries (*Jammâzah*), he having an urgent affair with Khâlid-al-Ḳasrî at Kûfah, in the year 120 (738), during the reign of the Omayyad Caliph Hishâm. Tabari, ii. 1627. (It will be remembered how Dick Turpin rode from London to York, 200 odd miles, in 18 hours: the rate is about the same.)

which lay a short distance south of Ḥâyil, the present chief town of Jabal Shammâr. The Baṣrah road went by Ḍarîyah, the older capital of what later became the Wahhâbî kingdom, the ruins of which town still exist a few miles to the west of Ar-Riyâḍ, the present chief town of Najd. From both the Kûfah and the Baṣrah Pilgrim ways there were branch roads, bifurcating to the right, leading direct to Medina[1].

From Baghdâd at the Kûfah Gate of the Round City a second high road branched westward, and going first to Muḥawwal kept along the bank of the 'Îsâ canal to Anbâr on the Euphrates, whence following up stream it passed Ḥadîthah, the last town in 'Irâḳ, and reached 'Ânah in Jazîrah. This is the first part of one of the roads (namely, by the Euphrates) going from Baghdâd to Syria, and it is given by Ibn Khurdâdbih and Ḳudâmah. The other road to Syria goes north along the Tigris by Mosul, and as far as Takrît lies in the 'Irâḳ province. This, which was the post-road, left the Baradân Gate of East Baghdâd and keeping up the left bank of the river through 'Ukbarâ and Sâmarrâ came to Takrît. It was here joined by the caravan road which, leaving the

[1] The Kûfah road to Mecca and Medina is given in I. K. 125. Kud. 185. I. R. 175. Ykb. 311. Muk. 107, 251. The Baṣrah road is given in I. K. 146. Kud. 190. I. R. 180, 182. Muk. 109, 251. It is worth noting that the older chief town of Najd is invariably written Ḍarîyah (with initial *Ḍâd*) by the Arab geographers. Ḥâjjî Khalfah is the first (J. N. 527) to give the modern pronunciation and spelling Dara'îyah (with initial *Dâl* and an *'Ayn*) though once or twice and in the Itinerary (J. N. 527, 543) he writes Ḍarîyah or Ḥiṣn Ḍarîyah. The geography of the Ḥijâz, and of Arabia in general lying north of the Dahnâ or Great Desert, has been fully worked out (from Arabic sources) by Professor F. Wüstenfeld, in a series of articles published in the *Abhandlungen der Königlichen Gesellschaft der Wissenschaft zu Göttingen*. These papers are provided with maps by Kiepert, and are well indexed; they include the following, of which I give the names in full, as they do not appear to be well known to English geographers. *Die von Medina auslaufenden Hauptstrassen* (vol. XI, 1862): *Die Wohnsitze und Wanderungen der Arabischen Stämme* (vol. XIV, 1869): *Die Strasse von Basra nach Mekka mit der Landschaft Dharija* (vol. XVI, 1871): *Das Gebiet von Medina* (vol. XVIII, 1873), which gives the Kûfah-Mecca Pilgrim road: *Bahrein und Jemama* (vol. XIX, 1874): lastly, *Geschichte der Stadt Medina* (vol. IX, 1860, and published separately), also vol. IV of *Chroniken der Stadt Mekka* (Leipzig, 1861) which contains a summary (in German) of the history of Mecca, with topographical notes.

Ḥarbîyah quarter in West Baghdâd, went up the Dujayl canal to Ḥarbâ, and thence by the palace grounds opposite Sâmarrâ passed along the Isḥâḳî canal to Takrît. This last is the road followed by Ibn Jubayr and Ibn Baṭûṭah[1].

Finally from the Khurâsân Gate of East Baghdâd started the great Khurâsân road which, crossing Persia, went, as already said, through Transoxiana, ultimately reaching the borders of China. This road is described in great detail, stage by stage, by Ibn Rustah; and almost all the other geographers give the distances along the various portions of this great highway, which is thus one of the best known to us of all the trunk roads[2].

[1] I. K. 72, 93. Kud. 214, 216, 217. Muk. 134. I. J. 232. I. B. ii. 132. Mst. 195.

[2] I. R. 163. Ykb. 269. I. K. 18. Kud. 197. Muk. 135. Mst. 193.

CHAPTER VI.

JAZÎRAH.

The three districts. The district of Diyâr Rabî'ah. Mosul, Nineveh, and the neighbouring towns. Great Zâb, Ḥadîthah, and Irbil. Little Zâb, Sinn, and Daḳûḳ. The Lesser Khâbûr, Ḥasanîyah, and 'Imâdîyah. Jazîrah Ibn 'Omar and Mount Jûdî. Naṣîbîn and Râs-al-'Ayn. Mârdîn and Dunaysir. The Hirmâs and the Khâbûr. 'Arabân and the Tharthâr river. Sinjâr and Ḥaḍr. Balad and Adhramah.

As already explained the Arabs named Upper Mesopotamia Al-Jazîrah, 'the Island' or 'Peninsula,' for its plains lay encompassed by the upper courses of the rivers Tigris and Euphrates. The province was generally divided into three districts called Diyâr Rabî'ah, Diyâr Muḍar, and Diyâr Bakr, after the Arab tribes of Rabî'ah, Muḍar, and Bakr respectively, who, in pre-Islamic days, had settled here under Sassanian rule, each receiving its appointed *Dâr* (plural *Diyâr*) or 'Habitation' to which the tribe had subsequently given its name. Of Diyâr Rabî'ah, Mosul on the Tigris was the chief town; of the district Diyâr Muḍar, Raḳḳah on the Euphrates was the capital; while Âmid on the upper course of the Tigris was the chief city of Diyâr Bakr, the northernmost of the three districts. Muḳaddasî, on the other hand, describes the Jazîrah province under the name of Iḳlîm Aḳûr, 'the Aḳûr Region'; the origin of the name is not clear, but Aḳûr would appear to have been the proper name at one time of the great plain of northern Mesopotamia.

A reference to the map shows that in Upper Mesopotamia the rivers Tigris and Euphrates receive their affluents almost exclusively on their left bank, that is flowing from the north-east or north. During the period of the middle-ages an exception

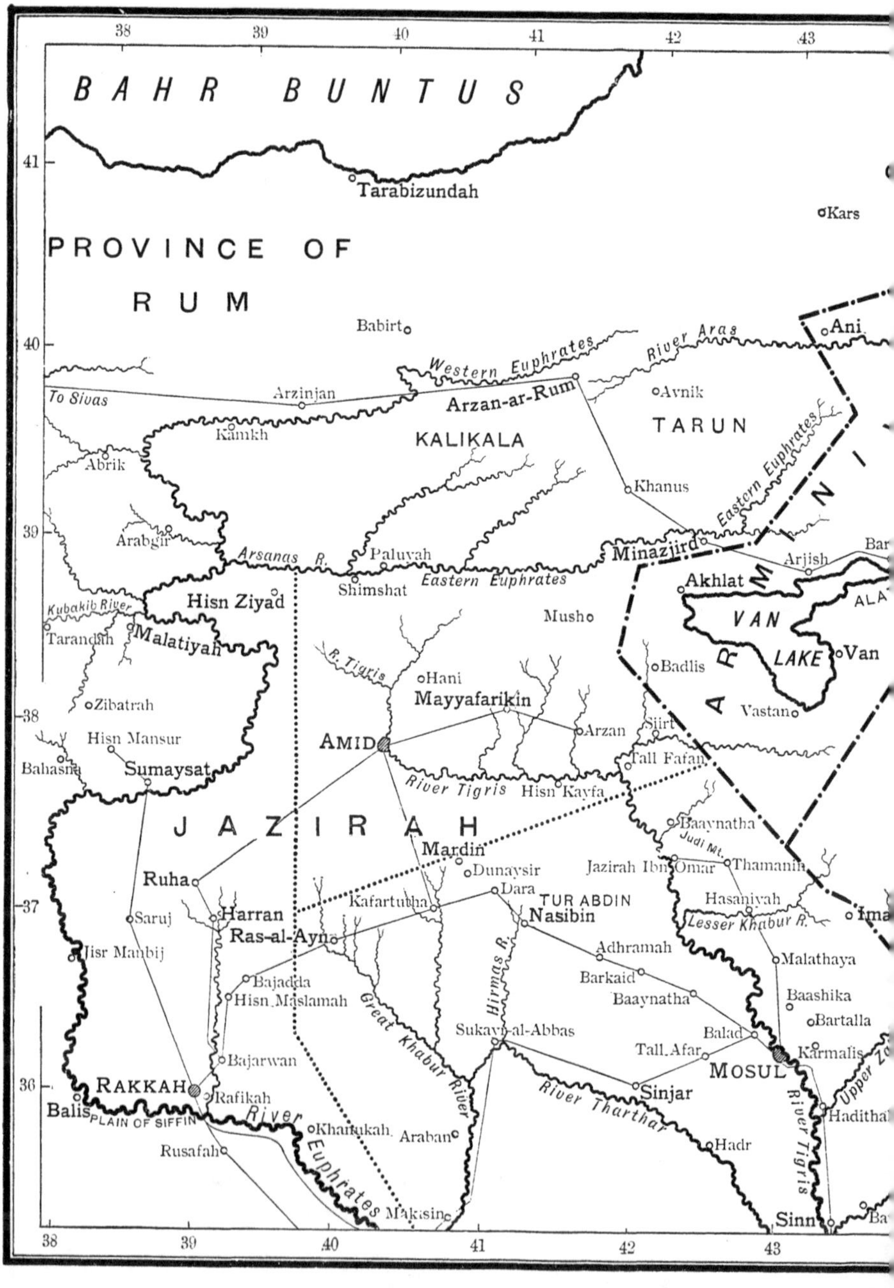

PROVINCES OF THE NORTH-WEST FRO

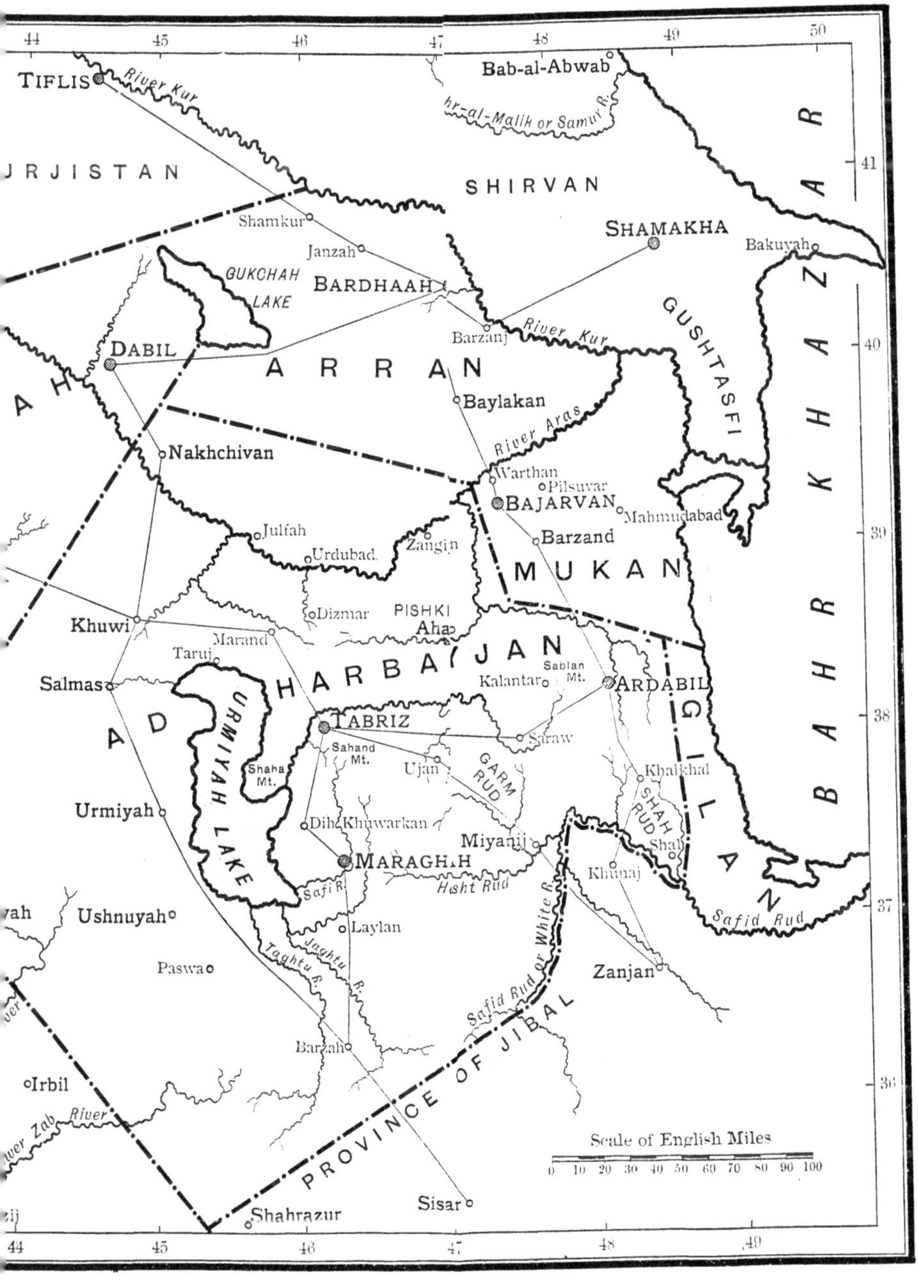

TIER, WITH JAZIRAH AND ADHARBAYJAN

occurs to this rule, namely in the drainage of the affluent of the (greater) Khâbûr, the Hirmâs river from Naṣîbîn. Just above its point of junction, the Hirmâs was dammed back at Sukayr-al-'Abbâs, and while a moiety passed on to join the Khâbûr which went to the Euphrates at Ḳarḳîsiyâ, the main stream of the Hirmâs flowed into the Tigris on its right bank at Takrît by the channel called the Nahr-ath-Tharthâr. Further, it will be seen that the limits of the three districts are determined by the water parting. Diyâr Bakr was the country watered by the Tigris from its source to the great bend south made by the river below Tall (the Hill of) Fâfân, with the land to the northward traversed by the numerous affluents of the Tigris which join its left bank west of Tall Fâfân. To the south-west, Diyâr Muḍar comprised all the lands along the Euphrates from Sumaysâṭ, where it left the mountain gorges, down to 'Ânah, with the plains watered by its affluent the river Balîkh, coming from Ḥarrân. Lastly Diyâr Rabî'ah was the district east of Muḍar; namely, of the (greater) Khâbûr coming from Râs-al-'Ayn, with the Hirmâs which, as we have seen, flowed eastward by the Tharthâr to the Tigris, also the lands on both banks of the Tigris from Tall Fâfân down to Takrît, namely those westward to Naṣîbîn, and those eastward which included the plains watered by the Lower and Upper Zâb and the Lesser Khâbûr river.

Mosul (Al-Mawṣil), the chief city of Diyâr Rabî'ah, stands on the western bank of the Tigris at the point where a series of loops in the river coalesce to form a single main stream, and Al-Mawṣil, meaning 'the confluence,' is said to take its name from this fact. In Sassanian times the city which existed here was called Bûdh Ardashîr. Under the Omayyads Mosul rose to importance, a bridge of boats was set across the Tigris, connecting the city on the western side with the ruins of Nineveh on the east bank, and Mosul became the capital of the Jazîrah province under Marwân II, the last of the Omayyad Caliphs, who also built here what afterwards came to be known as the Old Mosque[1].

Ibn Ḥawḳal who was at Mosul in 358 (969) describes it as a

[1] Muk. 136—138. I. K. 17. Yak. iv. 682—684. Mar. i. 84. Yâḳût gives the old Persian name of Mosul as Bawardashîr or Nawardashîr, but the latter form is undoubtedly a clerical error.

fine town with excellent markets, surrounded by fertile districts of which the most celebrated was that round Nînaway (Nineveh) where the prophet Yûnis (Jonah) was buried. In the 4th (10th) century the population consisted chiefly of Kurds, and the numerous districts round Mosul, occupying all Diyâr Rabî'ah, are carefully enumerated by Ibn Ḥawḳal. Muḳaddasî praises the numerous excellent hostelries of Mosul, and the town, he says, was extraordinarily well built, being in plan a semicircle, and about a third the size of Baṣrah. Its castle was named Al-Murabba'ah (the Square) and it stood on the affluent called the Nahr Zubaydah; within its precincts was held the Wednesday Market (Sûḳ-al-Arba'â) by which name also the castle was sometimes known. The Friday Mosque (that of Marwân II) stood a bow-shot from the Tigris, on a height to which steps led up. The roof of this building was vaulted in stone, and it had no doors to close the doorways going from the main building of the mosque into its court. The market streets of Mosul were for the most part roofed over, eight of the chief thoroughfares are named by Muḳaddasî, and the houses of the town stretched for a considerable distance along the Tigris bank. Muḳaddasî adds that formerly Mosul had borne the name of Khawlân: and that the Ḳaṣr-al-Khalîfah, 'the Palace of the Caliph,' stood on the opposite bank of the river, half a league from the town, overlooking Nineveh. This palace had of old been protected by strong ramparts, which the winds had overthrown, and the ruins, through which flowed the stream called the Nahr-al-Khawṣar, were when Muḳaddasî wrote occupied by fields.

In the year 580 (1184) Mosul was visited and described by Ibn Jubayr. Shortly before this date the famous Nûr-ad-Dîn, under whose banner Saladin began his career, had built the new Friday Mosque in the market place, but the old mosque of Marwân II still stood on the river bank, with its beautifully ornamented oratory and iron window-gratings. In the upper town was the great fortress, and the town walls with towers at intervals extended down to and along the river bank, a broad street connecting upper with lower Mosul. Beyond the walls were extensive suburbs with many small mosques, hostelries, and bath houses. The Mâristân (or hospital) was famous, also the great market buildings called the Ḳayṣarıyah[1], and there were also numerous colleges here. Ḳazwînî gives a list of the various Dayrs or Christian convents which were found in the vicinity of Mosul, and he notes especially the deep ditch and high walls of the Mosul fortress. All round the town were numerous gardens irrigated, he says, by waterwheels.

In regard to the Nineveh mounds, these were known from the time of Muḳaddasî as the Tall-at-Tawbah, 'the Hill of Repentance,' where the prophet Yûnis, Jonah, had sought to convert the people of Nineveh. The spot was marked by a mosque, round which, Muḳaddasî adds, were houses for pilgrims, built by Nâṣir-ad-Dawlah the Ḥamdânid prince, and half a league distant was a celebrated healing spring called 'Ayn Yûnis after the prophet Jonah, with a mosque adjacent, and here might be seen the Shajarah-al-Yaḳṭîn, namely 'the Tree of the Gourd' planted by the prophet himself. Yâḳût adds that most of the houses of Mosul were built of limestone or marble, with vaulted roofs, and that in the city might be seen the tomb of the prophet Jurjîs, or St George. In the 8th (14th) century Ibn Baṭûṭah passed through Mosul, which he describes as protected by a double wall and many high towers, 'like those of Dehlî.' The fortress was then known as Al-Ḥadbâ, 'the Hump-backed,' and in the new Friday Mosque (that of Nûr-ad-Dîn) was an octagonal marble basin with a fountain in its midst throwing up a jet of water a fathom high. A third Friday Mosque had recently been built overlooking the Tigris, and this is probably the building praised by Mustawfî, who says that the stone sculptured ornamentation of its oratory was so intricate that it might stand for wood-carving. In his day the circuit of Mosul measured a thousand paces, and he refers to the famous shrine of Jonah (Mashhad Yûnis) on the opposite bank of the Tigris, lying among the ruins of Nineveh[2].

[1] The Arabs, more especially those of the west, called the great buildings of a market, often used as a hostelry or caravanserai, Al-Ḳayṣarîyah, or Ḳaysârîyah, a term which they must have derived from the Greeks, though *Καισαρεία* does not occur, apparently, in the Byzantine historians, as applied to the Cæsarian, or royal market of a town. In any case the word seems hardly likely to have been taken by the Moslems from the name of the Cæsarion, the famous quarter of Alexandria; though this explanation is the one often given.

[2] I. H. 143—145. Muk. 138, 139, 146. I. J. 236—238. Yak. iv. 684. I. B. ii. 135. Kaz. ii. 247, 309. Mst. 165, 167.

A few miles to the east of Mosul lie the two small towns of Barṭallâ and Karmâlîs, which are mentioned by Yâḳût and Mustawfî, and Bâ'ashîḳâ is somewhat to the north of these, all three being of the dependencies of Mosul. Muḳaddasî mentions Bâ'ashîḳâ as noted for yielding a plant that cured scrofula and hæmorrhoids. It was a small town, Yâḳût adds, with a stream that worked many mills and irrigated its orchards, where olives, dates, and oranges grew abundantly. There was a large market here or Ḳayṣarîyah, with excellent bath houses. The Friday Mosque had a fine minaret, though in the 7th (13th) century most of the population were Christians. Barṭallâ lying a few miles south of Bâ'ashîḳâ was likewise counted as of the Nineveh district. It was, Yâḳût says, a place of great trade, mostly inhabited by Christians, though there was a fine mosque here, and many Moslems made the town their abode. The lettuces and greens of Barṭallâ were proverbial for their excellence, and Mustawfî praises its cotton crops. Karmâlîs, some miles further to the south again, had also a fine market according to Yâḳût, being a large village almost the size of a town, and much frequented by merchants. Mâr Juhaynah, or Marj (the meadow of) Juhaynah, was also near these places, but on the Tigris bank, being the first stage on the road from Mosul south to Baghdâd. Muḳaddasî describes it as having many pigeon towers. Its castle was strongly built of mortared stone, and a Friday Mosque stood in the midst of the town.

Between Mosul and Takrît the Tigris received, on its eastern bank, the waters of the two Zâbs, the one flowing in about a hundred miles above the other; and Ibn Ḥawḳal praises the magnificent fields occupying the broad lands lying between the two rivers. The upper or Greater Zâb rose in the mountains between Armenia and Adharbâyjân, and joined the Tigris at Ḥadîthah. The lower or Lesser Zâb, called also Majnûn, 'the Mad River,' from its impetuous current, flowed down from the Shahrazûr country, and came into the Tigris at Sinn. The country from which the Great Zâb flows is that known as Mush-takahar and Bâbghîsh according to Yâḳût, and its waters at first were red in colour, but afterwards ran clear. Al-Ḥadîthah, 'the New Town,' which stood a league above its junction with the Tigris (called Ḥadîthah of Mosul, to distinguish it from Ḥadîthah on the Euphrates already mentioned, p. 64), had been rebuilt by the last Omayyad Caliph, Marwân II, on a height overlooking the swampy plain; it was surrounded by famous hunting grounds, and had many gardens. The town was built in a semicircle, steps led up to it from the Tigris, and the Friday Mosque which was constructed of stone overlooked the river. Under the Sassanians the town was known as Nawkird, meaning in Persian likewise 'new town,' and before the rise of Mosul this had been the capital of the province[1].

The town of As-Sinn (the Tooth) lying one mile below the junction of the Lower Zâb according to Mas'ûdî, but above it with the Lesser Zâb flowing to the east according to Muḳaddasî, was in the middle-ages chiefly inhabited by Christians, and Yâḳût says there were many churches here. It was known as Sinn of Bârimmâ, to distinguish it from other towns of this name, the Bârimmâ chain of hills being cut through by the Tigris near this point. Sinn had in its market place a Friday Mosque, built of stone, and was surrounded by a wall. To the east of it, four leagues higher up the bank of the Lesser Zâb, stood the town of Bawâzîj (Madînat-al-Bawâzîj as Ibn Ḥawḳal gives the name) which however appears at the present day to have left no trace on the map. This also is the case with both Sinn and Ḥadîthah, and may be explained by the lower courses of both the Zâbs having much changed since the 4th (10th) century. Yâḳût refers to the town as Bawâzîj-al-Malik, 'of the King,' and in the 8th (14th) century it still existed, for Mustawfî describes it as paying 14,000 dînârs to the treasury of the Îl-Khâns.

South of Sinn the post-road to Sâmarrâ and Baghdâd kept along the left bank of the Tigris, passing first Bârimmâ, a hamlet lying under the hills of this name otherwise known as the Jabal Ḥumrîn, then coming to As-Sûdaḳâniyah, and finally reaching Jabiltâ (or Jabultâ) which appears to have been a mint city in 304 (916) lying on the east bank of the Tigris a little to the northward of Takrît. None of these small towns now appear on the map, but their positions are given very exactly in the Itineraries.

[1] Ist. 75. I. H. 147, 155. Muk. 139, 146. Yak. i. 446, 472, 567; ii. 168, 222, 552, 902; iv. 267. Mst. 165, 166, 214.

Rather more than a hundred miles due east of Sinn lies the town of Daḳûḳâ or Daḳûḳ—the name is generally written Tâûḳ or Tawûḳ in 'Alî of Yazd, as at the present day—which is frequently mentioned by Yâḳût and the later geographers. Mustawfî speaks of the river of Daḳûḳ (as he spells the name) which, rising in the Kurdistân mountains near Darband-i-Khalîfah (the Caliph's Pass), flowed out below the town of Daḳûḳ into the sandy plain, where, according to Mustawfî, there were most dangerous quicksands which swallowed up those who attempted to cross over. In flood time, he says, the Daḳûḳ river reached the Tigris, and its lower course is the stream now known as the Nahr-al-A'ẓam (the Great River): but in early times when the Nahrawân canal existed in its entirety, the spring floods of the Daḳûḳ river must have flowed into this. Mustawfî describes the town of Daḳûḳ as of medium size; it had a more healthy climate than that of Baghdâd, and near it were found naphtha springs. It is to be remarked that the place is not mentioned by the earlier Arab geographers[1].

Irbil, the ancient Arbela, lay in the plain between the Greater and Lesser Zâb, and is described by Yâḳût as a town much frequented by merchants. The castle, which crowned a hill, had a deep ditch and was in part enclosed by the town wall. A great market was held here, and the mosque, called Masjid-al-Kaff, 'of the Hand,' was celebrated for the mark of a man's palm on one of its stones. In the 7th (13th) century the market buildings had recently been restored, and great suburbs stretched beyond the city wall. Mustawfî praises the excellent crops, especially of cotton, that were produced by its lands. To the north of Mosul the city of 'Imâdîyah, near the head waters of the Upper Zâb, according to Mustawfî derived its name from its founder the Daylamite prince 'Imâd-ad-Dawlah who died in 338 (949). Other

[1] Ist. 75. I. H. 153. Mas. *Tanbih* 52. Kud. 214. Muk. 123. Yak. i. 464, 750; ii. 581; iii. 169. Mst. 139, 165, 220. A. Y. i. 660. Karkûk, not given by Yâḳût or the earlier geographers, is mentioned by 'Alî of Yazd (i. 661) as near Tâûḳ. In regard to Jabiltâ, or Jabultâ, on the Tigris opposite Takrît, it is to be remarked that this name has often been misread Ḥabiltâ (e.g. Muk. 135: the letters Ḥ and J being identical in Arabic script except for a diacritical point). The initial letter however is certainly J, for in Syriac the name frequently occurs under the form *Gebhiltâ*, and in this script G and Ḥ do not resemble one another.

authorities, however, ascribe 'Imâdîyah, or at any rate the restoration of that town in 537 (1142), to 'Imâd-ad-Dîn Zangi, father of that famous prince of Upper Mesopotamia, Nûr-ad-Dîn, under whom Saladin began his career. Yâkût reports that of old a castle had existed here held by the Kurds, and known under the name of Âshib. Mustawfî in the 8th (14th) century describes 'Imâdîyah as a town of considerable size.

In the neighbouring mountains were the head waters of the river Khâbûr-al-Ḥasanîyah, which flowed into the Tigris just north of the town of Faysâbûr, about 150 miles above Mosul. This river (not to be confounded with the Khâbûr of Râs-al-'Ayn) rose according to Yâkût in the district of Az-Zawzân, and at the town of Al-Ḥasanîyah it was spanned by a magnificent stone bridge, the remains of which still exist near the hamlet of Ḥasan Aghâ, which probably represents the older town. Ḥasanîyah, where there was a Friday Mosque, is described by Mukaddasî as a place of some importance, and one stage to the south of it on the road to Mosul was the small town of Ma'alathâyâ, where there was a Friday Mosque on a hill, the place being completely surrounded by gardens[1].

To the north of Faysâbûr is the important town of the Jazîrah (the Island), called Jazîrah Ibn 'Omar for distinction, after a certain Al-Ḥasan Ibn 'Omar of the tribe of Taghlîb, its founder; and the Tigris, as Yâkût explains, went half round the city in a semicircle, while a ditch filled with water on the land side made it an island. Ibn Ḥawkal in the 4th (10th) century describes Jazîrah as a walled town, whither the products of Armenia were brought for sale: its cheese and honey were famous. Its houses were of stone, and Mukaddasî adds that the mud at Jazîrah in winter time was phenomenal. Ibn Baṭûṭah who was here in the 8th (14th) century found it much ruined. The old mosque, however, stood in the market place, and the town wall, built of stone, still existed. Mustawfî adds that over a hundred villages were of its dependencies. Opposite Jazîrah Ibn 'Omar, on the west bank of the Tigris, was Bâzabdâ of the Bâkirdâ district, this representing the well-known

[1] Muk. 139. Kaz. ii. 192. Yak. i. 186; ii. 384; iii. 717, 931. Mst. 165, 166.

Roman fortress of Bezabda, but no description is given of the place.

From Jazîrah Ibn 'Omar Jabal Jûdî was visible to the eastward, with the Mosque of Noah on its summit, and Ḳarîyat-ath-Thamânîn (the Village of the Eighty) at the foot of the mountain. The Ḳurân (ch. xi. *v.* 46) states that 'the Ark rested upon Al-Jûdî,' which Moslem tradition identifies with this mountain in Upper Mesopotamia, and eighty of the companions of Noah are said to have built the village of Thamânîn named after their number. Muḳaddasî describes Thamânîn in the 4th (10th) century as a fair-sized city, and it lay one march to the north of Al-Ḥasanîyah; Mustawfî who calls it Sûḳ-Thamânîn—'the Market of the Eighty'—says that in his day it had fallen to ruin. Various affluents entered the Tigris on its left bank near Jazîrah Ibn 'Omar, and these are enumerated by Yâḳût, namely, the Yarnâ (or Yarnî) and the Bâ'aynâthâ (or Bâsânfâ as Ibn Serapion calls it), with a large village of the same name, above Jazîrah. Below this town, but to the north of the Khâbûr-al-Ḥasanîyah, and flowing down from the country of Az-Zawzân were the Al-Bûyâr and Dûshâ rivers[1].

On the western side of the Tigris, in the latitude of Jazîrah Ibn 'Omar, is the hilly district of Ṭûr 'Abdîn, 'the Mountain of (God's) Servants,' peopled by the Jacobites, in which the rivers Hirmâs and the Khâbûr of Naṣîbîn have their source.

Naṣîbîn, the Roman Nisibis, which Yâḳût describes as celebrated for its white roses and its forty thousand gardens, stood on the upper waters of the Hirmâs river, called by the Greek geographers the Saocoras or Mygdonius, and it is still one of the most important towns of Upper Mesopotamia. Ibn Ḥawḳal who was there in 358 (969) describes Naṣîbîn as the finest town of the Jazîrah province, and its neighbourhood produced the best barley and wheat crops. The hill above, from which its water came, was called the Jabal Bâlûsâ, the town was most pleasant to live in, and the only drawback was the fear of scorpions. It was more spacious than Moṣul, and Muḳaddasî praises both its fine baths, and the private houses. The market extended right across from gate to gate, a Friday Mosque stood in its midst, and a strong

[1] Ist. 78. I. H. 152, 153, 157. Muk. 139. I. S. 18. A. F. 55, 275. Yak. i. 466, 472; ii. 79, 144, 552, 957; iv. 1017. I. B. ii. 139. Mst. 165, 166.

fortress built of mortared stone protected the town. Naṣîbîn was visited by Ibn Jubayr in 580 (1184), who praises its gardens; in its Friday Mosque were two tanks, and a bridge crossed the river Hirmâs where it flowed by the town; also there was the hospital (Mâristân) and several colleges among other notable buildings. Ibn Baṭûṭah who was here in the 8th (14th) century describes Naṣîbîn as then for the most part in ruins, but its Friday Mosque was still standing with the two great tanks, and the gardens round the city produced the rose-water for which it was so celebrated. Mustawfî, who gives the circuit of the walls as 6500 paces, praises the grapes and other fruits grown here, and its wine, but the dampness of the climate, he says, made Naṣîbîn an unhealthy place. He, too, speaks of the excellence of its roses, also the abomination of the scorpions, which were equalled in virulence by the plague of gnats[1].

Râs-al-'Ayn, 'the Spring-head,' near the sources of the Khâbûr (the Roman Resaina, on the river Chaboras), was famous for its numerous springs, said to number 360 in all, and their waters made the surrounding country a great garden. Of these springs the 'Ayn-az-Zâhirîyah was supposed to be fathomless, and the stream flowing from this ran into the Khâbûr, by which pleasure-boats are described as travelling down from garden to garden from Râs-al-'Ayn to Ḳarḳîsiyâ on the Euphrates. Râs-al-'Ayn is described by Ibn Ḥawḳal as a walled town, having gardens and many mills within its circuit; and the arable fields stretched for 20 leagues beyond the houses. Muḳaddasî describes a small lake at the chief spring, two fathoms deep, but the water so clear that a silver piece could clearly be seen at the bottom. The buildings of Râs-al-'Ayn were of stone, well mortared, and Ibn Jubayr who passed through the town in 580 (1184) mentions its two Friday Mosques and the fine colleges and bath houses which stood along the banks of the Khâbûr. In his time the city apparently had no wall, though in the 8th (14th) century this must have been rebuilt, for Mustawfî describes it as 5000 paces in circuit. He adds that cotton, corn, and grapes were grown here abundantly.

[1] I. H. 140, 142, 143. I. S. 12. Muk. 140. I. J. 240. Yak. iii. 559; iv. 787. I. B. ii. 140. Mst. 167.

About half-way between Râs-al-'Ayn and Naṣîbîn, but more to the north, stood the great rock fortress of Mârdîn, overlooking the city of Dunaysîr in the plain below, some three leagues to the south. In the 4th (10th) century the castle of Mârdîn, called Al-Bâz (the Falcon), was the stronghold of the Ḥamdânid princes. The fortress crowned the hill-top, and on the southern side a suburb was built which by the 6th (12th) century had become very populous. Here there were many markets, some hostelries, and a few colleges, but all the buildings rose one above the other in steps, and the roads were stairs, each house having its cistern for storing rain water. Ibn Baṭûṭah, who visited Mârdîn in the 8th (14th) century, describes it as a fine town where much woollen stuff was woven. At that time the great fortress was known as Ḳal'at-ash-Shahbâ, 'the Grey Castle,' or Ḳal'at-i-Kûh, 'the Castle of the Hill.' Mustawfî describes Mârdîn as amply irrigated by the waters of the Ṣawr river, which flowed down from a hill of the same name in Ṭûr 'Abdîn, and this river ultimately joined the Khâbûr; he adds that corn, cotton, and abundant fruit was grown in all the neighbourhood.

Dunaysir, a few leagues distant (variously given as from 2 to 4, but its actual site appears to be unknown), was in the 7th (13th) century a great market town, and it was also known as Kûch Ḥiṣâr. Yâḳût writes that when he was a boy, that is to say at the close of the 6th (12th) century, Dunaysir had been merely a large village, but in 623 (1225) it was become a great city, with extensive markets. Ibn Jubayr who had passed through it in 580 (1184) describes it as unwalled, but it was then a meeting place for caravans, and a college had recently been built with numerous bath houses. Dârâ, lying a few miles to the eastward, which had been a great fortress in Roman days, is mentioned as a small town by Ibn Ḥawḳal, and Muḳaddasî describes how each house was supplied with water by an underground channel, these channels ultimately flowing into the tank of the Friday Mosque. The houses were all built, he says, of black stone, and well mortared. The town stood on a hill side, and Yâḳût states that it was famous for its *Maḥlab* or cherry-stone preserve, the gardens being most fruitful. When Ibn Baṭûṭah passed Dârâ in the 8th (14th) century, however, its fortress had already become an uninhabited ruin. Kafartûthâ, to the S.W. of Mârdîn and on its own small river, is described by Ibn Ḥawḳal as already a town of some importance in the 4th (10th) century, being at the junction of the high road coming down from Âmid. It was at that time a larger place than Dârâ, but in the 7th (13th) century Yâḳût refers to it as merely a large village[1].

The Greater Khâbûr from Râs-al-'Ayn received on its left bank the waters of the Mârdîn river, and below this again was joined by the Hirmâs coming from Naṣîbîn; but the major part of this latter stream, as already said, was diverted at the dam of Sukayr-al-'Abbâs, a short distance above the junction with the Khâbûr, into the Tharthâr channel. The Khâbûr now bearing the waters of three considerable streams, and—Mustawfî adds—further swelled by the confluence of 300 rivulets, flowed down south to Ḳarḳîsiyâ on the Euphrates, which is the chief town of the Diyâr Muḍar district and will be described presently. Before coming to this the river ran by the towns of 'Arâbân and Mâkisîn, which were of the Khâbûr lands and counted of Diyâr Rabî'ah province. 'Arbân or 'Arâbân, the ruins of which still exist, was in the 4th (10th) century a walled town where cotton stuffs were largely manufactured, cotton being grown in the surrounding country along the banks of the Khâbûr. Muḳaddasî speaks of 'Arâbân as standing on a high hill and surrounded by gardens. To the south of it, half-way to Ḳarḳîsiyâ, was the town of Mâkisîn (or Maykasîn) where a bridge of boats crossed the Khâbûr. Much cotton also was grown here, and near it lay the small lake of deep blue water called Al-Munkharik, about a third of an acre in extent and said to be unfathomable.

The source of the Hirmâs river is described as at a spring six leagues north of Naṣîbîn, where the water was dammed back by a masonry wall, clamped and with leaden joints. This, it was said, the Greeks had built, to preserve Naṣîbîn from being flooded, and the Caliph Mutawakkil at one time had commanded that it should be demolished, but finding the water beginning to overflow the city had promptly ordered the restoration of the wall. A hundred

[1] Baladhuri, 176. Ist. 73, 74. I. H. 143, 149, 152. Muk. 140. I. J. 242, 244. Kaz. ii. 172. Yak. ii. 516, 612, 733, 911; iii. 435; iv. 287, 390. Mst. 166, 205, 219. I. B. ii. 142. A. Y. i. 677.

miles or more south of Naṣîbîn was the dam or weir called Sukayr-al-'Abbâs, where in the 4th (10th) century there was a considerable town with a Friday Mosque and markets. This was at the head of the Tharthâr river, which, as already stated, flowed to the Tigris. At the present day its stream is so shrunken in volume that it no longer forms a waterway, and this shrinkage had already begun in the 7th (13th) century when Yâḳût wrote, for he reports that though when the rains were plentiful the flood still passed down its channel, in summer the bed was only marked by pools of water and brackish springs. Yâḳût had himself travelled along its course, and adds it was reported that in old times boats used to pass down this stream from the Khâbûr to the Tigris; and in those days a succession of villages lined its banks, where, when he wrote, there was only a desert to be seen.

In the plain of Sinjâr the river Tharthâr cut through the line of hills called the Jabal Ḥumrîn, otherwise the Jabal Bârimmâ, and received from the north a small stream which flowed down from the city of Sinjâr. This in the 4th (10th) century was a walled town, surrounded by a most fertile district. Muḳaddasî describes it as famous for its carpenters; oranges, lemons, and the date palm flourished abundantly here, and a large Friday Mosque stood in the midst of the town. Moslem tradition stated that the Ark first rested on the hill above Sinjâr during the Flood; but afterwards, continuing on its course, came finally to rest on Jabal Jûdî on the east side of the Tigris. Further, Yâḳût adds that Sinjâr was also famous as the birthplace of Sultan Sinjâr or Sanjar, the last of the great Saljûḳs, son of Malik Shâh. According to Ḳazwînî Sinjâr in the 7th (13th) century was remarkable for its bath houses, which had beautiful mosaic floors, and Ibn Baṭûṭah who passed through the place in the 8th (14th) century refers to its fine mosque. The town wall, 3200 paces in circuit, was built according to Mustawfî of mortared stone; most of the houses went step-fashion up the hill slope, and its gardens produced great quantities of grapes, olives, and sumach. Al-Ḥaḍr, the Roman Hatra, mentioned by Ibn Serapion, stood lower down the Tharthâr, about half-way between Sinjâr and where that river joined the Tigris near Takrît. At Ḥaḍr are still to be seen the remains of a great Parthian palace which Yâķût reports to have been built by a certain As-Sâṭirûn of squared stones, and there were many of its chambers whose ceilings and doors were likewise of stone slabs. Originally, he says, there had been sixty great towers, with nine turrets between each tower and its neighbour, while a palace stood over against each tower outside the walls[1].

The high road from Mosul to Naṣîbîn went up the right bank of the Tigris, and at Balad (corresponding with the place now known as Eski, or Old, Mosul), seven leagues from Mosul, the road bifurcated, the branch to the left hand going to Sinjâr by way of Tall A'far. Yâķût writes that Balad, where there was an Alid shrine, occupied the site of the old Persian town of Shahrâbâdh, and that the name of Balad was often written Balaṭ. Ibn Ḥawķal in the 4th (10th) century refers to Balad as a considerable city, and Muķaddasî tells us of its houses built of stone, well mortared, its good markets, and its Friday Mosque standing in the centre of the town. The neighbourhood produced sugar-cane and was very fertile. On the solitary hill of Tall A'far, one stage to the west, stood a castle, dominating a large suburb through which ran a stream. The castle was strongly fortified, Yâķût says, and the date palm grew in the surrounding district, which was known under the name of Al-Maḥlabîyah, from the *Maḥlab* perfume, or preserve, of cherry-stones chiefly made here.

The right-hand road at the bifurcation beyond Balad led to the town of Bâ'aynâthâ which Muķaddasî describes as lying in the midst of twenty-five fertile districts, the richest and pleasantest of all Mesopotamia, as he adds; and this Bâ'aynâthâ must not be confounded with 'the great village like a city' of the same name on the river which joins the Tigris to the north of Jazîrah Ibn 'Omar as mentioned on p. 94. Beyond Bâ'aynâthâ on the road to Naṣîbîn came Barķa'îd, a place evilly proverbial for the thieving ways of its people, practised against all strangers and their caravans. In the 3rd (9th) century it was a town of considerable size, with three gates, more than two hundred shops, and many

[1] The name of the town is written Sinjâr, with the last *a* long; the name of the Sultan is generally written Sanjar, with both vowels short. I. S. 12, 18. Ist. 73, 74. I. H. 139, 148, 150. Muk. 140, 141. Yak. i. 464, 921; ii. 281; iii. 109, 158; iv. 962. Mst. 166, 219. I. B. ii. 141. Kaz. ii. 263.

springs of excellent water. By the 7th (13th) century, however, though some traffic still passed through it, the evil reputation of its people had caused the place to be avoided by respectable travellers and it had fallen to the size of a village.

Adhramah, rather less than half-way between Barḳa'îd and Naṣîbîn, was a place of about the same size as Barḳa'îd; and its district was called Bayn-an-Nahrayn, 'Betwixt the Streams.' In the 3rd (9th) century it is stated that there had been a fine palace here, and a stone arched bridge crossed its stream. The little town then had double walls, surrounded by a deep ditch. Such at any rate is the description of the place left by the physician of the Caliph Mu'taḍid, who passed through it, when in attendance on the latter. In the 4th (10th) century Muḳaddasî describes Adhramah as a small place standing in the desert near some wells, and there were vaulted buildings round about these[1].

[1] Kud. 214. Ist. 73. I. H. 148, 149. Muk. 139, 140. Yak. i. 177, 472, 571, 715, 863 v. 428. Kaz. ii. 204.

CHAPTER VII.

JAZÎRAH (*continued*).

The district of Diyâr Muḍar. Raḳḳah and Râfiḳah. The river Balîkh and Ḥarrân, Edessa and Ḥiṣn-Maslamah. Ḳarḳîsiyâ. The Nahr Sa'îd, Raḥbah and Dâliyah. Ruṣâfah of Syria. 'Ânah. Bâlis, Jisr Manbij, and Sumaysâṭ. Sarûj. The district of Diyâr Bakr. Âmid, Ḥânî, and the source of the Tigris. Mayyâfarîḳîn and Arzan. Ḥiṣn Kayfâ and Tall Fâfân. Sâ'irt.

The district of Diyâr Muḍar, as already explained, lay along the banks of the Euphrates, and the chief town was Ar-Raḳḳah situated just above where the river Balîkh, coming down from the north, flows into the Euphrates. The site is that of the old Greek city of Callinicus or Nicephorium, for the Arab name Ar-Raḳḳah is merely descriptive; *Raḳḳah* being the term for the swampy land beside a river subject to periodical inundation, and as such Ar-Raḳḳah, 'the Morass,' is found elsewhere as a place-name, this particular Raḳḳah receiving the surname of As-Sawdâ, 'the Black,' for distinction.

In the 2nd (8th) century when the Abbasids had succeeded to the Caliphate, Raḳḳah, one of the chief cities of Upper Mesopotamia commanding the Syrian frontier, had to be secured, and for this purpose the Caliph Manṣûr in 155 (772) proceeded to build some 300 ells distant from Raḳḳah the town of Ar-Râfiḳah (the Companion or Fellow), which was garrisoned by Khurâsân troops entirely devoted to the new dynasty. Râfiḳah is said to have been laid out on the plan of Baghdâd, and was a round city. Hârûn-ar-Rashîd added to the town and built himself a palace here called the Ḳaṣr-as-Salâm (the Palace of Peace), for he at times resided in Raḳḳah, or Râfiḳah, when the climate of

Baghdâd was too hot. Soon the older town of Raḳḳah fell to ruin, new buildings covered all the intervening space, enclosing 'the Morass,' now a shallow lake, lying between Raḳḳah and Râfiḳah, and the name of Raḳḳah passed to Râfiḳah, which last, once the suburb, took the place of the older city, and lost its name in the process. Ibn Ḥawḳal in the 4th (10th) century, however, speaks of the twin cities of Raḳḳah and Râfiḳah, each with its own Friday Mosque, and he especially mentions the magnificent trees which surrounded the towns. Muḳaddasî describes only one town, namely Raḳḳah, as strongly fortified and having two gates; its markets were excellent and well supplied from the neighbouring villages; much traffic also centred here, and from the olive oil produced in the neighbourhood soap was manufactured. The Friday Mosque was, he says, a fine building standing in the Clothiers' market, and each of the great houses at Raḳḳah had its terraced roof. There were also excellent baths. Near by were the ruins of the old town, then known as Ar-Raḳḳah-al-Muḥtariḳah, 'Burnt Raḳḳah.' Mustawfî on the other hand speaks of Râfiḳah as still the name of a suburb, with its Friday Mosque standing in the Goldsmiths' market. Round this suburb grew mulberry and jujube trees, and a mosque stood near, overhanging the Euphrates bank.

On the right bank of the Euphrates opposite and above Raḳḳah was the celebrated plain of Ṣiffîn, which had been the battlefield between the partizans of the two Caliphs Mu'âwiyah and 'Alî. 'The Martyrs,' as the Shî'ahs called those who had fallen in the cause of 'Alî, had their shrines here, and Ibn Ḥawḳal, whose narrative is extended by Mustawfî, relates how miraculously, from afar off, each buried martyr was quite visible lying in his shroud underground, though, on coming up to the actual spot, no body could be perceived. Opposite the battlefield of Ṣiffîn on the north (left) bank of the Euphrates stands the fortress known as Ḳal'at Ja'bar, after its early possessor, an Arab of the Banî Numayr. Originally this castle had been called Dawsar. It is frequently mentioned in the later history of the Caliphate, and in the year 497 (1104) was taken possession of by the Franks from Edessa, during the time of the first Crusade. On its left bank below Raḳḳah the Euphrates receives the river Al-Balîkh, which the Greeks knew as the Bilecha. Its source was at a spring called the 'Ayn-adh-Dhahbânîyah lying to the north of Ḥarrân. The name of this spring is given variously by our authorities as Ad-Dahmânah or Adh-Dhahbânah, and Mustawfî (in Persian) writes of the Chashmah Dahânah[1].

The Balîkh took its course south and joined the Euphrates below Raḳḳah, passing by a number of important towns which were irrigated from it or from its tributaries. Ḥarrân (the ancient Carrhæ) near its source was famous as the home of the Sâbians (not identical with the Sabæans, but often confounded with them) who professed to hold the religion of Abraham, and tradition stated that Ḥarrân was the first city to be built after the Flood. Muḳaddasî describes Ḥarrân as a pleasant town protected by a fortress, built of stones so finely set as to recall the masonry of the walls of Jerusalem. It possessed a Friday Mosque. According to Ibn Jubayr, who passed through Ḥarrân in 580 (1184), the city itself was also surrounded by a stone wall, and he describes the mosque as having a large court with nineteen doors, while its cupola was supported on marble columns. The markets were roofed over with beams of wood, and the city possessed both a hospital and a college. Mustawfî adds that the circuit of the castle wall was 1350 paces. Three leagues to the south was to be seen the shrine (Mashhad) of Abraham, and the surrounding territory was fully irrigated by innumerable small canals.

Edessa, which the Arabs call Ar-Ruhâ (a corruption of the Greek name Callirrhoe), lay on the head-waters of one of the tributaries of the Balikh. The city is not held of much account by the Moslem geographers, for the majority of its population continued to be Christians, and the town was chiefly remarkable for its numerous churches, which Ibn Ḥawḳal estimates at more than 300 in number. Here originally had been preserved the famous relic known as 'the napkin of Jesus,' which had been given up by Moslem authorities to the Byzantines in 332 (944), in order to save Ruhâ from being stormed and plundered. Muḳaddasî in the latter part of the 4th (10th) century, after speaking of the

[1] Baladhuri, 179, 297. Ist. 75, 76. I. H. 153, 154. Muk. 141. I. S. 12. I. R. 90. I. K. 175. Yak. i. 734; ii. 621, 734; iv. 112, 164. Mst. 166, 219. Ibn-al-Athir, x. 253.

Friday Mosque, describes the magnificent cathedral of Edessa, celebrated as one of the four wonders of the world, whose vaulted ceiling was covered with mosaics. The Great Mosque of Al-Akṣâ at Jerusalem had been built, he says, on its plan. Mukaddasî adds that the city was well fortified. Notwithstanding its Arab garrison at the time of the first Crusade in 492 (1098), Edessa was taken by Baldwin, and during half-a-century remained a Latin principality. In 540 (1145), however, Zangî retook the city from Jocelin II, and after that date Ruhâ was in the hands of the Moslems. The ruins of its many handsome buildings might still be seen in the 8th (14th) century, and Mustawfî describes a great cupola of finely worked stone, rising beyond a court that was over 100 yards square. Ruhâ is more than once mentioned by 'Alî of Yazd in his account of the campaigns of Tîmûr, and it kept this name down to the beginning of the 9th (15th) century. After it passed into the possession of the Ottoman Turks its name was commonly pronounced Urfah, said to be a corruption of the Arabic Ar-Ruhâ, and as Urfah Edessa is known at the present day[1].

To the south of Ḥarrân, and lying some distance to the east

[1] Ist. 76. I. H. 154. Muk. 141, 147. I. J. 246. Yak. ii. 231, 591. A. Y. i. 662. Mst. 166. J. N. 443. In the matter of the famous napkin (*Mandîl*) of Christ once preserved at Edessa, this is one of the many Veronicas, but competent authorities are not agreed as to whether the Edessa Veronica is that now preserved in Rome, or the one shown at Genoa, and there are others. Our earliest Moslem authority, Mas'ûdî, who wrote in the very year when this famous relic had been delivered up to the Greek Emperor, calls it 'the napkin of Jesus of Nazareth, wherewith He had dried Himself after His baptism,' and Mas'ûdî mentions the year 332 (944) as that when the Byzantines got possession of it, to their great joy. Ibn Ḥawkal, writing in the same century, merely calls it 'the napkin of 'Îsâ, son of Mariyam, on whom be peace.' Ibn-al-Athîr in his chronicle under the year 331 (943) describes it as 'the napkin with which it was said the Messiah had wiped His face, whereby the likeness of His face was come thereon,' and he proceeds to relate how the Caliph Muttakî had been induced to give up this napkin to the Emperor of the Greeks in return for the release of many Moslem captives, and to save Ar-Ruhâ from assault and pillage. The Christian legend concerning the Edessa napkin, as given by Moses of Chorene, is that this relic was a portrait of Christ, wonderfully impressed on a cloth, which He had sent to Abgarus, King of Edessa. Mas. ii. 331. Ibn-al-Athir, viii. 302.

of the Balîkh river, was the small town of Bâjaddâ on the road to Râs-al-'Ayn. Its gardens were famous, and it was a dependency of Ḥiṣn Maslamah, which lay nearer to the Balîkh river. This great castle took its name from Maslamah, son of the Omayyad Caliph 'Abd-al-Malik, and it stood nine leagues south of Ḥarrân, lying about a mile and a half back from the actual river bank. From this point a canal brought water to the fortress to fill a cistern which Maslamah had caused to be dug here, 200 ells square by 20 deep, and lined throughout with stone. The cistern needed only to be filled once a year, and the canal served for irrigating the lands round Ḥiṣn Maslamah. The fortress buildings covered an area of a Jarîb (equivalent to a third of an acre), and its walls were fifty ells in height. To the south of Ḥiṣn Maslamah on the road to Raḳḳah, from which it was three leagues distant, stood Bâjarwân, which Ibn Ḥawḳal describes as having been a fine town, though in the 4th (10th) century already falling to ruin. Yâḳût, whose description of Ḥiṣn Maslamah has been given above, merely mentions Bâjarwân as a village of the Diyâr Muḍar district[1].

Some two hundred miles below Raḳḳah stands Ḳarḳîsiyâ, the ancient Circesium, on the left bank of the Tigris where, as already explained (p. 97), the moiety of the Khâbûr river flows in. Ibn Ḥawḳal describes it as a fine town surrounded by gardens; but Yâḳût and Mustawfî both refer to it as a smaller place than the neighbouring Raḥbah, which lay six leagues distant, standing back from the western side of the Euphrates. This Raḥbah—the name means the Square or Plain—was called for distinction Raḥbah-ash-Shâm, 'of Syria,' or Raḥbah Malik ibn Ṭawḳ after its founder, who had flourished during the reign of the Caliph Mamûn. Near it stood the small town of Ad-Dâliyah (the Waterwheel) and both places lay near the bank of a great loop canal, called the Nahr Sa'îd, which branched from the right bank of the Euphrates some distance above Ḳarḳîsiyâ and flowed back to it again above Dâliyah, which, like Raḥbah, was also known for distinction as Dâliyah of Malik ibn Ṭawḳ. The canal had been dug by Prince Sa'îd, son of the Omayyad Caliph 'Abd-al-Malik; he was a man of great piety, being known as Sa'îd-al-Khayr, 'the Good,' and was

[1] I. H. 156. Kud. 215. Yak. i. 453, 454, 734; ii. 278.

for some time Governor of Mosul. Raḥbah is described by Muḳaddasî as one of the largest towns on the Euphrates in Upper Mesopotamia. Its houses spread out in a great semicircle standing back to the desert border, it was well fortified, and had a large suburb. Dâliyah was much smaller, but still an important place, standing on an elevation and overlooking the west bank of the Euphrates.

In the desert between Raḥbah and Raḳḳah—and the ruins still exist four leagues south of the latter town—was Ruṣâfah (the Causeway), called Ruṣâfah-ash-Shâm—of Syria—or Ruṣâfah Hishâm, after its founder. The Caliph Hishâm, one of the many sons of 'Abd-al-Malik, built himself this palace in the desert as a place of safety to reside in at a time when the plague was raging throughout Syria. The spot had already been occupied by the Ghassanid princes before Islam, and there were ancient wells here, Yâḳût says, 120 ells deep. The physician Ibn Buṭlân, who wrote in 443 (1051), describes Ruṣâfah as possessing a church, said to have been built by the Emperor Constantine, the exterior of which was ornamented in gold mosaic work, and underneath was a crypt, as large as the church, with its roof supported on marble pillars. In the 5th (11th) century most of the inhabitants were still Christian, and they profitably combined brigandage with the convoying of caravans across the desert to Aleppo. On the eastern side of the Euphrates between Raḳḳah and Ḳarḳîsiyâ, two days' march above the latter town, was Al-Khânûḳah, a city of some size according to Ibn Ḥawḳal, and Yâḳût adds that in its vicinity was the territory of Al-Maḍiḳ.

Below Ḳarḳîsiyâ the only town of importance within the limits of the Jazîrah province was 'Ânah, the ancient Anatho, still found on the map, and mentioned by Ibn Serapion as on an island surrounded by the Euphrates. Ibn Ḥawḳal, however, describes this as formed by a creek branching off from the stream. Yâḳût adds that 'Ânah possessed a strong castle which overlooked the river, and here the Caliph Ḳâim found shelter in 450 (1058), when Basâsîrî the Daylamite, after taking possession of Baghdâd, had caused the public prayers to be read there in the name of the heterodox Fatimid Caliph of Cairo. Mustawfî says that in the 8th (14th) century 'Ânah was still a fine town, and famous for its palm-groves. The harbour of Al-Furḍah, called Furḍah Nu'm for distinction, lay due west of 'Ânah on the Euphrates, half-way to Ḳarḳîsiyâ, and probably marked the eastern bend of the Euphrates, but it is now no longer to be found on the map. This was an important station where the highway bifurcated, to the left-hand one road going direct across the desert by way of Ruṣâfah to Raḳḳah, while the right-hand road kept up stream along the river bank[1].

Above Raḳḳah there were three towns on the Euphrates, namely Bâlis, Jisr Manbij, and Sumaysâṭ, which were often counted as of Syria because they lay on the right or western bank of that river, though most authorities count them as belonging to Jazîrah. Bâlis lies due west of Raḳḳah, at the limit of the plain of Ṣiffîn, where the Euphrates after running south turns east. It was the Roman Barbalissus, the great river-port for Syria on the Euphrates, and hence the centre point of many caravan routes. Ibn Ḥawḳal describes Bâlis as having strong walls, with gardens lying between these and the Euphrates; of its lands the chief crops were wheat and barley. Though somewhat fallen to ruin, Muḳaddasî says, Bâlis was still populous in the 4th (10th) century; but Yâḳût reports that, by a change of bed, the Euphrates in the 7th (13th) century had come to flow more than four miles distant from the town, and Abu-l-Fidâ refers to Bâlis as a place that had long seen its best days.

Jisr Manbij, where a bridge of boats crossed the Euphrates, and the road led west up to Manbij (Hierapolis) of the Aleppo province, was a place of great importance during the middle-ages. The bridge was protected by a great fortress, and below this a small town stood on the Euphrates bank. The fortress was known as Ḳal'at-an-Najm, 'the Castle of the Star,' from its height on the hill, and it was also called Ḥiṣn Manbij, 'the Manbij Fortress.' When Ibn Jubayr passed Ḳal'at-an-Najm, coming from Ḥarrân in 580 (1184), he speaks of the market which was held below its walls. Abu-l-Fidâ says that the fort had been rebuilt by Sultan Nûr-ad-Dîn, son of Zangî, and its garrison freely harassed the neighbouring

[1] Ist. 77, 78. I. H. 155, 156. Muk. 142. Baladhuri, 179, 180, 332. I. S. 10, 14. Yak. ii. 394, 538, 764, 784, 955; iii. 595, 876; iv. 65, 560, 840. Mst. 139, 166.

towns occupied by the Crusaders. Ḳazwînî, writing in the latter half of the 7th (13th) century, gives a long account of the frauds practised by sharpers here who, getting acquainted with rich travellers passing Ḳal'at-an-Najm, by means of games of hazard, aided by confederates, would win all their money and possessions. The play ran so high that, according to Ḳazwînî, the stranger was often left 'with nothing but his drawers (*sârawîl*) of all his clothes or former possessions.' The sharpers, indeed, would sometimes hold the victim himself in pawn, until his companions could be induced to buy him off.

Sumaysâṭ, the Roman Samosata, was still higher up the Euphrates, and lay on the right or north bank of the great river, which here runs west. It was a very strong fortress. Mas'ûdî states that Sumaysâṭ was also known as Ḳal'at-aṭ-Ṭîn, 'the Clay Castle,' and Yâḳût reports that in the 7th (13th) century one of its quarters was exclusively inhabited by Armenians. Finally to complete the list of towns of the Muḍar district Sarûj is to be mentioned, which lies about half-way on the direct road from Raḳḳah north, across the desert plain, to Sumaysâṭ; this road forming the chord of the great semicircular sweep followed by the Euphrates. Sarûj was also on the caravan road from Ḥarrân and Edessa to Jisr Manbij, and is described by Ibn Ḥawḳal as a fine city, surrounded by fertile districts, a description which Yâḳût, adding nothing further, corroborates[1].

The cities of Diyâr Bakr, the smallest of the three districts into which the Jazîrah province was divided, lay exclusively on, or to the north of, the upper course of the Tigris. The chief town of the district was Âmid, sometimes written Ḥâmid, the Roman Amida. In later times the city was generally known under the name of the district, as it is at the present day, being called Diyâr Bakr, or else Ḳârâ Âmid (Black Âmid) from the colour of the stone used here.

The town stood on the right or west bank of the Tigris, and a hill 100 fathoms in height dominated it. Ibn Ḥawḳal states that its walls were built of black mill-stones. Muḳaddasî describes its strong fortifications as being like those of Antioch, the outer walls,

[1] Ist. 62, 76, 78. I. H. 119, 120, 154, 157. Muk. 155. Mas. i. 215. I. J. 250. Yak. i. 477; iii. 85, 151; iv. 165. A. F. 233, 269. Kaz. ii. 160.

battlemented and with gates, being separated from the inner fortifications by a clear space, afterwards occupied by the suburbs. There were springs of water within the town and Muḳaddasî also remarks on the black stone of which, and on which, he says the city was built. Âmid possessed a fine Friday Mosque, and its walls were pierced by five chief gates, namely the Water gate, the Mountain gate, the Bâb-ar-Rûm (the Greek gate), the Hill gate, and the Postern gate (Bâb-as-Sirr) used in time of war. The line of fortified walls included the hill in their circuit, and in the 4th (10th) century Muḳaddasî says that the Moslems possessed no stronger or better fortress than Âmid on their frontier against the Greek Empire.

Nâṣir-i-Khusraw the Persian pilgrim passed through Âmid in 438 (1046), and has left a careful description of the city as he saw it. The town was 2000 paces in length and in breadth, and the wall built of black stone surrounded the hill overlooking it. This wall was 20 yards in height and 10 yards broad, no mortar was used in its construction, but each stone block was, Nâṣir estimates, of the weight of 1000 *man* (equivalent to about three tons). At every hundred yards along the wall was built a semicircular tower, and the crest had battlements of the aforesaid black stone, while stone gangways at intervals led up to the ramparts from within the circuit. There were four iron gates, facing the cardinal points; namely, to the east the Tigris gate, to the north the Armenian gate (Bâb-al-Arman), to the west the Greek gate, and to the south the Hill gate (Bâb-at-Tall). Beyond the city wall ran the outer wall, ten yards in height, also of black stone, a suburb occupying the space between the two, in a ring that was fifteen yards across. This outer wall also had battlements, and a gangway along it for the defence, and there were here four iron gates corresponding with those of the inner wall. Âmid, Nâṣir adds, was one of the strongest places he had seen.

In the centre of the town a great spring of water, sufficient to turn five mills, gushed out; the water was excellent, and its overflow irrigated the neighbouring gardens. The Friday Mosque was a beautiful building, of black stone like the rest of the town, with a great gable roof and containing over 200 columns, each a monolith, every two connected by an arch, which supported in turn a row of dwarf columns under the roof line. The ceiling was of carved wood, coloured and varnished. In the mosque court was a round stone basin, from the midst of which a brass jet shot up a column of clear water, which kept the level within the basin always the same. Near the mosque stood a great church, built of stone and paved with marble, the walls finely sculptured; and leading to its sanctuary Nâṣir saw an iron gate of lattice-work, so beautifully wrought that never had he seen the equal thereof.

This description of the magnificence of Âmid is borne out by what the anonymous annotator of the Paris MS. of Ibn Ḥawḳal writes, who was here in 534 (1140). He notes that its markets were well built and full of merchandise. In the 7th (13th) century Yâḳût and Ḳazwînî repeat much of the foregoing description, and the latter speaks of Âmid as then covering a great half-circle of ground, with the Tigris flowing to the eastward, and surrounded on the other side by magnificent gardens. Mustawfî in the following century writes of it as a medium-sized town, paying the Îl-Khâns a revenue of 3000 gold pieces. At the close of this century Âmid was taken by Tîmûr[1].

To the north of Âmid, and near one of the eastern arms of the upper Tigris, stands the town of Ḥânî, which is said by Yâḳût to be famous for the iron mine in its neighbourhood, which produced much metal for export. Ḥânî is also mentioned by Mustawfî. Some distance to the west of Ḥânî lies the chief source of the Tigris, which Muḳaddasî describes as flowing with a rush of green water out of a dark cave. At first, he says, the stream is small, and only of sufficient volume to turn a single mill-wheel; but many affluents soon join and swell the current, the uppermost of these being the Nahr-adh-Dhîb (the Wolf River), apparently identical with the Nahr-al-Kilâb (the River of Dogs) referred to by Yâḳût, which came down from the hills near Shimshâṭ, to the north of Ḥânî. The source of the Tigris, according to Yâḳût, was distant two and a half days' journey from Âmid, at a place known as Halûras, 'where 'Alî, the

[1] Ist. 75. I. H. 150, 151. Muk. 140. N. K. 8. Yak. i. 66. Kaz. ii. 331. Mst. 165. A. Y. i. 682.

Armenian, obtained martyrdom,' and he too speaks of the dark cavern from which its waters gushed forth. The names of many other affluents are mentioned both by Muḳaddasî and Yâḳût, whose accounts are not quite easy to reconcile, and probably the names of these streams varied considerably between the 4th and the 7th (10th and 13th) centuries.

Some distance below Âmid the Tigris turns due east at a right angle, and then from the north receives a stream called the Nahr-ar-Rams or the Nahr Ṣalb. A more important affluent, however, is the river coming down from the north of Mayyâfarîḳîn, a tributary of which flowed by that city. This is the river Sâtîdamâ, or Sâtîdamâd, one branch of which was called the Wâdî-az-Zûr flowing from the district of Al-Ḳalḳ, while the Sâtîdamâ river itself had its head-waters in the Darb-al-Kilâb—'the Dogs' Pass'—so called, Yâḳût says, from a famous massacre of the Greeks, 'when these were all killed like dogs,' which the Persian army effected in the reign of King Anûshirwân, some time before the birth of the prophet Muḥammad. This river Sâtîdamâ, which is mentioned by Ibn Serapion, is that which Muḳaddasî names the Nahr-al-Masûliyât, and is now known as the Batman Ṣû, one of whose affluents, as already said, flows down from Mayyâfarîḳîn[1].

The Arabic Mayyâfarîḳîn appears to be a corruption of the Aramaic name Maypharkath, or the Armenian Moufargin, and it is identical with the Greek town called Martyropolis. Muḳaddasî in the 4th (10th) century describes it as a fine city, surrounded by a stone wall, with battlements and a deep ditch, beyond which stretched extensive suburbs. Its mosque was well built, but Muḳaddasî remarks that its gardens were scanty. Mayyâfarîḳîn was visited by Nâṣir-i-Khusraw in 438 (1046), who speaks of the town as surrounded by a wall built of great white stones, each of 500 *man* weight (about a ton and a half), and while all Âmid, as already said, was of black stone, in every building at Mayyâfarîḳîn the stones used were notably white. The town wall was then new, it had good battlements and at every 50 yards rose a white stone tower. The city had but one gateway, opening to the west,

[1] I. S. 17, 18. Muk. 144. Yak. ii. 188, 551, 552, 563, 956; iii. 7, 413; iv. 300, 979. Mst. 165.

and this possessed a solid iron door, no wood having been used in its construction. There was according to Nâṣir a fine mosque within the city, also a second Friday Mosque in the suburb outside, standing in the midst of the markets, and beyond lay many gardens. He adds that at a short distance to the north of Mayyâfarîḳîn stood a second town called Muḥdathah, 'the New Town,' with its own Friday Mosque, bath houses, and markets; while four leagues further distant was the city of Naṣrîyah, lately founded by the Mirdâsid Amîr Naṣr, surnamed Shibl-ad-Dawlah.

Both Yâḳût and Ḳazwînî give a long account of various churches, of the three towers, and the eight town gates, which had existed of old at Mayyâfarîḳîn—the Greek name of which, Yâḳût says, was Madûrṣâlâ, meaning 'the City of the Martyrs.' These buildings dated from the days of the Emperor Theodosius, and some of their remains, especially those of an ancient church built, it was said, 'in the time of the Messiah,' might still be seen in the 7th (13th) century. Thus there was in particular, on the summit of the south-western tower of the town wall, a great cross, set up to face Jerusalem, and this cross, it was reported, was the work of the same craftsman who had made the great cross that adorned the pinnacle of the Church of the Resurrection in Jerusalem, the two crosses being alike, and wonderful to behold. Further, in the Jews' quarter of Mayyâfarîḳîn near the Synagogue, was to be seen a black marble basin, in which was kept a glass belt (possibly a phylactery), wherein was preserved some of the blood of Joshua the son of Nun, this having been brought hither from Rome, and to touch it was a sovereign remedy against all disease. In the 8th (14th) century under the Mongols Mayyâfarîḳîn was still an important place, and Mustawfî praises its excellent climate and abundant fruits[1].

Arzan, a short distance to the east of Mayyâfarîḳîn, stood on the western side of the river called the Nahr, or Wâdî, as-Sarbaṭ. Arzan had a great castle, well fortified, and it was visited in 438 (1046) by Nâṣir-i-Khusraw. He writes of it as a flourishing place with excellent markets, being surrounded by fertile and well irrigated gardens. Yâḳût describes Arzan (which must not be

[1] I. H. 151. Muk. 140. N. K. 7. Yak. iv. 703—707. Kaz. ii. 379. Mst. 167.

confounded with Arzan-ar-Rûm or Erzerum which will be noticed in the next chapter) as in his day gone to ruin; but Mustawfî in the 8th (14th) century, who generally spells the name Arzanah, speaks of it as though it were still a flourishing place.

On the southern bank of the Euphrates, between where the two rivers from Mayyâfarîkîn and Arzan flow in from the north, stands the castle called Ḥiṣn Kayfâ, or Kîfâ, which the Greeks called Kiphas or Cephe. Muḳaddasî describes the place as a strongly fortified castle, and the markets of its suburbs were plentifully supplied. There were, he adds, many churches here, and the anonymous annotator of the MS. of Ibn Ḥawḳal, already referred to, writing in the 6th (12th) century, speaks of the great stone bridge which crossed the Tigris here, and which had been restored by the Amîr Fakhr-ad-Dîn Ḳârâ Arslân in the year 510 (1116). Below the castle, at that time, was a populous suburb, with many markets and hostelries, the houses being well built of mortared stone. The surrounding district was fertile, but the climate was bad, and the plague was often rife during the summer heats. Yâḳût, who had been at Ḥiṣn Kayfâ, says that suburbs had formerly existed here on both banks of the Tigris, and he considered the great bridge as one of the finest works he had seen. It consisted of a single great arch, which rose above two smaller arches, and these, presumably by a central pier, divided the bed of the Tigris. In the next century Mustawfî describes Ḥiṣn Kayfâ as a large town, but for the most part gone to ruin, though still inhabited by a numerous population.

The hill known as Tall Fâfân, with a town of this name at its foot, stood on the northern or left bank of the Tigris, some 50 miles east of Ḥiṣn Kayfâ, where the river makes its great bend south. The town, Muḳaddasî writes, in the 4th (10th) century was surrounded by gardens, its markets were well provisioned, and though the houses were mostly clay-built, the market streets were roofed over. The river which joins the Tigris at Tall Fâfân comes down from Badlîs (Bitlis), rising in the mountains of Armenia to the south-west of Lake Vân. This river is joined by a great affluent rising to the south of the lake, which Muḳaddasî and Yâḳût name the Wâdî-ar-Razm, and the Tigris below the junction of their united streams became navigable for boats. On the banks of the river Razm, north of Tall Fâfân, just above where the Badlîs river runs in, stands the town of Sâ'irt, also written Sî'ird and Is'îrt, which was often counted as of Armenia. Yâḳût more than once refers to it, but gives no description; Mustawfî, however, speaks of Sâ'ird as a large town, famed for the excellent copper vessels made by its smiths; and the drinking cups from here were exported far and wide. Near Is'îrt, according to Ḳazwînî, was the small town of Ḥîzân, where alone in all Mesopotamia the chestnut-tree (Shâh-balûṭ) grew abundantly[1].

[1] Ist. 76. I. H. 152. Muk. 141, 145. N. K. 7. Yak. i. 205; ii. 277, 552, 776; iii. 68, 854. Kaz. ii. 241. Mst. 165, 166. The name of the river Razm is variously given in the MSS. as Zarm, Razb, or Zarb, and the true pronunciation is unknown.

CHAPTER VIII.

THE UPPER EUPHRATES.

The Eastern Euphrates or Arsanâs. Milâsgird and Mûsh. Shimshâṭ and Ḥiṣn Ziyâd or Kharpût. The Western Euphrates. Arzan-ar-Rûm or Ḳâlîḳalâ. Arzanjân and Kamkh. The castle of Abrîḳ or Tephrike. Malaṭiyah and Ṭarandah. Zibaṭrah and Ḥadath. Ḥiṣn Manṣûr, Bahasnâ and the Sanjah bridge. Products of Upper Mesopotamia. The high roads.

The cities and districts lying along the banks of the Eastern and Western upper Euphrates (for the great river had two head-streams) were generally counted as dependent on northern Mesopotamia, and are often included in the Jazîrah province. The Eastern Euphrates, the southernmost of the two branches of the river, and by some geographers counted as the main source, is the Arsanias Flumen of Tacitus and Pliny. In the 4th (10th) century Ibn Serapion still calls this the Nahr Arsanâs, and the same name is given to it by Yâḳût as in use in the 7th (13th) century, who refers to the extreme coldness of its waters. At the present day it is generally known to the Turks as the Murâd Ṣû, being so named, it is commonly said, in honour of Sultan Murâd IV, who conquered Baghdâd in 1048 (1638).

The Arsanâs took its rise in the Ṭarûn country, a name the Armenians write Daron, and the Greeks knew of as Taronites, which includes the mountains lying to the north of Lake Vân. The first place of importance on the Arsanâs was the town of Malâzkird, which in the various dialects of this region was also known as Minâzjird, Manzikart, and Milâsgird. In the 4th (10th) century Muḳaddasî describes Malâzkird as a strong fortress with a mosque in its market street, the place being surrounded by many gardens. In 463 (1071) Manzikart, as the Greeks called it, was the field of the decisive battle between the Byzantines and Moslems, when the Emperor Romanus IV (Diogenes) was taken prisoner by the Saljûḳs, this leading up to their conquest and permanent settlement in Asia Minor. Yâḳût more than once refers to Minâzjird or Minâzkird, and Mustawfî, who gives the name as Malâzjird, praises its strong castle, its excellent climate, and its fertile lands. The town of Mûsh to the south of the Arsanâs, in the great plain on the west of Lake Vân, is often counted as of Armenia. It is mentioned by Yâḳût, and Mustawfî describes it as having excellent pasture lands, watered by streams that flowed north to the Eastern Euphrates and south to the Tigris. The town was in his day in ruins[1].

The Arsanâs received on its right bank two affluents coming down from the north, and the Ḳâlîḳalâ country. These affluents are important as they enable us to fix the approximate position of Shimshâṭ, a town of some note, which has disappeared from the map, and which has often been confounded with Sumaysâṭ on the Euphrates already mentioned (p. 108). Ibn Serapion states that the first affluent was the Nahr-adh-Dhîb, 'the Wolf River,' which rising in Ḳâlîḳalâ fell into the Arsanâs a short distance above Shimshâṭ; the second was the Salḳiṭ river, which rose in the mountains called Jabal Marûr (or Mazûr) and joined the Arsanâs one mile below Shimshâṭ. A reference to the map shows that these two streams are those now known respectively as the Gunek Ṣû and the Perî Chay; the Ḳâlîḳalâ country representing the mountain region lying between the Arsanâs and the Western Euphrates, and to the west of the Ṭarûn country.

Shamshâṭ (Shimshâṭ) was much the most important place on the Arsanâs, which Ibn Serapion also refers to as the river of Shimshâṭ, and the town appears to have stood on the southern or left bank of the river. Shamshâṭ is undoubtedly the Arsamosata of the Greeks, and Yâḳût—who particularly remarks that it is not

[1] I. S. 11. Kud. 246, 251. Muk. 376. Yak. i. 207; iv. 648, 682. Mst. 165, 167. Ḥâjjî Khalfah (J. N. 426) in 1010 (1600) is apparently our earliest authority for the Eastern Euphrates being called the Murâd Ṣû, and as his work was apparently written before the reign of Sultan Murâd IV, this goes to prove that the stream was not called after that monarch, as is commonly said.

to be confounded with Sumaysâṭ—says that Shamshâṭ lay between Pâlûyah (modern Pâlû) and Ḥiṣn Ziyâd (modern Kharpût). In the 7th (13th) century when Yâḳût wrote, Shamshâṭ was already in ruins, but the data above given by Ibn Serapion and Yâḳût enable us to fix its position within narrow limits. The fortress of Ḥiṣn Ziyâd, which Ibn Khurdâdbih mentions as situated at no great distance from Shamshâṭ, was on the authority of Yâḳût the Arab name for the Armenian Khartabirt, now more generally called Kharpût. Mustawfî gives the spelling Kharbirt, but adds no details, referring to it merely as a large town enjoying a good climate. In this district Balâdhurî and other early authorities mention the bridge of Yaghrâ, which crossed a stream that was probably some tributary of the Arsanâs, and this bridge (*Jisr*) lay 10 miles distant from Shamshâṭ; its exact position, however, is unknown. Then about a hundred miles to the westward of Shamshâṭ the Arsanâs or Eastern Euphrates finally mingles its stream with the Western Euphrates[1].

The Western Euphrates has generally been considered the main branch of the great river, and it is that now commonly known to the Turks as the Ḳârâ Ṣû (Black Water), and this is the Nahr-al-Furât of Ibn Serapion. According to him it took its rise in the mountains called Jabal Aḳradkhis (the name is apparently written Afradkhis by Mas'ûdî, and other variants occur) which are of the Ḳâlîḳalâ country to the north of Erzerum. This important town, which the Arabs called Arzan-ar-Rûm or Arḍ-ar-Rûm (the Land of the Romans), the Armenians knew as Karin, and the Greeks as Theodosiopolis. It is the Moslem city of Ḳâlîḳalâ, and the chief place in this district. The origin of the name Ḳâlîḳalâ, so frequently mentioned by all the earlier Arab geographers, appears to be unknown, but all agree that this was the country in which the Western Euphrates, the Araxes river, and the affluents of the Arsanâs took their rise. Of the town of Erzerum the earlier Arab geographers afford no details, except to state that it was a great city: Mustawfî speaks of there being many fine churches here, one especially with a dome whose circle was fifty ells in diameter. Opposite this

[1] I. S. 10, 13, 30. I. K. 123. Baladhuri, 139. Yak. ii. 276, 417; iii. 319. Mst. 262.

church was a mosque built on the model of the Ka'bah at Mecca. Ibn Baṭûṭah, who was in Arz-ar-Rûm (as he writes the name) in 733 (1333), describes it as a large city, belonging to the Sultan of 'Irâḳ, for the most part in ruins, but still famous for its gardens, and three rivers ran through its suburbs. Eight leagues to the east of Arzan-ar-Rûm, on the summit of a mountain and near one of the head-streams of the Araxes, is Avnîk, a great fortress, of which Mustawfî says that the town at its foot was named Abaskhûr (or Abshakhûr). It belonged to Arzan-ar-Rûm, and Yâḳût adds that the district was called Bâsîn. At the close of the 8th (14th) century Tîmûr took Avnîk after a long siege, and it is frequently mentioned in the history of his campaigns.

Some 200 miles west of Arzan-ar-Rûm and on the right or north bank of the Euphrates, is the town of Arzanjân, which Yâḳût says was more often called Arzingân. He speaks of it as a fine town well provisioned, in his day inhabited for the most part by Armenians, who openly drank wine to the scandal of their Moslem fellow-citizens. Mustawfî adds that its walls had been restored by the Saljûḳ Sultan 'Alâ-ad-Dîn Kayḳubâd at the close of the 7th (13th) century, and that they were built of well-cut jointed stone masonry. Arzanjân had an excellent climate, its lands producing corn, cotton, and grapes in abundance. Ibn Baṭûṭah who passed through here in 733 (1333) writes of it as mostly inhabited by Turkish-speaking Armenians, who were Moslems. In the neighbourhood were copper mines, and the brass work of the native smiths was famous; the markets were good and much cloth was woven in the town. Bâbirt to the north of Arzanjân is mentioned by Yâḳût as a considerable town, mostly peopled by Armenians; but Mustawfî adds that in his day it had much diminished in importance. The fortress of Kamkh (or Kamakh) lay on the Western Euphrates a day's journey below Arzanjân, on the left or south bank of the river. It is frequently mentioned by Ibn Serapion and the earlier Arab geographers, and was the Greek Kamacha. Mustawfî describes it as a great castle, with a town below on the river bank, and many fertile villages were of its dependencies[1].

[1] I. S. 10. I. R. 89. I. K. 174. Mas. i. 214. *Tanbih* 52. Yak. i. 205,

Sixty miles or more to the west of Kamkh the Euphrates, which from Erzerum has flowed westward, makes a great bend and takes its course south, and it here receives on its right bank the river called by Ibn Serapion the Nahr Abrîk, from the castle of Abrîk which is on its upper course. This is the stream now known as the Chaltah Irmak, which comes down from Divrîk or Dîvrîgî. In Mustawfî and Ibn Bîbî the name is given as Difrîgî, which the Byzantines wrote Tephrike (the form Aphrike also occurs in the Greek MSS.), and the earlier Arab geographers shortened this to Abrîk. The place was celebrated at the close of the 3rd (9th) century as the great stronghold of the Paulicians, a curious sect of Eastern Christians whose Manichæan beliefs caused them to be ruthlessly persecuted by the orthodox Emperors of Constantinople. The Paulicians, whose name the Arab writers give under the form of Al-Baylakânî, took possession of Tephrike, fortified it, and countenanced or aided by the Caliphs, for some years successfully defied the armies of Constantinople. Kudâmah and Mas'ûdî, who are nearly contemporary authorities, both refer to the castle of Abrîk as 'the capital of the Baylakânî'; and 'Alî of Herat (quoted by Yâkût) writing in the 7th (13th) century has left a curious account of a great cave and a church near Al-Abrûk (as he spells the name) where were preserved the bodies of certain martyrs, which he considered to be those of the Seven Sleepers of Ephesus.

A short distance to the south of the Chaltah Irmak and Divrîk, the Sârîchîchek Sû joins the Euphrates, on which stands the fortress of 'Arabgîr. This place does not appear to be mentioned by any of the earlier Arab geographers, though Ibn Bîbî in his Saljûk chronicle of the 8th (14th) century names it more than once; also under the form Arabraces it is found in the Byzantine chronicles. 'Arabgîr in any case does not represent Abrîk and Tephrike, as has been sometimes erroneously urged. Apparently the earliest occurrence of the name of 'Arabgîr or 'Arabkîr in any Moslem geographer is to be found in the Turkish *Jahân Numâ* of Hâjjî Khalfah at the beginning of the 11th (17th) century.

206, 408, 444; iii. 860; iv. 19, 304. Kaz. ii. 370. Mst. 162, 163. A. Y. i. 691; ii. 252, 403. I. B. ii. 293, 294.

He also mentions Dîvrîkî (as the town is now called), but unfortunately we have no description of the old Paulician stronghold[1].

Malaṭiyah, which the Greeks called Melitene, was in early days one of the most important fortresses of the Moslem frontier against the Byzantines. Balâdhurî states that its garrison held the bridge, three miles distant from the fort, where the high road crossed the Ḳubâḳib river near its junction with the Euphrates. The Ḳubâḳib is the river known to the Greeks as the Melas, and called by the Turks at the present time the Tukhmah Ṣû, and it rises far to the west of Malaṭiyah in the mountains from which the Jayḥân, the ancient Pyramus, flows south-west (as will be noticed in the next chapter) to the Mediterranean in the Bay of Alexandretta. Except for the Arsanâs the river Ḳubâḳib is by far the most important of the many affluents of the upper Euphrates, and the Ḳubâḳib itself has many tributaries that are duly named by Ibn Serapion. The city of Malaṭiyah was rebuilt by order of the Caliph Manṣûr in 139 (756), who provided it with a fine mosque, and he garrisoned it with 4000 men. Iṣṭakhrî describes it in the 4th (10th) century as a large town surrounded by hills on which grew vines, almonds, and nut-trees, for its lands produced the crops of both the hot and the cold regions. It was more than once taken by the Byzantines and retaken by the Moslems, and Yâḳût in the 7th (13th) century counts it as of the Greek country. Mustawfî in the next century speaks of Malaṭiyah as a fine town with a strong fortress. Its pasture lands were famous, corn, cotton, and abundant fruit being grown in the neighbourhood. On a mountain peak near Malaṭiyah was the convent called Dayr Barṣûmâ, which Ḳazwînî describes as greatly venerated by the Christians, and as inhabited by many monks.

The fortress of Ṭarandah, the modern Darandah—under which form it is mentioned in the *Jahân Numâ*—lay on the upper waters of the Ḳubâḳib, three marches above Malaṭiyah. A Moslem garrison was placed here, to hold the pass, as early as the year 83 (702), but the post was subsequently abandoned in 100 (719) by order of the Caliph 'Omar II. In the Byzantine chronicles

[1] I. S. 11, 31. Kud. 254. Mas. viii. 74. *Tanbih* 151, 183. Yak. i. 87. Ibn Bibi, 210, 318. Mst. 162. J. N. 624. *Cf.* also *J. R. A. S.* 1895, p. 740, and the corrections given in *J. R. A. S.* 1896, p. 733.

this place is frequently mentioned as Taranta, and in the 3rd (9th) century it was one of the strongest of the Paulician fortresses[1].

The river Ḳubâḳib had an important tributary, the Nahr Ḳarâḳîs, which joined it from the south, and on the upper waters of the Ḳarâḳîs stood the great fortress of Zibaṭrah, which the Byzantines called Sozopetra or Zapetra, the ruins of which are probably those of Vîrân Shahr, some leagues to the south of Malaṭiyah on the river Sulṭân Ṣû, the modern name of the Ḳarâḳîs. Balâdhurî and Iṣṭakhrî both speak of Zibaṭrah as a great fortress on the Greek frontier, many times dismantled by the Byzantines and rebuilt by the Caliph Manṣûr and later by Mamûn. Yâḳût and other authorities couple together the names of Zibaṭrah and the fortress Al-Ḥadath, which will be noticed presently. In the Arab and Byzantine chronicles Zibaṭrah or Sozopetra is famous for its capture by the Emperor Theophilus, and again for its recapture by the Caliph Mu'taṣim in his great expedition against 'Amûrîyah, which will be mentioned in the next chapter. Zibaṭrah long continued a place of importance, but Abu-l-Fidâ who visited it in the year 715 (1315) describes the fortress as then a ruin. The line of the old walls could at this time barely be traced, and its fields were completely wasted, so that Abu-l-Fidâ found excellent hunting in the oak woods near the formerly well-cultivated lands, the hares here being, he says, of a size not met with elsewhere. He describes the place as two marches south of Malaṭiyah and the same distance from Ḥiṣn Manṣûr, which will be noticed below[2].

The fortress of Al-Ḥadath, the Byzantine Adata, was taken by the Moslems in the reign of the Caliph 'Omar, and is frequently mentioned in the chronicles. The word *Ḥadath* in Arabic means

[1] I. S. 10, 12, 13. Baladhuri, 185, 187. Ist. 62. I. H. 120. Yak. iv. 26, 633. Mst. 163. Kaz. ii. 356. J. N. 624. The modern town of Malaṭiyah lies two leagues distant to the south of the medieval fortress. The ruins of the old town are at Eskî-Shahr, a league from the ancient bridge, called Ḳirkgoz, crossing the Tukhmah Ṣû immediately above its junction with the Euphrates.

[2] I. S. 13. Baladhuri, 191. Ist. 63. Yak. ii. 914. A. F. 234. The identification of the sites of Zibaṭrah and Ḥadath are discussed by Mr J. G. C. Anderson in the *Classical Review* for April, 1896, in his paper on *The Campaign of Basil I against the Paulicians in* 872 A.D.

'news,' and more especially 'bad news,' and Balâdhurî says that the road thither, of old called Darb-al-Ḥadath, 'the Road of Bad News,' was changed to Darb-as-Salâmah, 'the Road of Safety,' after the capture of the fortress by the Moslems. Darb-as-Salâmah, however, as will be mentioned in the following chapter, is more generally the name given to the Constantinople road, going by the Cilician Gates. There was a mosque at Ḥadath, and the town was rebuilt by the Caliph Mahdî in 162 (779), and again restored by Hârûn-ar-Rashîd, who kept a garrison here of 2000 men. Iṣṭakhrî mentions its fertile lands, and relates how this frontier fortress had been taken and retaken many times alternately by Byzantines and Moslems. According to Yâḳût and others Al-Ḥadath was called Al-Ḥamrâ, 'the Red,' because of the colour of the ground thereabout, and the castle stood on a hill called Al-Uḥaydab, 'the Little Hump-back.' In 343 (954), after many vicissitudes, it was finally taken from the Greeks and rebuilt by Sayf-ad-Dawlah the Ḥamdânid, and in 545 (1150) it passed into the hands of Mas'ûd, son of Ḳilij Arslân the Saljûḳ.

The river near which Ḥadath stood was called the Jûrîth or Ḥûrîth; this Ibn Serapion, in error, gives as an affluent of the Ḳubâḳib (the Malaṭiyah river), but Yâḳût, who writes the name Ḥûrîth, rightly says that it was a tributary of the Nahr Jayḥân, the Pyramus. Ibn Serapion records that the source of the Ḥadath river was at a spring called 'Ayn Zanîthâ, and that before passing Ḥadath it ran through a series of small lakes; further, that the Jûrîth river (as he writes the name) was joined by the river Al-'Arjân, whose sources were in the Jabal-ar-Rîsh, the town of Ḥadath being supplied by water-channels from the 'Arjân river, to which they again returned. To supplement this Abu-l-Fidâ states that Ḥadath lay twelve miles distant from a place on the main stream of the Jayḥân where that river was crossed at 'the Ford of the Alid.' The exact site of Ḥadath has not been identified, but there is little doubt that it protected the pass going from Mar'ash (Germanicia) to Al-Bustân (Arabissus), and that it lay on the banks of the present Âḳ Ṣû, near Iniklî, the Âḳ Ṣû being in fact one of the head-waters of the Jayḥân[1].

[1] Baladhuri, 189—191. I. S. 14. Ist. 62. I. H. 120. Yak. ii. 218; iv. 838. A. F. 263.

Each of the two fortresses of Ḥiṣn Manṣûr and Bahasnâ (which exist to the present day) lies on its own river, and both these are right-bank affluents of the Euphrates, joining it successively below Sumaysâṭ. Ḥiṣn Manṣûr, in modern days more often called Adiamân, was by the Byzantines called Perrhe. It took its name from its builder, Manṣûr of the tribe of Ḳays, who was commander of this frontier station during the reign of the last Omayyad Caliph, Marwân II, having been killed in 141 (758). Ḥiṣn Manṣûr was re-fortified by Hârûn-ar-Rashîd during the Caliphate of his father Mahdî, and it is described by Ibn Ḥawḳal as a small town with a Friday Mosque. Its fields were well irrigated, but Ibn Ḥawḳal writes that the fate of this place, like other frontier fortresses, was to be ravaged and dismantled alternately by the Byzantines and the Moslems. Yâḳût adds that the town had a wall with three gates and a ditch outside; and that in its midst stood the fortress defended by a double wall. When Abu-l-Fidâ wrote in the 8th (14th) century Ḥiṣn Manṣûr was a ruin, though the fields round it were still cultivated.

The Nahr-al-Azraḳ (the Blue River) passed down to the north-west of Ḥiṣn Manṣûr, this fortress occupying the tablelands above the Euphrates, which flowed along their southern border. The fortress of Bahasnâ, which the crusading chronicles call Behesdin, lies to the west of Ḥiṣn Manṣûr, and its district was called Kaysûm. Bahasnâ stood on a hill-top, and had a Friday Mosque in the town below, where there were excellent markets, the surrounding country being very fertile. Yâḳût speaks of it as an impregnable castle. The neighbouring Sanjah river, which appears to be that which the Greeks called Singas, had on its banks the small town of Sanjah, near which the stream was crossed by a celebrated bridge, built of dressed stone, with well-set arches of beautiful workmanship. This bridge, the Ḳanṭarah Sanjah, was one of the wonders of the world according to Ibn Ḥawḳal. Yâḳût, who speaks of the Sanjah and the Kaysûm rivers, reporting both as affluents of the Euphrates, describes this great bridge as being of a single arch, going from bank to bank, and over 200 paces in span. It was built, he adds, of huge well-dressed blocks of stone, each block being ten ells long and five high, the width not being shown, and it had been constructed, he affirms, by aid of a talisman[1].

In the matter of trade, the province of Jazîrah or Upper Mesopotamia produced little. Muḳaddasî gives us a list and the items are chiefly the natural products of the land. Mosul, the capital, exported grain, honey, charcoal, cheese, butter, the sumach fruit and pomegranate pips, manna, salted meat, and the ṭirrîkh fish; also iron, and for artificers' work knives, arrows, chains, and goblets. The district of Sinjâr produced almonds, pomegranates, sumach fruit, and sugar-cane; Naṣîbîn, walnuts; Raḳḳah, olive oil, soap, and reeds for pens. Raḥbah was famous for its quinces; Ḥarrân for its honey and the preserve called Ḳubbayt; Jazîrah Ibn 'Omar for nuts, almonds, and butter, also excellent horses were reared on its pastures. Ḥasanîyah on the Little Khâbûr (on the east bank of the Tigris) produced cheese, partridges, fowls, and fruit preserve; the neighbouring Ma'alathâyâ, charcoal, grapes and other fresh fruits, salted meat, hemp seed and hemp stuffs; and finally Âmid in Diyâr Bakr was famous for its woollen and linen fabrics[2].

The high roads of Upper Mesopotamia are in continuation

[1] Baladhuri, 192. Ist. 62. I. H. 120. Yak. i. 770; ii. 278; iii. 162, 860. A. F. 265, 269. The Sanjah bridge is always given as one of the four wonders of the world—the other three are the church at Edessa already mentioned, the Pharos at Alexandria, and the Great Mosque at Damascus (Yak. ii. 591). It is curious that Muḳaddasî on two occasions confounds this bridge over the Sanjah, which last by all accounts was a right-bank affluent of the Euphrates joining it near Sumaysâṭ, with the no less remarkable bridge at Al-Ḥasanîyah, which was built over the Lesser Khâbûr, an affluent of the Tigris (Muk. 139, 147, and see above, p. 93). The stream now known as the Bolam Ṣû which, after being joined by the Kâkhtah Chay, falls into the Euphrates from the north a short distance above Sumaysâṭ, is apparently the Nahr Sanjah of the Arab geographers; and the great bridge, so famous as one of the wonders of the world, still exists. It was built by Vespasian, and by a single arch of 112 feet span crosses the Bolam Ṣû just above the junction of the Kâkhtah Chay. It is described as 'one of the most splendid monuments of the Roman period in existence,' and an illustration of it will be found in the *Geographical Journal* for October, 1896, p. 323; also, with more detail, in Humann and Puchstein, *Reisen in Kleinasien*, plates 41, 42, and 43.

[2] Muk. 145, 146.

of those of 'Irâķ. The post-road from Baghdâd to Mosul, going up the eastern bank of the Tigris, entered the Jazîrah province at Takrît; it continued on the left bank of the river, going straight to Jabultâ, whence by way of Sinn and Ḥadîthah Mosul was reached. This road is given by our earlier Arab authorities and by Mustawfî[1].

From Mosul the post-road, changing to the right or western bank of the Tigris, went up to Balad, where it bifurcated, the left road going by Sinjâr to Ķarķîsiyâ on the Euphrates, the right through Naṣîbîn to Kafartûthâ, where again it bifurcated, the right leading to Âmid, the left by Râs-al-'Ayn down to Raķķah on the Euphrates. This main road from Mosul to Âmid is given by Ibn Khurdâdbih and Ķudâmah, also—but in marches—by Muķaddasî; and the same authorities give the cross roads to the Euphrates. Muķaddasî also gives the marches from Mosul straight to Jazîrah Ibn 'Omar by Ḥasanîyah, and he mentions the road from Âmid by Arzan to Badlîs near Lake Vân[2].

The post-road up the Euphrates kept along its right or western bank, from Alûsah passing 'Ânah to the river harbour of Al-Furḍah. Here it bifurcated, one road running beside the Euphrates up to Fâsh opposite Ķarķîsiyâ, and thence still along the western side of the river to Raķķah; while the left road of the bifurcation at Furḍah went straight across the desert through Ruṣâfah to Raķķah, thus avoiding the windings of the Euphrates. Ruṣâfah, further, was an important station, for here two roads went off to the west across the Syrian desert, namely to Damascus and to Ḥimṣ (Emessa). At Ķarķîsiyâ and Raķķah, as already said, branch roads came in, one from Mosul *viâ* Sinjâr, the other from Naṣîbîn *viâ* Râs-al-'Ayn and Bajarwân; while from Raķķah by Bajarwân a road went through Ḥarrân and Ruhâ (Edessa) to Âmid.

Lastly from Raķķah, *viâ* Sarûj, the direct road, avoiding the great bend of the Euphrates, reached Sumaysâṭ; whence the various distances to Ḥiṣn Manṣûr, Malaṭiyah, Kamkh and the other fortresses are mentioned in round numbers. Unfortunately,

[1] I. K. 93. Ķud. 214. Muk. 135, 148, 149. Mst. 195.
[2] I. K. 95, 96. Ķud. 214, 215, 216. Muk. 149, 150.

however, these last distances are not given with sufficient exactness to be of much use in fixing the positions of Ḥadath and Zibaṭrah, about which there is some question, though Muḳaddasî often adds some useful indications even as regards these outlying frontier forts[1].

[1] I. K. 96, 97, 98. Kud. 215, 216, 217. Muk. 149, 150.

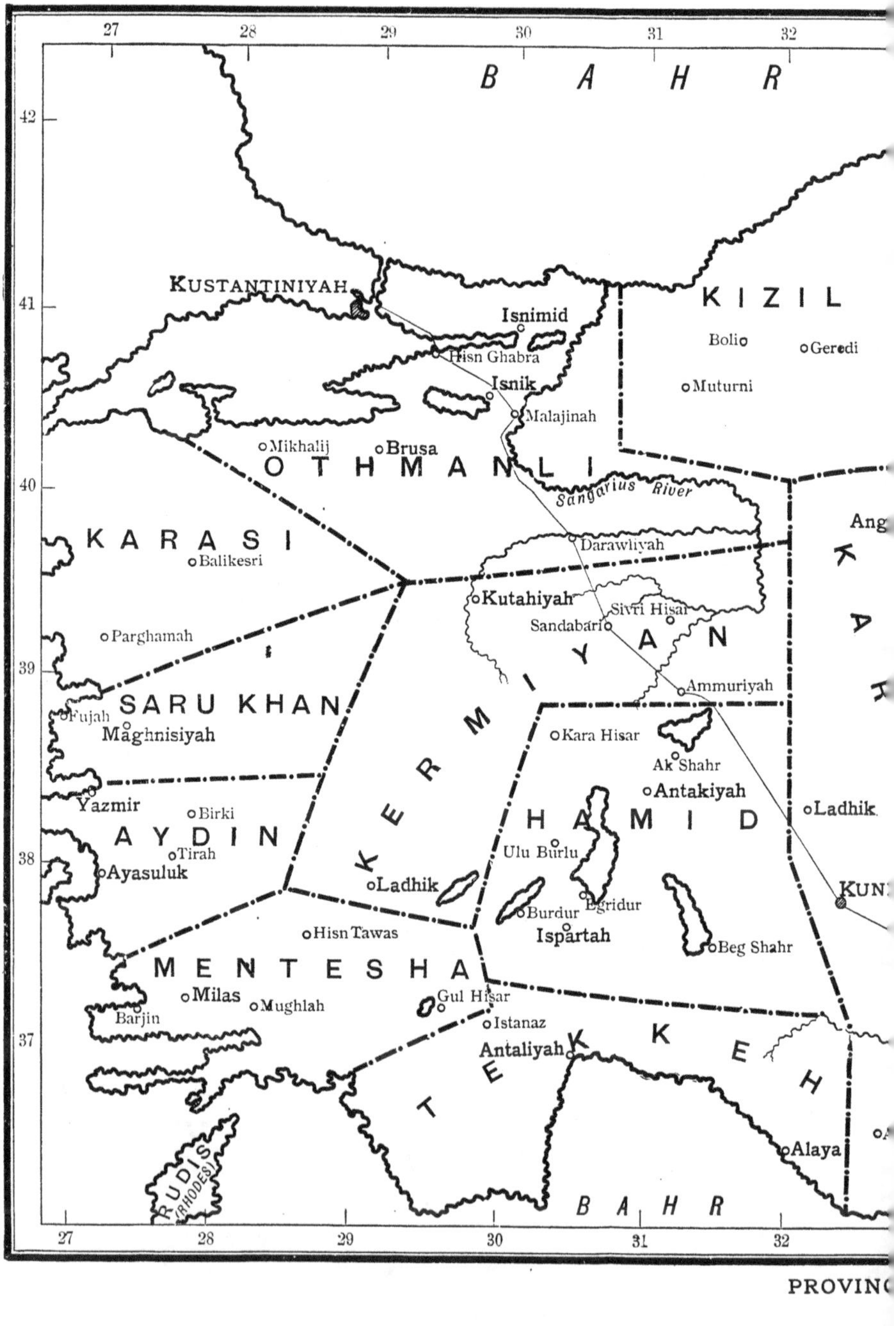
B A H R
KUSTANTINIYAH
Isnimid
Hisn Ghabra
Isnik
Malajinah
KIZIL
Bolio
Geredi
Muturni
Mikhalij
Brusa
O T H M A N L I
Sangarius River
K A R A S I
Balikesri
Darawliyah
Kutahiyah
Sivri Hisar
Sandabari
Parghamah
SARU KHAN
Fujah
Maghnisiyah
Ammuriyah
K E R M I Y A N
Kara Hisar
Ak Shahr
Antakiyah
Yazmir
Birki
A Y D I N
Tirah
Ayasuluk
H A M I D
Ulu Burlu
Ladhik
Burdur
Egridur
Ispartah
Beg Shahr
Hisn Tawas
M E N T E S H A
Milas
Mughlah
Barjin
Gul Hisar
Istanaz
Antaliyah
T E K K E H
Alaya
RUDIS (RHODES)
B A H R
Ang
Ladhik
KUN
27
28
29
30
31
32
42
41
40
39
38
37

PROVINC

MAP IV

VINCE OF RUM

CHAPTER IX.

RÛM OR ASIA MINOR.

Bilâd-ar-Rûm or the Greek country. The line of fortresses from Malaṭiyah to Ṭarsûs. The two chief passes across the Taurus. The Constantinople high road by the Cilician Gates. Trebizond. Three sieges of Constantinople. Moslem raids into Asia Minor. The sack of Amorion by Mu'taṣim. Invasion of Asia Minor by the Saljûḳs. The kingdom of Little Armenia. The Crusaders. The chief towns of the Saljûḳ Sultanate of Rûm.

The provinces of the Byzantine empire were known collectively to the Moslems as Bilâd-ar-Rûm, 'the Lands of the Greeks'; the term 'Rûm' standing for the Romaioi or Romans, being in early Moslem times the equivalent for 'Christian,' whether Greek or Latin. The Mediterranean too, was generally known as the Bahr-ar-Rûm, 'the Roman Sea.' Then Bilâd-ar-Rûm, abbreviated to Rûm, in course of time came more especially to be the name of the Christian provinces nearest to the Moslem frontier, and hence became the usual Arab name for Asia Minor, which great province at the close of the 5th (11th) century finally passed under the rule of Islam when it was overrun by the Saljûḳs.

Unfortunately, for lack of contemporary authorities, we are extremely ill-informed concerning the details of the history and historical geography of Asia Minor during the middle-ages—whether under Christian or Moslem rule[1]. The earlier Arab

[1] The *Historical Geography of Asia Minor* by Professor W. M. Ramsay (referred to as *H. G. A. M.*) contains an admirable summary of all that is at present known on the subject, and is indispensable to any one who wishes to gain a clear understanding of this knotty problem. The present chapter owes far more to this work than appears from the citations in the notes, and reference

geographers not unnaturally knew little of the country that was in their day a province of the Roman empire, and after it had come under the rule of the Saljûḳ Turks our Moslem authorities unfortunately almost entirely neglect this outlying province of Islam. No systematic description of it, such as we possess of the other provinces, therefore has come down to us, and the first complete account of Moslem Asia Minor is that written by Ḥâjjî Khalfah, which only dates from the beginning of the 11th (17th) century, when for nearly two hundred years this province had formed part of the Ottoman empire[1].

Under the Omayyads, as under the Abbasid Caliphs down to rather more than a century and a half before the final overthrow of their dynasty by the Mongols, the frontier line between the Moslems and the Byzantines was formed by the great ranges of the Taurus and Anti-Taurus. Here a long line of fortresses (called Ath-Thughûr in Arabic), stretching from Malaṭiyah on the upper Euphrates to Tarsus near the sea-coast of the Mediterranean, served to mark and guard the frontier; these, turn and turn about, being taken and retaken by Byzantines and Moslems as the tide of war ebbed or flowed. This line of fortresses was commonly divided into two groups—those guarding Mesopotamia (Thughûr-al-Jazîrah) to the north-east, and those guarding Syria (Thughûr-ash-Shâm) to the south-west. Of the former were Malaṭiyah, Zibaṭrah, Ḥiṣn Manṣûr, Bahasnâ, Al-Ḥadath, which have been already described in the previous chapter, next Mar'ash, Al-Hârûnîyah, Al-Kanîsah and 'Ayn Zarbah. Of the latter group lying near the northern coast-line of the bay of Iskandarîyah (Alexandretta), and protecting Syria, were Al-Maṣṣîṣah, Adhanah, and Ṭarsûs.

Mar'ash, the Byzantine Marasion, and it is said occupying the site of Germanicia, was rebuilt by the Caliph Mu'âwiyah in the 1st (7th) century; under the later Omayyads it was strongly

must be made to Professor Ramsay's important papers in the *Geographical Journal* for September, 1902, p. 257, and October, 1903, p. 357.

[1] In the eastern part of the Mediterranean the islands of Cyprus (Ḳubrus) and Rhodes (Rûdis) were both well known to the Arabs, the first having been raided by the Moslems as early as the year 28 (648) under the leadership of Mu'âwiyah, afterwards Caliph. No geographical details, however, are given. Baladhuri, 153, 236. Yak. ii. 832; iv. 29.

fortified, and a large Moslem population settled here, for whose use a Friday Mosque was built. It was re-fortified by Hârûn-ar-Rashîd with double walls and a ditch. Its inner castle, according to Yâkût, was known as Al-Marwânî, being so called after Marwân II, the last Omayyad Caliph. In 490 (1097) Mar'ash was captured by the Crusaders under Godfrey de Bouillon, and subsequently became an important town of Little Armenia (to be described later), remaining for the most part in Christian hands till the fall of that kingdom. The fortress of 'Ayn Zarbah, which the Crusaders knew as Anazarbus, still exists. It was rebuilt and well fortified by Hârûn-ar-Rashîd in 180 (796), and the place is described by Iṣṭakhrî as lying in a plain where palm-trees grew, the surrounding lands being very fertile, while the city had fine walls and its prosperity in the 4th (10th) century was considerable. About the middle of this century Sayf-ad-Dawlah the Ḥamdânid prince spent, it is said, three million dirhams (about £120,000) on its fortification, but it was taken more than once by the Greeks from the Moslems. Then at the close of the next century the Crusaders captured it and left it a ruin; afterwards it formed part of the dominions of the king of Little Armenia. Abu-l-Fidâ describes the town as lying at the base of a hill crowned by a strong castle, it being one day's march south of Sîs, and south of it, he adds, flowed the Jayḥân river. The name 'Ayn Zarbah had in the 8th (14th) century become corrupted into Nâwarzâ.

The exact positions of Al-Hârûnîyah and Al-Kanîsah are unknown, but they lay in the hill country between Mar'ash and 'Ayn Zarbah. Hârûnîyah, which was one march to the west of Mar'ash and considered as its outlying bulwark, took its name from its founder Hârûn-ar-Rashîd who built it in 183 (799). The fortress lay in a valley to the west of the Lukkâm mountains, a name by which the Moslem geographers roughly indicate the chain of the Anti-Taurus. Ibn Ḥawḳal appears to have visited it, for he says the hamlet was populous and the fort had been strongly built, but had been ruined by the Byzantines. This was in 348 (959), when, according to Yâḳût, one thousand five hundred Moslems, men and women, were taken captive. Subsequently Hârûnîyah was rebuilt by Sayf-ad-Dawlah the Ḥamdânid, but again the Christians took it, after which it remained a possession of the king of Little Armenia. Kanîsah, the full name being Kanîsah-as-Sawdâ, 'the Black Church,' was a very ancient fortress built of black stones, and by the Greeks, says Balâdhurî, who adds that Hârûn-ar-Rashîd had it strongly fortified and well garrisoned. It possessed a Friday Mosque and apparently lay to the south of the Jayḥân, for Iṣṭakhrî describes it as 'at some distance from the sea-shore.' Abu-l-Fidâ adds that it was only 12 miles from Hârûnîyah; being in his day included like the latter place in the kingdom of Little Armenia.

Another fortress of this neighbourhood was that known to the Arabs under the name of Al-Muthaḳḳab, 'the Pierced'; so called, according to Yâḳût, 'because it stands among the mountains, all of which are *pierced* as though with great openings.' Its exact site appears to be unknown, but it stood not far from Al-Kanîsah, being at the foot of the Lukkâm mountains, near the sea-shore, and in the vicinity of Maṣṣîṣah. The fortress was built by the Omayyad Caliph Hishâm; others say by 'Omar II; and a Ḳurân, written by the hand of 'Omar II, the most pious of the Omayyad Caliphs, was according to Ibn Ḥawḳal preserved here. Further, Balâdhurî states that when the engineers first came to dig the ditch at Ḥiṣn-al-Muthaḳḳab, they found buried in the earth a human leg, but of such monstrous size that it was considered a portent, and it was forthwith despatched to the Caliph Hishâm as a unique gift[1].

The three cities of Al-Maṣṣîṣah (Mopsuestia), Adhanah (Adana) and Ṭarsûs (Tarsus), all of Greek foundation, still exist. Al-Maṣṣîṣah lies on the Nahr Jayḥân (the river Pyramus). It was conquered by 'Abd-Allah, son of the Omayyad Caliph 'Abd-al-Malik, in the 1st (7th) century, who rebuilt its fortifications and established a strong garrison here. A mosque was erected on the summit of the hill, and the church in the fortress was turned into a granary. A suburb or second town was built shortly afterwards on the other bank of the Jayḥân, called Kafarbayyâ, where the Caliph Omar II founded a second mosque and dug a great cistern. A third quarter, lying to the east of the Jayḥân, was built by the

[1] Ist. 55, 63. I. H. 108, 121. Baladhuri, 166, 171, 188. Mas. i. 26; viii. 295. Yak. i. 927; iii. 761; iv. 314, 498, 945. A. F. 235, 251.

last Omayyad Caliph Marwân II, and named Al-Khuṣûṣ; he surrounded it by a wall with a ditch, and wooden doors closed its gateways. Under the Abbasids the Caliph Manṣûr turned an ancient temple into a Friday Mosque, making it thrice as large as the older mosque of 'Omar II. Hârûn-ar-Rashîd rebuilt Kafarbayyâ, and its mosque was further enlarged by Mamûn. The two quarters of Kafarbayyâ and Maṣṣîṣah proper were connected by a stone bridge across the Jayḥân; the town bore the title of Al-Ma'mûrîyah, 'the Populous,' or 'Well-built,' said to have been bestowed upon it by the Caliph Manṣûr, who restored Maṣṣîṣah after it had been partially destroyed by earthquake in 139 (756). At a later date Maṣṣîṣah, like its neighbours, passed into the possession of the kings of Little Armenia.

The adjacent city of Adhanah lay on the Nahr Sayḥân (the river Sarus), and on the road thither from Maṣṣîṣah was the great bridge which dated from the time of Justinian, but was restored in the year 125 (743) and called Jisr-al-Walîd after the Omayyad Caliph Walîd. This bridge was again restored in 225 (840) by the Abbasid Caliph Mu'taṣim. Adhanah had been in part rebuilt in 141 (758) by Manṣûr, and Iṣṭakhrî describes it as a very pleasant city, lying to the west of the Sayḥân, well fortified and populous. The fortress was on the eastern bank of the river, and was connected with the town by a bridge of a single arch, according to Yâḳût, and Adhanah itself was defended by a wall with eight gates and a deep ditch beyond it.

The rivers Sarus and Pyramus were known to the Moslems respectively as the Nahr Sayḥân and the Nahr Jayḥân. In early days they were the frontier rivers of the lands of Islam towards the Greek country. As such on the analogy, or in imitation, of the more famous Oxus and Jaxartes of Central Asia, which latter were called the Jayḥûn and the Sayḥûn by the Arab geographers, as will be more fully explained later, the rivers Pyramus and Sarus were named the Jayḥân and Sayḥân. Both had their sources in the highlands lying to the north of Little Armenia, and the Jayḥân—which Abu-l-Fidâ compares for size to the Euphrates, adding that in his day the name was commonly pronounced Jahân—after passing Maṣṣîṣah flowed out to the Mediterranean in the Bay of Ayâs to the north of the port of Al-Mallûn (Mallus, later Malo). The Sayḥân was of lesser size, and Adhanah was the only important town on its banks. It was however famous for the great bridge, already mentioned, and both the Jayḥân and Sayḥân, as reported by Mas'ûdî, were held to have been of the rivers of Paradise[1].

The most important, however, of all the frontier fortresses was Ṭarsûs (Tarsus), where a great army of both horse and foot was kept in early times, for Tarsus commanded the southern entrance of the celebrated pass across the Taurus known as the Cilician Gates. Ibn Ḥawḳal states that Tarsus was surrounded by a double stone wall, and garrisoned by 100,000 horse-soldiers; he adds, 'between this city and the Greek lands rises a high mountain range, an offshoot of the Jabal-al-Lukkâm, which stands as a barrier between the two worlds of Islam and Christendom.' Ibn Ḥawḳal explains that the great garrison he saw here in 367 (978) was made up for the most part of volunteers coming from all the provinces of Islam to aid in fighting against the Byzantines, 'and the reason thereof,' he adds, 'is this, that from all the great towns within the borders of Persia and Mesopotamia, and Arabia, Syria, Egypt, and Marocco, there is no city but has in Tarsus a hostelry (*Dâr*) for its townsmen, where the warriors for the Faith (*Ghâzî*) from each particular country live. And when they have once reached Tarsus they settle there and remain to serve in the garrison; among them prayer and worship are most diligently performed; from all hands funds are sent to them, and they receive alms rich and plentiful, also there is hardly a Sultan who does not send hither some auxiliary troops.'

Already under the earlier Abbasid Caliphs, namely Mahdî and Hârûn-ar-Rashîd, Tarsus had been carefully re-fortified and well

[1] Baladhuri, 165, 166, 168. Ist. 63, 64. I. H. 122. Mas. ii. 356; viii. 295. Yak. i. 179; ii. 82; iv. 558, 579. A. F. 50. The names of both rivers are occasionally, but incorrectly, written Sayḥûn and Jayḥûn, like their Central Asian prototypes. In the matter of the ancient mouth of the Sarus, it is worth noting that Ibn Serapion (MS. *folio* 44 *a*) states that in his day, at the beginning of the 4th (10th) century, the Sayḥân (Sarus) flowed into the Jayḥân (Pyramus) five leagues above Maṣṣîṣah, having but one mouth to the sea with the Jayḥân. At the present day the Sayḥân has its separate mouth to the westward near Marsinah, but the old bed may still be traced. See the *Geographical Journal* for Oct. 1903, p. 410.

garrisoned at first with 8000 troops; and from the celebrated Bâb-al-Jiḥâd, 'the Gate of the Holy War,' the yearly expeditions against the Christians were wont to set forth. The Caliph Mamûn, who had died at the neighbouring Badhandûn (Podandos), was buried at Tarsus, on the left-hand side of the great Friday Mosque. Through the city ran the Nahr-al-Baradân (the river Cydnus); the double walls of the town were pierced by six gates, and outside was a deep ditch. Tarsus, Yâḳût adds, remained the frontier city of Islam until the year 354 (965), when the Emperor Niḳfûr, Nicephorus Phocas, having conquered many of the frontier fortresses, laid siege to Tarsus and took it by capitulation. Among the Moslems, those who could left the city; those who remained were forced to pay the capitation tax. The mosques were all destroyed 'and Niḳfûr burnt all the Ḳurâns, further he took all the arms away from the arsenals, and Ṭarsûs with all the country round has remained in the hands of the Christians to this day of the year 623 (1226).'

The ancient Cydnus river, as already said, was generally known as the Nahr-al-Baradân or Baradâ, and Ibn-al-Faḳîh states it was also called the river Al-Ghadbân. It rose in the hill country to the north of Tarsus in a mountain known as Al-Aḳra', 'the Bald,' and flowed into the Mediterranean not far from the later mouth of the Sayḥân. To the westward, one march from Tarsus, the frontier in early times was marked by the river Lamos, which the Arabs called the Nahr-al-Lamis, and here the ransoming of Moslem and Christian captives periodically took place. Beyond this was the Greek town of Salûḳîyah (Seleucia of Cilicia) which in later times, under the Turks, came to be known as Selefkeh[1].

The line of the Taurus was traversed by many passes, but two more especially were used by the Moslems in their annual raids into the Byzantine country. The first, to the north-east, was the Darb-al-Ḥadath which led from Mar'ash north to Abulustân, a town in later times known as Al-Bustân (Byzantine Ablastha and the Greek Arabissus), this pass being defended by the great fortress of Ḥadath (Adata) already noticed in the last chapter. The

[1] I. H. 122. I. F. 116. Baladhuri, 169. Mas. i. 264; vii. 2; viii. 72. Yak. i. 553, 558; iii. 526. Tabari, iii. 1237. In Ibn-al-Athir (vi. 340) the name of the Lamos river is incorrectly printed as Nahr-as-Sinn.

second, and most frequently used pass in early times was that of the Cilician Gates, leading north from Tarsus, and through this went the high road to Constantinople. This road, which was traversed by the post-couriers, and periodically by the embassies passing between the Cæsar and the Caliph, in addition to being followed more or less exactly in innumerable raiding expeditions whether of the Moslems or the Christians, is carefully described by Ibn Khurdâdbih writing in 250 (864), and his account has been copied by many later writers. It was known in its southern part as the Darb-as-Salâmah, 'the Pass of Safety,' and threaded the Pylæ Ciliciæ—the celebrated Cilician Gates.

The account is as follows. Many of the places of course cannot now be exactly identified, but the names are added where possible in brackets. Ibn Khurdâdbih writes:—From Ṭarsûs it is six miles to Al-'Ullayḳ and thence 12 to Ar-Rahwah ('the Water-meadow,' probably the ancient Mopsukrene) and Al-Jawzât, then seven miles on to Al-Jardaḳûb, and again seven to Al-Badhandûn (Podandos, the modern Bozanti), where is the spring called Râḳah near which the Caliph Mamûn died. And then on from Badhandûn it is 10 miles to the (northern) end of the pass (of the Cilician Gates) at Luluah (Loulon) of Mu'askar-al-Malik, 'the King's Camp,' near the hot springs, and here is Aṣ-Ṣafṣâf, 'the Willows' (near Faustinopolis), also Ḥiṣn-aṣ-Ṣaḳâlibah, 'the Fortress of the Sclavonians.' From the King's Camp (where the Pylæ Ciliciæ end) it is 12 miles to the Wâdî-aṭ-Ṭarfâ, 'the Tamarisk Valley,' thence 20 to Minâ, thence 12 to the river of Hiraḳlah (later Arâkliyah, the Greek Heraclia), the town which Hârûn-ar-Rashîd took by storm. From Hiraḳlah it is eight miles to the city of Al-Libn, thence 15 to Râs-al-Ghâbah, 'the Beginning of the Forest,' thence 16 to Al-Maskanîn, thence 12 to 'Ayn Burghûth, 'the Spring of Bugs,' thence 18 to Nahr-al-Aḥsâ, 'the Underground River,' and thence 18 miles on to the suburb of Ḳûniyah (Iconium). From Ḳûniyah it is 15 miles to Al-'Alamayn, 'the Double Sign-posts,' thence 20 to Abrumasânah, thence 12 to Wâdî-al-Jawz, 'Nut River,' and 12 miles on to 'Ammûriyah (Amorion). But there is another route also going from Al-'Alamayn, 'the Double Sign-posts' aforesaid, to 'Ammûriyah; namely from Al-'Alamayn 15 miles to the villages of Naṣr the Cretan, thence 10 to the head of the lake of

Al-Bâsiliyûn (lake of the Forty Martyrs), thence 10 to As-Sind, thence 18 to Ḥiṣn Sinâdah (the fortress of Synades), thence 25 to Maghl, and then 30 miles on to the forest at 'Ammûriyah.

From 'Ammûriyah (Amorion) it is 15 miles to the villages of Al-Ḥarrâb, and two on to the river Sâgharî (the Sangarius) of 'Ammûriyah; thence 12 to Al-'Ilj, 'the Barbarian,' and thence 15 to Falâmî-al-Ghâbah, 'Falâmî of the Forest,' then 12 to Ḥiṣn-al-Yahûd, 'the Jews' Fortress,' and 18 miles on to Sandâbarî (Santabaris), 35 miles beyond which lies the Meadow of the King's Asses at Darawliyah (Dorylæum). From Darawliyah it is 15 miles to the fortress of Gharûbulî, and three on to Kanâis-al-Malik, 'the King's Churches' (the Basilica of Anna Comnena), then 25 miles to At-Tulûl, 'the Hills,' and 15 to Al-Akwâr, whence in 15 miles you reach Malajînah (Malagina). From here it is five miles to Iṣṭabl-al-Malik, 'the King's Stables,' and 30 on to Ḥiṣn-al-Ghabrâ, 'the Dusty Fortress' (namely Kibotos, whence the ferry goes over to Aigialos), and thence it is 24 miles on to Al-Khalîj, 'the Strait' (which is the Bosporus of Constantinople). And over against (namely south of) the fortress of Al-Ghabrâ is Niḳîyah (Nicæa). This ends the account in Ibn Khurdâdbih of the Constantinople road[1].

Off the line of the great high road to Constantinople, the earlier Arab writers had but very incorrect notions of the geography of Asia Minor;—as is shown, for instance, by the confusion which Ibn Ḥawḳal makes between the two very distinct rivers Alis and Ṣâghirah, the Halys and Sangarius. The names of a number of Greek towns appear, in an Arabicized form, in the

[1] I. K. 100—102, 110, 113. Some other variants of this route are given by Ibn Khurdâdbih (pp. 102 and 103), for which the distances have been added by Idrîsî (Jaubert, ii. 308, 309), and compare especially Ramsay, *H. G. A. M.* pp. 236 and 445. Professor Ramsay (see *Geographical Journal* for Oct. 1903, p. 383) has identified the famous fortress of the Sclavonians (Ḥiṣn-aṣ-Ṣaḳâlibah) with the ruins of the Byzantine fortress, built of black marble, and now known as Anasha-Ḳal'ahsi, which is perched high on the mountain overlooking, from the south, the vale of Bozanti (Badhandûn, Podandos). The Byzantine castle of Loulon, which the Arabs called Luluah, 'the Pearl,' he has also identified (*loc. cit.* pp. 401 and 404, where a photograph of the place is given). It lay to the north, above Aṣ-Ṣafṣâf, 'the Willows,' which marked the settlement in the valley below, where the Greek town of Faustinopolis had stood.

earlier chronicles, and these names for the most part recur, but in an altered form after the Turkish conquest; the Arab authors, however, have unfortunately left no descriptions of these towns. Their identity is not disputed, and we have, to name but a few, Aṭ-Ṭawânah (Tyana), Dabâsah (Thebasa), Malaḳûbiyah (Malacopia), Hiraḳlah (Heraclia), Lâdhiḳ (Laodicea), Ḳayṣarîyah (Cæsarea Mazaka, of Cappadocia), Anṭâkiyah (Antioch of Pisidia), Ḳuṭiyah (Cotyæum), Anḳurah (Angora), Afsûs (Ephesus), Abidûs (Abydos) and Niḳmûdiyah (Nicomedia), with some others.

Trebizond, written Ṭarâbazandah or Aṭrabazandah, according to Ibn Ḥawḳal, was the chief port by which goods from Constantinople, in early Abbasid times, were brought for sale to Moslems. Arab merchants or their agents took the goods thence across the mountains to Malaṭiyah and other towns on the upper Euphrates. The carrying trade was in the hands of Armenians, according to Ibn Ḥawḳal, but many Moslem merchants, he adds, resided permanently at Trebizond. Greek linen and woollen stuffs are more especially mentioned and Roman brocades, all of which were brought by sea from the Khalîj or Bosporus. The fame and importance of Trebizond at this time is also proved by the Black Sea being then commonly known as the Sea of Trebizond (Baḥr Ṭarâbazandah). Its official name, however, was the Baḥr Bunṭus or Punṭush, the Greek Pontos, which by a clerical error (from the misplacing of the diacritical points of the Arabic character) had from a very early time been incorrectly written and pronounced Nîṭus or Nîṭush, under which form the name is still often quoted by Persian and Turkish writers, and the mistake is now become so stereotyped as to be beyond recall[1].

Although so little topographical information is recorded in the Arab writers about the towns of Asia Minor previous to the Saljûḳ conquest in the latter half of the 5th (11th) century, the Moslems must have had ample practical acquaintance with much of the country; for almost yearly, and often twice a year in spring and autumn, under the Omayyads and the earlier Abbasids, raids

[1] I. H. 129, 132, 245, 246. I. K. 103. Baladhuri, 161. Tabari, iii. 709, 710. A. F. 34. Yak. i. 401, 499. Mas. i. 260. The Black Sea is also occasionally called the Baḥr-al-Khazar, the Sea of the Khazars, a name more generally applied only to the Caspian. I. K. 103.

were made across the Taurus passes into the Greek country, and their ultimate object was ever the capture of Constantinople. Three times, in fact, under the Omayyad Caliphs was Constantinople besieged by Moslem armies, but the result was in each case disastrous to the assailants, which is hardly to be wondered at, seeing that the Bosporus, measuring in a direct line across the mountainous plateau of Asia Minor, is over 450 miles from Tarsus, the base of the Arab attack.

These three famous sieges are: the first in the year 32 (652), under the reign of 'Othmân, when Mu'âwiyah the future Caliph raided across Asia Minor and attempted to take Constantinople, first by assault, and then by siege, which last he had to raise when news came of the murder of the Caliph 'Othmân. The events which followed soon led to the foundation of the Omayyad dynasty. The second siege was in 49 (669), when Mu'âwiyah, established as Caliph, sent his son and successor Yazîd against the Emperor Constantine IV; but the generals were incapable, the Moslem army suffered a crushing defeat, and Yazîd, succeeding to the Caliphate on his father's death, had to return home. The third and best known attempt against Constantinople was the great siege lasting, off and on, for many years in the reign of the Caliph Sulaymân, who sent his brother Maslamah in 96 (715) against Leo the Isaurian. Of this campaign, which again ended in a defeat for the Moslems, we have very full accounts both from the Arab and the Greek chroniclers; and it was in these wars that 'Abd-Allah, surnamed Al-Baṭṭâl, 'the Champion,' made himself famous, who long after, among the Turks, came to be regarded as their national hero, the invincible warrior of Islam.

In spite of frequent defeat and disaster the raids continued, year by year, with a brief interlude while the Abbasids were establishing themselves in power, till more than a century after the date when the latter, having supplanted the Omayyads, became Caliphs; and though again to besiege Constantinople was beyond their power, they raided, sacked, and burnt again and again throughout Asia Minor. One of the most famous of these expeditions was that of the Caliph Mu'taṣim, son of Hârûn-ar-Rashîd, in 223 (838) against 'Ammûriyah (Amorion), described as the most splendid city of the East, 'the strongest fortress of the Bilâd-ar-Rûm and the very eye of the Christians,' which none the less was plundered and burnt to the ground by the Caliph, who returned unmolested, laden with the spoils[1].

The division of Asia Minor into Themes, under the Byzantine Emperors, has been carefully described by Ibn Khurdâdbih, and his account is of use in correcting the confused details given by Constantine Porphyrogenitus. This however need not be discussed here, as it belongs of right to the geography of the Byzantine empire. Besides the towns already mentioned the Arab writers, when recounting the Moslem expeditions across the frontier, notice a number of places which, either from the vagueness of the statement or the ambiguity in the name, can now hardly be identified. Thus Marj-al-Uskuf, 'the Bishop's Meadow,' is frequently mentioned, which from one of the itineraries given by Ibn Khurdâdbih lay some distance west of Podandos. Al-Maṭmûrah[2], or (in the plural) Al-Maṭâmîr, 'the Cellars,' or 'Grottos,' also frequently occurs, and must be sought for in the neighbourhood of Malacopia. Dhû-l-Kulâ' (the Strong Castle), otherwise spelt Dhu-l-Ḳilâ' (the Castle of the Rocks), was a famous fortress, which Balâdhurî states was called 'the Fortress

[1] The long list of Moslem raids into Asia Minor, from Arab sources, has been fully worked out and annotated by Mr E. W. Brooks in his papers 'The Arabs in Asia Minor, 641 to 750' (published in the *Journal of Hellenic Studies*, vol. XVIII, 1898) and 'Byzantines and Arabs in the Time of the early Abbasids, 750 to 813' (published part i. in the *English Historical Review* for October, 1900, and part ii. in the January number, 1901). The great siege of Constantinople during the Caliphate of Sulaymân he has separately treated of in the *Journal of Hellenic Studies* (vol. XIX, 1899) in a paper on 'The Campaign of 716—718 from Arabic sources.' From the Byzantine side this famous siege has been fully discussed by Professor J. B. Bury, *History of the Later Roman Empire*, ii. 401. The Moslems called Constantinople Al-Ḳusṭanṭinîyah, but in regard to the Byzantine name, from which the modern Turkish Istambûl is said to be derived, it is worth noting that Mas'ûdî, in the early part of the 4th (10th) century, writes (*Tanbîh* p. 138) that the Greeks in his day spoke of their capital as Bûlin (i.e. Polin—for πόλις, 'the city'), also as *Istan-Bûlin* (εἰς τὴν πόλιν), and he notes that they did not generally use the name Constantinople (Al-Ḳusṭanṭinîyah), as did the Arabs.

[2] *Mazmorra* in Sp. 'a dungeon' = Scotch *Massamora* (*v. The Antiquary*, ch. xxxiii, note).

of the Stars' by the Greeks, which would seem to identify it with Sideropolis in Cappadocia.

The town of Luluah (the Pearl), as the Arabs, to give the name a meaning, called the Byzantine Loulon, stood as already mentioned at the northern end of the pass of the Cilician Gates. Still further north was Tyana (Ṭawânah or Ṭuwânah), which for a time Hârûn-ar-Rashîd strongly garrisoned and where a mosque was built. The town or fortress called Ṣafṣâf, 'the Willows,' was on the Constantinople road near Luluah, probably as already said (p. 134) at the site of Faustinopolis, while immediately to the south of Podandos was the fortress of the Sclavonians (Ḥiṣn-aṣ-Ṣaḳâlibah) already mentioned, where according to Balâdhurî certain Sclavonians who had deserted from the Byzantines were quartered to guard the pass by Marwân II, the last of the Omayyad Caliphs[1].

After the year 223 (838) the date of the Caliph Mu'taṣim's famous expedition against Amorion, the Moslem raids into the Greek country became less frequent, for the recurrent disorders at Baghdâd left the Abbasid Caliphs less and less free to think of invading the Byzantine territory. Still, from the middle of the 3rd (9th) century to the 5th (11th) century, many of the great semi-independent vassals of the Caliph led Moslem armies across the passes, and at different times the line of the frontier varied considerably, backwards and forwards, though speaking generally it may be stated that no land was ever permanently held by the Moslems beyond the Taurus.

The rise, however, of the Saljûḳ Turks in the 5th (11th) century, which followed the epoch of the Crusades, entirely changed the face of affairs in Asia Minor. In the spring of the year 463 (1071) Alp Arslân the Saljûḳ gained the battle of Malasjird (Manzikart), completely routing the Byzantine forces, and taking the Emperor Romanus Diogenes prisoner. Moreover, previously to this, in 456 (1064), Alp Arslân had taken Ânî, the capital of Christian Armenia, an event which broke up the older

[1] For the themes see 'Arabic lists of the Byzantine themes; by E. W. Brooks,' in the *Journal of Hellenic Studies*, vol. XXI, 1901. I. K. 102, 105. Baladhuri, 150, 170. Tabari, iii. 710, 1237. Ibn-al-Athir, vi. 341. Ramsay, *H. G. A. M.* 340, 354, 356.

Armenian kingdom of the Bagratids, and led to the founding by Rupen, their kinsman, of the kingdom of Little Armenia in the Taurus country. The result of the battle of Malasjird was that Alp Arslân sent his cousin Sulaymân, son of Ḳutlumish, into Asia Minor; and then the Saljûḳs permanently settled down, after their nomadic fashion, in all the high plateau lands forming the centre of the province, and the kingdom of Rûm became from henceforth one of the lands of Islam. In their first flush of victory the Saljûḳs had raided so far west as Nicæa, which for a short time they held, making it temporarily their capital. From here they were driven back by the first Crusade, and retiring to the central plateau, Iconium or Ḳûniyah, which was conquered by them in 477 (1084), became and remained the centre of their government[1].

The line of the Saljûḳ Sultans of Ḳûniyah lasted over two centuries, from 470 (1077) to 700 (1300), but their real power was ended by the Mongol conquest of Ḳûniyah in 655 (1257), the year previous to the fall of Baghdâd. The establishment of the Saljûḳs in the plateau of Asia Minor was coincident with the rise of the Christian kingdom of Little Armenia in the Taurus

[1] Ibn-al-Athir, x. 25, 44; xii. 125. J. N. 621. On the battle of Manzikart see *History of the Art of War* by C. Oman, pp. 216—221. The history of the Saljûḳs in Rûm, and their successors the ten Turkoman Amîrs, ending in the establishment of the Ottoman Sultans, is unfortunately the most obscure period in all the Moslem annals. The Persian historians Mîrkhwând and Khwând-amîr have nothing to add to the bald summary on the Saljûḳs of Rûm given by Mustawfî in his *Târîkh-i-Guzîdah*. Perhaps the fullest account of the dynasty is that given by Ibn Khaldûn in his *Universal History* (volume v. pp. 162—175): but this is in fact little more than a list of names and dates. The Chronicle of Ibn Bîbî, lately published by Professor Houtsma, unfortunately begins only with the reign of Ḳilij Arslân II, in the year 551 (1156), and regarding the first seventy years of Saljûḳ rule, when they were conquering and establishing themselves in Asia Minor, we know next to nothing. The battle of Manzikart is the only great victory that is alluded to, all the fighting that resulted in the ejection of the Byzantines from the high lands of Asia Minor passes unrecorded. Also there is no mention of a treaty, which must have been made, formally or informally, between the Byzantines and the Saljûḳs after Manzikart. For a summary of all that is known of the Turkoman Amîrs who succeeded to the Sultans of Rûm see Professor Lane-Poole, 'The successors of the Saljûḳs in Asia Minor' in the *J. R. A. S.* for 1882, p. 773.

country. Sîs, otherwise called Sîsîyah, soon after 473 (1080) became the capital of Rupen, the founder of the new dynasty. After a century Leo took the title of king in 594 (1198), and the kings of Little Armenia, weathering the Mongol invasion, only came to an end in 743 (1342). From Sîs the kingdom grew to include all the mountainous country watered by the Sayḥân and Jayḥân rivers, down to the Mediterranean, with the cities of Maṣṣîṣah, Adhanah, and Ṭarsûs, as well as much of the coast-line to the west of Ṭarsûs. Sîs, or Sîsîyah, the ancient Flaviopolis, under the early Abbasids had been counted an outlying fortress of 'Ayn Zarbah, and its walls were rebuilt by the Caliph Mutawakkil, grandson of Hârûn-ar-Rashîd. It was afterwards taken by the Byzantines, and when Abu-l-Fidâ wrote in 721 (1321) he alludes to it as having been recently rebuilt by Leo II (Ibn Lâwûn), surnamed the Great, king of Little Armenia. Its castle, surrounded by a triple wall, crowned the hill, and the gardens descended to the river, which was an affluent of the Jayḥân. Yâḳût adds that, in his day, Sîs was the commonly used form of the name.

To the west and north of this kingdom of Little Armenia stretched the territories of the Saljûḳ Sultans, and during the first hundred years of their occupation of the plateau lands of Asia Minor this province was three times traversed by the armies of the Crusades. The first Crusade in 490 (1097) resulted in the expulsion of Ḳilij Arslân I (son and successor of Sulaymân, the first Sultan of Rûm) from Nicæa, and the rabble of the Crusaders passing by Ḳûniyah regained the sea at Tarsus, and took ship for Palestine. In the second Crusade Louis VII of France defeated Sultan Mas'ûd (son of Ḳilij Arslân) on the banks of the Meander in 542 (1147), but the Franks in their passage onward to the port of Anṭâliyah suffered great losses in the mountain country. In the third Crusade the Emperor Frederick Barbarossa is said in 586 (1190) to have captured Ḳûniyah, the Saljûḳ capital, from Ḳilij Arslân II (son of Mas'ûd), but marching onward Barbarossa was accidentally drowned in a river near Salûḳiyah (Seleucia of Cilicia), possibly in the Lamos or Nahr-al-Lamis, already mentioned (p. 133), where under the earlier Abbasids Moslem and Christian captives were exchanged or ransomed.

The extent of the country governed by the Saljûḳ Sultans of Rûm varied of course at different times, according to the waning or recovered power of the Byzantine empire, the growth of the Christian kingdom of Little Armenia, and the condition of the neighbouring Moslem principalities, which the Crusaders had in part overcome, and where for a time Frank princes ruled over Moslem subjects. The chief towns of the Saljûḳ Sultanate in Rûm as it existed in 587 (1191) are made known to us by the division of his dominions which Ḳilij Arslân II made in that year among his eleven sons. Ḳûniyah (Iconium), as already stated, was the capital, and the second city of the Sultanate was Ḳayṣariyah (Cæsarea Mazaka). Malaṭiyah (Melitene) was the chief town of the eastern province on the Euphrates boundary. To the north Sîvâs (Sebastia), Nakîsâr (or Nîksâr, the older Neo-Cæsarea), Tûḳât and Amâsiyah (Amasia) each became the appanage of a Saljûḳ prince, likewise Angûriyah (Angora) to the north-west, and on the western border Burughlû, probably identical with the modern Ulû Burlû, lying to the west of the Egridûr lake. On the southern frontier, lying eastwards of Ḳûniyah, the chief towns were Arâkliyah (Heraclia), Nakîdah or Nigdah, and Abulustân, later called Al-Bustân (Arabissus).

Sultan 'Alâ-ad-Dîn, who succeeded in 616 (1219) and was the grandson of Ḳilij Arslân II, extended his rule north and south from the shores of the Black Sea to the Mediterranean. He took Sînûb (Sinope) on the former, and on the southern coast made a great harbour at 'Alâyâ—named after him—where the slips for ship-building and remains of other constructions connected with the great navy of the Saljûḳs may still be seen: and on the north-west he extended his power to the town of Ṣârî Bûlî. His reign was made famous by the writings of the great Ṣûfî poet Jalâl-ad-Dîn Rûmî, who lived and died at Ḳûniyah. Thirty years after the death of 'Alâ-ad-Dîn, which occurred in 634 (1237), the Mongol armies broke up the power of the Saljûḳs; the four last Sultans were in fact merely governors under the Îl-Khâns of Persia, and in the year 700 (1300) the province of Rûm was divided up among the ten Turkoman Amîrs, who originally had been the vassals of the Saljûḳ Sultans[1].

[1] Baladhuri, 170. Yak. iii. 217. A. F. 257. Ibn Bibi, 5. J. N. 621,

622. Idrîsî, who wrote in 548 (1153), and who, according to his own testimony (Jaubert, ii. 300), was at Amorion and visited the cave of the Seven Sleepers in 510 (1116), is the one Moslem geographer who gives us an account of Asia Minor in the time of the Saljûḳs. Unfortunately his text has come down to us in a most corrupt form. He gives a number of routes, traversing Asia Minor in all directions, which are very difficult to plot out, for the names of intermediate places are for the most part unrecognisable, though the terminal stages are beyond dispute. Idrisi, ii. 305—318. The limits of the Saljûḳ kingdom have been clearly traced by Professor Ramsay (*H. G. A. M.* pp. 78, 382, 384), and a description of the Great Mosques and other buildings of the Saljûḳ Sultans will be found in a series of papers by M. C. Huart entitled 'Épigraphie Arabe d'Asie Mineur,' in the *Revue Sémitique*, 1894, pp. 61, 120, 235, 324, and 1895, pp. 73, 175, 214, 344; and in the *Journal Asiatique* for 1901, i. 343, also by M. F. Grenard, 'Monuments Seljoukides de Sivas etc.,' *J. As.* 1900, ii. 451. See further a paper by Professor Ramsay, with remarks of Sir C. Wilson and others, in the *Geographical Journal* for September, 1902, p. 257.

CHAPTER X.

RÛM (*continued*).

The ten Turkoman Amirates. Ibn Baṭûṭah and Mustawfî. Ḳaysârîyah, and Sîvâs. The Sultan of Mesopotamia. The Amîr of Ḳaramân. Ḳûniyah. The Amîr of Tekkeh, 'Alâyâ, and Anṭâliyah. The Amîr of Ḥamîd, Egridûr. The Amîr of Germiyân, Kutâhîyah, and Sîvrî Ḥiṣâr. The Amîr of Menteshâ, Mîlâs. The Amîr of Aydîn, Ephesus, and Smyrna. The Amîr of Ṣârûkhân, Magnesia. The Amîr of Ḳarâsî, Pergamos. The 'Othmânlî territory, Brusâ. The Amîr of Ḳizil Aḥmadlî: Sinûb.

The limits of the ten Turkoman Amirates of the 8th (14th) century very roughly corresponded with the following ancient Greek provinces of Asia Minor. Ḳarâmân or Ḳaramân, the largest, was the older Lycaonia; on the Mediterranean coast Tekkeh included Lycia and Pamphylia; inland Ḥamîd corresponded with Pisidia and Isauria; Kermiyân or Germiyân with Phrygia; and on the coast of the Black Sea Ḳizil Aḥmadlî, sometimes called Isfandiyâr, had been Paphlagonia. On the Ægean shores Menteshâ was the older Caria; Aydîn and Ṣârûkhân combined were the kingdom of Lydia; Ḳarâsî was Mysia; and lastly the 'Othmânlî territory (of those Ottomans who ultimately conquered all the other nine provinces) was at first only the small province of Phrygia Epictetus, backed by the high lands of Bythia which the 'Othmanlîs had recently conquered from the Byzantines.

Of the state of Asia Minor under these Turkoman Amîrs we possess an extremely curious account in the travels of Ibn Baṭûṭah the Berber, who landing from Syria at 'Alâyâ, in 733 (1333), visited many of the petty courts on his way to Ṣinûb (Sinope), where he took ship across the Black Sea to the Crimea. Unfortunately, a part of his account appears to be missing. From

'Alâyâ he journeyed along the sea-shore to Anṭâliyah, and then struck north across the hills to Egridûr in Ḥamîd, on the lake of that name. From here by a devious road through Lâdhiḳ (Laodicea ad Lycum) he travelled to Mîlâs in Menteshâ, and thence right across Asia Minor diagonally, by Ḳûniyah and Ḳaysâriyah, to Sîvâs and Arzan-ar-Rûm. Here a lacuna occurs, for the next town mentioned is Birkî in Aydîn, whence Ayâ Sulûḳ (Ephesus) was visited. Finally, going north and east, Ibn Baṭûṭah takes Brusâ and other towns on his road to the Black Sea coast at Ṣinûb (Sinope). His contemporary Mustawfî, in the chapter of his Geography on Rûm, has added some details to the description of towns given by Ibn Baṭûṭah. Mustawfî, however, though writing in 740 (1340) works on earlier sources, and his information gives the state of Rûm under the later Saljûḳs, rather than the country as it existed when the ten Amîrs had established their power.

At the beginning of the 9th (15th) century the irruption of Tîmûr into Asia Minor temporarily altered the course of affairs, and threw back the rising Ottoman power for a quarter of a century. The account of his campaigns given by 'Alî of Yazd again adds something to our knowledge of the country, some further details also being given in the pages of the Turkish *Jahân Numâ*, which, though written in the beginning of the 11th (17th) century, when the Ottoman power had long been established in Asia Minor, makes mention of the chief monuments left by the Saljûḳ Sultans.

Before describing the ten provinces, already named, of the Turkoman Amîrs, some account must be given of the towns lying to the eastward of the boundary of Ḳarâmân, which may be taken as marked by the lower course of the Halys (the Ḳizil Irmâḳ of the Turks) continued by a line going south to the Jayḥân. East of this boundary Asia Minor in the 8th (14th) century belonged to the Îl-Khâns, the Mongol princes who ruled in Mesopotamia and Persia, and sent hither their governors to keep the peace among the smaller hordes of Turkoman nomads who had settled down in this country after the great Mongol invasion. The chief city east of the Ḳaramân frontier was Ḳayṣariyah (also spelt Ḳaysârîyah, namely Cæsarea Mazaka, of Cappadocia), which under the Saljûķs had been the second city of Rûm, and which indeed Ķazwînî names as their capital. Here among other shrines might be seen the Friday Mosque dedicated to the hero of Omayyad days, Al-Baṭṭâl. Mustawfî describes Ķayṣariyah as surrounded by the stone walls built by Sultan 'Alâ-ad-Dîn the Saljûķ; it was a great town with a castle and lay at the base of Mount Arjâish (Argæus). Mount Arjâish, Mustawfî adds, was an extremely high mountain, its summit never being free from snow, and from it many streams descended. At its foot lay Davalû, a place which will be mentioned below. On the summit of the mountain might be seen a great church. In Ķayṣârîyah stood the famous and greatly venerated shrine of Muḥammad ibn Ḥanafîyah, a son of the Caliph 'Alî, and when Ibn Baṭûṭah visited Ķaysârîyah (as he writes the name) the city was occupied by a strong garrison in the pay of the Mongol Sultan of 'Irâķ. In the beginning of the 9th (15th) century Ķayṣariyah was the first great city in Asia Minor occupied by the armies of Tîmûr.

Abulustân (Arabissus) to the east of Ķayṣariyah, the frontier fortress of Byzantine times, is also mentioned in the conquests of Tîmûr; and Mustawfî speaks of Abulustân as a medium-sized town. In the *Jahân Numâ* the modern spelling Al-Bustân (with the signification of 'the Garden') is given. Ķîrshahr (Byzantine Justinianopolis Mokissus), about 80 miles west of Ķayṣariyah, was a place of great importance and is frequently mentioned in the account of the campaigns of Tîmûr. Mustawfî describes Ķîrshahr as a large town with fine buildings, and in the *Jahân Numâ* it is counted as one of the cities of Ķaramân. Amâṣiyah or Amâsiyah (Amasia) under the Saljûķs had been one of their centres of government; and Mustawfî relates that it had been rebuilt by Sultan 'Alâ-ad-Dîn. Ibn Baṭûṭah, who passed through it, describes it as a great city with broad streets and fine markets, surrounded by splendid gardens irrigated by means of waterwheels erected along the river. In his day it was under the Sultan of Mesopotamia, and not far distant from it was the town of Sûnusâ (spelt Ṣûnîsâ in the *Jahân Numâ*) with a population of fanatical Shî'ahs. To the north of Amâsiyah lies Lâdîķ (Laodicea Pontica), a place of importance under the Saljûķs, and frequently mentioned in the chronicle of Ibn Bîbî. The port of Samsûn (or Ṣâmṣûn, the Greek Amysos) is described by Mustawfî as a great harbour for ships, and already by the latter part of the 8th (14th) century it was growing rich on the trade diverted to it from the older port of Sanûb or Ṣinûb (Sinope)[1].

Nîksâr (or Nakîsâr, the Greek Neo-Cæsarea) had been an important place under the Saljûḳs, and is frequently mentioned by Ibn Bîbî; Mustawfî describes it as a medium-sized town, with many gardens producing much fruit. Tûḳât (also spelt Dûḳâṭ) lies to the west of Nîksâr on the road to Amâsiyah, and was one of the great governments under the Saljûḳs; further west again lies Zîlah, mentioned by Ibn Bîbî and later authorities. The city of Sîvâs (Sebastia), on the Ḳizil Irmâḳ (Halys), had been rebuilt by Sultan 'Alâ-ad-Dîn, who used hewn stone for all the new masonry. Mustawfî reports that the place was famous for its woollen stuffs, which were largely exported; it had a cold climate, but cotton was grown here, as well as much grain. Ibn Baṭûṭah speaks of Sîvâs as the largest city in the province ruled by the Sultan of Mesopotamia. Here were a Government House, fine streets and excellent markets, and a great Madrasah or college.

Mustawfî gives an account of the high road which went west from Sîvâs to Persia: two stages led to Zârah, a town of some importance, and two more to Âḳ Shahr (White Town), a place frequently mentioned in the Saljûḳ chronicle. North-west of Âḳ Shahr lies Ḳarâ Ḥiṣâr (the Black Fortress) which is often referred to by Ibn Bîbî, who calls it Ḳarâ Ḥiṣâr Dawlah—'of the State'—to distinguish this fortress, which is referred to also by Mustawfî, from other places of like name. In the *Jahân Numâ* it is called Ḳarâ Ḥiṣâr Shâbîn, from the alum (*Shâb*) mines that lie near it. From Âḳ Shahr the high road to Persia went on in three stages to Arzanjân, and thence it was the like distance to Arzan-ar-Rûm. From here the way went south in three stages to Khanûs (or Khûnâs as Ibn Bîbî writes the name, Khinis being the modern form), whence it was 10 leagues to Malâsjird (Manzikart), this being eight leagues distant from Arjîsh on the lake of Vân[2].

[1] Kaz. ii. 371. I. B. ii. 287, 289, 292. Ibn Bibi, 26, 308. Mst. 162, 163, 164, 202. A. Y. ii. 270, 416, 417. J. N. 599, 615, 620, 622, 623.

[2] Ibn Bibi, 26, 292, 308. I. B. ii. 289. Mst. 161, 163, 164, 199. J. N. 424, 622, 623.

The province of Ḳaramân (or Ḳarâmân), the largest of the ten Amirates, took its name from the Turkoman tribe which had settled in this region, and the capital was Lârandah, also called Ḳaramân after the province. Lârandah dated from Byzantine days, and Ibn Baṭûṭah who visited it in the 8th (14th) century, and spells the name Al-Lârandah, describes it as a fine town standing in the midst of gardens, abundantly supplied with water. At the close of the century it was taken and plundered by the troops of Tîmûr, but afterwards regained its former prosperity. To the south of Lârandah is Armanâk, which is spoken of by Mustawfî as having been formerly a large city, though in the 8th (14th) century it had sunk to the condition of a provincial town. It is also mentioned in the *Jahân Numâ*, together with Selefkeh, the older Arabic Salûḳiyah (Seleucia of Cilicia). Under the Ottoman rule these places were included in the province called Îch Îlî, which in Turkish signifies 'the Interior Land,' and as this description is hardly applicable to the province in question, which lies along the coast, it has been suggested that Îch Îlî is in reality only a corruption, truncated, of the older Greek name *Cilicia*.

Ḳûniyah (Iconium), as already stated, had been the Saljûḳ capital, but under the Ḳaramân Amîrs it sank to a city of the second rank. Mustawfî relates that the town possessed a great Aywân, or hall, in the palace which had been built by Sultan Ḳilij Arslân, by whom also the castle had been founded. At a later date 'Alâ-ad-Dîn had built, or restored, the town walls, making them of cut stone, 30 ells in height, with a ditch 20 ells deep outside. The walls were 10,000 paces in circuit, they were pierced by twelve gates, each having a great castellated gateway. Abundant water was brought down from a neighbouring hill, to be stored at one of the city gates in a great tank under a dome, whence over 300 conduits distributed it throughout the city. The neighbourhood of Ḳûniyah was renowned for its gardens, famous for yellow plums, and immense quantities of cotton and corn were grown in the fields around the town.

Mustawfî adds that in his day much of Ḳûniyah was in ruin, though the suburb immediately below the castle had a large population. In the city was the tomb of the great mystic, the Ṣûfî poet Jalâl-ad-Dîn Rûmî, already mentioned, which was an object of pilgrimage. This shrine is noticed by Ibn Baṭûṭah, who praises the fine buildings and abundant water-supply of Ḳûniyah. He speaks of its gardens and the apricots grown here, called *Ḳamar-ad-Dîn* (Moon of Faith), which were exported largely to Syria. The streets were broad and the markets abundantly supplied, each trade keeping to its own quarter. Ibn Bîbî in his Saljûḳ chronicle incidentally mentions the names of three of the gates of Ḳûniyah, namely, the Gate of the Horse Bâzâr, the Gate of the Assay-house, and the Gate of the Aḥmad bridge.

The fortress of Ḳarâ Ḥiṣâr of Ḳûniyah lies at some distance to the east of Ḳûniyah, and is mentioned by Mustawfî who says that it was built by one Bahrâm Shâh. Beyond this is Hiraḳlah (Heraclea), a name which in later times appears as Arâklîyah, and is frequently mentioned in the *Jahân Numâ*. To the north of Ḳûniyah is Lâdîḳ Sûkhtah, the Burnt Lâdîḳ (Laodicea Combusta, the Greek Katakekaumena), which Ibn Bîbî speaks of as the Village of Lâdîḳ to distinguish it from the other towns called Laodicea (Pontica and Ad Lycum). The *Jahân Numâ* refers to Laodicea Combusta as Yurgân Lâdîḳ, otherwise called Lâdhiḳiyah of Karamân[1].

In the northern part of the Ḳaramân province is Angora (Greek Ancyra), the name of which is spelt by the earlier Arabic authorities Anḳurah, and by later Persian and Turkish authors Angûriyah. Mustawfî speaks of it as a town possessing a cold climate; much corn, cotton, and fruit being grown in the neighbourhood. It is famous in history as the place where in 804 (1402) Tîmûr defeated in a pitched battle, and took prisoner, the Ottoman Sultan Bayazîd Ilderim. Kûsh Ḥiṣâr, or Kûch Ḥiṣâr, on the eastern border of the great Salt Lake, is mentioned by Mustawfî as a medium-sized town, and its name also occurs in the *Jahân Numâ*. Some distance east of the southern end of the lake stands Âḳ Sarây (the White Palace) built by Sultan Ḳilij Arslân II in 566 (1171), and described by Mustawfî as a fine town surrounded by fruitful lands. Âḳ Sarâ (as Ibn Baṭûṭah spelt the name) stood on three streams, and its gardens were magnificent, also there were many vineyards within the walls.

[1] I. B. ii. 281, 284. Mst. 162, 163. A. Y. ii. 458. J. N. 611, 615, 616. Ibn Bibi, 8, 9, 287, 324.

The townspeople in the 8th (14th) century made excellent carpets from the wool of their sheep, and these carpets were largely exported to Syria, Egypt, and Mesopotamia. Ibn Baṭûṭah adds that in his day Âḳ Sarâ was in the government of the Sultan of Mesopotamia.

Some fifty miles east of Âḳ Sarâ is Malanḳûbiyah (Malacopia), which is mentioned by Mustawfî as a place of importance in the 8th (14th) century. To the north of this is another Ḳarâ Ḥiṣâr, described by Mustawfî as of the Nigdah district, and east of this again is Davalû (in the *Jahân Numâ* the name is written Davahlû), a place already spoken of as at the foot of Mount Arjâish. It occurs more than once in the history of Ibn Bîbî in connection with Ḳayṣariyah. Mustawfî describes Davalû as a town of medium size, and its walls had been rebuilt by Sultan 'Alâ-ad-Dîn the Saljûḳ. South of Malanḳûbiyah is Nigdah (in Ibn Bîbî written Nakîdah) which had taken the place of the earlier Ṭuwânah (Tyanah), having been built by Sultan 'Alâ-ad-Dîn. Nigdah is described by Mustawfî as a medium-sized town, and Ibn Baṭûṭah, who passed through it, notes that the greater part was already in ruin. It lay, he adds, in the territories of the Sultan of Mesopotamia; its stream was called the Nahr-al-Aswad, 'the Black River,' and was crossed by three stone bridges. The gardens of Nigdah were most fruitful; and waterwheels were employed for their irrigation. To the south of Nigdah was Luluah (Loulon), frequently mentioned by Ibn Bîbî, a great fortress which, as already said, marked the northern end of the pass of the Cilician Gates. In the 8th (14th) century Mustawfî describes Luluah as a small town, surrounded by excellent pasture lands. It had a cold climate, and in the neighbourhood there were famous hunting grounds[1].

In the territories of the Amîr of Tekkeh the most important towns appear to have been 'Alâyâ and Anṭâliyah, famed for their harbours. The first, as already mentioned, had been founded by the Saljûḳ Sultan 'Alâ-ad-Dîn on the site of the ruins of Coracesium. Ibn Baṭûṭah landed here from Syria in 733 (1333), and describes 'Alâyâ as at that time the great port for the trade with Alexandria.

[1] Ibn Bibi, 5, 34, 44, 279, 314. I. B. ii. 285, 286. Mst. 162, 163, 164, 202. Yak. iv. 635. A. Y. ii. 429. J. N. 617, 620.

In the upper town, very strongly built by 'Alâ-ad-Dîn, was the castle, which Ibn Baṭûṭah carefully examined; but in his day 'Alâyâ appears to have belonged to the Sultan of Ḳaramân.

Anṭâliyah, the second harbour, lying a hundred miles to the westward of 'Alâyâ, at the head of the bay, was famous as the usual place of re-embarkation of the Crusaders for Palestine. It was a fine town, and was known to Yâḳût as the chief port of Rûm, being strongly fortified and surrounded by fruitful lands, with many vineyards. Here Sultan Ḳilij Arslân the Saljûḳ had built himself a palace on the hill overlooking the sea, and here, too, Ibn Baṭûṭah found many Christian merchants settled, especially down at the Mînâ or port, their quarter being shut off by a wall, and each trade, he adds, had its own street in the markets. There was a Jews' quarter also, and the Moslems lived in their own part of the city, where stood the mosque and Madrasah (college). Anṭâliyah, the name of which occurs in the Crusading chronicles as Satalia or Attaleia, is frequently mentioned in the campaigns of Tîmûr under the form 'Adâliyah. To the west of it, also mentioned by 'Alî of Yazd, is Istânûs, a town whose name in the *Jahân Numâ* is written Istanâz[1].

To the north of Tekkeh the Amîr of Ḥamîd owned the country round the four lakes of Egridûr, Burdûr, Beg Shahr, and Âḳ Shahr. Under the Saljûḳs, according to Ibn Bîbî, the seat of government had been at Burughlû, apparently identical with the later Ulû Burlû (to the west of the Egridûr lake), the Byzantine Sozopolis or Apollonia. Anṭâkiyah (Antioch of Pisidia), which in the earlier Moslem chronicles is frequently referred to, in Turkish times took the name of Yalâvâch, and was situate in the plain between the lakes of Egridûr and Âḳ Shahr. The chief town of the province, according to Mustawfî in the 8th (14th) century, appears to have been the city of Egridûr (the ancient Prostanna) at the southern end of the lake of that name. Ibn Baṭûṭah describes it as a great place, well built, with fine markets, surrounded by abundantly watered gardens; and the lake (he adds) was traversed by the boats of the merchants, who thus transported

[1] In the New Testament Attalia is mentioned in Acts xiv. 25. Yak. i. 388. I. B. ii. 257, 258. J. N. 611, 638, 639. A. Y. ii. 447, 449.

their goods to neighbouring places, and traded with the towns on the shores of the Âḳ Shahr and Beg Shahr lakes.

The town of Beg Shahr (or Bey Shahr, Karallia of the Byzantines) at the foot of its lake, according to the *Jahân Numâ*, had been founded by Sultan 'Alâ-ad-Dîn the Saljûḳ. It had a stone wall with two gates, a Friday Mosque, and fine baths; also a market at a place called Alarghah. To the west of Egridûr lies Burdûr, on the lake of the same name, a small town, according to Ibn Baṭûṭah, with many streams and gardens, protected by a castle on the neighbouring hill. Ispârtah, south of Egridûr, is given in the *Jahân Numâ* as the capital of Ḥamîd in later times. Ibn Baṭûṭah writes the name Sabartâ, and describes it as a well-built city of many gardens, protected by a castle. This represents the Byzantine town of Baris, and Sparta is the common pronunciation of the present day[1].

The lake of Âḳ Shahr is that which Ibn Khurdâdbih (see above, p. 135) calls Bâsiliyûn, and which the Byzantines knew as the Lake of the Forty Martyrs. To the west of it is the great castle of Ḳarâ Ḥiṣâr, which in connection with Âḳ Shahr is frequently mentioned in the campaigns of Tîmûr. At Âḳ Shahr, according to 'Alî of Yazd, the unfortunate Sultan of the 'Othmânlîs, Bayazîd Ilderim, whom Tîmûr had defeated at Angora, died broken-heated in 805 (1403), and both this Âḳ Shahr and this Ḳarâ Ḥiṣâr are mentioned by Mustawfî among the many celebrated places of those names. This Ḳarâ Ḥiṣâr, now surnamed Afyûn from the quantity of opium grown round it, marks the site of the Greek town of Prymnessos or Akroenos, and local tradition asserts that Al-Baṭṭâl, the champion of the earlier Omayyad wars against the Byzantines, was killed in battle near here. Ṭabarî, however, our earliest authority, only says that in the year 122 (740) 'Abd-Allah-al-Baṭṭâl was slain in the Greek country, and no indication of the place is given[2].

[1] Sabartâ or Ispârtah is the corruption of the Greek εἰς Βάριδα: *cf.* footnote, p. 157, on Izmîd and Iznîḳ (Nicomedia and Nicæa).

[2] Ibn Bibi, 5, 212, 251, 283. I. B. ii. 265, 266. Mst. 162, 163, 164. J. N. 618, 639, 640, 641. A. Y. ii. 457, 458, 489, 492. Ramsay, *H. G. A. M.* 87, 139, 396, 401, 406. Tabari, ii. 1716. The tomb of Al-Baṭṭâl is given in the *Jahân Numâ* (p. 642) as existing in the 11th (17th) century at Sîdî Ghâzî, more than fifty miles north of Ḳarâ Ḥiṣâr to the east of Kûtâhiyah. At the present day it is shown at Ḳîrshahr. In regard to Antioch of Pisidia there was

North and west of the Ḥamîd province was the country governed by the Amîr of Kermiyân, or Germiyân, whose capital was at Kûtâhiyah (Cotyæum). The Arab chroniclers wrote the name, as already mentioned, Ḳuṭiyah; but the Byzantine town must early have fallen to ruin, and according to the *Jahân Numâ* it was the Sultan of Germiyân to whom the later medieval town of Kûtâhiyah owed its foundation. Ibn Baṭûṭah refers to it as inhabited by robbers. At the close of the 8th (14th) century the place is frequently mentioned in the campaigns of Tîmûr, he for a time having made it his head-quarters. A hundred miles east of Kûtâhiyah, near the upper affluents of the Sangarius, stands the great fortress of Sîvrî Ḥiṣâr, where Tîmûr also for a time had his head-quarters. The name in Turkish means 'the Pointed Castle' (Ḳazwînî spells it Sîbrî Ḥiṣâr), and it stands above the site of the Roman Pessinus, which afterwards was renamed Justinianopolis Palia. Ḳazwînî reports that in the 7th (13th) century there was a famous church here called Bay'at Kamnânûs, and if animals suffering from stricture were seven times led round this church, the stricture would yield and they then recovered their health.

South of Sîvrî Ḥiṣâr lies 'Ammûriyah (Amorion, at the modern Assar Ḳal'ah), already spoken of (p. 137), which Mustawfî refers to as if in the 8th (14th) century it were still a place of importance. For some unexplained reason the common people, he adds, called it Angûriyah or Angûrah (Angora), and this strange misnomer is repeated in the *Jahân Numâ*, only that according to the latter authority it was Angûriyah, Angora, that was commonly called 'Ammûriyah. In the south-eastern part of Germiyân is Lâdhiḳ (Laodicea ad Lycum), which the Turks called Denizlû, 'Many Waters,' from its abundant streams; the place is now known as Eskî Ḥiṣâr (Old Fort). Ibn Baṭûṭah describes it

at all times a tendency in the earlier Arab chronicles to confound this with other places of the same name, and especially with Antioch of Syria. Ya'ḳûbî in his *History* (i. 177) refers to Anṭâkiyah-al-Muḥtariḳah, 'Burnt Antioch,' by which apparently the town of Pisidia is meant. The same author (ii. 285) speaks further of a raid made in the year 49 (669), and then mentions 'Black Antioch' (Anṭâkiyah-as-Sawdâ), by which name possibly Antioch of Isauria is intended.

as a great city, with seven mosques for the Friday prayers, and excellent markets. The Greek women of Lâdhiḳ wove cotton stuffs, which they afterwards embroidered finely with gold, and these embroideries were famous for their wear. In the *Jahân Numâ* the older form of the name is given as Lâdhiḳiyah[1].

In the province governed by the Menteshâ Amîr, Ibn Baṭûṭah visited the three neighbouring cities of Mughlah, Mîlâs, and Barjîn. The Amîr lived at Mughlah (the older Mobolla), the capital, according to the *Jahân Numâ*, which Ibn Baṭûṭah describes as a fine town. Mîlâs (Mylasa, or Melisos) was also a great city with gardens, much fruit, and plentiful streams. Barjîn (Bargylia, now known as Assarlik), a few miles from Mîlâs, was a newly built town, standing on a hill-top, with a fine mosque and good houses. In the eastern part of Menteshâ, Ibn Baṭûṭah visited Kul Ḥiṣâr, which under the name of Gul is described by Mustawfî as a medium-sized town, and it is also spoken of in the campaigns of Tîmûr. Ibn Baṭûṭah describes it as surrounded on all sides by the waters of the little lake on which it stood, this being almost entirely overgrown with reeds. A single road by a causeway led to the town across the lake, and the castle, which was very strong, crowned a hill rising immediately above the town. In the north of Menteshâ was the castle of Ḥiṣn Ṭawâs, at the present time called Daonas, a day and a half distant from Lâdhiḳ (Laodicea ad Lycum). Ibn Baṭûṭah describes Ṭawâs as a great fortress with a walled town below it. Tradition stated that Suhayb, a celebrated Companion of the prophet Muhammad, had been born here[2].

North of Menteshâ was the territory of the Amîr of Aydîn, of which Tîrah (Teira) was the capital. Ibn Baṭûṭah, who visited the Amîr of Aydîn here, says it was a fine city with many gardens and abundant streams. He also passed through Birkî (Pyrgion), one march north of Tîrah, of which he praises the magnificent trees. The city of Aydîn or Guzel Ḥiṣâr occupies the site of the Byzantine Tralleis, and was a town of secondary importance. Ephesus, on the coast, was well known to the earlier Arab

[1] Kaz. ii. 359. I. B. ii. 270, 271, 457. Mst. 162. A. Y. ii. 448, 449. J. N. 631, 632, 634, 643.

[2] I. B. ii. 269, 277, 278, 279, 280. Mst. 163. J. N. 638. A. Y. ii. 448.

geographers as Afasûs, or Abasûs, and was famous as the place where might be seen the Cave of the Seven Sleepers referred to in the Ḳurân (ch. XVIII, v. 8). In later times the town came to be known as Ayâsulûḳ (also written Ayâthulûkh or Ayâsalîgh), a corruption of the Greek *Agiou Theologou*, and so called from the great church to Saint John Theologos, built here by the Emperor Justinian. This church was visited by Ibn Baṭûṭah when he was here in 733 (1333). He describes it as constructed of great stones, each ten ells in length, carefully hewn. Another church had, on the Moslem conquest, become the Friday Mosque, and this was a most beautiful building, the walls being faced with divers coloured marbles, while the pavement was of white marble, and the roof, which was formed of eleven domes, was covered with lead. Ibn Baṭûṭah states that Ayâsulûḳ in his day had fifteen gates, a river (the Cayster) flowed past it to the sea, and the city was surrounded by jasmine gardens and vineyards.

The other great port of Aydîn was Smyrna, called by the Turks Azmîr or Yazmîr, which was taken by Tîmûr from the Knights Hospitallers in the beginning of the 9th (15th) century. Ibn Baṭûṭah, who was here in 733 (1333), describes it as then for the most part in ruin; there was a great castle on the hill hard by, and from this port, he adds, the Amîr of Aydîn was wont to send out ships to harass the Byzantines, and plunder the neighbouring Christian towns. Of these last was Fûjah (or Fûchah, Phocia) on the coast of the province of Ṣârûkhân, mentioned later on in the time of Tîmûr as a Moslem castle, but which Ibn Baṭûṭah writes of in his travels as then in the hands 'of the infidels,' namely the Genoese. The capital of Ṣârûkhân was Maghnîsiyah (or Maghnîsiyâ, Magnesia) which he speaks of as a fine city standing on the hill-side, surrounded by many gardens with abundant streams, and here the Amîr of Sârûkhân held his court. In the campaigns of Tîmûr the province round Maghnî Siyâh (as the name was then written) is called Saruhân-Ilî[1].

North of Ṣârûkhân was the territory of the Karâsî (or

[1] I. B. ii. 295, 307, 308, 309, 312. A. Y. ii. 466, 468, 470, 480. J. N. 634, 636, 637. Ramsay, *H. G. A. M.* 110, 228. Yak. i. 91; ii. 806. The legend of the Cave of the Seven Sleepers of Ephesus I have already discussed in *Palestine under the Moslems*, p. 274.

Ḳarah-Sî) Amîr, whose capitals were Bâlîkesrî and Barghamah (Pergamos). Ibn Baṭûṭah, who visited Pergamos in 733 (1333), describes it as a city for the most part in ruin, but defended by a huge castle perched on a hill-top near by. Bâlîkesrî, which he also visited, was a well-built and populous town with excellent markets. There was, however, no Friday Mosque here at this date, though the Sultan of Ḳarâsî, Dumûr (or Tîmûr) Khân, generally lived here, and his father had built Bâlîkesrî. At a later period the town is frequently mentioned in the campaigns of Tîmûr.

From Bâlîkesrî Ibn Baṭûṭah travelled on to Bruṣâ, at that time the capital of the 'Othmânlî state, which already had begun to overshadow and absorb all the other Turkoman Amirates. Bruṣâ or Brûsah (Prusa) was already a great city, with fine markets and broad streets. The town was surrounded by extensive gardens, and within the city was a great tank where the water was collected for distribution to all the houses. At Bruṣâ there was a hospital, with one ward for men and another for women, where the sick were attended to and supplied gratis with all necessities, and there was also a hot bath. The 'Othmânlî Sultan whom Ibn Baṭûṭah visited was Orkhân (grandfather of that Bayazîd Ilderim, already mentioned as defeated at the beginning of the following century by Tîmûr), and the chief monument of his capital was the tomb of Sultan 'Othmân, his father, who was buried in what had formerly been a church.

Mikhâlîj (Miletopolis, which the Byzantines called Michaelitze), lying about 50 miles west of Bruṣâ, is frequently mentioned in the campaigns of Tîmûr, and in the *Jahân Numâ*. The most important town of the Ottoman territory in 733 (1333), however, was Nicæa, which had been taken from the Byzantines by Sultan Orkhân. Nicæa, which the earlier Arab geographers called Nîḳiyah, the Turks knew as Yaznîḳ or Iznîḳ. Ibn Baṭûṭah describes the lake of Yaznîḳ as covered with reeds. At the eastern end of it the town stood, and was entered by a single causeway across the waters, so narrow that only one horseman at a time could approach. The town itself he describes as much in ruin, but its circuit enclosed many gardens; it was surrounded by four separate walls with a water ditch dug between every two, traversed by drawbridges. To the north of Nicæa lies Nicomedia, which the earlier Arab authorities knew as Niḳmûdiyah; the Turks called it Iznekmîd, as the *Jahân Numâ* writes the name, shortened later to Izmîd, which is that now in use. No description of this town is given by Ibn Baṭûṭah or our other authorities[1].

The province of Ḳizil Aḥmadlî lay along the coast of the Black Sea from the neighbourhood of the Bosporus to Sinope. Travelling from Yaznîḳ, after passing the river Sangarius, which the Turks called Saḳarî, the first large town which Ibn Baṭûṭah came to was Muṭurnî or Mûdurnî (modern Mudurlû, and the ancient Modrene) which he speaks of as a place of considerable size; it is also mentioned in the *Jahân Numâ*. The town of Bûlî (Claudiopolis), to the north-east of Muṭurnî, Ibn Baṭûṭah describes as standing on a river of some volume; and Kerehdeh (or Geredî) Bûlî, one march to the east of this, was a fine large city in a plain, with good markets and broad streets, each separate nation among its people having a distinct quarter. Geredî Bûlî in 733 (1333) was the residence of the Amîr, and appears to have been then the chief town of Ḳizil Aḥmadlî.

In the eastern part of the province stands Ḳaṣṭamûniyah (or Ḳaṣṭamûnî, for Castamon) which Mustawfî describes as a medium-sized town. Ibn Baṭûṭah speaks of it as one of the largest cities which he visited in Asia Minor, and provisions, he notes, were here both cheap and abundant. To the north-east of it lay the great port of Ṣanûb (or Sînûb, Sinope), where he took ship for the Crimea, and from his description we learn how Sinope was surrounded on three sides by the sea, the town being entered by a single gate to the east. It was a beautiful and populous harbour and strongly defended. A fine Friday Mosque was to be seen here, the dome supported on marble pillars; and a place of

[1] Iznekmîd is a corruption of the Byzantine *εἰς Νικομήδειαν*: Iznîḳ of *εἰς Νίκαιαν*. I. B. ii. 315, 316, 317, 322. A. Y. ii. 466. J. N. 631, 656, 661, 662. Ramsay, *H. G. A. M.* 179. The picture Ibn Baṭûṭah gives of Sultan Orkhân, the founder of the celebrated corps of the Janizaries, is very curious. Ibn Baṭûṭah states that this chief was already the most powerful of all the Turkoman Amîrs. He possessed a hundred castles, and never stayed a month in any one town, being always out campaigning and inspecting his frontiers.

popular veneration was the reputed tomb of Bilâl the Abyssinian, the Companion of the prophet Muḥammad, and his Muezzin who had been the first to call the Moslems to prayer.

The Byzantine city of Gangra Germanicopolis, which lies some 50 miles south of Ḳasṭamûnî, the Turks called Kânḳrî. In the earlier Arab chronicles the name is given as Khanjarah, and a great raid was made by the Moslems in the reign of the Omayyad Caliph Hishâm as far into the Greek lands as this town. Ḳazwînî, who spells the name Ghanjarah, says that it stood on a river called the Nahr Maḳlûb, 'the stream which was turned over,'—because unlike other rivers it ran from south to north. He adds that in 442 (1050) Ghanjarah was almost entirely destroyed by an earthquake. Finally, to complete the list of towns in the Ḳizil Aḥmadlî province, Kûch Ḥiṣâr, which is named in the *Jahân Numâ*, must be mentioned. It lies about midway between Ḳasṭamûnî and Kânḳrî, and possibly is the Kûsh Ḥiṣâr of Mustawfî already noticed (p. 149), and there identified with the city of the same name on the great Salt Lake[1].

In regard to the high roads traversing Asia Minor, except for the road from Tarṣus to Constantinople (given p. 134), and the road east from Sîvâs towards Tabrîz (given p. 147), no itineraries that are of any use are forthcoming. In the *Jahân Numâ*[2] a certain number of roads are mentioned that radiated from Sîvâs as a centre, and along these the names of various villages and post-stations are set down, many of which may still be found on the map. Unfortunately the distances are in most cases omitted, and hence the amount of information to be derived from these routes is not of much account.

[1] Mst. 163, 164. I. B. ii. 325, 332, 336, 338, 341, 348. J. N. 645, 646, 648, 649, 651, 652. Yak. ii. 475. Kaz. ii. 368. Tabari, ii. 1236.

[2] J. N. 627, 628.

CHAPTER XI.

ADHARBÂYJÂN.

The lake of Urmiyah. Tabrîz. Sarâv. Marâghah and its rivers. Pasawâ and Ushnuh. Urmiyah city and Salmâs; Khoi and Marand. Nakhchivân. Bridges over the Araxes. Mount Sablân. Ardabîl and Âhar. The Safîd Rûd and its affluents. Miyânij. Khalkhâl and Fîrûzâbâd. The Shâl river and Shâh Rûd district.

The mountainous province of Adharbâyjân—the name of which is pronounced Azarbîjân[1] in modern Persian—was of much less importance under the Caliphate than it became in the later middle-ages after the Mongol invasion. In the earlier period it lay off the line of traffic, which passed by the Khurâsân road through the Jibâl province (Media); and the remoteness of Adharbâyjân was also increased, according to Muḳaddasî, by the fact that over seventy languages or dialects were spoken among its mountains and high plains, while none of the cities were of any very considerable size.

In successive epochs different towns rose one after another to the position of the provincial capital. At first, with the earlier Abbasids, it was Ardabîl; then, under the later Caliphs, Tabrîz took the first position, but after the Mongol invasion for a time gave place to Marâghah. Tabrîz, however, soon regained its pre-eminence under the Îl-Khâns, but again under the first

[1] See Map III. p. 87. The older form of the name in Persian was Adharbâdhagân, a name which the Greeks corrupted to Atropatene. Muḳaddasî (p. 373) describes Adharbâyjân, Arrân and Armenia as forming part of a single great province, which he designates as the Iḳlîm-ar-Riḥâb, 'the region of the high plains'—in distinction to the mountains (Jibâl) of Media, and the lowlands (Aḳûr) of Mesopotamia.

Ṣafavid kings was eclipsed by Ardabîl. At a later date, in the 11th (17th) century, when Isfahân was made the capital of all Persia by Shâh 'Abbâs and Ardabîl fell to decay, Tabrîz was reinstated once more in the position of chief city of Adharbâyjân, and so remains to the present day, being now by far the most important town in the north-western part of Persia.

The most remarkable natural feature of the province is the Lake of Urmiyah, the largest permanent sheet of water in Persia, being over 80 miles long from north to south and a third of this across in its broadest part. It lies to the west of Tabrîz, and takes its name from the town of Urmiyah which lies on its western shore. Our authorities give the lake a variety of names. In the Zend Avesta it is called Chaechasta, and this, the old Persian form, is retained in Chîchast, the name by which the lake is referred to in the *Shâh Nâmah*, and which was still in use as late as the times of Mustawfî. Mas'ûdî and Ibn Ḥawḳal in the 4th (10th) century call it the Buḥayrah Kabûdhân, a name derived from the Armenian and meaning 'the Blue Lake' (*gaboid* being 'blue' in that language). Iṣṭakhrî calls it the lake of Urmiyah (being followed in this by Muḳaddasî), otherwise the Buḥayrah-ash-Shurât, 'the Lake of the Schismatics,' from the heterodox beliefs of the various peoples inhabiting its shores, and he describes its waters as very salt. It was, he adds, in those days covered with boats trafficking between Urmiyah and Marâghah, and on its shores were many most fertile districts.

In the middle of the lake was an island, called the Kabûdhân island by Ibn Serapion, with a small town, inhabited by boatmen. Its waters were full of fish according to Iṣṭakhrî (Ibn Ḥawḳal, on the contrary, says there were none), and there was a curious fish found here known as the Water-dog (*Kalb-al-Mâ*); in winter time storms raised great waves, and the navigation was very dangerous. By Abu-l-Fidâ the lake is referred to as the Buḥayrah Tilâ—but the latter name is of unknown signification; Ḳazwînî speaks of the salt and the Tûtiyâ (tutty of zinc) which were produced here and largely exported. Mustawfî who, as already said, more generally writes of it as the Chîchast lake, also calls it the Daryâ-i-Shûr, 'the Salt Lake,' or else refers to it as the lake of Ṭarûj or Ṭasûj, from the name of an important town on its northern shore.

He and Ḥâfiẓ Abrû both refer to the island (a peninsula, when the waters are low) of Shâhâ, where there was a great castle crowning a hill, the burial-place of Hûlâgû and other of the Mongol princes. The fortress of Shâhâ is mentioned in the 3rd (9th) century, for Ibn Mashkuwayh when relating the events of the Caliphate of Mutawakkil, grandson of Hârûn-ar-Rashîd, speaks of Shâhâ and Yakdur, two castles then held by rebel chieftains of these parts. In the 7th (13th) century Hûlâgû rebuilt the castle of Shâhâ—which Ḥâfiẓ Abrû calls the Ḳal'ah-i-Tilâ of the Urmiyah lake—and stored here all his treasures, the plunder of Baghdâd and the provinces of the Caliphate. This castle subsequently becoming his burial-place it was known in Persian as Gûr Ḳal'ah, 'the Castle of the Tomb,' and when Ḥâfiẓ Abrû wrote in the time of Tîmûr it was entirely uninhabited[1].

The city of Tabrîz lies some thirty miles east from the lake shore on a river which debouches near the Shâhâ island or peninsula. Tabrîz appears to have been a mere village till the 3rd (9th) century, when in the reign of Mutawakkil a certain Ibn-ar-Rawâd settled here, he and his brother and son building themselves palaces and afterwards enclosing with a wall the town which gathered round these. A late tradition indeed refers the foundation of Tabrîz to Zubaydah, the wife of Hârûn-ar-Rashîd, but the earlier chronicles give no support to this statement, moreover it is nowhere recorded that this princess ever visited Adharbâyjân. Muḳaddasî in the 4th (10th) century describes Tabrîz as a fine town, with a Friday Mosque, well watered by numerous streams, and surrounded by fruitful orchards. Yâḳût who was here in 610 (1213) speaks of it as at that time the chief town of Adharbâyjân, Ḳazwînî adding that it was famous for its 'Attâbî (or tabby) silks, its velvets and woven stuffs. The Mongols

[1] The name Urmiyah is now commonly pronounced Urûmiyah, and this is the spelling given by Ibn Serapion, MS. f. 25 *a*. Ist. 181, 189. I. H. 239, 247. Muk. 375, 380. Mas. i. 97. A. F. 42. Yak. i. 513. Kaz. ii. 194. Mst. 226. Hfz. 27 *a*. Ibn Mashkuwayh, 539. In the *Shâh Nâmah* (Turner Macan, Calcutta, 1829), p. 1860, line 4, and p. 1927, line 6 from below, for *Khanjast* (a clerical error), 'Chîchast' is to be read, the two names only differing by a shifting of the diacritical points.

who captured Tabrîz in 618 (1221) were promptly bought off, and the city thus escaped the usual sack; and, as already said, under the subsequent Îl-Khân dynasty it became the largest town of these parts.

Mustawfî gives a long account of Tabrîz. Twice, he says, it had been destroyed by earthquakes and rebuilt, namely in 244 (858), and in 434 (1043) when 40,000 of its inhabitants perished. After being finally restored it was surrounded by a wall 6000 paces in circuit, with ten gates, and continued thus till the 8th (14th) century, when Ghâzân Khân began to build great suburbs beyond the older wall, surrounding these in turn by a new wall. This, which was pierced by six gates, included the hill of Valiyân in its circuit, and measured 25,000 paces round. Mustawfî gives the names of the inner and outer gates of Tabrîz (the MSS. vary considerably in these), and he states that Ghâzân Khân was buried in 703 (1303) in the great suburb of Shâm, which he had laid out. His successors added many fine mosques and erected public buildings within the city and in the suburb of Rashîdî, which occupied the slopes of the hill of Valiyân. The orchards of Tabrîz were watered by the river Mihrân Rûd, which rose in Mount Sahand lying to the south of the city. Round Tabrîz lay seven districts, called for the most part after their respective streams. These names, with the villages adjacent, are given in detail by Mustawfî, but the readings of the many proper names are very uncertain. Ibn Baṭûṭah, who visited Tabrîz in the year 730 (1330), speaks of the Shâm quarter lying outside the town, with its fine college built by Ghâzân Khân and the oratory. He entered the city by the Baghdâd gate, and notes the market of Ghâzân, and the jewellers' market where an abundance of precious stones was offered for sale. Near by was the musk and ambergris market. The Friday Mosque, he says, had been built by the Wazîr 'Alî Shâh of Gîlân; its court was paved with marble, and to the tank a channel brought water. The walls were faced with enamelled tile-work (Kâshânî-ware), and to right and left of the mosque stood, on the one side an oratory, and on the other a college[1].

The two rivers, called respectively the Mihrân Rûd, which ran

[1] Muk. 378. Yak. i. 822. Kaz. ii. 227. Mst. 153—155. J. N. 380. I. B. ii. 129.

through the suburbs of Tabrîz, and the Sard Rûd (the Cold River), flowing to the south-west, which like the first named took its rise in Mount Sahand to the south of Tabrîz, both joined the Sarâv river at a short distance to the north of the city. The Sarâv Rûd, which was also called the Sarkhâb river, rose in the mountains of Sablân Kûh, which lay 200 miles to the eastward of Tabrîz, overhanging Ardabîl. After a long and winding course, passing through successive salt marshes and receiving many affluents, the Sarâv river flowed out into the Urmiyah lake at a point about 40 miles to the westward of the city of Tabrîz. The two mountains of Sahand and Sablân, and the rivers that flowed down from them, are described in much detail by Mustawfî. The town of Sarâv or Sarâb, which gave its name to the river, lies on the road from Tabrîz to Ardabîl, and according to Mustawfî was surrounded by the four districts of Warzand, Darand, Barâghûsh, and Saḳhîr. The earlier Arab geographers spell the name of the town Sarât (for Sarâb), and Ibn Ḥawḳal describes it as a fine place with many mills, surrounded by fields and orchards where much corn and fruit was grown. In Sarât were found numerous hostelries and excellent markets. Yâḳût, who spells the name Sarâv or Sarv, speaks of it as having been ruined by the Mongol invasion of the year 617 (1220), when most of its inhabitants were slaughtered. It had however recovered when Mustawfî wrote a century later; he adds that it lay three days' march from Tabrîz and two from Ardabîl.

On a left (south) bank affluent of the Sarâv river stood the town of Awjân or Ûjân, which was ten leagues from Tabrîz on the road to Miyânah. Yâḳût who had been here in the 7th (13th) century describes Ûjân as a walled town with an excellent market. It had, however, been ruined by the Mongols, and in Mustawfî's day was rebuilt by Ghâzân Khân, who at one time had resided here. He renamed it Shahr-i-Islâm, 'the City of Islam,' and enclosed it with a wall 3000 paces in circuit built of mortared stones. The surrounding districts were very fertile, growing cotton, corn, and much fruit. Its river, called the Âb-i-Ûjân, rose in an eastern spur of Mount Sahand. To the south-west of this mountain, and about 60 miles from Tabrîz, being four leagues from the shore of the lake, was the great village of Dâkharraḳân, as Ibn Ḥawḳal and the Arab geographers spell the name, which the Persians write Dih Khuwârḳân. Yâḳût gives Dih Nakhîrjân as an alternative reading, explaining this as meaning the village (*Dih*) of Nakhîrjân, treasurer of Chosroes, king of Persia. Mustawfî describes it as a small town, surrounded by dependencies and eight villages, where much fruit and corn was grown[1].

The city of Marâghah stood about 70 miles south of Tabrîz, on the river Ṣâfî, which flowed south down to it from Mount Sahand, and then turned west to reach the lake. Marâghah, an abbreviation for Ḳariyat-al-Marâghah, 'the Village of the Pastures,' is said to have been called Afrâzah Rûdh by the Persians. In the 4th (10th) century Marâghah is described by Ibn Ḥawḳal as a town of the size of Ardabîl, at that time the chief city of Ādharbâyjân; he adds further that Marâghah had already even then been for a time the provincial capital, where the government treasury and offices were stationed, before they were permanently transferred to Ardabîl. Marâghah was a most pleasant town, surrounded by a wall beyond which lay fruitful orchards. It was famous for a particular kind of perfumed melon grown here, green outside and red within, which tasted of honey. Muḳaddasî speaks of its castle and fortifications, with a great suburb lying outside these. Yâḳût records that its fortifications were built under Hârûn-ar-Rashîd and restored by the Caliph Mamûn.

Under the earlier Mongols, as we have already seen, Marâghah became the capital of Adharbâyjân, and Mustawfî describes it as a great city surrounded by numerous and fertile districts, some of which he names, amply watered by many streams. Outside Marâghah stood the great observatory built by the astronomer Nâṣir-ad-Dîn of Ṭûs, where by order of Hûlâgû the celebrated Îl-Khânî tables had been calculated and published. The observatory, of which the ruins still exist, was however already dilapidated when Mustawfî wrote in the 8th (14th) century. Ḳazwînî mentions the castle, called Ruwîn Diz, which lay three leagues distant from Marâghah, having a stream flowing on either

[1] Ist. 190. I. H. 248, 253. Yak. i. 131, 198; ii. 425, 636; iii. 64. Mst. 155, 158, 204, 205, 217, 218.

side of it, and within the castle a famous garden called Umîdâbâd with its own cistern to irrigate it. A league from here stood the village of Janbadhaķ, with a hot spring, of which many wonders were related.

The Ṣâfî river, which flowed into the lake to the west of Marâghah, mingled its waters in flood-time with those of the Jaghtû river and its affluent the Taghtû, both of which as described by Mustawfî rose in the Kurdistân mountains; and the whole of the southern shore of the lake at their outflow was a great swamp. Here surrounded by tortuous streams stood the small town of Laylân (or Naylân), among fruitful orchards, and inhabited in the time of Mustawfî by Mongols. Some way to the south of Laylân, according to the distances given in the Itineraries, was the village of Barzah, where the road coming up from Sîsâr (in the Jibâl province) bifurcated. To the right one way went on north-east to Marâghah; while to the left, and by the west of the lake, lay the way to Urmiyah.

Fifty miles from the southern shore of the lake was Baswâ, by the Persians pronounced Pasawâ, which Yâķût had visited, and he states that in his day the inhabitants were mostly robbers. Mustawfî praises its fruitful orchards, and to the north-west of it lay the town of Ushnuh, which in the time of Ibn Ḥawķal was inhabited by Kurds. In the 4th (10th) century Ushnuh did a great trade in horses and cattle with the neighbouring towns of Mesopotamia, especially Mosul: its lands were very fertile and its sheep pastures were famous. Yâķût, who had visited it, speaks of its fine gardens, and Mustawfî, who spells the name Ushnûyah, describes it as a medium-sized town of the mountain region which he calls Dih Kiyâhân[1].

The city of Urmiyah, which gave its name to the lake, lay at a short distance from its western shore. Tradition proclaimed Urmiyah to have been the birth-place of Zardûsht or Zoroaster. The town, according to Ibn Ḥawķal in the 4th (10th) century, was of the same size as Marâghah, being a pleasant place and surrounded by vineyards; its markets were well supplied with merchandise, among the rest being the clothiers' market,

[1] Ist. 181. I. H. 238, 239. Muk. 377. Yak. i. 284, 564, 626; iv. 476. Kaz. ii. 350, 358. Mst. 158, 159, 218.

where stood the Friday Mosque. Urmiyah was fortified and defended by a castle, and a stream flowed through it down to the lake, which was about a league distant. In the 8th (14th) century it had grown to be a large place, its wall measuring 10,000 paces in circuit, and a score of villages were of its dependencies. On the high road north of Urmiyah, and at some distance back from the north-western corner of the lake, is Salmâs. Muḳaddasî describes this as a fine town with good markets and a Friday Mosque built of stone; the population of the place in the 4th (10th) century was of Kurd origin. Yâḳût says that in the 7th (13th) century Salmâs lay for the most part in ruin; but the Wazîr 'Alî Shâh, Mustawfî writes, rebuilt its walls 8000 paces in circuit during the following century, in the reign of Ghâzân Khân, the Mongol, and the town had then regained its former importance. Its climate was cold, and a river which rose in the mountains to the west passed through it to the lake.

On the northern shore of the lake was the town called Ṭarûj or Ṭasûj, which is apparently identical with the modern Tursah. Mustawfî, as already said, often speaks of the Salt Lake of Ṭasûj or Ṭarûj, and the town therefore shared with Urmiyah the honour of giving its name to this sheet of water. In the 8th (14th) century Ṭasûj must have been an important place, it was warmer than Tabrîz and damper, being so near the lake, and it was surrounded by gardens and orchards. To the north-east of Salmâs lies Khawî, pronounced Khoi, on a stream that flows north to the river Aras (Araxes). Khawî was a strongly fortified and flourishing town according to Yâḳût and Ḳazwînî, surrounded by fertile lands and famous for its excellent brocades. There was also a spring here which had the reputation of being hot in winter and cold in summer. Mustawfî says that the *enceinte* of its town walls measured 6500 paces, and that its people were a white-skinned race like the Khatâi (Chinese); eighty villages were of its dependencies.

The town of Marand which lay to the east of Khoi, on the banks of a stream which was a right bank affluent of the Khoi river, is described by Muḳaddasî in the 4th (10th) century as a small fortress with a mosque, and a market in the suburb, which was surrounded by gardens. Yâḳût says that it was ruined by the Kurds who had carried off most of its inhabitants after plundering the town. Its river according to Mustawfî was called the Zûlû (or Zakvîr), and a part of it was said to flow for four leagues underground. Mustawfî states that in his day Marand was only half its former size, but was still famous for the rearing of the Ḳirmiz-worm (cochineal), used for making the red dye, and that round the town were 60 villages that were of its dependencies[1].

Nakhchivân, or Naḳjawân, to the north of the Aras river, was generally counted as of Adharbâyjân. It is identical with Nashawâ of the Arab geographers, and is often mentioned in the Itineraries, but no description of the town is given. Nakhchivân rose to importance under the Mongols, and Mustawfî describes it as a large town built of brick. Near it, to the eastward, was the fortress of Alanjik, and to the north rose the snow-clad mountain called Mâst Kûh. In Nakhchivân stood the dome built by Ḍiyâ-al-Mulk, son of Niẓâm-al-Mulk, the great Wazîr of Malik Shâh the Saljûḳ, and 'Alî of Yazd describes the famous bridge of Ḍiyâ-al-Mulk (the ruins of which still exist) which crossed the Aras at the fortress of Karkar on the road to Marand, about 15 miles from Nakhchivân.

A little lower down on the Aras is Julfah, otherwise written Jûlâhah, which was destroyed by Shâh 'Abbâs of Persia in 1014 (1605), when he transported all its Armenian inhabitants to the new suburb which he built to the south of Isfahân and named Julfah from the older Julfah on the Araxes. Among other towns on the banks of the Aras river Mustawfî mentions Urdûbâd (which still exists), near where a river joins the Aras from the south, on whose banks stood the castle of Dizmâr, which is also mentioned by Yâḳût. Still lower down the Aras lay the town of Zangiyân in the Murdân Na'îm district, where a second bridge, still in existence, crosses the Araxes. This is called the Pûl-i-Khudâ-Âfarîn in Persian, 'the Bridge of Praising God,' which Mustawfî says had been built by one of the Companions of the prophet Muḥammad in the year 15 (636). The Murdân (or Murâd) Na'îm territory comprised in its circuit over 30 villages[2].

[1] Ist. 181. I. H. 239. Muk. 377. Kaz. i. 180; ii. 354. Yak. i. 218; ii. 502; iii. 120; iv. 503. Mst. 156—159, 218.

[2] Yak. iv. 262, 767, 784. Mst. 157, 159, 206. A. Y. i. 398, 399; ii. 573.

The city of Ardabîl stood on the upper waters of the river called Andarâb by Mustawfî, and the Ardabîl river, after being joined lower down on its left bank by the Âhar river, flowed into the Araxes some way below the bridge of Khudâ-Âfarîn. The rivers of Ardabîl and Âhar rose on the eastern and western slopes, respectively, of the great mountain called Sablân Kûh, which overhangs Ardabîl, and from whose southern slopes the Sarâv river, as already mentioned, takes its course westward to the Urmiyah lake. Mount Sablân is mentioned in the 4th (10th) century by Ibn Ḥawḳal, who erroneously considered it as higher than Damâvand, some miles to the north of Tihrân. Its slopes were covered with trees, and here stood villages and many towns, which are enumerated by Mustawfî. The mountain, he adds, was visible 50 leagues away, its summit being always covered with snow, while near the top was a spring the surface of which remained always frozen. Near Mount Sablân also were two other peaks, Kûh Sarâhand north of Âhar, and Siyâh Kûh (the Black Mountain), which last towered above Kalantar, a small town with a castle which stood among woods, with a river flowing through its many cornfields.

Ardabîl, as already said, was the capital city of Adharbâyjân in the 4th (10th) century. It is described by Iṣṭakhrî as walled, and measuring two-thirds of a league across every way; the houses were of burnt brick and clay, and at that time troops were kept here in garrison. Its dependencies were extremely fertile, and the Ardabîl honey was famous. Muḳaddasî speaks of the fortress, and the markets of Ardabîl were in four cross-streets, with the Friday Mosque standing at the intersection point. Outside the town was an extensive suburb. In 617 (1220) Ardabîl was sacked by the Mongols and left a ruin; but just before this, when Yâḳût was here, it was a most populous city. Ardabîl had been known anciently by the Persian name of Bâdhân Fîrûz. When Mustawfî wrote in the 8th (14th) century, though no longer the chief town of Adharbâyjân, it had recovered much of its former splendour; and in the 10th (16th) century, as already stated, it became for a time the capital of the whole of Persia under the newly founded dynasty of the Ṣafavids, before they removed, first to Tabrîz and afterwards to Isfahân.

Âhar which lies 150 miles west of Ardabîl, on the Âhar river, is named in the lists of the earlier Arab geographers, and described by Yâḳût as a well-built city, to the north of which lay Mount Sarâhand. It was surrounded by many small towns standing on the hill-slopes, the names of which are recorded by both Yâḳût and Mustawfî, but these are difficult now to recognise or identify. The surrounding district was known as Pîshkîn (at the present day Mîshkîn), from the name of the ruling family who flourished here in the 8th (14th) century. The town of Pîshkîn lay one march from Âhar, and originally had been known as Varavî. The river Andarâb, just above where the Âhar river joined it, Mustawfî says, was crossed by a fine bridge that had been built by 'Alî Shâh, the Wazîr of Ghâzân Khân the Mongol[1].

The Safîd Rûd, or White River, with its many affluents, drained all the south-eastern part of Adharbâyjân. Its main stream for most of its length formed the frontier dividing Adharbâyjân from the Jibâl province, and the river finally flowed out to the Caspian Sea through the province of Gîlân. Iṣṭakhrî and other Arab writers give the name as the Sabîd-rûdh. Mustawfî says that in his time it was known to the Mongols as the Hûlân Mûlân (more exactly Ulân Mören), which in Mongolian means 'Red River'; and at the present day part of the Safîd Rûd is known as Ḳizil Uzen, which in Turkish also signifies 'Red Stream.' Mustawfî writes that the Safîd Rûd rose in the highlands of Kurdistân, in a mountain called Panj Angûsht (in Persian) or Besh Parmaḳ (in Turkish), and both names mean 'the Five Fingers.' Flowing north the Safîd Rûd first received the Zanjân river on its right bank, coming from the city of that name, which will be described in a later chapter; then on its left bank there flowed in the Miyânij river, formed by the confluence of many streams coming down from the west. North of Miyânij the Safîd Rûd turned west, receiving on its left bank the united streams of the Sanjîdah and Gadîv rivers coming down from Khalkhâl to the south of Ardabîl, and next the Shâl river from the Shâh Rûd district of Khalkhâl. Below this on its right bank, and coming from the Jibâl province (as will be described in

[1] Ist. 181. I. H. 237, 238, 240, 266. Muk. 374, 377. Yak. i. 197, 367, 409, 461; iv. 918. Mst. 156, 157, 204, 205, 217.

Chapter XV), the Ṭârum river joins the Safîd Rûd, and next the river Shâh Rûd (not to be confused with the district of Shâh Rûd just named) coming from the country of the Assassins, and then finally, after piercing the mountain barrier, the Safîd Rûd reaches the Caspian Sea at Kawtam in the province of Gîlân.

The Miyânij river, as already said, was the most important left bank affluent of the Safîd Rûd. It came from the west, rising in the country south of Ûjân (see p. 163), and in the Garm Rûd district received on its left bank the waters of the Garm Rûd (Hot River), a stream which rose in the hills to the south of Sarâv. Below the town of Miyânij the main stream receives on its right bank the waters of the Hasht Rûd, 'the Eight Streams,' which have their sources in the hills to the east of Marâghah; and, in the time of Mustawfî, where the Hasht Rûd joined the Miyânij river, there spanned it a great masonry bridge of thirty-two arches.

The town of Miyânij or Miyânah, 'the Middle Place,' which stands at the junction of all these streams, was an important centre from the earliest times. Ibn Hawkal writes of it as very populous in the 4th (10th) century, and its district—in later times known under the name of the Garm Rûd—produced great quantities of fruit. Muḳaddasî, who gives the modern form of the name Miyânah, praises its store of goods, and Yâḳût, who had visited it in the 6th (12th) century, extols its situation. In the following century, when Mustawfî wrote, it had sunk to the size of a large village, but was still an important stage on the road system inaugurated by the Mongols. The climate was hot, and insect pests were numerous (the Miyânah bug at the present day is a terror to travellers), but the Garm Rûd district comprised over a hundred fertile villages, and much corn was grown.

The three rivers called Sanjîdah, Gadîv (or Kadpû, in the *Jahân Numâ*), and Shâl, joined the Safîd Rûd from the north, coming down from the Khalkhâl district. Khalkhâl was also the name of the chief town of this district, the position of which is given in the Itinerary as 12 leagues south of Ardabîl. Fîrûzâbâd, situated at the summit of the pass, where there was a boiling spring bubbling up in the midst of the snow-clad peaks, according to Mustawfî had in former times been the residence of the governor, but when it fell into ruin Khalkhâl city took its place. The exact position of Fîrûzâbâd, however, cannot now be fixed. The small towns of Kalûr and Shâl, which are still to be found on the map, were of the Shâh Rûd district, and lay on the Shâl river (now called the Lesser Shâh Rûd) which rose in the Shâl hills. Mustawfî mentions a number of other places in Khalkhâl, the names of which, however, cannot now be identified[1].

The few products of Adharbâyjân will be described at the end of the next chapter; and the summary of the high roads through this province must be deferred to the conclusion of Chapter XV, after describing the Jibâl province, for these roads all start from various points on the great Khurâsân road which traverses the latter province.

[1] Ist. 189. I. H. 246, 253. Muk. 378. Yak. i. 239; iv. 710. Mst. 156, 158, 198, 215, 218. J. N. 384, 388.

CHAPTER XII.

GÎLÂN AND THE NORTH-WEST PROVINCES.

The Gîlâns. Daylam and the Ṭâlish districts. Barvân, Dûlâb, and Khashm. Lâhîjân, Rasht, and other towns of Gîlân. The district of Mûghân. Bajarvân and Barzand. Maḥmûdâbâd. Warthân. The province of Arrân. Bardhâ'ah. Baylaḳân. Ganjah and Shamkûr. The rivers Kur and Aras. The province of Shirvân. Shamâkhî. Bâkûyah and Bâb-al-Abwâb. The province of Gurjistân, or Georgia. Tiflîs and Ḳarṣ. The province of Armenia. Dabîl or Duwîn. The lake of Vân. Akhlâṭ, Arjîsh, Vân, and Bitlîs. Products of the northern provinces.

The Safîd Rûd, as described in the last chapter, after traversing the chain of the Alburz mountains by a tortuous course, flows into the Caspian Sea at the western end of its southern shore, and here forms a delta with marshlands of some breadth backed by the mountain chain. This delta of the Safîd Rûd, with the great amphitheatre of forest-clad foot-hills surrounding it on the south and west, is the small province of Gîlân, which the Arabs called Jîl or Jîlân, and which comprised three very different districts[1].

The alluvial delta lands are those more especially called Jîl or Jîlân by the Arab geographers, who when referring to the whole province often give the name in the plural form, Jîlânât, 'the Gîlâns,' which may then be taken to include the mountain districts. To the south and west, the mountain range bordering on the districts of Ṭâliḳân and Ṭârum in the Jibâl province, was the Daylam country, generally also given in the plural form as Ad-Daylamân; and this country became famous in history as the original home of the Buyids, or Daylamites, whose chiefs were masters of Baghdâd, and of the Caliphate for the most part,

[1] For Gîlân see Map v, at the beginning of the following chapter.

during the 4th (10th) century. The narrow strip of shore and mountain slope, running north from the south-west corner of the Caspian, and facing east over that sea, is the Ṭâlish country, a name which Yâḳût gives under the plural form Ṭâlishân or Ṭîlshân. To the east, on the Ṭabaristân frontier, was the mountain range of Ar-Rûbanj, beyond which came the hill district belonging to the great Ḳârin family, whose chiefs had from time immemorial been rulers of these fastnesses, as will be further mentioned in Chapter XXVI.

When Muḳaddasî wrote in the 4th (10th) century, and the Buyid supremacy was at its height, all Gîlân, together with the mountain provinces to the eastward and along the shore of the Caspian, namely, Ṭabaristân, Jurjân, and Ḳûmis, were included in the province of Daylam, but in later times these eastern provinces came to be counted as separate. Afterwards the name of Daylam itself for the most part fell out of use, and the lowlands of the Safîd Rûd delta gave their name to the whole of the adjacent district, which was commonly known as the Jîlân province. More exactly, however, Jîlân was the coast district, while Daylam was the mountain region overhanging it, and at different times either of these names in turn might be taken commonly to include the whole province lying round the south-western corner of the Caspian Sea[1].

The chief city of Daylam is said to have been called Rudhbâr, but its situation is unknown. Muḳaddasî on the other hand says the capital was known as Barvân, but unfortunately it no longer exists and none of the Itineraries give its exact position. Barvân, Muḳaddasî adds, had neither good houses nor good markets, and possessed no Friday Mosque. Where the governor resided was called the Shahrastân, and the merchants living here were wealthy, so that it was a flourishing town. Of Jîlân, Muḳaddasî gives Dûlâb as the chief town, which he describes as a fine place, its houses being well built of stone; the market was excellent, and a Friday Mosque stood in it. According to Abu-l-Fidâ Dûlâb is

[1] Ist. 204, 205, 206. I. H. 267, 268. Muk. 353. Yak. i. 174, 812; ii. 179, 711; iii. 571. Mst. 147, 191. A. F. 426. The name of Tâlish is written with either the soft *t*, or the hard Arabic *ṭ*; and in the plural as Tâlishân or Ṭîlshân, also Ṭawâlish in Mustawfî.

the same as Kaskar, and in the only Itinerary of this country that has come down to us, Muḳaddasî gives Dûlâb as lying four marches from Baylamân, a small town like a farmstead according to Abu-l-Fidâ, which appears to have been one of the chief places in the Ṭâlish country. Two marches from the Safîd Rûd, and four from Baylamân, was the town of Khashm, the residence of the Alid chief (the Dâʿî or Missioner), who in the latter half of the 3rd (9th) century ruled these provinces as an independent (heretical) sovereign, who did not acknowledge the Caliph. Muḳaddasî describes Khashm as having a fine market and a Friday Mosque near the chief's palace. A river ran through the town, which was crossed by a remarkable bridge of boats. The identification and situation of all these early towns is exceedingly uncertain[1].

In the 8th (14th) century the chief towns of Gîlân, according to Mustawfî, were Lâhîjân and Fûmin. Abu-l-Fidâ also mentions Lâhîjân, which lies to the eastward of the mouth of the Safîd Rûd. It was then a fair-sized town; much silk was manufactured here and the district grew rice and corn, also oranges and shaddocks with other fruits of a hot region. Kawtam or Kûtam, nearer the mouth of the Safîd Rûd, was the harbour for ships coming from other parts of the Caspian; it is mentioned by Yâḳût and Abu-l-Fidâ, having been a place of much commerce in the 8th (14th) century, and the town lay one day's march from the actual shore of the Caspian. Fûmin with its district lies further inland, and to the west of the Safîd Rûd. It is counted as the chief town of the mountain region of Daylam, and Mustawfî writes of it as a large place standing in a fertile district growing much corn and rice. Silk was also produced and manufactured here.

Mustawfî is one of the earliest authorities to describe Rasht, now the capital of Gîlân, but none of the Arab geographers appear even to name it. He notices its warm damp climate, cotton and silk being both largely produced for export, and the place was already in his time of some size and importance. To the westward of Rasht extends, at the present day, the district of

[1] Ist. 204, 205. Muk. 355, 360, 373. A. F. 429 (where, in error, Baylamân is printed *Bîmân*). Yak. ii. 831. For the Dâʿî dynasty of Alids (Ḥasanids), see G. Melgunof, *Das südliche Ufer des Caspischen Meeres*, p. 53.

Tûlim, and Mustawfî gives this as the name of an important town in the 8th (13th) century. According to Abu-l-Fidâ it was the chief city of the Jîlân or lowlands; its districts were very fertile, corn, cotton, rice, oranges, shaddocks and lemons being grown for export. Shaft, or Shaftah, is the name of a town mentioned in similar terms by Mustawfî, though at present only the Shaft district exists, which lies to the southward of Rasht. Finally, as of Gîlân, Mustawfî mentions the little town of Iṣfahbad, which Yâḳût spells Isfahbudhân, adding that it stood two miles distant from the coast of the Caspian, but not otherwise indicating its position; corn, rice, and a little fruit were grown here, and in the neighbouring district were near a hundred villages. The name of the township came from the Iṣfahbads or Ispahbids, who had been the semi-independent kings of this country under the Sassanians, and who, nominally converted to Islam, continued to rule as princes in Ṭabaristân under the earlier Caliphs[1].

Mûghân.

Mûghân, Mughkân, or Mûḳân[2] is the name of the great swampy plain which stretches from the base of Mount Sablân to the east coast of the Caspian Sea, lying south of the mouth of the river Aras, and north of the mountains of Ṭâlish. It was sometimes counted as part of the Adharbâyjân province, but more often formed a separate district.

The capital of Mûghân in the 4th (10th) century was a city of the same name, the position of which it is difficult to fix. Muḳaddasî speaks of Mûḳân city as lying on two rivers, with gardens all round, and as almost of the size of Tabrîz. From his description it is not improbable that this Mûḳân city was identical with Bajarvân, which Mustawfî names as the older capital of the district, and which in his day had already gone to ruin. The position of Bajarvân he gives in his Itineraries as four leagues north of Barzand, a name which is still found on the map. Further, Moslem tradition connected Bajarvân with the Fountain of Life, said to have been discovered near here by the prophet Khiḍr,

[1] Yak. i. 298; iv. 316. A. F. 426, 429. Mst. 191, 192. J. N. 343, 344.

[2] For Mûghân and the north-west frontier provinces see Map III, p. 87.

otherwise Elias. As already stated, to the south of Bajarvân lay Barzand, which is described as a great city by Ibn Hawḳal, and Muḳaddasî praises its markets, where goods from all the surrounding regions were collected for exportation, for this was the commercial centre of the district. Mustawfî mentions both Bajarvân and Barzand as sunk to be mere villages in his time; the climate in the surrounding districts was hot, and much corn was grown[1].

In the Mûḳân plain Mustawfî names the three towns of Pîlsuvâr, Maḥmûdâbâd, and Hamshahrah. Pîlsuvâr, which stood on a stream coming down from Bajarvân, lay at a distance of eight leagues from the latter place, and it is said to have been so called after the Amîr Pîl-Suwâr sent here by the Buyids, whose name signified 'great rider or soldier.' Maḥmûdâbâd in the plain of Gâvbârî, near the Caspian, was twelve leagues beyond Pîlsuvâr, and Mustawfî adds that it had been built by Ghâzân Khân the Mongol. The neighbouring Hamshahrah was two leagues from the coast, and originally had been known as Abrashahr, or Bûshahrah, having been founded, says Mustawfî, by Farhad, son of Gûdarz, 'whom they identify with Nebuchadnezzar.' To the north of Bajarvân, in earlier times, was Balkhâb, described as a populous village with guard-houses and hostelries for travellers; and beyond this stage on the northern high road, and upon the south bank of the Aras, was Warthân, at the crossing into the Arrân country. In the 4th (10th) century Warthân was a walled city with markets and much merchandise, having a suburb without its gates. The place was very populous, standing in a plain two leagues from the river bank, and its Friday Mosque was in the suburb; further, tradition averred that Warthân had been built by order of Zubaydah, wife of Hârûn-ar-Rashîd[2].

Arrân.

The provinces of Arrân, Shirvân, Georgia and Armenia, which for the most part lay north of the river Araxes, were hardly counted among the lands of Islam, and hence are but perfunctorily described by the Arab geographers. From early days Moslems

[1] I. H. 251. Muk. 376, 378. Yak. i. 454, 562; iv. 686. Mst. 159, 160, 198. J. N. 392.

[2] I. H. 251. Muk. 376. Yak. iv. 919. Mst. 160, 198. J. N. 393.

lived here, and governors were appointed at various times by the Caliphs, but the majority of the population continued to be Christian until near the close of the middle-ages. Hence it was not till the resettlement subsequent to the Mongol invasion, and more especially after the many campaigns which Tîmûr waged in Georgia at the close of the 8th (14th) century, when these lands came to be permanently settled by the Turks, that Islam became the dominant faith.

The province of Arrân is included in the great triangle of land lying to the west of the junction point of the rivers Cyrus and Araxes—the Kur and the Aras of the Arabs—and it is thus 'between the two rivers' (Bayn-an-Nahrayn) as Mustawfî calls it. The earlier Arab geographers write the name Al-Rân (pronounced *Ar-Rân*) to give it the appearance of an Arabic word, and the capital town in the 4th (10th) century was Bardhâ'ah, the ruins of which still exist. Bardhâ'ah, later written Bardâ', Ibn Ḥawḳal describes in the 4th (10th) century as measuring a league across, and it was by far the largest city of these parts. It was built in the form of a square, was protected by a fortress, and stood about three leagues from the Kur river, on the bank of its affluent the Tharthûr. Near by the town, in the Kur, was caught the fish called Sarmâhî (otherwise Shûr-mâhî in Persian, *salt-fish*), which after being salted was exported to all neighbouring towns. This fish was also found in the Aras river near Warthân. The fertile district round Bardhâ'ah was known by the name of Al-Andarâb, where villages with continuous gardens and orchards, a day's journey across in every direction, produced abundant fruits, especially chestnuts, filberts, and figs. In these parts also the silkworm was reared.

A great market was held every Sunday outside Bardhâ'ah at the Bâb-al-Akrâd, 'the gate of the Kurds'; and the market-place stretched a league in length. It was called locally Al-Kurkî (from the Greek *Kuriakos*, 'the Lord's day'), and Sunday, we are told, was here commonly known as the day of the Kurkî. Bardhâ'ah further had a fine Friday Mosque, the roof of which was supported on wooden pillars, its walls being of burnt brick covered with stucco. Also there were many ḥammâms, or hot-baths; and in Omayyad times the Treasury of the province was kept at

Bardhâ'ah. In the 7th (13th) century, when Yâḳût wrote, Bardhâ'ah had already fallen to ruin, though Mustawfî in the following century still refers to it as a considerable town on the river Tharthûr. At the crossing of the Kur, probably below the junction of the Tharthûr, and 18 leagues, counted as a day's march, on the direct road from Bardhâ'ah to Shâmâkhî in Shirvân, was the town of Barzanj, much frequented by merchants, where goods were stored for import and export[1].

The city of Baylaḳân, known in Armenian as Phaidagaran, became the capital of Arrân after the decay of Bardhâ'ah. Though all traces of the town have now apparently disappeared, its approximate position is clearly given by the Arab itineraries. Baylaḳân lay 14 leagues south of Bardhâ'ah and seven or nine leagues north of the Aras on the road up from Barzand, and it still existed as a great place in the 9th (15th) century. Ibn Ḥawḳal in the 4th (10th) century describes it as a fine city, watered by streams with many mills, and surrounded by gardens and orchards. It was celebrated for a particular kind of syrup made here. In the year 617 (1220) Baylaḳân was stormed by the Mongols, who, finding no stones in the surrounding plain for their mangonels, cut down the plane trees, sawed the trunks into blocks, and shot these against the walls and houses of the city, which was subsequently plundered and burnt. The population, however, after a time returned, rebuilt their houses, and the place regained its former prosperity. At the close of the 8th (14th) century it was besieged and taken by Tîmûr, who afterwards caused it to be rebuilt, and a canal was dug from the river Aras, six leagues in length and 15 ells in width, by which the new town was well supplied with water. This canal was known as the Barlâsî, from the Barlâs tribe, from which Tîmûr was sprung.

Two other cities of Arrân are also mentioned, both of which lie to the north-west of Bardhâ'ah, on the road to Tiflîs. The first of these is Ganjah (now better known as Elizabetpol), which the Arab geographers write Janzah, and its river is called by Ḳazwînî the Ḳirdḳâs. Further to the north-west again lay Shamkûr, the ruins of which still exist, and this town in the 3rd

[1] Ist. 182, 183, 187, 188. I. H. 240, 241, 244, 251. Muk. 374, 375. Yak. i. 558, 562. Mst. 160. Kaz. ii. 344.

(9th) century was known as Mutawakkilîyah, from having been rebuilt by orders of the Caliph Mutawakkil in the year 240 (854)[1].

The two rivers bounding the province of Arrân, which the Greeks knew as the Araxes and the Cyrus, are called by the Arabs the Nahr-ar-Rass (or Aras) and the Nahr-al-Kurr (or Kur). The Aras rises in the Ḳâlîḳalâ country of western Armenia, and after passing along the northern frontiers of Adharbâyjân joins the river Kur (according to Mustawfî) in the Ḳarâbâgh country in the eastern part of Arrân. The river Kur rises in the mountains west of Tiflîs in Georgia, namely, in the country of the Khazars, which comprised the districts of Abkhâs and Allân. Passing Tiflîs the Kur flows down to Shamkûr, and here, according to Mustawfî, sends off a branch, or canal, which ends in the great Shamkûr swamp or lake. The Kur, after being joined by the Aras river some distance below Bardhâ'ah, flows out to the Caspian in the Gushtâsfî district[2].

Shirvân.

Beyond the Kur river, and along the Caspian where the Caucasus range sinks to the sea, is the Shirvân province, of which the capital was Ash-Shamâkhiyah, now called Shâmâkhî or Shâmâkhâ. In the 4th (10th) century Muḳaddasî describes this as a stone-built town, at the foot of the mountains, surrounded by gardens. Its governor, the ruler of the province, was called the Shirvân Shâh. Much corn was grown here, and in the neighbourhood, according to Moslem tradition as reported by Mustawfî, was to be seen both the Rock of Moses (referred to in the Ḳurân, XVIII. 62) and the site of the Fountain of Life, already mentioned as also localised in Bajarvân. Two other towns of the Shirvân

[1] No trace of the ruins of Baylaḳân appear on the Russian ordnance map. I. K. 122. Kud. 213. Ist. 187, 189. I. H. 244, 251. Muk. 376. Yak. i. 797; iii. 322. Kaz. ii. 345, 351. A. Y. ii. 543, 545. Mst. 160.

[2] In the *Jahân Numâ* (396, 397) a long description of both the Aras and the Kur, with their various affluents, is given. This serves to correct Mustawfî, also to elucidate the campaigns of Tîmûr in Georgia, though many of the names of towns cannot now be identified. Ist. 189. I. H. 246. Muk. 379. Kaz. i. 184; ii. 331. Mst. 213, 215.

province are mentioned by Muḳaddasî and other early authorities, the sites of which have not been fixed, namely Shâbarân, chiefly inhabited by Christians, which is said to have stood 20 leagues distant from Darband, and the city of Shirvân, which lay in the plain, having a Friday Mosque in its market-place. The latter was three days' march from the capital Shâmâkhî on the road to Darband.

The northernmost place in Shirvân was Bâb-al-Abwâb, 'the Gate of Gates,' as the Arabs called Darband, the famous port on the Caspian. Ibn Ḥawḳal says that in the 4th (10th) century the town was larger than Ardabîl, then the capital of Adharbâyjân. The harbour was protected by two moles, stretching out into the sea, and at their extremity was a water-gate, closed by chains, so that no ship could go out or in except by permission. These moles were built of blocks of stone fastened by lead joints. A stone wall enclosed the town, and it had two gates, the Great Gate and the Little Gate, besides the Water Gate aforesaid; and the walls had towers. The linen stuffs which were made in Darband were largely exported, also saffron from the neighbouring countryside.

There was a fine mosque in the market-place of Bâb-al-Abwâb, which was here the frontier town of Islam, for the place in early days was surrounded by infidel folk. Yâḳût gives a long account of the various tribes inhabiting the mountains and highlands of the Caucasus to the westward, among which he says that seventy different languages were spoken, and no man could understand that of his neighbour. Of these the Khazars, from whom the Caspian Sea, generally called the Baḥr-al-Khazar, took its name, were the most important. Yâḳût also describes the great wall which ran along the hill-crests westward from Darband, built to keep out the Barbarians, which had been erected, it was said, by King Anûshirvân of Persia in the sixth century A. D. The river Samûr, which flows into the Caspian a short distance to the south of Darband, is described by Muḳaddasî under the name of the Nahr-al-Malik, 'the King's River,' otherwise the Nahr-as-Samûr, and there was a bridge of boats (*Jisr*) across it, some 20 leagues from Darband, on the road coming up from Shâmâkhî.

The port of Bâkûh, or Bâkûyah (modern Bâkû), lies south of

Darband, and Iṣṭakhrî refers to its well-known naphtha springs. Yâḳût and others describe these in detail, the produce was worth a thousand dirhams (£40) a day; the naphtha flowed continuously, and all the ground was on fire round and about. Mustawfî speaks of the castle of Bâkûyah, which being high placed above the town kept it in shadow at midday. To the south of Bâkûh was the Gushtâsfî district, near the mouth of the Kur river, from which its lands were watered by a canal, much corn and cotton being grown here. Lastly, in the mountains near Darband was the fortress of Ḳabalah, where according to Muḳaddasî there was a mosque on a hill. Ḳabalah is more than once mentioned in the campaigns of Tîmûr, Mustawfî adding that both silk and corn are of its produce[1].

Gurjistân.

Gurjistân, which we call Georgia, and Abkhâs, otherwise Abkhasia, were lands that only became Moslem districts after the campaign of Tîmûr in these parts, at the close of the 8th (14th) century. Tiflîs, the capital of Gurjistân, on the upper waters of the river Kur, was, however, well known to the geographers of the 4th (10th) century. Ibn Ḥawḳal describes it as possessing double walls, strongly fortified, with three gates. There were natural hot-baths in Tiflîs where hot springs gushed out in the river bed, and the surrounding country was extremely fertile. The town lay on both banks of the Kur, and a bridge of boats, Muḳaddasî writes, connected the two quarters.

The neighbouring district of Abkhâs, or Abkhâz, was according to Muḳaddasî to be counted as of the Jabal-al-Ḳabḳ, the Caucasus. Here stood the village of Jonah, Ḳariyat Yûnis, inhabited by Moslems, and round this were the tribes of the Gurj (Georgians), Allân, and others. Many rivers flowed down from the mountain of Alburz, according to Mustawfî, who further mentions Ḳarṣ as one of the chief towns of Georgia[2].

[1] Ist. 184, 189, 190. I. H. 241, 251. Muk. 376, 379, 381. Yak. i. 437, 477; iii. 225, 282, 317; iv. 32. Mst. 159—161. Kaz. ii. 389. A. Y. i. 406.

[2] Ist. 185. I. H. 242. Muk. 375—377. Mst. 161, 202. Yak. i. 78, 350, 857. Mustawfî always writes of *Jibâl Alburz*, 'the Alburz mountains,' in

Armenia.

Great Armenia (spelt Armîniyah, in Arabic) was divided into Inner and Outer, and though mostly inhabited by Christians, was brought under Moslem rule at an early period. The country lay comprised within the great knot of mountains lying between the lake of Vân and the Gukchah lake, and from these highlands the Aras river and the two branches of the Euphrates took their rise.

The capital of Moslem Armenia in early times was Dabîl, otherwise called Duwîn or Tovin, now marked by a small village to the south of Erivan, near the Aras river. In the 4th (10th) century Dabîl was a larger town than Ardabîl, and was the chief place in Inner Armenia. It was a walled town, having three gates, and a Friday Mosque stood here side by side with the church. Mount Ararat, with its double peak, towered above Dabîl to the south, across the Araxes. As already said (p. 94) Moslem tradition identified Jabal Jûdî, in Upper Mesopotamia, as the mount on whose summit the Ark of Noah had come to rest. Ararat, in Armenia, they called Jabal-al-Ḥârith (of 'the Labourer' or 'Ploughman,' or else Al-Ḥârith was taken as the proper name of a pre-Islamic Arab who had settled in these parts). The lesser peak of Ararat was called Al-Ḥuwayrith, 'Little Ḥârith,' and Iṣṭakhrî says that both summits were always covered with snow, and they were not to be scaled by reason of their great height and steepness. The people of Dabîl cut firewood on their slopes, and hunted the abundant game here, and Muḳaddasî adds that a thousand hamlets were situated among the spurs flanking the great mountain. The wool stuffs of Dabîl, dyed red with the ḳirmiz insect, were famous. In the 4th (10th) century Muḳaddasî describes Dabîl as peopled by Kurds, and the Christians, he says, had the upper hand. Outside the town was

the plural, meaning the range; but he uses the term vaguely, and only one part of these corresponded with the Caucasus chain. At the present day Alburz, generally pronounced Elburz, or Elbruz, is the name of the highest mountain peak of the Caucasus; and in Persia Alburz is now used to designate the great range of mountains (of which Damâvand is the highest peak) lying to the north of Ṭihrân.

a great suburb surrounded by gardens. Ânî, the older capital of Christian Armenia, which was taken and sacked in 456 (1064) by Alp Arslân the Saljûḳ, is mentioned by Mustawfî as a town in the mountains where much fruit was grown. At some distance to the north-east of Dabîl lies the sweet-water lake, called Gukchah Tangîz (the Blue Lake) by 'Alî of Yazd; this, however, does not appear to be named by any earlier Moslem authority than Mustawfî[1].

The lake of Vân, or of Arjîsh as it is called by the earlier authorities, was naturally the best known of the Armenian lakes, having on its shores the cities of Akhlâṭ, Arjîsh, Vân, and Vasṭân. Iṣṭakhrî describes it as twenty leagues in length, and it was celebrated for the fish called ṭirrîkh (of the herring kind and still caught here in immense numbers) which after being salted was in the 7th (13th) century exported to Mesopotamia, and even to the furthest parts of Khurâsân, for Yâḳût says he bought some of this salt fish in Balkh. The waters of the lake were salt and bitter. Akhlâṭ, or Khilâṭ, at the western end of the lake, was one of the largest cities of Armenia. Mustawfî describes it as standing in a plain, surrounded by gardens, and dominated by a fortress. The Friday Mosque stood in the market-place. The cold here was severe in winter, but the town was very populous; it stood on the banks of a small stream across which was a bridge; and Mustawfî praises the gardens of the neighbouring district. Above Akhlâṭ was the great mountain called Kûh Sîpân, visible, says Mustawfî, fifty leagues away, and its summit was always snow-clad.

Arjîsh, a town on the northern shore of the lake, to which it frequently gave its name, according to Mustawfî, had been strongly fortified by the Wazîr 'Alî Shâh by order of Ghâzân Khân in the 8th (14th) century, and the country round was famous for its corn lands. Further to the east was the town of Bârkîrî, or Bahargîrî, near Band-i-Mâhî (the Fish Dam), on the road from Arjîsh to Khuwî (Khoi) in Adharbâyjân, and it is described by Mustawfî as having a strong castle crowning a hill. Its river came down from the Alâṭâḳ pastures, where the Îl-Khân, Arghûn, had built his great summer palace in the midst of

[1] Ist. 188, 191. I. H. 244. Muk. 374, 377, 380. Yak. ii. 183, 549. Mst. 126, 161, 164. A. Y. i. 414, 415; ii. 378. Ibn al-Athir, x. 25.

carefully preserved hunting grounds. The city of Vân, which at the present day gives its name to the lake, stands near its eastern shore; but we have no description of it. The fortress of Vasṭâm or Vasṭân lies on the south shore and is spoken of by Mustawfî, in the 8th (14th) century, as having a large town near it. Finally near the south-western corner of the lake lies Badlîs (Bitlis), described by Muḳaddasî as situated in a deep gorge where two streams met. A castle built of stone protected the town, and according to Yâḳût the apples grown in its district were so excellent as to be largely exported to all neighbouring lands[1].

The products of these northern provinces were few, and the manufactures consisted chiefly of stuffs dyed red with the ḳirmiz, an insect that fed on the oak trees growing throughout Adharbâyjân, and gave its name to the 'cramoisie' silks, being the origin of our words 'crimson' and 'carmine.' Ibn Ḥawḳal and Muḳaddasî both describe the ḳirmiz. The former says it was a worm like the silkworm, spinning for itself a cocoon exactly like the silkworm's cocoon; Muḳaddasî, on the other hand, writes that the ḳirmiz insect, or worm, was found on the earth, and that the women went out to gather it up, and afterwards dried it in an oven on brass pans. Silk, goat's-hair stuffs, linen, and wool were dyed with it, and the colour was famous in all lands. Armenia in general was also noted for its girdles, ribbed coverlets, carpets, rugs, cushions and veils; these commodities with figs, walnuts, and the salted ṭirrîkh fish from lake Vân already noticed, were the chief exports, and might all be found in great store at Dabîl. The town of Bardhâʿah was also celebrated for the silk produced in its neighbourhood, and from the countryside, as from Bâb-al-Abwâb, great numbers of mules were obtained for export; while lastly from the latter port, otherwise called Darband, came slaves brought thither from out of the northern lands[2].

[1] Ist. 188, 190. I. H. 245, 248. Muk. 377. Yak. i. 526; ii. 457. Kaz. ii. 352. Mst. 164, 165, 205, 226. J. N. 411, 412. A. Y. i. 685, 688.

[2] I. H. 244. Muk. 380, 381.

PROVINCES OF JIBAL AND JILAN, W

MAZANDARAN, KUMIS, AND JURJAN

CHAPTER XIII.

JIBÂL.

The province of Al-Jibâl, or 'Irâḳ 'Ajam, with its four districts. Ḳirmasîn or Kirmânshâhân. Bîsutûn and its sculptures. Kanguvâr. Dînavar. Shahrazûr. Ḥulwân. The great Khurâsân road. Kirind. Kurdistân under the Saljûḳs. Bahâr. Jamjamâl. Alânî and Alîshtar. Hamadân and its districts. Darguzîn. Kharaḳânayn and the northern Âvah. Nihâvand. Karaj of Rûdrâvar, and Karaj of Abu-Dulaf. Farâhân.

The broad mountain region, which the Greeks called Media, stretching across from the Mesopotamian plains on the west to the great desert of Persia on the east, was known to the Arab geographers as the province of Al-Jibâl, 'the Mountains.' This name afterwards fell out of use, and during the 6th (12th) century under the later Saljûḳs, the province came by a misnomer to be called 'Irâḳ 'Ajamî, which means Persian 'Irâḳ, being so named to distinguish it from the older 'Irâḳ of the Arabs, which was Lower Mesopotamia[1].

How this change in the name came about would appear to have been as follows. Al-'Irâḳ, as already said (Chapter II, p. 25, note), besides being the Moslem denomination for the lower half of Mesopotamia, was commonly, but in the dual form, applied

[1] *'Ajam* or *'Ajamî* is the name originally applied by the Arabs to a 'foreigner,' or non-Arab, as the Greeks used the term Barbarian. Since the Persians were the first foreigners with whom the Arabs came into contact 'Ajam and 'Ajamî soon became specialised to mean 'the Persian foreigner,' and as the equivalent of 'Persian' is in use at the present time. *Jibâl* is in Arabic the plural of *Jabal*, 'a hill.' Abu-l-Fidâ (p. 408) has the double name; he writes 'Bilâd-al-Jabal (Provinces of the Mountain) which is called by the people 'Irâḳ-al-'Ajam (Persian 'Irâḳ).'

by the Arabs to the two chief provincial cities, Kûfah and Basrah, which hence were known as Al-'Irâkayn—meaning 'the Two (capitals of) 'Irâk.' This was the older and classical usage; but in the latter part of the 5th (11th) century the Saljûks had come to rule over all western Persia, having their capital at Hamadân, and they also governed Mesopotamia, where the Abbasid Caliph resided. From him they received the title of Sultân of the Two 'Irâks, which seemed fitting to their case, and the second of the two 'Irâks soon came to be understood as meaning the province of Jibâl, where the Saljuk prince more especially resided, which thus by the vulgar came to be known for distinction as Persian 'Irâk. This is the account of the matter given by Yâkût, who states that the Persians in his day, but incorrectly and as a modern usage, called the province Persian 'Irâk. Yâkût himself uses the older name of Al-Jibâl, for which his contemporary Kazvînî, writing also in Arabic, gives the Persian equivalent of Kuhistân (the Mountain province). The name Jibâl, however, apparently became completely obsolete after the Mongol conquest, and Mustawfî in the 8th (14th) century nowhere uses it. He divides the older Jibâl province into two parts, the smaller being Kurdistân on the west, the larger moiety Persian 'Irâk on the east; and the name of 'Irâk is still in use at the present day, for that part of the older Jibâl province which lies south-west of Tihrân is now locally known as the 'Irâk district[1].

Four great cities—Kirmîsîn (later Kirmânshâh), Hamadân, Ray, and Isfahân—were from early days the chief towns of the four quarters of this province. In Buyid times, namely in the 4th (10th) century, according to Ibn Hawkal, the offices of the government were at Ray; at the close of the next century Hamadân became the capital under the Persian Saljûks; but at all times Isfahân would appear to have been the largest and generally the most flourishing city of the Jibâl province. In the present work it will be found convenient to describe the province as divided into the dependencies of its four great cities, and to begin with the western quarter, that dependent on Kirmânshâh, which since the days of the Saljûks has been commonly known as Kurdistân, signifying the land of the Kurds.

[1] Yak. ii. 15. Kaz. ii. 228. Mst. 141.

The capital city of Kirmânshâhân, a name generally curtailed to Kirmânshâh, was by the earlier Arabs known as Ḳirmîsîn (written also Ḳirmâsîn and Ḳirmâshîn). In the 4th (10th) century it is described by Ibn Ḥawḳal as a pleasant town surrounded by trees, with running waters, where fruit was cheap and all commodities abundant. Muḳaddasî, who is the first to mention the Persian name of Kirmânshâhân, adds that there was a Great Mosque in the market-place, and that 'Aḍud-ad-Dawlah the Buyid had built himself a fine palace here in the main street of the city. Ḳazvînî in the 7th (13th) century speaks of Ḳirmîsîn as standing close to Kirmânshâhân, as though these were twin cities; Yâḳût, who gives both names, says little of the town, confining himself to a description of the sculptures and ruins on the neighbouring mountain of Bihistân. The Mongol invasion in the 7th (13th) century effected the ruin of Kirmânshâh, which Mustawfî in the following century describes as reduced in his day to the size of a village, the name of which 'in books' was, he says, still written Ḳirmîsîn (since his time become obsolete), and he too is chiefly concerned with describing the Bihistân or Bîsutûn sculptures.

These are on the side, and at the foot of the great mountain of black rocks, about a day's march to the east of Kirmânshâh, near the Khurâsân road, and they consist of remains dating from the Achæmenian kings (5th century B.C.) and the Sassanians (7th century A.D.). They are described in the 4th (10th) century by Iṣṭakhrî and Ibn Ḥawḳal, who write the name of the mountain Bîhistûn and Bîsutûn, adding that the sculptures were to be found near the village of Sâsâniyân, doubtless the same village which Mustawfî in the 8th (14th) century called Vasṭâm or Basṭâm and which is now known as Ṭâḳ-i-Bustân, 'the Garden Arch.' Here the well-known sculpture of Darius receiving the tributary kings, with the trilingual cuneiform inscription, is referred to by Ibn Ḥawḳal, who describes it as being 'the representation in stone of a school-house, with the master and the boys; further (he adds) in the school-master's hand is an instrument like a strap wherewith to beat: also there be cauldrons as used in a kitchen sculptured in stone.' In regard to the Sassanian sculptures, added over a thousand years later, these are chiefly in and about a grotto, where there is a spring of water gushing out at the foot of the great mountain and, according to Ibn Ḥawḳal, repeated by all later Persian authorities, they represent King Khusraw Parvîz on his celebrated horse Shibdâz (or Shabdîz), while above him stands the beautiful Queen Shîrîn, her portrait adorning the roof of the grotto aforesaid. Somewhat defaced, these sculptures exist at the present day, and have been more than once figured and described. Yâḳût who quotes the travels of Ibn Muhalhal, 4th (10th) century, and Mustawfî, give in some detail the popular legends of their time. The story of Khusraw and Shîrîn, and of her lover the sculptor Farhâd who in despair slew himself, will be found localised in many of the neighbouring places; the incidents are well known, both from the *Shâh Nâmah* of Firdûsî, and from Niẓâmî's great poem (which Mustawfî quotes) called the 'Loves of Khusraw and Shîrîn[1].'

Overhanging Kirmânshâh to the north, and on the left hand of one travelling along the great Khurâsân road, was the isolated hill called Sinn Sumayrah, 'Sumayrah's Tooth,' whence the northern road started leading to Dînavar and the Adharbâyjân province. 'Sumayrah's Tooth' was so called from an Arab woman of that name, celebrated for her projecting teeth, and the Moslems gave the hill this nickname in jest, as they marched past it to the conquest of Nihâvand. Eastward beyond Bîsutûn, on the great Khurâsân road, lies the village of Siḥnah, as mentioned by Iṣṭakhrî, and still existing though not to be confused with the modern town of Siḥnah to be spoken of later. Beyond Siḥnah village lies Kanguvâr, which the Arabs called Ḳaṣr-al-Luṣûṣ, 'the Robbers' Castle,' from the evil ways of the inhabitants, who at the time of the first Moslem conquest stole all the baggage animals of the army sent against Nihâvand. There was here, according to Ibn Rustah and others, a great arched building standing on a platform, and dating from the days of Khusraw Parvîz, being constructed with columns and of mortared brickwork. The town of Kanguvâr was of considerable size, and had a Friday Mosque

[1] I. R. 166. Ykb. 270. Ist. 195, 203. I. H. 256, 265, 266. Muk. 284, 393. Kaz. ii. 290. Yak. iii. 250; iv. 69. Mst. 168, 203. J. N. 451. Bihistân is the older form. Bîsutûn, meaning 'without pillars' in Persian, i.e. unsupported, is probably the result of popular etymology.

built by Mûnis the chamberlain of the Caliph Muḳtadir. Yâḳût asserts that the platform, where the Sassanian buildings stood, was 20 ells above the ground level, and Mustawfî adds that the great stones for its construction had been brought from the mountain of Bîsutûn[1].

About 25 miles to the westward of Kanguvâr are the ruins of Dînavar, which in the 4th (10th) century was the capital of the small independent dynasty named after Ḥasanawayh, or Ḥasanûyah, the Kurdish chief of the dominant tribe settled in these parts. At the time of the Moslem conquest of Persia Dînavar had received the name of Mâh-al-Kûfah, 'because (as Ya'ḳûbî writes) its revenues were apportioned to the payment of the state pensions of the inhabitants of Kûfah'; and Mâh Kûfah for a time became the common name for the city and its surrounding territory. Ibn Ḥawḳal in the 4th (10th) century describes Dînavar as two-thirds the size of Hamadân, and the population as more urbane and better mannered than the Hamadân people. Muḳaddasî adds that the markets were well built, the surrounding gardens being very fruitful. The Great Mosque, which had been built by Ḥasanawayh, stood in the market-place, and over the pulpit rose a fine dome that was ornamented with sculptures. Dînavar was still an inhabited town when Mustawfî wrote in the 8th (14th) century; the climate was temperate, water plentiful, corn and grapes being abundantly grown. The place probably fell to its present state of ruin after the conquest of Tîmûr, who according to 'Alî of Yazd left some of his troops in garrison here.

Probably in the neighbourhood of Dînavar, but the site appears to be as yet unknown, stood the great castle of Sarmâj, described by Yâḳût as impregnable, being built of hewn stones by Ḥasanawayh, who died here in 369 (979), after a glorious reign, according to Ibn-al-Athîr, of nearly fifty years. In the next century Sarmâj was taken after a four years' siege in 441 (1049) by Tughril Beg the Saljûḳ, who, however, had to bring together an army of 100,000 men before he could force his brother Yunnâl out of this almost impregnable stronghold[2].

[1] Ist. 196. I. H. 256. I. R. 167. Muk. 393. Yak. iii. 50, 169; iv. 120, 381. The name of the village is spelt either Ṣiḥnah or Siḥnah. Mst. 168.

[2] Ykb. 171. I. H. 260. Muk. 394. Mst. 167. Yak. iii. 82. A. Y. ii.

About sixty miles north of the ruins of Dînavar stands at the present day the important town of Siḥnah, which is the modern capital of the Persian province of Kurdistân, though under this name it is not mentioned by any of the medieval Arab or Persian geographers. In the position of the modern Siḥnah, however, according to the itineraries of Ibn Khurdâdbih and Ḳudâmah, stood, during the middle-ages, the city of Sîsar, a name which Yâḳût rightly says means in Persian 'Thirty Heads.' The neighbourhood of Sîsar abounded in springs and was known as the Sad-Khâniyah—'the Hundred Houses' or Heads of Water—from the number of these springs. The Caliph Amîn had built a fortress here, which his more celebrated brother Mamûn had garrisoned, taking into his pay the Kurdish tribes who held the surrounding pastures, and using them in the civil war against his brother, whom he deprived later on of the Caliphate. Sîsar was counted as one of the 24 sub-districts of Hamadân; and it is possible that the modern name of Siḥnah may be merely a corruption of Sad-Khâniyah, shortened to Sî-Khânah, 'Thirty Houses,' but of this there is no direct evidence.

Four marches north-west of Dînavar was the town of Shahrazûr, standing in the district of the same name. Ibn Ḥawḳal, in the 4th (10th) century, mentions Shahrazûr as a walled and fortified town inhabited by Kurds, whose tribes he names; they occupied all the surrounding region, which was most fruitful. The traveller Ibn Muhalhal (as quoted by Yâḳût) describes in the 4th (10th) century the many towns and villages of this district, and the chief town, he says, was known among the Persians as Nîm-Râh, or 'the Half-way House,' because it stood at the middle stage between Madâin (Ctesiphon) and Shîz, the two great fire-temples of Sassanian times. The neighbouring mountains were called Sha'rân and Zalam, where according to Ḳazvînî a species of

530. Ibn-al-Athir, viii. 518, 519; ix. 380. According to Yâḳût (iv. 405) the Persian word *Mâh* is synonymous with Ḳaṣbah (chief town) in Arabic. The prefix *Mâh*, which occurs in the older names for Dînavar and Nihâvand, is in Old Persian *Mada*, and as a place-name is radically the same word which has come down to us, through the Greeks, in the form of Media and the Medes. The ruins of Dînavar have been lately visited, and are described by De Morgan, *Mission en Perse*, ii. 95, 96.

grain was grown that was deemed a powerful aphrodisiac. The Kurds in this region, when Ibn Muhalhal visited the place, numbered 60,000 tents, and when Mustawfî wrote in the 8th (14th) century Shahrazûr was still a flourishing town, and inhabited by Kurds[1].

The great Khurâsân road, which, as already described in our first chapter, went eastwards from Baghdâd to the uttermost limits of Moslem lands, after crossing the Mesopotamian plain entered the mountainous region of Persia at Ḥulwân, a town of the Jibâl province, which however was sometimes counted as of Arabian 'Irâḳ. Ibn Ḥawḳal says that in the 4th (10th) century Ḥulwân was half the size of Dînavar, and its houses were built of both stone and clay bricks. Though the climate was hot, dates, pomegranates, and figs growing abundantly, snow could all the summer through be found on the mountains two leagues above the city. Muḳaddasî adds that there was an old castle in the town within which stood the mosque, and the city wall had eight gates, the names of which he enumerates. Outside the town stood a synagogue of the Jews, much venerated by them, which was built of squared stones set in mortar. In the 7th (13th) century, when Ḳazvînî wrote, Ḥulwân was already in ruins, but famous for its sulphur springs. In the next century Mustawfî praises its crops, but says that the town stood desolate, except for divers shrines of Moslem saints, though the surrounding territory comprised thirty villages.

Along the Khurâsân road, and four leagues above Ḥulwân towards Kirind, lay Mâdharûstân, where according to Yâḳût might be seen a great arched building surmounting a platform. This had formed part of the palace of the Sassanian king Bahrâm Gûr, who laid out a paradise round it that, in Yâḳût's days, had long gone to ruin. Six leagues beyond this comes the town of Kirind, which is apparently first mentioned by Mustawfî in the 8th (14th) century; he couples Kirind with the neighbouring village called Khûshân, which however has now completely disappeared, though Mustawfî describes it as in his day more

1 I. K. 120. Kud. 212. I. H. 263, 265. Yak. iii. 216, 340; iv. 988. Kaz. ii. 266. Mst. 167. The district of Shahrazûr still keeps the name, the old city stood where are the ruins now known as Yasîn Tappah.

populous even than Kirind. These two places lie together at the head of the Ḥulwân pass, in a fertile plain, and correspond in position—for as already said neither are mentioned by the earlier Arab geographers—with the station of Marj-al-Ḳal'ah (the Meadow Castle), which Ibn Ḥawḳal describes as a great walled town surrounded by populous and fertile districts. Ya'ḳûbî states that in these pastures the Abbasid Caliph kept his stud of horses. Four leagues beyond these pastures the high road passed Ṭazar, where, according to Muḳaddasî, might be seen the remains of a palace of the Chosroes, built Yâḳût records by one Khusrûjird, son of Shahân. Ṭazar had good markets, and it appears to be identical with Ḳaṣr Yazîd (Yazîd's palace or castle), mentioned by other authorities. Six leagues beyond Ṭazar again was Az-Zubaydîyah, 'a fine healthy place' according to Ibn Ḥawḳal, the position of which on the high road shows it to be identical with the present village of Hârûnâbâd. Here the Khurâsân road turns east, and crossing the plain of Mâyidasht (or Mâhîdasht) runs direct to Kirmânshâh. The Mâyidasht plain is described by Mustawfî as in his day dotted with some fifty villages, surrounded by excellent pasture lands that were well watered from the neighbouring hills. In this region was the castle of Harsîn with a small town at its base, which still exists, lying about 20 miles to the south-east of Kirmânshâh[1].

As regards the origin of the Kurdistân province, it is stated that about the middle of the 6th (12th) century Sulṭân Sanjar the Saljûḳ divided off the western part of the Jibâl province, namely the region which was dependent on Kirmânshâh, and giving it the name of Kurdistân put it under the government of his nephew Sulaymân Shâh, surnamed Abûh (or Ayûh), who, at a later period —that is from 554 to 556 (1159 to 1161)—succeeded his uncle as chief of the house of Saljûḳ and Sultan of the Two 'Irâḳs. This is the account given by Mustawfî, who states that under Sulaymân Shâh Kurdistân flourished greatly, and its revenues then amounted to two million gold dînârs (equivalent to about a million sterling),

[1] I. H. 168, 256, 262. I. R. 165. Ykb. 270. Muk. 123, 135, 393. Kaz. ii. 239, 302. Mst. 138, 168. Yak. iii. 537; iv. 382. J. N. 450. The ruins of Ḥulwân exist at the village now called Sar-i-Pul (Bridge-head), where a bridge crossed the stream.

which was near ten times the sum yielded by the province in the 8th (14th) century under the Mongols, when Mustawfî was their revenue officer. Sulaymân Shâh made Bahâr—a town that still exists, lying some eight miles to the north of Hamadân—his capital; and here there was a strong castle. In Mongol times a second capital was built, by Uljaytû Sulṭân, at Sulṭânâbâd Jamjamâl (or Chamchamâl) near the foot of the Bîsutûn mountain, and this town Mustawfî describes as standing in a rich country where much corn was grown. Of Jamjamâl, or Chamchamâl, the position is given in his itineraries (four leagues from Siḥnah village, and six from Kirmânshâh) and its ruins still exist, being marked on the map at the spot indicated. The town is frequently mentioned by 'Alî of Yazd when describing the marches of Tîmûr through Kurdistân.

Among other towns which occur in the description of the campaigns of Tîmûr, and which are noticed by Mustawfî, are Darband Tâj Khâtûn, 'a medium-sized town now for the most part in ruin,' and Darband Zangî, a smaller place, which had good pasture grounds with a temperate climate. Both towns apparently have disappeared from the map; but Darband means 'a pass,' and from 'Alî of Yazd, who writes the name of the first as Darband-Tâshî-Khâtûn, these two Darbands would appear to have stood on the western frontier of Kurdistân (between Shahrazûr and Ḥulwân), among the hills that here dominate the plains of Mesopotamia.

Mustawfî also mentions four other towns in Kurdistân, namely Alânî, Alîshtar, Khuftiyân, and Darbîl, as important places in his day, but it is not easy now to identify their sites. Alânî, for which some MSS. give the reading Alâbî, in the 8th (14th) century was presumably one of the chief towns of the province, though no other authority but Mustawfî appears to mention it. Its lands grew wheat crops, it had a good climate, well-watered pastures lying round it, and there were well-stocked hunting grounds in the neighbourhood. At Alîshtar also was an ancient fire-temple called Ardahish (Arûkhsh or Arakhash). Unfortunately none of the Itineraries give its position; but the plain of Alîshtar still exists, and probably one of its ruined sites is the town mentioned by Mustawfî. It is doubtless identical with the town of Lîshtar or Lâshtar mentioned by Ibn Ḥawḳal and others as lying 10 leagues south-west of Nihâvand, being 12 leagues north of Sâbûrkhwâst. On the other hand the reading of the name Alîshtar is, it must be admitted, extremely doubtful; many of the best MSS., also the Turkish *Jahân Numâ*, give Al-Bashr, and a variety of other forms occur. Nothing is known of Khuftiyân (for which the *Jahân Numâ* gives Ḥaḳshiyân, and the MSS. a variety of readings) except that it was a strong castle surrounded by villages lying on the banks of the Zâb river; but whether this was the Upper or the Lower Zâb is not indicated. Its site is unknown and the same is the case with Darbîl (or Dizbîl), 'a medium-sized town with a good climate,' the position of which is not even approximately indicated by Mustawfî. This concludes his notice of the Kurdistân district[1].

Hamadân (which name the Arabs wrote Hamadhân)[2] is the ancient Ecbatana, the capital of the province of Media. Ibn Ḥawḳal in the 4th (10th) century describes Hamadhân as a large fine city, over a league square, which had been rebuilt since the Moslem conquest. Its walls had four gates, and without them was a suburb. There was much merchandise in its markets, and the surrounding district was very fertile, producing large crops, more especially saffron. Muḳaddasî adds that the town possessed three rows of markets, and that in one of these stood the Great Mosque, a very old structure. Yâḳût, who has some notes on Hamadân, written shortly before it was laid in ruins by the Mongol invasion of 617 (1220), states that there were twenty-four Rustâḳs, or sub-districts, dependent on the city, and these he enumerates. The list is again given by Mustawfî in the following century, who adds thereto the names of the villages of each district; most of them however it is impossible now to identify. Mustawfî describes the city, in the 8th (14th) century, as measuring two leagues across, in the centre of which stood the ancient castle, built of clay, called the Shahristân. This ancient citadel of Hamadân like that of Isfahân—to be noticed later—is named Sârûḳ by Ibn Faḳîh, but the meaning of the word is not explained. The goldsmiths' market

[1] I. H. 259, 264. Yak. i. 276; iii. 5. Mst. 167, 192. A. Y. i. 584, 585, 599, 640. J. N. 450. Neither Bahâr, Alânî, Khuftiyân, Darbîl, nor the two Darbands, are mentioned by any of the earlier Arab geographers.

[2] *Hamadhân* represents the *Hagmatâna* of the Achæmenian inscriptions, which the Greeks wrote Ecbatana.

in Hamadân was famous, built on the site of the former village of Zamîn Dîh; and the city walls measured 12,000 paces in circuit. Originally, says Mustawfî, Hamadân comprised five cities, namely Ḳal'ah Kabrît, 'Sulphur Castle,' Ḳal'ah Mâkîn, Girdlâkh, Khurshîd, and Kurasht. He adds, 'this last, formerly a large town, is now entirely ruined.' Of Hamadân, too, were the following five great districts, with their villages; namely, Farîvâr near the city, next Azmadîn, Sharâmîn, and A'lam; with, lastly, the district of Sard Rûd and Barhand Rûd. It must, however, be added that the readings of these names are uncertain, as the MSS. vary considerably[1].

Three leagues from Hamadân (but in what direction is not stated, and the name does not appear on the map), in the village of Juhastah, stood the ruins of the ancient castle of King Bahrâm Gûr, described by Ibn Faḳîh. It was a huge structure, with halls, passages, and chambers, in part cut out of the live rock. At the four corners were sculptured female figures, and along one face of the building ran an inscription in Old Persian (*Farsîyah*) commemorating the conquests of the Chosroes. Half a league distant from this palace was a hill, where was to be seen the so-called Antelope's tomb (Nâûs-aẓ-Ẓabîyah), and Ibn Faḳîh gives a long anecdote concerning King Bahrâm Gûr and his mistress, and of the many gazelles that he slew in the neighbouring plain, and how he finally put his mistress to death here for her insolent remarks in disparagement of his shooting.

To the south-west of Hamadân rises the great mountain of Alvand, or Arvand as Yâḳût writes the word, and this form of the name appears as the mint city on silver dirhams of Abu-Sa'îd, the Mongol Îl-Khân, dated 729 (1329). Mustawfî gives a long account of Kûh Alvand, which he says was thirty leagues in circuit, its summit always being covered with snow. There was an abundant spring of water on the topmost peak, which issued from a sort of building cut in the rock, and forty-two other streams, he adds, gushed from the various spurs of the mountain. Travelling west from Hamadân, after crossing the Alvand pass, on the high road

[1] J. H. 256, 260. Muk. 391. I. F. 219. Yak. iv. 988. Mst. 151, 152. The Turkish *Jahân Numâ* (p. 300) repeats the enumeration of districts and villages from Mustawfî.

to Kanguvâr, stands Asadâbâd, which Ibn Ḥawḳal describes as a populous city; and Muḳaddasî adds that a league distant from it was to be seen the arch (*Aywân*), in a building which Yâḳût refers to as the Maṭâbikh-al-Kisrâ, 'the Kitchens of Chosroes.' Asadâbâd had a mosque, and good markets; its district was very fertile and produced honey. Mustawfî says that 35 villages were of its dependencies[1].

The plain in which Hamadân stands drains to the north and east, its numerous streams uniting to form the head-waters of the river Gâvmâhâ (or Gâvmâsâ) whose course will be described later when speaking of the Ḳum river. To the north of Hamadân lies the district of Darguzîn, and north of this again that of Kharraḳân. Mustawfî in the 8th (14th) century writes of Darguzîn as a considerable town, formerly a village, the capital of the A'lam district, mentioned on the previous page as one of the five dependent on Hamadân. The A'lam district, he adds—and Yâḳût confirms him—was wrongly called Al-Amr by the Persians: it was a high plateau lying between Hamadân and Zanjân, where grapes, cotton, and corn grew abundantly. Kharraḳân, more often called Kharraḳânayn, 'the two Kharraḳâns,' lay north of the A'lam district. It comprised many villages, which Mustawfî enumerates (but the readings in the MSS. are uncertain), and the chief town which still exists was Âvah, or Âbah of Hamadân, so named to distinguish it from Âvah of Sâvah, which will be noticed later. This, the northern Âvah, sometimes also written Âvâ, is mentioned by Yâḳût, and it is referred to as early as the 4th (10th) century by Muḳaddasî. The Kharraḳân river, according to Mustawfî, during the spring freshets poured its waters into the stream of the Khushk Rûd which ultimately lost itself in the great desert in the Ray district. In the summer time, however, the Kharraḳân river never flowed beyond the boundaries of its own immediate district, its waters drying up in irrigation channels[2].

The city of Nihâvand, lying about forty miles south of Hamadân, was an important place dating from Sassanian times. After the first Moslem conquest, which was effected by the troops from

[1] I. H. 256. I. F. 255. Muk. 393. Yak. i. 225, 245; iv. 110, 733. Kaz. ii. 236, 311. Mst. 152, 202.

[2] Muk. 25, 51, 386. Yak. i. 316, 408. Mst. 152, 217. J. N. 301, 305.

Baṣrah more particularly, the town and its district received the name of Mâh-al-Baṣrah, for its revenues were allotted to the payment of pensions in Baṣrah, just as those of Dînavar were paid to Kûfah (see above, p. 189). Ibn Ḥawḳal in the 4th (10th) century speaks of the rich merchandise sold in its markets, whither the saffron of the neighbouring district of Rûdhrâvar was brought for distribution. Nihâvand had then two Great Mosques, the old and the new. Yâḳût adds the tradition that many Arabs coming from Baṣrah had settled here in early days; and the city was famous for the manufacture of perfumes. Mustawfî in the 8th (14th) century states that in his day the population consisted mostly of the Kurdish tribesmen; much cotton was grown in the neighbouring districts, three of which in particular he names, Malâir, Isfîdhân, and Jahûḳ. About half-way between Hamadân and Nihâvand lay the rich district of Rûdhrâvar, so famous for its saffron, of which district the chief city was Karaj, possessing a fine mosque. The district was three leagues across, and comprised 93 villages according to Yâḳût. Mustawfî generally spells the name Rûdârûd, and of its towns he mentions Sarkân and Tuvî, both of which still exist; and Tuvî, at the present day, is the name commonly given to the district[1].

To the eastward of Nihâvand lay the district of the two Îghârs (Al-Îghârayn) of which the capital was also called Karaj, known for distinction as Karaj of Abu Dulaf. The exact site of this Karaj is unknown, but from the distances given in the Itineraries, and from the fact stated by Mustawfî that the town lay beneath the Râsmand mountains (almost certainly to be identified with the present range called Râsband), its site must be sought for near the head-waters of the stream which flows past Sârûk to join the modern Ḳarâ Ṣû. Ibn Ḥawḳal in the 4th (10th) century speaks of Karaj as smaller than Burûjird, but it was a place of importance, built on a height. The houses of the town covered a space of over two leagues, and there were two markets, one at the Bâb Masjid-al-Jâmi', 'the gate of the Great Mosque,' the other situated at the opposite town gate opening

[1] I. R. 166. I. H. 258, 259, 262. Muk. 393. Yak. ii. 832; iv. 251, 827. Mst. 152, 153. The ruins of Karaj of Rûdhrâvar are doubtless those described by De Morgan, *Mission en Perse* (ii. 136), which he names Rûdîlâvar.

beyond what was known as 'the great plain.' Baths were numerous and the houses were well built, mostly of clay bricks; the gardens were few, but those round the town limits were very fertile. Abu Dulaf, from whom the place took its distinguishing name, had been a celebrated general, also a poet at the court of Hârûn-ar-Rashîd and his son Mamûn. Abu Dulaf together with his descendants settled in this district, which with that lying round Burj, 12 leagues distant towards Isfahân, had been granted to them as Îghârs—that is 'fiefs in perpetuity,' paying a fixed yearly tribute to the Caliph, but free of all other taxes. Yâkût states that the Persians pronounced the name of Karaj *Karah*, and Farrazîn was the name of a castle not far from the gate of Karaj. Mustawfî, who refers to the river as the River of Karah—the Karah Rûd—says that the Râsmand mountain here rose above the plain to the north. At the foot of the mountain was an abundant spring of water, called the fountain of King Kay-Khusraw, which irrigated the neighbouring pasture lands, six leagues long by three wide, known as the Margzâr of Kîtû, which lay under the protection of the Farrazîn castle. The Râsmand mountain is described as a black rock towering up like the hill of Bîsutûn, with glens at its base, and it was ten leagues in circuit. The site of Burj, the second city of the Îghârayn, has not yet been identified. Its position, however, is approximately known. Ibn Ḥawkal speaks of it as a fine well-conditioned town, and tells us that it lay on the high road towards Isfahân, some 12 leagues distant from Karaj[1].

Lower down the Karaj river, and to the north of Karaj of Abu Dulaf, is the town of Sârûk of the Farâhân district, noticed by Yâkût and Mustawfî, being counted by them as belonging to Hamadân. Dawlatâbâd, which still exists, is mentioned as a prominent place of the neighbourhood; and there was a salt marsh near here, formed by a lake, measuring four leagues square, which when dried up by the summer heats produced excellent salt for export. This lake, according to Mustawfî, the Mongols named Jaghân Nâûr, meaning 'Salt Lake.' It is doubtless

[1] I. H. 258, 262. Muk. 394. Yak. i. 420, 548; iii. 873; iv. 250, 270. Mst. 151, 204.

identical with the present lake of Tualâ. Lastly, to the south-east of Hamadân, and about half-way between that city and Nihâvand, lies the small town of Râmîn, which is noticed by Yâḳût as of this district, but it is not further described by any other authority[1].

[1] Yak. iii. 867, 887; iv. 683. Mst. 151. At the present day the chief town of this district, now famous for its carpets, is Sulṭânâbâd, founded by Fatḥ 'Alî Shâh at the beginning of the nineteenth century; it is commonly known as Shahr-i-Naw (New Town).

CHAPTER XIV.

JIBÂL (*continued*).

Little Lur. Burûjird. Khurramâbâd. Shâpûrkhwâst. Sîrawân and Ṣaymarah. Isfahân and its districts. Fîrûzân; Fârifân and the river Zandah Rûd. Ardistân. Kâshân. Ḳum, Gulpaygân, and the Ḳum river. Âvah and Sâvah. The river Gâvmâhâ.

South of Hamadân lies Luristân, the district of the Lur tribes, kinsmen of the Kurds, and this mountainous region is divided by its rivers into two parts, Great Lur to the south and Little Lur to the north. The district of Little Lur is separated from Great Lur by the main stream of the Upper Kârûn, and the towns of Great Lur will be more conveniently described in the chapter on Khûzistân, although the district of Great Lur also is by some authorities regarded as forming part of 'Irâḳ 'Ajam.

The chief towns of Little Lur, as enumerated by Mustawfî in the 8th (14th) century, were Burûjird, Khurramâbâd, and Shâpûrkhwâst. Burûjird is described by Ibn Ḥawḳal in the 4th (10th) century as a fine city, measuring over half a league across. Its fruits were exported to Karaj, much saffron was grown, and its importance increased after Ḥamûlah, the Wazîr of the Abu Dulaf family just mentioned, built the Friday Mosque here. When Mustawfî wrote, in the 8th (14th) century, there were two mosques, the old and the new; but the town, he says, was then already falling to ruin. 'Alî of Yazd, who always writes the name Vurûjird, frequently refers to it in describing the campaigns of Tîmûr, by whose orders the castle, called the Ḳal'ah Armiyân, was restored[1].

[1] I. H. 258, 262. Yak. i. 596; ii. 737. Mst. 151. A. Y. i. 587; ii. 515.

The name of Khurramâbâd, since the time of Tîmûr the most important place in Little Lur after Burûjird, does not occur in any of the Arab geographers of the middle-ages; and it has often been suggested that Khurramâbâd was identical with the town of Shâpûrkhwâst, a place frequently mentioned in earlier days. That this, however, is not the case, is proved by the mention, separately, of both towns by Mustawfî, who further indicates the position of Shâpûrkhwâst. Khurramâbâd, when Mustawfî wrote in the 8th (14th) century, was a fine town, though already partly in ruin. The date palm produced abundantly here, and he adds that this was the only place in the hill country where it grew, excepting Ṣaymarah: but this statement cannot be accepted as quite exact.

In regard to Shâpûrkhwâst, which the Arab geographers wrote Sâbûrkhwâst, this also had been a town famous for its dates since the time of Ibn Ḥawḳal. In the 4th (10th) century Sâbûrkhwâst with Burûjird and Nihâvand came under the power of Ḥasanawayh, the Kurdish chief who had established his government at Dînavar (see above, p. 189), and at Dizbaz, the castle of Sâbûrkhwâst, which rivalled Sarmâj for strength, Badr, son of Ḥasanawayh, kept his treasures, which in 414 (1023) fell into the hands of the Buyids. During the 5th (11th) century Sâbûrkhwâst is frequently mentioned in the chronicles relating to the doings of the Saljûḳs, and in 499 (1106) the Atabeg Mankûbars came into possession of the city, together with Nihâvand and Lîshtar (Alîshtar). Writing in the early part of the 8th (14th) century Mustawfî (in the *Guzîdah*) gives the information that in his day there were, in Little Lur, three populous cities, namely Burûjird, Khurramâbâd, and Shâpûrkhwâst (as he spells it in Persian). He relates that, 'this last, though once a great city, and very populous, being full of people of various nations and the capital of the kingdom, is now reduced to become a provincial town'; and in regard to its position he states that beyond (south) of Burûjird, 'the road (coming from Nihâvand and going to Isfahân) branches to the right to Shâpûrkhwâst,' while to the left (eastward) the main road went on to Karaj of Abu Dulaf. These details are in accordance with the accounts given by Ibn Ḥawḳal and Muḳaddasî; for the former states that from Nihâvand it was 10 leagues (south) to Lâshtar, and thence 12 on to Sâbûrkhwâst, from which it was counted 30 leagues to (Great) Lur—that is to say the plains lying north of Dizful which will be noticed later in Chapter XVI. Muḳaddasî adds that from Sâbûr-khwâst to Karaj of Abu Dulaf was four marches, it being the same from Sâbûrkhwâst to Lur[1].

To the west of Little Lur, and on the frontier of Arabian 'Irâḳ, lay the two districts of Mâsabadhân and Mihrajânḳudhaḳ, of which the chief towns were, respectively, Sîrawân and Ṣaymarah. The ruins of both towns still exist, and Mâsabadhân is in use as the name of the region to the south of the Mâyidasht plain. Sîrawân (or As-Sîrawân) was, according to Ibn Ḥawḳal, a small town, its houses built of mortared stone, not unlike Mosul. It produced the fruits of both hot and cold regions, especially nuts and melons, the latter of the celebrated kind known as Dastabûyah; moreover the date palm, as already said, flourished here. Ḳazvînî refers to mines of salt, sulphur, vitriol, and borax as being found in the Mâsabadhân district. Situated some fifty miles to the eastward, Ṣaymarah was not unlike Sîrawân, and it remained a populous town to a later date than the latter, its position being better chosen. The Mihrajânḳudhaḳ district lying round it was celebrated in the 4th (10th) century for great fertility; and Muḳaddasî refers to its numerous population. 'Dates and olives, nuts and snow are all found here abundantly,' Yâḳût writes, and on the road between Ṣaymarah and the neighbouring hamlet of Tarḥân was a wonderful bridge, 'twice as great as the bridge between Ḥulwân and Khâniḳîn.' When Mustawfî wrote in the 8th (14th) century Ṣaymarah, though already falling to ruin, was still a fine town, and the surrounding country was celebrated for its date-groves[2].

At the south-eastern corner of the Jibâl province, and not far distant from the borders of the Great Desert, stands Isfahân (the

[1] I. H. 259, 264. Muk. 401. Yak. ii. 572; iii. 4, 82, 225. Ibn-al-Athir, ix. 174; x. 274. Mst. 151, 195; also *Guzîdah* (Gantin), I. 622, and MS. f. 159 *b*, giving the paragraph on Lesser Lur, at the end of section xi of chapter IV, immediately preceding the section treating of the Mongols. The name is variously spelt Sâbûrkhwâst, Shâburkhast, and Shâpûrkhwâst. The exact site of the ruins has not been identified.

[2] I. H. 263, 264. Muk. 394. Ykb. 269. Kaz. ii. 172. Yak. iii. 443, 525. Mst. 151.

name being spelt Iṣbahân by the Arabs and by the Persians Ispahân), which from the earliest times must have been a place of importance, on account of the fertility of its lands which are watered by the abundant stream of the Zâyindah Rûd. At the present day Isfahân and its suburbs occupy both banks of the river, but in the middle-ages the inhabited quarters lay only on the northern or left bank of the Zâyindah Rûd. Here there were two cities side by side; namely, to the east Jay, otherwise called Shahristânah[1], girt by a wall with a hundred towers; and two miles to the westward of this Al-Yahûdîyah, 'the Jew Town,' double the size of Jay, taking its name, so tradition asserted, from the Jews who had been settled here in the time of Nebuchadnezzar.

Ibn Rustah, at the close of the 3rd (9th) century, describes the city of Jay as measuring half a league across, and covering an area of 2000 Jarîbs (about 600 acres). There were four gates, Bâb Khawr or 'of the Creek,' otherwise Bâb Zarîn Rûd, for this was the earlier spelling of the name of the river; then Bâb Asfîj, Bâb Ṭîrah, and the Yahûdîyah Gate. Ibn Rustah enumerates the number of towers on the wall between each gate, and he also gives the space in ells. In Jay was an ancient building like a fortress called Sârûḳ, the name likewise of the Hamadân citadel, as above stated, which Ibn Rustah says dated from before the Flood. Ibn Ḥawḳal and Muḳaddasî in the next century describe both Jay and Yahûdîyah. In each city was a Great Mosque for the Friday prayers; and Yahûdîyah alone equalled Hamadân in size, being indeed the largest city in the Jibâl province, Ray only possibly excepted. Isfahân was already a great commercial centre, and its silks, especially the 'Attâbî (tabby stuffs), and its cottons, were largely exported. Saffron and all kinds of fruit grew well in its districts, which were the broadest and richest of the whole Jibâl. Al-Yahûdîyah, according to Muḳaddasî, had been originally settled by the Jews in the time of Nebuchadnezzar because its climate resembled that of Jerusalem. The town, which he reports had twelve gates (*Darb*), was built mostly of unburnt brick, and it had both open and covered markets. The

[1] Shahristân, or Shahristânah, means, in Persian, 'the Township,' and is a common name for the capital city.

Great Mosque was in one of the markets, built with round columns, having a minaret on the Ḳiblah (Mecca) side, 70 ells in height. The neighbouring township of Jay, a couple of miles to the eastward, was according to Muḳaddasî called Al-Madînah, 'the City,' the Arabic equivalent of Shahristânah, and immediately below its ancient fortress, in the 4th (10th) century, the river was crossed by a bridge of boats.

In 444 (1052) Isfahân was visited by the Persian traveller Nâṣir-i-Khusraw, who describes it as the largest city in all Persian-speaking lands that he had seen. There were two hundred bankers, and fifty caravanserais; and the town was surrounded by a wall said to be three and a half leagues in circuit, with battlements and a gangway running along the summit. The Great Mosque was a magnificent building, and the money-changers' market a sight to be seen, and each of the other numerous markets was shut off by its own gate. When Yâḳût wrote, in the beginning of the 7th (13th) century, both Yahûdîyah and Jay had fallen to ruin; and of the two the latter was then the more populous. He further speaks of the Great Mosque in Jay built by the Caliph Manṣûr Râshid, who, as the chronicles relate, having been deposed by his uncle Muḥammad Muḳtafî in 530 (1136), was afterwards killed in battle and brought to be buried outside the gate of Isfahân. Yahûdîyah, however, after the Mongol invasion, recovered a part of its former glory, and was a populous thriving city when Abu-l-Fidâ wrote in 721 (1321), having, he says, the suburb of Shahristân a mile distant to the eastward, which occupied part of the older site of Jay.

His contemporary, Mustawfî, gives us a long account of Isfahân and its districts, mentioning the names of many places that still exist; and his description proves that Yahûdîyah of medieval times is the city of Isfahân as described by Chardin at the close of the 17th century, when it had become the capital of Persia under Shâh 'Abbâs, the past glories of which are to be seen at the present day. According to Mustawfî the city walls, 21,000 paces in circuit, dated from the 4th (10th) century, having been built by 'Aḍud-ad-Dawlah the Buyid. The area of Isfahân had formerly been occupied by four villages, whose names survived in the town quarters, namely, Karrân (the Karrân Gate is given by

Chardin as opening on the east side), Kûshk, Jûbarah (this was the eastern quarter when Chardin wrote, and the Jûbarah Gate was to the north-east), and Dardasht (the gate of this name lay to the north, and the Dardasht quarter was to the north-west). Mustawfî writes that the most populous quarter under the Saljûḳs had been that known as Julbârah (the Gulbâr quarter of Chardin, round the present Maydân-i-Kuhnah or 'Old Square'), where stood the College and Tomb of Sulṭân Muḥammad the Saljûḳ, and here might be seen a block of stone weighing 10,000 *mans* (equivalent, perhaps, to a little less than 32 tons weight), this being a great idol, carried off by the Sulṭân from India, and set up before the college gate[1].

When Tîmûr conquered Isfahân at the close of the 8th (14th) century, the name of the citadel which he occupied is given as Ḳal'ah Ṭabarik (the latter word meaning a 'hillock' in the Persian dialect), and the ruins of this castle, which still exist, are described by Chardin as standing outside the Dardasht Gate. Further we are informed that Malik Shâh the Saljûḳ erected another strong castle—the Shâh-Diz, 'the Royal Fort'—on the summit of a mountain close to Isfahân in the year 500 (1107), and Ḳazvînî adds a long anecdote relating the circumstances that brought about its foundation. At the beginning of the 10th (16th) century, Persia came under the rule of Shâh Ismâîl the Ṣafavid, and at the close of the century Shâh 'Abbâs the Great transferred his capital from Ardabîl to Isfahân, whither he also removed the whole Armenian population of Julfah on the river Aras, settling them in a new quarter of the city which he founded on the southern or right bank of the Zâyindah Rûd. Shâh 'Abbâs also added other new quarters and suburbs to Isfahân, but north of the river, all of which are minutely described by Chardin, who lived at Isfahân for many years during the latter half of the 17th century A.D.[2]

[1] History, however, does not record that this Sulṭân Muḥammad—he reigned from 498 to 511 (1104—1117) and was a son of Malik Shâh—made any conquests in India; possibly Mustawfî has mistaken him for Maḥmûd of Ghaznah.

[2] I. R. 160, 162. I. H. 161. Muk. 386, 387, 388, 389. N. K. 93. Yak. i. 295; ii. 181; iii. 246; iv. 452, 1045. A. F. 411. Mst. 142. A. Y.

The eight districts round Isfahân, which Mustawfî carefully enumerates with their villages, still exist, and the same names appear in Ya'ḳûbî and other early authorities of the 3rd and 4th (9th and 10th) centuries. Four of these districts lie to the north of the river, while the other four are on its right bank to the southward. Beginning with the north bank, the home district, that immediately round the city, was called Jay, the name of the older town to the eastward. The Marbîn district was to the west of Isfahân, and here stood an ancient fire-temple built by the mythical king Tahmurath, surnamed Dîv Band, 'the demon binder.' To the north-west, at some distance from the city gates, lay the Burkhwâr district, of which Jaz (modern Gaz) was the largest village; while to the north-east was the district called Ḳahâb, the fourth on the northern river bank. South of the Zâyindah Rûd, and to the south-east of the old Shahristânah city, was the district of Baraân, with the Rûdasht district beyond it lying further down the river, of which last the chief centre was Fârifân, a large town in the 8th (14th) century, though now only a village, standing near the great Gâv-Khânah swamp. The Karârij district is south of Baraân; and westward of this, higher up the right bank of the Zâyindah Rûd, is the great Khânlanjân district, the last of the four to the south of the river, of which the chief town was Fîrûzân. Of this city no trace apparently remains, but it was a considerable town 'in two parts' in the 8th (14th) century, situated on the Zâyindah Rûd, and Ibn Baṭûṭah, who passed through it, says it lay six leagues distant from Isfahân. The Khânlanjân district was already famous in the 4th (10th) century for its plentiful fruits and the fertility of its lands. Its name is often written Khâlanjân or Khûlanjân, and it was also known as Khân-al-Abrâr, 'the Caravanserai of the Benefactors.' As the name of a town Khânlanjân is doubtless identical with Fîrûzân aforesaid, and in the Itineraries this is the first stage southward from Isfahân on the western road to Shîrâz. In the 5th (11th) century Nâṣir-

i. 431. Kaz. ii. 265. The description of Isfahân fills volume VIII (see especially pp. 122, 126, 147, 153, 212, 227, 229, for passages referred to) of the *Voyages du Chevalier Chardin en Perse* (Amsterdam, 1711). For modern Isfahân see Houtum-Schindler, *Eastern Persian 'Irâḳ* (1897), pp. 18, 19, 120, 122.

i-Khusraw passed through Khânlanjân on his way to Isfahân, and noticed on the city gate an inscription bearing the name of Tughril Beg the Saljûḳ[1].

The main stream of the Isfahân river, at the present time generally called the Zandah Rûd, is known as the Zâyindah Rûd or the Zarîn-Rûdh to our various authorities, though this last name is now generally given to a tributary river. The main stream, in its upper reach, was named the Jûy-Sard, 'the Cold River,' and this rose in the Zardah-Kûh, 'the Yellow Mountains' —still so called from their yellow limestone cliffs—30 leagues west of Isfahân, not far from the head-waters of the Dujayl or Kârûn river of Khûzistân; and here, according to Mustawfî, were also the Ashkahrân mountains, which marked the frontiers of Great Lur. Below the town of Fîrûzân in Khânlanjân, the Zandah Rûd receives an affluent, almost equal to its main stream in volume, which comes down from near Gulpaygân (Jurbâdhaḳân); then after passing Isfahân, and irrigating its eight districts, the Zandah Rûd somewhat to the eastward of Rûdasht flows finally into the swamp of Gâv-Khânah on the borders of the Great Desert. According to popular belief, which is mentioned already by Ibn Khurdâdbih in the 3rd (9th) century, the river, after sinking into this swamp, rose again to the ground surface 90 leagues away in Kirmân, thence reaching the sea; but Mustawfî not unnaturally discredits the story, because of the high mountains lying between Isfahân and Kirmân, and though he states that it was said that bits of reeds thrown into the Gâv-Khânah marsh reappeared in Kirmân, he adds 'but this account is incredible[2].'

Nâyin, which lay to the north of the Gâv-Khânah swamp on the border of the Great Desert, and the towns to the south-east

[1] I. K. 20, 58. I. R. 152. Kud. 197. I. H. 201. Ykb. 275. Muk. 389, 458. Yak. i. 294; ii. 394; iii. 839. Mst. 143, for the most part reproduced in J. N. 291. I. B. ii. 42. N. K. 92. Khânlanjân is famous also as the place of refuge of Firdûsî, when he fled from the wrath of Sultan Maḥmûd of Ghaznah. An account of his reception by the governor of Khânlanjân is given in a copy of the *Shâh Nâmah* preserved in the British Museum (Or. 1403, f. 518 *a*), of which the text and translation are given by C. Schéfer in his edition of Nâṣir-i-Khusraw (Appendix iv. p. 298).

[2] I. R. 152. I. K. 20. Mst. 201, 202, 214.

of it towards Yazd, were all included in the province of Fârs during the middle-ages, as will be explained in Chapter XVIII, but Ardistân, some miles north-west of Nâyin, was counted as of the Jibâl province. As early as the 4th (10th) century, Iṣṭakhrî describes Ardistân as a walled city, a mile across, with five gates and well fortified. The Friday Mosque stood in the centre of the town, and much silk was manufactured here, chiefly for export. At Zuvârah, to the north-east of Ardistân, some ancient ruins were attributed to King Anûshirwân the Just, and Muḳaddasî adds that the soil of Ardistân was white, 'like wheat flour, whence its name,' for *Ard* in Persian meaning 'meal,' Ardistân would have the signification of 'the place like flour.' The ruins are referred to under the name of Uzvârah by Yâḳût, who states that there were many vaulted buildings, also the remains of a fire-temple that had become the castle of Ardistân, and here according to tradition Anûshirwân had been born. Mustawfî however, who spells the name Zuvârah, attributes all these remains, including the fire-temple, to King Bahmân, son of Isfandiyâr; and records that the town, which stood close to the desert, had round it 30 villages, giving as a tradition that these had been built by Dastan, brother of the hero Rustam.

On the desert border between Ardistân and Kâshân were the Kargas Kûh, 'the Vulture Hills,' which Muḳaddasî describes as the highest mountains in the Great Desert of Persia. The neighbouring Siyâh Kûh, 'Black Hills,' were of almost equal height and ruggedness:—'black evil-looking mountains'; and both, says Iṣṭakhrî, were famous hiding-places for robbers. In a valley of the Vulture Hills was a fine spring called the Âb-i-Bandah, which gushed out from a cleft that was completely enclosed by rocks. About half-way between the Kargas Kûh and the Siyâh Kûh on the desert road, stood the caravanserai called Dayr-al-Jiṣṣ, 'Gypsum Convent,' a strong place, built entirely of burnt brick and shut by iron gates. In this hostel, according to Iṣṭakhrî, guides for the desert routes were to be found, stationed here by order of the Sultan. Further, great tanks had been constructed here for storing water, which Muḳaddasî relates were never allowed to go out of repair, and there were shops in the caravanserai for the sale of provisions. Mustawfî describes the Kargas Kûh as standing solitary, being joined to no other range, and some ten leagues in circuit. In their rocky heights the vultures nested, and the ibex (*wa'l*), that could live long without water, was found here in great numbers. To the west of Ardistân is the town of Naṭanz, or Naṭanzah, which appears to be mentioned by no Arab geographer before the time of Yâḳût. Mustawfî states that its castle was called Washâḳ, after one who was governor of Naṭanz, though originally this castle had borne the name of Kamart. Close to Naṭanz also was the large village of Ṭarḳ, almost a town says Yâḳût, and here according to Ḳazvînî the people were celebrated for their skill in carving bowls out of ivory and ebony; these being largely exported[1].

The city of Kâshân is mentioned by Iṣṭakhrî 'as a pleasant town, clay built, like Ḳum.' The earlier Arab geographers always spell the name Ḳâshân (with the dotted ḳ). The place became famous throughout the east for its tile-work, which took the name Ḳâshî (for Ḳâshânî), this being still the common term for the well-known enamelled blue and green tiles so much used in mosque decoration. According to Muḳaddasî Ḳâshân was the reverse of famous for its scorpions; and Yâḳût, who refers to the beautiful green bowls of Ḳâshî-ware which were in his day largely exported, speaks of the population as all fanatical Shî'ahs of the Imâmite sect. Mustawfî asserts that Kâshân had originally been built by Zubaydah, the wife of Hârûn-ar-Rashîd; and he praises the palace of Fîn, lying near Kâshân, for its tanks and water-courses, which were supplied by the river from Kuhrûd. The Kâshân river, which in summer went dry before reaching the town limits, in spring often endangered the city with its floods, which passing on were lost in the neighbouring desert.

The city of Ḳum (more correctly spelt Ḳumm according to Arab orthography), to the north of Kâshân, is now famous among the Shî'ahs for its shrine, said to mark the tomb of Fâṭimah, sister of the sixth Imâm 'Alî-ar-Riḍâ, a contemporary of Hârûn-ar-Rashîd, whom they assert to have died here of poison on her way to join her brother in Khurâsân. Already in the 4th (10th) century Ibn Ḥawḳal describes Ḳum as peopled by Shî'ahs; it was then a

[1] Ist. 202, 228, 230, 231. I. H. 288—291. Muk. 390, 490, 491. Yak. i. 198; iii. 531; iv. 793. Mst. 150, 151, 206. J. N. 299.

walled town, with fertile gardens round it, celebrated for pistachio nuts and filberts. The ancient name of Ḳum according to Yâḳût had been Kumandân, curtailed by the Arabs to Ḳumm. The remains of a Persian fortress were, he says, still to be seen among the ruins of the town, and an ancient stone bridge crossed the river which separated the older site from the Moslem town. Mustawfî states that the walls of Ḳum measured 10,000 paces in circuit, and, like Âvah, the place was celebrated for its numerous ice-houses excavated in the ground; also for its cypress trees, and for vines which produced the famous red grapes. When Mustawfî wrote in the 8th (14th) century most part of Ḳum lay in ruins, and it is to be remarked that neither he nor any earlier authorities make any mention of the tomb of Fâṭimah, although the city is always noted as being a centre of the Shî'ah sect[1].

The river of Ḳum rose in the Gulpaygân district near the mountains of Khânsâr, as Mustawfî writes the name, and these ranges are the watershed between the Ḳum river and the left-bank tributary of the Isfahân river already mentioned. Jurbâdh-aḳân is the Arab name for Gulpaygân, of which the older form was Gurbâyigân, and Mustawfî explains the name to mean 'the place of roses,' writing it *Gul-âbâd-ikân* and goes on to praise its fertility and the excellent water, 50 villages being of its dependencies. Muḳaddasî refers to Jurbâdhaḳân as lying about half-way between Karaj of Abu Dulaf and Isfahân, and the village of Khânsâr which gave its name to the district, Yâḳût adds, was of its neighbourhood. The town of Dalîjân lies further down the Ḳum river; and according to Yâḳût the name was pronounced Dulayjân or Dulaygân. Formerly it had been a flourishing place, but when Mustawfî wrote it had fallen to ruin. After passing the city of Ḳum, the Ḳum river joined the waters of the great stream coming down from Hamadân, called the river Gâvmâhâ, or Gâvmâsâ, which itself a short distance above Ḳum had received on its right bank the Âvah river, and on its left bank the river passing Sâvah. All these streams branched to form many water channels, and intermingling by cross canals finally became lost in the Great Desert to the north-east of Ḳum.

[1] Ist. 201. I. H. 264. Muk. 390. Yak. iv. 15, 175. Mst. 150, 217. J. N. 305.

The town of Âvah (called Âvah of Sâvah to distinguish it from Âvah near Hamadân, see p. 196) lay a short distance to the west of Ḳum. The Âvah river took its rise in Tafrîsh, which Mustawfî describes as a district 'that on all sides was only approached by passes,' and the country here was very fertile, with many villages. The town of Âvah is mentioned by Muḳaddasî, who names it Âvâ or Âvah of Ray; and Yâḳût, who speaks of it as a village or a small town, writes the name Âbah, adding that its population were ardent Shî'ahs. In the 8th (14th) century Mustawfî describes Âvah as enclosed by a wall a thousand paces in circuit, and there were pits for storing ice, which were famous, for ice was much in demand during the summer heats; but the bread here was very bad. Between Âvah and Ḳum, he describes an isolated hill, called Kûh Namak Lawn (Salt Mountain), where the earth was everywhere mixed with salt. To reach the summit was impossible on account of the friable nature of the ground; no snow either would remain on its sides, and the salt was too bitter to be used by man. This hill was three leagues in circuit, and so high as to be visible at a distance of 10 leagues[1].

The city of Sâvah, lying midway between Hamadân and Ray on the great caravan road which traversed Persia (the Khurâsân road), was a place of importance as early as the 4th (10th) century, when Ibn Ḥawḳal describes it as noted for its camels and camel-drivers, both much in demand throughout the land by pilgrims and travellers. Muḳaddasî adds that the town was fortified, that there were fine baths here, and that the Friday Mosque stood near the high road, and at some distance from the market. The people of Sâvah were Sunnîs, and Yâḳût writes that in his day they were perpetually at feud with their neighbours of Âvah, who were Shî'ahs. Sâvah suffered severely at the hands of the Mongols in 617 (1220), who plundered the town, slaying most of its inhabitants; and among other buildings burning the great library, which Yâḳût had seen, and describes as having had no equal throughout all Persian 'Irâḳ. This library is also referred to by Ḳazvînî, who says it was housed in the Great Mosque, and contained, besides books on all subjects, a set of astrolabes and globes for the study

[1] Ist. 195, 198. Muk. 25, 51, 257, 386, 402. Yak. i. 57; ii. 46, 392, 584. Mst. 147, 150, 206, 216.

of astronomy. In the town was a hospital, as well as many colleges and caravanserais; and at the gate of the mosque was a mighty arch, recalling the arch of the Chosroes at Madâin.

In Moslem legend Sâvah was famous for the great lake which had been here before the days of Islâm, and which had suddenly dried up on the night of the birth of the prophet Muḥammad; 'the water sinking down into the earth in joy at the good news,' as Mustawfî writes. He adds that in his day the walls of Sâvah had been recently rebuilt of burnt brick, being then 6200 ells in circuit. Four leagues to the west of Sâvah was the shrine of the prophet Samuel, and when Mustawfî wrote the population of the town had nearly all become Shî'ahs. He mentions the names of many of the surrounding villages, and adds that corn, cotton, and pomegranates were grown abundantly throughout the district.

The Sâvah river was called the Muzdaḳân, from a town of this name which stood on its banks. This stream rose at Sâmân, a large village on the border of the Kharraḳân district of Hamadân (see p. 196), lying in a rich country producing corn and grapes. From Sâmân the river came to Muzdaḳân (also spelt Muṣdaḳân), a town which Mustawfî describes as 3000 paces in circuit, with a cold climate, being in the hill country. Yâḳût speaks of a celebrated Rubâṭ—guard-house or monastery—at Muzdaḳân, where many Ṣûfîs had their abode; and the town was a stage on the great caravan road crossing Persia. After passing through Sâvah, Mustawfî tells us, the Muzdaḳân river divided, part of its waters sinking underground into a great pit, while a moiety joined the Gâvmâhâ.

The long river called the Gâvmâhâ (or Gâvmâsâ as some MSS. write the name), which Mustawfî carefully describes for us, is now known as the Ḳârâ Ṣû—Black Water—along a part of its course. It had its head-waters, as already said, in the Hamadân plain, where divers streams came down from Asadâbâd, the Alvand mountain, and the Farîvâr district. Flowing first northward and then bending sharply to the east, it received from the south a great affluent, the river rising near Karaj of Abu Dulaf. Beyond Sâvah and Âvah, where it received the two other affluents we have previously described, a great dam was built across the river to retain its waters for irrigation purposes during the summer droughts. The Gâvmâhâ eventually mingled its stream with the river of Ḳum coming from Gulpaygân, and Mustawfî adds that their surplus waters after passing a place called Haftâd Pulân, 'Eighty Bridges,' finally escaped and were lost in the Great Desert. The Gâvmâhâ river was to its district, says Mustawfî, what the Zandah Rûd was to Isfahân, being the chief fountain of its riches and prosperity. It is to be remarked that none of the earlier Arab geographers make mention of this river[1].

[1] I. H. 258. Muk. 392. Yak. iii. 24; iv. 520. Kaz. ii. 258. Mst. 148, 149, 152, 217. The dam on the Gâvmâhâ was built by Shams-ad-Dîn, prime minister (Ṣâḥib-Dîvân) of Sulṭân Aḥmad, son of Hûlâgû, the third Îl-Khân of Persia.

CHAPTER XV.

JIBÂL (*continued*).

Ray. Varâmîn and Ṭihrân. Ḳazvîn and the castle of Alamût. Zanjân. Sulṭânîyah. Shîz or Satûrîḳ. Khûnaj. The districts of Ṭâliḳân and Ṭârum. The castle of Shamîrân. The trade and products of the Jibâl province. The high roads of Jibâl, Adharbâyjân and the frontier provinces of the north-west.

At the north-eastern corner of the Jibâl province stood Ray, more correctly spelt Rayy, which the Arab geographers always write with the article Ar-Rayy, the name representing the Greek Rhages. In the 4th (10th) century Ray appears to have been the chief of the four capital cities of the Jibâl province; 'except for Baghdâd, indeed, it is the finest city of the whole east,' Ibn Ḥawḳal writes, 'though Naysâbûr in Khurâsân is more spacious,' and Ray covered at that time an area of a league and a half square. Officially, during the Abbasid Caliphate, Ray was known as Muḥammadîyah, in honour of Muḥammad, afterwards the Caliph Mahdî, who had lived here during the reign of his father Manṣûr, and had rebuilt much of the city. His son Hârûn-ar-Rashîd was born here, and under its official title of Muḥammadîyah it became the chief mint city of the province, this name occurring on many of the Abbasid coins.

In Ray the houses were mostly built of clay, but burnt bricks were also largely used. The town was strongly fortified, and Ibn Ḥawḳal mentions five gates; the gate of the Bâṭâḳ Arch opening (S.W.) on the Baghdâd road, Bâb Balîsân (N.W.) towards Ḳazvîn, Bâb Kûhak (N.E.) towards Ṭabaristân, Bâb Hishâm (E.) on the Khurâsân road, and Bâb Sîn (S.) towards Ḳum. The markets of the city lay at, and outside, these gates, and the most frequented were in the suburbs of Sârbânân and Ar-Rûdhah, where shops, and warehouses filled with merchandise, extended along both sides of the main thoroughfare for a great distance. Two rivers, according to Ibn Ḥawḳal, brought water to Ray, one called Sûrḳanâ running past the Rûdhah suburb; and the other, the river Al-Jîlânî, flowing through Sârbânân. Yâḳût also mentions the Nahr Mûsâ (River of Mûsâ), coming down from the mountains of Daylam, which may therefore be identical with the Jîlânî or Gîlân river, aforesaid. Muḳaddasî refers to two great buildings in Ray, one the Dâr-al-Baṭṭîkh, 'the water-melon house,' a name commonly given to the city fruit-market, the other the Dâr-al-Kuttub, or library, lying below Rûdhah in a khân (caravanserai), where, however, there were not many books, according to his account.

In the 4th (10th) century both Ibn Ḥawḳal and Muḳaddasî speak of Ray as already much gone to ruin, the chief traffic then being in the suburbs of the older town. High above the Great Mosque, which Yâḳût states was built by the Caliph Mahdî and finished in 158 (775), was the castle, which stood on the summit of a steep hill, of which Ibn Rustah writes that 'from its top you overlook all the roofs of Ray.' The account of Ray given in Yâḳût is not very clear, but he quotes, in one part of his work, an old topographical description of the town, which is to the following effect. The Inner City, where the mosque and the Government House stood, was the quarter surrounded by a ditch, and this was generally known as Al-Madînah, 'the City' proper. The Outer City was that part more especially known as Al-Muḥammadîyah, which at first had been a fortified suburb. It crowned the summit of the hill overlooking the lower (or inner) town, and according to the information quoted by Yâḳût its castle was known as Az-Zubaydîyah (some MSS. give the name as Az-Zaybandî), which had been the palace of Prince Mahdî when he was quartered in Ray. Afterwards this became the prison, and it was rebuilt in 278 (891). Further, there was another castle in Ray called the Ḳal'ah-al-Farrukhân, also known as Al-Jawsaḳ, 'the Kiosque,' and during the 4th (10th) century Fakhr-ad-Dawlah the Buyid, who disliked the old palace on the hill-top, built himself a great house in the midst of gardens, which was afterwards known as Fakhrâbâd[1].

The most celebrated in early days of the many fertile districts round Ray were the following:—Rûdhah (or Ar-Rûdhah), with a large village of the same name beyond the city suburb; Varâmîn, which afterwards took the place of Ray as the chief city of this part of the Jibâl province; Pashâvíyah, still existing under the form Fashâvîyah; lastly, Ḳûsîn and Dîzah, with the districts of Al-Kaṣrân, 'the Two Palaces'—the outer and the inner—Dîzah being the name of two large villages or towns lying one day's journey from Ray, to wit, Dîzah of Ḳaṣrân, and Dîzah of Varâmîn. All these hamlets according to Ibn Ḥawḳal, with some others that he names, were like small towns, each with a population of over 10,000 men. In the year 617 (1220) Ray was taken, plundered, and burnt by the Mongol hordes, and from this great calamity it never recovered. Yâḳût, who passed through the place at this time, states that the city walls alone remained intact, most of the houses being reduced to ruin. Many of these had originally been built of burnt brick, faced with blue enamelled tiles, which Yâḳût describes as 'varnished smooth like the surface of a bowl.' The Shâfi'ite suburb, the smallest of the city quarters, alone had escaped the Mongols, the quarters of the Ḥanbalites and of the Shî'ahs having been completely ruined[2].

From its state of utter ruin Ghâzân Khân the Mongol, by imperial decree, according to Mustawfî, attempted to restore Ray, ordering the city to be rebuilt and repeopled. The attempt, however, failed, for the population had already shifted to the neighbouring towns of Varâmîn and Ṭihrân, more especially the former, which, having a better climate than the older Ray, had become at the beginning of the 8th (14th) century the most flourishing city of the district. The ruins of Varâmîn lie at some distance to the south of Ray, while to the north of the city, Mustawfî says, was the hill of Ṭabarik—presumably not that on which the castle

[1] Ykb. 275. I. R. 168. I. H. 265, 269, 270. Muk. 390, 391. Yak. ii. 153, 894, 895; iii. 855; iv. 431. Whether or not the fortress of Ray built by Mahdî was called Zubaydîyah (if this indeed be the true reading) after the future wife of his son Hârûn-ar-Rashîd is not clear.

[2] I. H. 270, 289. Yak. ii. 572, 833, 893, 894.

already mentioned as built by the Caliph Mahdî had stood—where a silver mine was worked at much profit to the state. This castle of Ṭabarik, according to the chronicle of Ẓahîr ad Dîn, was founded by Manûchahr the Ziyârid at the beginning of the 5th (11th) century. Yâḳût states that it was destroyed in 588 (1192) by Ṭughril II, the last Saljûḳ Sulṭân of 'Irâḳ, and a long account is given of the siege of this famous stronghold. The Ṭabarik hill, he adds, lay on the right of the Khurâsân road to a traveller leaving Ray, while the Hill of Ray (presumably the site of the castle built by Mahdî) lay to the left of one leaving the city gate. Mustawfî describes the shrine of the Imâm Zâdah 'Abd-al-'Aẓîm as situated close to Ray, and this Mashhad, or place of martyrdom, is still the most venerated sanctuary of modern Ṭihrân; the saint being a certain Ḥusayn, son of 'Alî-ar-Riḍâ, the eighth Imâm.

One of the famous districts near Ray was called Shahriyâr, and Mustawfî incidentally mentions a castle (Ḳal'ah) of this name as lying to the north of the city. In later times this castle must have become important, for Shahriyâr or Ray-Shahriyâr is the name which 'Alî of Yazd, when describing the campaigns of Tîmûr, gives to Ray. Varâmîn, as already said, was then the chief centre of population, but this town in the beginning of the 9th (15th) century was itself already falling to ruin. At a later time its place was taken by Ṭihrân, which in the 7th (13th) century is merely mentioned as one of the largest villages of Ray. The early Ṭihrân (also spelt Tihrân with the soft *t*) had many half-underground houses, 'like Jerboa holes' according to Ḳazvînî, and the people of its twelve wards were always fighting, each ward against the other. Mustawfî in the next century describes Ṭihrân as a medium-sized town; but it was not till long after, namely at the close of the 12th (18th) century, that the city was made the capital of Persia by Âḳâ Muḥammad Shâh, founder of the Ḳajar dynasty[1].

The rivers that water the plain in which Ray, Varâmîn, and Ṭihrân stand, flow thence to the neighbouring border of the Great

[1] Kaz. ii. 228, 250. Mst. 143, 144, 205. Yak. iii. 507, 564. A. Y. i. 583, 586, 597. Ẓahîr-ad-Dîn (Dorn, *Muhammadanische Quellen*, i. p. 15 of the Persian text) states that *Ṭabarik* means 'a hillock,' being the diminutive of *Ṭabar* which signifies 'a hill or mountain' in the Ṭabaristân dialect. Ṭabarik of Isfahân has been noticed on p. 205.

Desert and there are lost. One of the chief streams was the Nahr Mûsâ already mentioned, along whose bank lay many villages; further, Mustawfî speaks of the river Karaj, which was crossed by a bridge of a single arch known as the Pul-i-Khâtûn, 'the Lady's Bridge,' and so called, it was said, in memory of the lady Zubaydah, wife of Hârûn-ar-Rashîd. The ruins of this bridge still exist not far from Ṭihrân. Ḳazvînî also mentions the Nahr Sûrîn, whose waters were carefully avoided by the Shî'ah population of Ray, because the body of the murdered Yaḥyâ, grandson of 'Alî Zayn-al-'Âbidîn the fourth Imâm, had been washed in it, and thus polluted the stream for evermore. The chief river of Ray, however, according to Mustawfî, was the Jâyij Rûd, which, rising in the Jâyij range under Damâvand, divided into forty channels on reaching the plain of Ray.

On the western border of this plain lies the district of Sâûj Bulâgh—meaning 'Cold Springs' in the Turkish dialect—which is described by Mustawfî as having been an important place under the Saljûḳs. In the time of the Mongols it paid revenues to the amount of 12,000 dînârs, and the chief among its numerous villages was Sunḳurâbâd (which still exists), an important stage on the itinerary given by Mustawfî. Sâûj Bulâgh district was watered by the Garm Rûd, which, rising in the mountains to the east of Ḳazvîn, irrigated the districts of Ray and Shahriyâr, where it was joined by many streams from the mountain range to the north before such of its waters as were not used up in irrigation channels were absorbed by the Great Desert[1].

Ḳazvîn (otherwise Ḳazwîn) lies about a hundred miles northwest of Ṭihrân, immediately below the great mountain chain, and from the earliest times was an important place, guarding the passes that led across the Ṭabaristân province to the shores of the Caspian. The mountain region to the north-west had in early times formed part of the district of Daylam (already described in Chapter XII) which for a time was semi-independent, not having been brought under the government of the Abbasids. During this period Ḳazvîn was the chief fortress against these fierce infidels, and was strongly garrisoned by Moslem troops. Already in the times of the

[1] Kaz. i. 181. Mst. 144, 148, 196, 216: and see British Museum MS. Add. 23,543, f. 179*b*. J. N. 292, 304.

Omayyad Caliphs, Muḥammad, son of Ḥajjâj—the latter being the celebrated governor of Arabian 'Irâḳ—had been sent by his father at the head of an army against the infidels of the Daylam mountains. This Muḥammad had halted at Ḳazvîn, and built here the first Friday Mosque, which Yâḳût describes as standing near the gate of the palace of the Banî Junayd. It was called the Masjid-ath-Thawr, 'the Bull Mosque,' and was the chief mosque of the city till the days of Hârûn-ar-Rashîd. Ibn Ḥawḳal in the 4th (10th) century describes Ḳazvîn as consisting of a double city, one without, the other within, and there were two Friday Mosques in the central town, which was like a fortress. Its lands were very fertile, and the houses of the city covered an area of a square mile. The people were brave and warlike, and it was from this city that the Abbasid Caliphs were wont to despatch punitive expeditions into Ṭâliḳân and Daylam.

The two chief rivers of Ḳazvîn, according to Ya'ḳûbî, were the Wâdî-al-Kabîr (the Great Stream), and the Wâdî Sayram. There were the remains of many fire-temples in this neighbourhood, and Muḳaddasî praises the grapes grown in the gardens round the place. Of the double town the two quarters were called the Madînah Mûsâ and the Madînah Mubârak, otherwise the Mubârakîyah. The Caliph Hâdî (elder brother of Hârûn-ar-Rashîd), whose name was Mûsâ, had built here the town quarter named after him, Madînah Mûsâ. This was during the Caliphate of his father Mahdî; and afterwards Hârûn-ar-Rashîd (who succeeded Hâdî) on his way to Khurâsân had halted in Ḳazvîn, where he laid the foundations of the new mosque and built the city walls. Mubârak the Turk, a freedman either of the Caliph Mamûn or of Mu'taṣim, was the builder of the Mubârakîyah fortress at Mubârakâbâd, otherwise called the city of Mubârak.

Throughout the middle-ages Ḳazvîn continued to be a flourishing town, but at the beginning of the 7th (13th) century it was laid in ruins by the Mongols. A hundred years later, Mustawfî, who was himself a native of Ḳazvîn, gives a long account of the place, derived in part from local traditions. He states that on the site of later Ḳazvîn there had stood an ancient Persian city, built by King Shâpûr and called Shâd Shâpûr—'the Joy of Sapor.' Near its ruins the two Moslem cities of Madînah

Mûsâ and Mubârakâbâd (Mubârak, he says, was a freedman of the Caliph Hâdî) were subsequently built, and Hârûn-ar-Rashîd surrounded all three towns by a great fortified wall. This wall was only completed in 254 (868) by the Turk commander Mûsâ ibn Bughâ in the reign of the Caliph Mu'tazz; and it was afterwards rebuilt in burnt brick by Ṣadr-ad-Dîn, the Wazîr of the Saljûḳ Sulṭân Arslân II, in 572 (1176). Mustawfî further states that 300 villages were of the dependencies of Ḳazvîn, and of these the most important were Fârisjîn and Sagsâbâd, both mentioned in his itinerary. He also names a number of streams which irrigated the Ḳazvîn territory, namely the Kharûd, with the Buh Rûd and Kardân Rûd both flowing from Ṭâliḳân, and the Turkân Rûd coming from the Kharraḳân district (see p. 196). According to Ḳazvînî the streams that watered the gardens of the city were the Daraj river on the east, and the Atrak river on the west; and the same author also names a number of towns and villages that were situated in the plain, and in the hill country overlooking Ḳazvîn[1].

Dastuvâ (or Dastabâ) under the Omayyads holds the position of a mint city, and is the name of a great district, of which Yazdâbâd was the chief village. In Omayyad times Dastuvâ had belonged in part to Ray, in part to Hamadân, and we are told that the direct post-road from Ray to the Adharbâyjân province lay through it, avoiding Ḳazvîn. The name is no longer found on the map, but Dastabâ must have been to the south of Ḳazvîn, of which city in later days, under the Abbasid Caliphs, it came to be counted as a dependency.

To the north-west of Ḳazvîn, on the summits of the mountains dividing this district from that of Rûdbâr, which lay along the

[1] I. H. 259, 263, 271. Ykb. 271. I. K. 57. Muk. 391. Yak. iv. 88, 89, 454, 455. Kaz. ii. 190, 193, 194, 196, 244, 274, 275, 290. Mst. 145, 146, 196, 217. As his name implies, Ḳazvînî (like Mustawfî) was a native of Ḳazvîn, and Mustawfî in his history (the *Guzîdah*) has left a long account of his birth-place, which M. Barbier de Meynard has translated in the *Journal Asiatique* for 1857, ii. p. 257. Ḳazvînî (ii. 291) gives a rough ground plan of the town, which is figured in concentric circles of walls. The inner circle was the Shahristân, and this was surrounded by the great city (Al-Madînah-al-'Uẓmâ), which in turn was enclosed by gardens, depicted as encircled by arable fields; the latter traversed by the two rivers.

river Shâh Rûd in Ṭabaristân, stood the famous castles of the Assassins (Ismâîlians), fifty in total number Mustawfî says, of which Alamût was the capital and Maymûn Diz the strongest fortress. The name Alamût is said to mean 'the eagle's nest' or 'the eagle's find' in the Ṭabaristân dialect, and the first to build a castle here was a Daylamite king whose hunting eagle had by chance once perched on the crag. Ḳazvînî, who doubtless knew the place well, describes the castle as surrounded by deep and wide ravines, cutting it off from all communication with the neighbouring mountain spurs, and rendering it impregnable, for it was beyond bow-shot or even the bolts from a mangonel. Alamût lay six leagues distant from Ḳazvîn, and its later fortress was built by the 'Alid missioner Ḥasan, surnamed Ad-Dâ'î-ilâ-l-Ḥaḳḳ, in 246 (860). In 483 (1090)—or 446 (1054) according to Ḳazvînî—it came into the possession of Ḥasan Sabâḥ, surnamed the Old Man of the Mountain, and for 171 years was the chief stronghold of his followers. Alamût was taken and dismantled in 654 (1256) by order of Hûlâgû Khân the Mongol, and, after its fall, the remaining castles of the Assassins were quickly captured and razed to the ground. Its supposed site has been visited by various travellers, and the remains of many other fortresses, said to be those of the Ismâîlians, still exist in the mountains to the north of Ḳazvîn[1].

Abhar and Zanjân, two cities often named together, lay on the high road west of Ḳazvîn, and were famous from early times. Ibn Ḥawḳal in the 4th (10th) century mentions Abhar as peopled by Kurds, its fields were very fertile and well watered, corn being largely grown here. It was protected by a strong castle built upon

[1] Kaz. ii. 200. Mst. 147. In the *Guzîdah* (chapter IV, section ix, part 2) Mustawfî gives the history of the Ismâîlians or Assassins in Persia; and this has been translated, with notes, by Defrémery, in the *Journal Asiatique* (1849, i. 26). He gives in a list (p. 48) the names of the Ismâîlian fortresses taken and destroyed by order of Hûlâgû, but the position of most of these is unknown. Girdkûh and Lanbasar were the last strongholds to fall. Alamût, however, appears not to have been entirely destroyed by Hûlâgû, or perhaps it was rebuilt later, for it served as a state prison under Shâh Sulaymân the Ṣafavî, as is mentioned by Chardin (*Voyage en Perse*, x. 20). In the last century Colonel Monteith visited the ruins, and has described them in the *J. R. G. S.* for 1833 (p. 15).

a great platform, and Ḳazvînî reports that it was famous for its water-mills, also for the so-called 'Abbâsî pear grown here, in shape like an orange and very sweet. According to Yâḳût the Persians pronounced the name Avhar. Mustawfî records that the fortress was rebuilt under the Saljûḳs by the Atabeg Bahâ-ad-Dîn Ḥaydar, and hence was known as the Ḥaydarîyah. The city walls measured 5500 paces in circuit, and the Abhar river, after watering the district, flowed towards Ḳazvîn, becoming lost in the desert plain. The city of Zanjân lay about 50 miles to the north-west of Abhar, and on the Zanjân river, which flowed west to the Safîd Rûd. Zanjân is described by Ibn Ḥawḳal as larger than Abhar; and it was on the high road into Adharbâyjân. The Persians, Yâḳût says, pronounce the name Zangân, and Mustawfî states that the place was founded by King Ardashîr Bâbgân, being first named Shahîn. Zanjân had been ruined during the Mongol invasion; its walls, however, were still 10,000 paces in circuit, the district was most fertile, and its revenues amounted to 20,000 dînârs. Mustawfî adds that the language talked here, in the beginning of the 8th (14th) century, was still 'almost pure Pahlavî,' by which a local Persian dialect is doubtless indicated[1].

About half-way between Abhar and Zanjân, in the centre of the great plain forming the watershed between rivers flowing west to the Safîd Rûd and east to the Great Desert, lie the ruins of the Mongol city of Sulṭânîyah, which, founded by Arghûn Khân, was completed by Uljaytû Sulṭân in 704 (1305) and made the capital city of the Îl-Khân dynasty. Abu-l-Fidâ states that its Mongol name was Kungurlân, and according to Mustawfî nine cities were of its dependencies. Its walls were 30,000 paces in circuit, and in the central fortification stood the great sepulchre of Uljaytû, adorned with many carvings in stone. The ruins of this domed tomb (or mosque) still exist, but of the city nothing now remains, although Mustawfî says that in his day Sulṭânîyah contained finer buildings than any other town in Persia, Tabrîz alone excepted. On the Abhar road five leagues east of Sulṭânîyah lay the village of Ḳuhûd, 'which the Mongols call Ṣâin Ḳal'ah,' Mustawfî writes, and under the latter name—'Ṣâin's Fortress'—

[1] I. H. 258, 271, 274. Muk. 378, 392. Kaz. ii. 191. Yak. i. 104; ii. 573, 574, 948; iv. 1017. Mst. 146, 147, 217.

the place still exists, Ṣâin, otherwise called Bâtû Khân, being the grandson of Changîz Khân. The strong castle of Sarjahân stood on the mountain spurs half-way between Ṣâin Ḳal'ah and Sulṭânîyah. From the latter it was distant five leagues, and it crowned a hill-top overlooking the great plains which extended thence eastward to Abhar and Ḳazvîn. Yâḳût describes Sarjahân, which was of the Ṭârum district, as one of the strongest fortresses that he had seen; but when Mustawfî wrote it was in ruins, the result of the Mongol invasion, its munitions of war and garrison having been transferred to Ṣâin Ḳal'ah.

To the west of Sulṭânîyah lay the two small neighbouring towns of Suhravard and Sujâs, which were still of some importance when Mustawfî was here in the 8th (14th) century, though now entirely gone to ruin. Ibn Ḥawḳal writes in the 4th (10th) century that Suhravard with its Kurdish population was then as large as Shahrazûr, it was a walled town and well fortified, lying to the south of Zanjân on the road to Hamadân. Sujâs, or Sijâs, lay close to Suhravard, and Mustawfî describes both places as having been ruined during the Mongol invasion, so that in his day they were merely large and populous villages. The surrounding districts were called Jarûd and Anjarûd (at the present day they are known under the names of Ijarûd and Angurân), and Sujâs lay five leagues west of Sulṭânîyah in the midst of more than a hundred villages settled by Mongols. In the mountain near was the grave of Arghûn Khân, made a *Kurûgh* or 'inviolate sanctuary' after the custom of the Mongols, and his daughter Uljaytû Khâtûn had built here a khânḳâh or convent for Darvishes[1].

On the western border of the Jibâl province, near one of the head-streams of the Safîd Rûd, are the remarkable ruins called Takht-i-Sulaymân—'Solomon's Throne'—at the present day, with a little lake or pool which is always kept full by a natural syphon, however much water may be drawn off. These ruins

[1] I. H. 258, 263. Kaz. ii. 261. Yak. iii. 40, 70, 203. A. F. 407. Mst. 144, 145, 148, 149, 196. Both Sujâs and Suhravard have apparently now disappeared from the map; though Sir H. Rawlinson writes (*J. R. G. S.* 1840, p. 66) that Sujâs was in his time a small village lying 24 miles S.E. of Zanjân he further adds that Suhravard is 'now lost.'

have been identified with the city of Ash-Shîz, mentioned by the early Arab geographers, which Mustawfî also describes under the name of Satûrîḳ. At Shîz, Ibn Khurdâdbih, writing in the 3rd (9th) century, describes the great fire-temple, so much honoured by the Magians, which bore the name of Adharjushnas. Hither, walking on foot all the way from Madâin (Ctesiphon), and halting at the half-way stage of Shahrazûr, already noticed p. 190, each of the Sassanian Chosroes was bound to come as a pilgrim immediately after his accession to the crown; for according to one tradition Shîz was the birth-place of Zoroaster. Yâḳût reports that the Persian name was Jîs, otherwise Gazn, of which Shîz was an Arab corruption. He then quotes a long account from Ibn Muhalhal, who in 331 (943) wrote a description of Shîz, which he had visited in search of gold mines said to exist in its mountains. The town walls of Shîz, he states, surrounded a lake, that was unfathomable, about a Jarîb (one third of an acre) in extent, and whose waters always kept the same level though seven streams continually flowed from it, and these streams had the property of producing petrifaction on objects laid in their waters. Ibn Muhalhal also describes the fire-temple, from which the sacred fire was taken to all the other temples throughout Persia; and for seven hundred years, he says, the sacred fire had never been extinguished in Shîz. The same place is described by Mustawfî who gives it as the chief town of the Anjarûd district, and adds that the Mongols called it Satûrîḳ. He describes a great palace here, originally built, report said, by King Kay-Khusraw, the court of which was occupied by a bottomless pool or small lake that always maintained its level, although a stream perpetually flowed from it, while if the stream were dammed back the pool did not overflow. Mustawfî relates that Abaḳah Khân the Mongol had built himself a palace here, for there were excellent pasture grounds in the neighbourhood[1].

In the north-western angle of the Jibâl province, on the high road from Zanjân to Ardabîl, lay the important commercial town of Khûnaj, according to Ibn Ḥawḳal noted already in the 4th (10th)

[1] I. K. 119. I. F. 286. Kaz. ii. 267. Yak. iii. 353. Mst. 148. Sir H. Rawlinson (*J. R. G. S.* 1840, p. 65) would identify Takht-i-Sulaymân or Shîz with the northern Ecbatana of the Greek writers.

century for its fine breed of horses, sheep, and oxen. Yâķût, who had visited the town, gives the alternative spelling of Khûnâ, but he adds that it was more generally called Kâghadh Kunân, 'the Paper Factory'—for the people augured evil of the name Khûnâ which signified 'bloody' in Persian. Mustawfî, who in his itinerary gives the position of Kâghadh Kunân as lying six leagues south of the Safîd Rûd, and fourteen north of Zanjân on the direct road to Ardabîl, says that during the Mongol invasion it had been ruined, and was, when he wrote, merely the size of a village. The stream that watered its lands was a tributary of the Safîd Rûd. Excellent paper, however, was still manufactured, and the Mongols who had settled in the place gave it the name of Mughûlîyah, 'the Mongol Camp.' The exact site of Khûnaj has not, apparently, been identified.

Along the southern slope of the great range dividing the Jibâl province from Daylam and Ṭabaristân to the north, were the three districts of Pushkil-Ḍarrah, Ṭâliķân, and Ṭârum, of which the last two overlap, the names often being used indifferently, one for the other. These districts were each divided into Upper and Lower, the Upper region being of the mountains, and as such counted to be of the Daylam province. Pushkil-Darrah, according to Mustawfî, lay to the west of Ķazvîn, and south of Ṭâliķân. It comprised forty villages whose revenues had formerly gone to the up-keep of the Friday Mosque in Ķazvîn. The name Ṭâliķân—the district lying between the Sulṭânîyah plains and the northern mountain range—has disappeared from the map, but Aṭ-Ṭâliķân (as it is generally written) is frequently mentioned by the earlier Arab geographers. Muķaddasî refers to it as a most populous and fertile region; and expresses his wonder that the Sulṭân (the Governor of Daylam) does not live here instead of in the mountain valleys, 'but his people will not have it,' he adds. Ķazvînî refers to the abundant olives and fine pomegranates grown in Ṭâliķân, and Yâķût names some of its villages. Of these last Mustawfî gives a long list, but the majority of them it is impossible now to identify on the present map. He considered that most part of the Ṭâliķân region belonged rather to Gîlân.

To the north of Zanjân, likewise along the foot of the hill spurs, lies the Ṭârum district, which with the Arab geographers is generally found in the dual form Aṭ-Ṭârumayn, 'the Two Ṭârums,' Lower and Upper, the latter being entirely of the Daylam country. As already said, the Ṭârum river was a right-bank affluent of the Safîd Rûd, and its many tributaries irrigated this fertile district. Yâḳût, who spells the name Târum or Tarm (with the unemphatic t), says that there was no great city here, but in history the land was famous for the memory of the Vahsûdân family, and the last of these native chiefs had been dispossessed by Rukn-ad-Dawlah the Buyid. Mustawfî mentions Fîrûzâbâd as the capital town of Lower Ṭârum, Andar (or Aydî) being the chief place in Upper Ṭârum, with the fortress called Ḳal'ah Tâj, and he names five districts, each comprising numerous villages.

As being in Lower Ṭârum, but the position is nowhere given, Mustawfî mentions the great castle of Shamîrân, or Samîrân as the name is spelt by Yâḳût, who had himself visited its ruins. Yâḳût quotes also a long account from Ibn Muhalhal, who passed through Samîrân in about the year 331 (943), when it was counted as one of the chief strongholds of the Daylamite kings, and contained (he writes) 2850 and odd houses, large with small. Fakhr-ad-Dawlah the Buyid took the place in 379 (989), dispossessing the last of the Vahsûdân family, a child, whose mother the Buyid chief married. At about this date Muḳaddasî, who spells the name of the castle Samîrûm, describes it as being of the Salârvand district, and on its walls were 'lions of gold, and the sun and the moon,' though its houses were built but of mud-brick. In the middle of the next century the Persian traveller Nâṣir-i-Khusraw visited Samîrân on his pilgrimage to Mecca. This was in 438 (1046) and he describes it as the capital of Ṭârum in Daylam. It apparently lay three leagues west of the junction of the Shâh Rûd with the Safîd Rûd on the high road to Sarâv in Adharbâyjân. Above the lower town was an immense fortress, crowning a rock with its triple wall, garrisoned by a thousand men, water being obtained by an underground conduit. Yâḳût, who saw Samîrân in the earlier years of the 7th (13th) century, found it a ruin, the result of an order of the chief of the Assassins at Alamût. The remains were those of a mighty fortress, 'a mother of castles,' and it was situated on a great river that flowed from the mountains of Ṭârum. Its site, however, does not appear to have been identified by any modern traveller. Another fortress of this district is also mentioned by Yâḳût, bearing the name of Ḳilâṭ, which was situated in the Ṭârum mountains, on the frontier of Daylam between Ḳazvîn and Khalkhâl. It occupied the summit of a mountain, and below, on the river bank, where a masonry bridge of many arches crossed the stream, was a suburb with excellent markets. Yâḳût states that this castle had belonged to the chief of the Assassins at Alamût, but like Samîrân its site as yet remains unidentified[1].

In the matter of the manufactures and products of some of the chief towns of the Jibâl province Muḳaddasî gives us a succinct account. He says that Ray exported various kinds of stuffs, especially those known as *Munayyar*. Cotton was spun here and dyed blue, and the striped cloaks of Ray were famous. Needles, combs, and great bowls were made for export, the last two articles, according to Ḳazvînî, being made from the fine-grained hard wood known as khalanj, which came from the Ṭabaristân forests. Ray also was famous for its melons and peaches, and for a kind of saponaceous clay, much used in washing the head.

In Ḳazvîn well-made clothes were to be bought, also leathern sacks used on journeys as wallets. Bows for archery were exported, also the calamint herb. Ḳum was noted for its chairs, bridles, stirrups, and various stuffs; much saffron, too, came from its district. Ḳâshân exported a kind of dried immature date; also tarragon. Isfahân was famous for its overcloaks; and a special kind of salted meat was made for export; further, the Isfahân padlocks were renowned. Hamadân and its neighbourhood produced cheese, and much saffron; and the skins of foxes and martens were exported. Tin is named as found near here, and various stuffs, as well as good boots, were made in the city. Finally from Dînavar came famous cheeses[2].

The chief highway through the Jibâl province was part of the great caravan road, commonly called the Khurâsân road, which, as already described in the introductory chapter, went from Baghdâd to Transoxiana and the farther east. Entering the

[1] I. H. 253. Muk. 360. Yak. i. 63, 811; ii. 499, 500; iii. 148, 492, 533; iv. 156. Kaz. ii. 268. Mst. 149, 150, 198, 217. J. N. 297. N. K. 5.

[2] Muk. 395, 396. Kaz. ii. 250.

province at Ḥulwân this high road passed through it diagonally, coming first to Ḳirmîsîn (or Kirmânshâh), then to Hamadân, from which town Sâvah was the next point, thence finally north to Ray, beyond which it passed eastward out of the Jibâl province into Ḳûmis, and through this to Khurâsân. Of the Khurâsân road, the fullest of the early descriptions, as already explained, is that given by Ibn Rustah at the close of the 3rd (beginning of the 10th) century, who, stage by stage, mentions all the streams and bridges crossed by the road, also whether it ascends or descends or runs across level ground, further naming the various villages and towns that are passed. We have, besides, four other early accounts of this road, the last by Muḳaddasî, who gives the distances by the day's march (*Marḥalah*).

After the Mongol conquest and the establishment of the dynasty of the Îl-Khâns in Persia, Sulṭânîyah became the capital, and hence the centre of the road system. In the itineraries of Mustawfî, therefore, instead of starting from Baghdâd and going east, the roads start from Sulṭânîyah, and towards Baghdâd the reverse direction is of course followed. From Ḥulwân to Hamadân (to revert to the older order of the route) the stages are however practically the same in both systems. But from Hamadân, instead of going by Sâvah to Ray, the Mongol high road goes north direct to Sulṭânîyah across the Darguzîn and Kharraḳân districts. No great towns, however, are passed, and the stages on the road, as given by Mustawfî, being names of villages, are all extremely uncertain[1].

From near Kirmânshâh, at the hill called 'Sumayrah's Tooth,' Sinn Sumayrah (see p. 188), the road to Marâghah in Adharbâyjân and the north turns off from the great Khurâsân road, running first to Dînavar and thence to Sîsar (probably identical with the modern Siḥnah town, see p. 190) and the Jibâl frontier. This route, of which the continuation through Adharbâyjân will be described presently, is given by both Ḳudâmah and Ibn Khurdâdbih, and the earlier portions of it are found in Ibn Ḥawḳal. From Kirmânshâh (Ḳirmîsîn), from Kanguvâr and from Hamadân, roads branched to the right, going south-east to Nihâvand,

[1] I. R. 165—169. I. K. 19—22. Kud. 198—200. I. H. 256—258. Muk. 400—402. Mst. 192.

whence, and from Hamadân direct, the way went by Burûjird to Karaj of Abu Dulaf and thence on to Isfahân. Mustawfî gives the stages from Kanguvâr to Nihâvand and then on by a devious route to Isfahân; while from Karaj Muḳaddasî gives the direct road to Ray going *viâ* Âvah and Varâmîn[1].

The present high road from Isfahân to Ṭihrân (past Ray) goes up through Ḳâshân and Ḳum; but in the earlier middle-ages the caravan route kept more to the east and nearer the desert border, sending off branches to the left westward, in turn, to Ḳâshân and to Ḳum. Muḳaddasî, however, at the close of the 4th (10th) century, already gives the route direct through Ḳâshân and Ḳum, as it goes nowadays. In Mustawfî the road after passing these two towns turned to the left through Âvah to Sâvah, whence Sulṭânîyah was reached, the great high road from this new capital to Ray being joined at the stage of Sûmghân, as will be described in the next paragraph[2].

The number of marches between the towns to the west of Ray on the high road to the Adharbâyjân province is given by Ibn Ḥawḳal and others, also those from Zanjân north to Ardabîl. The stages on this route, however, are found in fullest detail in Mustawfî. Between Sulṭânîyah and Ray the road passed through Abhar to Fârisjîn, leaving Ḳazvîn to the north, and thence reached a stage called Sûmghân (the reading of this name is uncertain), where it bifurcated. The Khurâsân high road went straight onward by the shrine of 'Abd-al-'Aẓîm to Ray, and thence to Varâmîn; while branching to the right southwards, the Isfahân road went first to Sagzâbâd (or Sagziâbâd), and thence on to Sâvah as already described[3].

Of the roads through Adharbâyjân, in early times under the Caliphate, as already noticed, the great northern branch starting from the Khurâsân road at Hamadân went to Sîsar, and thence on to Barzah in Adharbâyjân, 60 miles south of the Urmîyah lake, where it bifurcated[4]. To the right the main road passed to

[1] I. K. 119, 120. Kud. 199, 200, 212. I. H. 256, 257, 258. Muk. 401, 402. Mst. 195.

[2] I. R. 190, 191. I. K. 58, 59. I. H. 289, 290. Muk. 491. Mst. 199.

[3] I. H. 252, 258. Muk. 383. Mst. 196, 198, 199.

[4] See Map III, p. 87.

the east of the lake by Marâghah to Tabrîz, and thence east through Sarâv to Ardabîl. The left branch at the bifurcation at Barzah kept to the west of the lake, going by Urmîyah city to Khuwî, and thence by Nakhchivân (Nashawâ) to Dabîl, the capital of Armenia. From Tabrîz there was the cross-road by Marand to Khuwî, and thence on by Arjîsh to Khilâṭ at the western end of the Vân lake. This last section is given by Iṣṭakhrî and Muḳaddasî only[1].

From Ardabîl, north, the road went across the Mûghân district to Warthân, where the Araxes was crossed, and thence by Baylaḳân to Bardhâ'ah. From this town one road went by Shamkûr north-westwards up the Kur river to Tiflîs in Georgia; while to the right by Barzanj, at the crossing of the Kur, another road led to Shamâkhâ, the capital of Shirvân, and thence on to Bâb-al-Abwâb, otherwise Darband. A road from Dabîl, the capital of Armenia, to Bardhâ'ah is also given by Muḳaddasî and others, but the stages are not easy to identify[2].

The Mongol road system which went through Adharbâyjân to the north-western frontiers, as described by Mustawfî in the 8th (14th) century, started from the new capital Sulṭânîyah, and at Zanjân bifurcated. To the right, the northern branch passed through Khûnaj or Kâghadh Kunân, crossed the Safîd Rûd, and by Khalkhâl city came to Ardabîl, from whence Bajarvân, the capital of Mûghân, was reached. From Zanjân, and crossing the Safîd Rûd by a stone bridge (called the Ḳanṭarah Sabîd Rûdh), this road is also given in part by Iṣṭakhrî and Ibn Ḥawḳal, with a cross-road from Miyânij. Continuing on from Bajarvân Mustawfî first notices the branch road, east, to Maḥmûdâbâd, and then mentions the stages on the main road, which went from Bajarvân by Bardhâ'ah and Shamkûr to Tiflîs.

Returning to the bifurcation at Zanjân, the left branch, as described by Mustawfî, went up to Miyânij in Adharbâyjân, and thence by Ûjân to Tabrîz, following the line given (in the contrary direction) by the earlier Arab geographers. From Tabrîz Mustawfî likewise gives the road on to Arjîsh on the lake of Vân,

[1] I. K. 119—121. Kud. 212, 213. Ist. 194. I. H. 252—254. Muk. 382, 383.

[2] I. K. 121, 122. Kud. 213. Ist. 192, 193. I. H. 251. Muk. 381.

whence, bearing away from the left road along the lake shore to Khilâṭ, he records the distances going north-west to Malâsjird, and on by Arzan-ar-Rûm (Erzerum) through Arzanjân to Sîvâs, the capital of the Saljûḳ province of Rûm. Finally, starting from Tabrîz and going north-east, Mustawfî gives the cross-road to Bajarvân, which went by Âhar, crossing two passes; and along this line, he tells us, the Wazîr 'Alî Shâh had recently built a number of Rubâṭs or guard-houses[1].

[1] Mst. 198, 199. Ist. 194. I. H. 252.

CHAPTER XVI.

KHÛZISTÂN.

The Dujayl or Kârûn river. Khûzistân and Ahwâz. Tustar or Shustar. The Great Weir. The Masruḳân canal. 'Askar Mukram. Junday Shâpûr. Dizfûl. Sûs and the Karkhah river. Başinnâ and Mattûth. Ḳarḳûb and Dûr-ar-Râsibî. Ḥawîzah and Nahr Tîrâ. Dawraḳ and the Surraḳ district. Ḥiṣn Mahdî. The Dujayl estuary. Râmhurmuz and the Zuṭṭ district. Territory of Great Lur. Îdhaj or Mâl-Amîr. Sûsan. Lurdagân. Trade and products of Khûzistân. The high roads.

The province of Khûzistân comprises all the alluvial lands of the river Kârûn, known to the Arabs as the Dujayl of Al-Ahwâz, with its many affluents[1]. This river was called the Dujayl (Little Tigris) of Al-Ahwâz, past which city it flowed, in order to distinguish it from the Dujayl canal of the Tigris to the north of Baghdâd. Khûzistân means 'the Land of the Khûz,' a name otherwise written Ḥûz or Hûz; and the plural of *Hûz*, in Arabic, is *Ahwâz*, which was the capital city, Al-Ahwâz being the shortened form of Sûḳ-al-Ahwâz, 'the Market of the Hûz people.' The name Khûzistân for the province is now become almost obsolete, and at the present day this district of Persia is known as 'Arabistân, 'the Arab Province.' Its great river, too, is no longer called the Dujayl, being now known as the Kârûn, a name which is said to be a corruption of *Kûh Rang*, 'the Coloured Hills,' namely the mountains from which it descends; the name Kârûn, however, appears to have been unknown to the medieval Arab or Persian geographers.

The upper waters of the Dujayl or Kârûn river ramify

[1] For Khûzistân see Map II, p. 25.

through the gorges of the district of Greater Lur, and its affluents come down from Lesser Lur and the Kurdistân mountains. The source of the Dujayl is in the Ḳûh Zard, 'the Yellow Mountain' (see p. 207); from which, on the other versant, the main stream of the river Zandah Rûd flows towards Isfahân. The Dujayl river after a long and winding course through the gorges, with many minor affluents on either bank, comes to the city of Tustar, which Mustawfî in the 8th (14th) century counts as the capital of Khûzistân, whence he calls the river the Dujayl of Tustar. At Tustar the stream bifurcates, but coming together again at 'Askar Mukram, thence flows past Ahwâz, where it is joined by the Junday Sâbûr or Dizfûl river. The Dizfûl takes its course from Burûjird in Lesser Lur (see p. 200), and its upper waters were known as the Ḳar'ah (or Ḳaw'ah). After being joined by another river, called the Kazkî, the main stream flowed past the city of Dizfûl to join the Dujayl, as we have seen. Another great affluent of the Dujayl ran further to the westward, namely the river of Sûs, otherwise called the Karkhah. This rose in the mountains of Lesser Lur, and was joined by the Kûlkû, also by the river of Khurramâbâd. After a long course these united streams, flowing down past the city of Sûs, came to the Ḥawîzah country to the west of Ahwâz and finally joined the Dujayl. At some distance below the junction of these affluents, the Dujayl river became a great tidal estuary, through which, to the eastward of the estuary of the Tigris (already described in Chapter II) the combined waters of the Khûzistân rivers found their way out to the Persian Gulf[1].

Al-Ahwâz, the capital of the province, had originally been known by the name of Hurmuz-Shahr (variously given in the MSS. as Hurmuz Awshîr and Hurmuz-Ardashîr), this being the Persian name. Muḳaddasî describes the town as having suffered greatly during the rebellion of the Zanj in the 3rd (9th) century, and their chief for a time had made it his place of residence. In the following century it was in part rebuilt by the Buyid prince 'Aḍud-ad-Dawlah; and Muḳaddasî writes of it as possessing in his day many great warehouses, where merchandise was collected

[1] I. S. 32. I. R. 90, 91. Yak. ii. 496, 555. Mst. 204, 214, 215, 216. J. N. 286.

from the inland towns and stored, before being transferred to Baṣrah for final sale and export.

In those days Ahwâz consisted of two quarters; one, the eastern, on the river bank, was the main quarter of the town, containing the great markets and the Friday Mosque, and it was connected by a bridge with the island in the Dujayl river, on which stood the western quarter of the city. This bridge, built of kiln-burnt bricks, and known as the Ḳanṭarah Hinduwân, had been restored by 'Aḍud-ad-Dawlah, and on it stood a mosque overlooking the river, which near the town had many waterwheels along its banks. The main stream of the Dujayl flowed past the further, or western side of the island, and a little below Ahwâz a great weir (Shâdhurwân), built on the rocks, dammed back the waters, raising them for irrigation purposes. Three canals, used for watering the lands round the town, left the river above the weir, in which sluices regulated the level for supply and when opened in flood-time saved the city from inundation. The climate of Ahwâz, according to Muḳaddasî, was execrable, hot winds blew all day, and by night sleep became impossible by reason of innumerable mosquitoes and bugs, which 'bite like wolves,' he tells us, adding that the noise of the waters rushing over the weir had prevented him from resting, being plainly audible all over the town. Snakes and scorpions, he says, infested the neighbouring plain, which in many parts was a salt marsh, and the rice-flour bread on which the population fed was most indigestible[1].

In complete contrast to the evil-famed city of Ahwâz was the second capital of Khûzistân, called Tustar by the Arabs, and Shustar, or Shushtar, by the Persians. This as the crow flies lay about 60 miles north of Ahwâz, but perhaps double that distance by water, on account of the windings of the Dujayl river. Muḳaddasî records Tustar as surrounded by gardens, where grapes, oranges, and dates grew abundantly, and no town of Khûzistân, he says, was more beautiful or pleasanter to live in, though he admits that the heat was extreme in summer. The markets of Tustar were abundantly supplied; brocades, with

[1] Ist. 88. I. H. 171. Muk. 406, 410. Yak. i. 410—413; iv. 969. Mst. 169.

embroidered cotton stuffs of all kinds were made here, the brocade (Dîbâj) of Tustar being most famous. The Friday Mosque stood in the middle of the cloth-merchants' market; and the fullers' quarter, down by the river, was a fine place.

In the year 260 A.D. the Roman Emperor Valerian fell a prisoner into the hands of King Shâpûr (Sapor I), the second monarch of the Sassanian dynasty, and during his seven years' captivity, according to the Persian historians, had been employed to build the Great Weir (Shâdhurwân) across the Dujayl immediately below Tustar. This was held by the Arabs to be one of the wonders of the world, and the remains of it still exist at the present day. The bed of the stream to the west of Tustar was paved, and the weir held back the water, enabling a part of the full river to be diverted above Tustar into an artificial channel turning off eastwards, which rejoined the Dujayl river many miles lower down after irrigating the lands through which it passed. The weir of Tustar is given by the older authorities as measuring nearly a mile across, and according to Muḳaddasî a bridge of boats (*Jisr*) stretched over it, carrying the high road which went west from Tustar towards 'Irâḳ. At the present day an ancient bridge of many small arches, over a quarter of a mile in length, carries the road across the weir, but this does not appear to have existed in the earlier middle-ages. Mustawfî in the 8th (14th) century describes the city of Shustar as having four gates, and it was protected by a strong fortress. His contemporary Ibn Baṭûṭah calls the Dujayl (or Kârûn) the Nahr-al-Azraḳ, 'the Blue River,' and speaks of the bridge of boats, 'like those at Baghdâd and Ḥillah,' which crossed the river west of the town from the Dizfûl Gate. He describes at some length the various shrines at the place, which, when he was there, was, he reports, an extremely flourishing town[1].

The Great Weir at Tustar, as already said, was built to raise the water sufficiently high for a canal to be taken from the Dujayl

[1] Ist. 89, 92. I. H. 172, 174, 175. Muk. 405, 409. Yak. i. 847. Mst. 168. I. B. ii. 24. The story of Valerian, and the building of the Great Weir by Sapor I, is narrated by Ṭabarî (i. 827), who, with unusual accuracy, gives the name of the Roman Emperor as Alariyânûs (the Greek form is Οὐαλεριανός). Mas'ûdî (ii. 184) in error gives these events under the reign of Sapor II.

above the city, which should water the lands to the eastward. This canal, now called the Âb-i-Gargar, was in the earlier middle-ages known as the Masruḳân or Mashruḳân, and according to Ibn Muhalhal—a traveller of the 4th (10th) century, quoted by Yâḳût—its waters were white, while those of the main stream of the Dujayl were red in colour. The main stream of the Dujayl (called at the present day the Shuṭayṭ, or 'Little River,' in the reach immediately below Shustar) is rejoined by the Masruḳân branch some 25 miles south of Shustar, at a point near the ruins of Band-i-Ḳir. These mark the site of the city called 'Askar Mukram, which, throughout the middle-ages, was the most important town on the Masruḳân, and the canal throughout its course passed through and irrigated lands planted with sugar-canes, the finest, it was said, in all Khûzistân.

In the early part of the 9th (15th) century, Ḥâfiẓ Abrû and 'Alî of Yazd, writing after the time of Tîmûr, refer to these water-ways under the following names: the moiety of the main stream of the Dujayl, which passed off to the eastward above Shustar (the Masruḳân, or Âb-i-Gargar), was then called the Dû Dânikah or 'Two Sixths'; while the major part of the Dujayl, which went over the weir to the west of the town, was known as the Chahâr Dânikah or 'Four Sixths.' At the present day a canal, called the Mînaw, is diverted south-east from the main stream, and passing through a tunnel under the rock on which the castle of Shustar stands, irrigates the high-lying lands to the south of the city. This channel is the Dashtâbâd canal mentioned by Mustawfî; and it is referred to by Ḥâfiẓ Abrû, who says that the Chahâr Dânikah was divided near the city into two streams, of which only one re-united below with the Dû Dânikah (or Masruḳân). According to tradition the Masruḳân had been originally dug by Ardashîr Bâbgân, founder of the Sassanian monarchy. Mustawfî mentions the city of Masruḳân as standing on the canal bank; and south of this, as already said, at a point half-way between Tustar and Ahwâz, the Masruḳân stream poured back into the Dujayl near the city of 'Askar Mukram.

The Masruḳân district was famous for a particularly excellent kind of date, as well as for the sugar-cane already alluded to.

'Askar Mukram took its name from the camp ('Askar) of Mukram, an Arab commander sent into Khûzistân by Ḥajjâj, the celebrated viceroy of 'Irâḳ under the Omayyads, to put down a revolt. Mukram encamped near the ruins of a Persian town originally called Rustam Kuwâd, a name corrupted by the Arabs into Rustaḳubâdh; and this afterwards became known as 'Askar Mukram, a new city having sprung up on the site of the Arab camp. At the present day the name of 'Askar Mukram has disappeared from the map, but its site is marked by the ruins known as Band-i-Ḳîr, 'the Bitumen Dyke,' where the Âb-i-Gargar (the Masruḳân) runs into the Kârûn. In the 4th (10th) century 'Askar Mukram was a town occupying both banks of the Masruḳân canal, the western quarter being the larger, and this was connected with the other side by two great bridges of boats. The city had well-built markets, which, with the Friday Mosque, stood in the western quarter, but a great drawback to the place was the number of particularly venomous scorpions that were found there. According to Mustawfî the older Persian town had been called Burj Shâpûr, after King Sapor II, who had rebuilt and enlarged it; Mustawfî states that it was in his day commonly called Lashkar, meaning 'the Camp' in Persian, being when he wrote, in the 8th (14th) century, accounted as the healthiest of all the towns of Khûzistân.

According to Ibn Serapion, and other early authorities, the Masruḳân channel, in the 4th (10th) century, did not flow back into the Dujayl at 'Askar Mukram, but took its separate course, running parallel with the Dujayl main stream, down to the tidal estuary. Further, Ibn Ḥawḳal, in the previous century, describes how he himself travelled down the bed of the Masruḳân, at a season of low water, going by this route from 'Askar Mukram to Ahwâz; the first six leagues were, he says, by boat, the remaining four being completed on horseback in the dry bed of the canal. The old course of the lower part of the Masruḳân cannot now be followed, for in this alluvial country the lapse of a thousand years has completely changed the face of the land. Below Ahwâz city, in the 3rd (9th) century, began the broad reach of the Dujayl called the Nahr-as-Sidrah—'the Lotus Canal'—which, after receiving many affluents, ended at Ḥiṣn Mahdî, near the head of the Kârûn tidal estuary[1].

Eight leagues north-west of Tustar, on the road to Dizfûl, lie the ruins now called Shâhâbâd, which mark the site of the city of Junday Sâbûr, or Jundî Shâpûr. Under the Sassanians Junday Sâbûr had been the capital city of Khûzistân, and as late as the time of the Caliph Manṣûr it was famous for the great medical school founded here by the Christian physician Bukht-Yishû‘, who, followed by his sons and grandsons, stood high in favour with more than one of the Abbasid Caliphs. The neighbourhood was celebrated also for the sugar that it produced, which was exported thence to Khurâsân and the further east, though already by the 4th (10th) century Muḳaddasî speaks of Junday Sâbûr as falling to ruin, on account of the inroads of the Kurds. Its embroideries, however, were famous, and rice was largely grown; and in the town was to be seen the tomb of Ya‘ḳûb, son of Layth the Ṣaffârid, who having made this city his capital, died here in 265 (878). Mustawfî in the 8th (14th) century describes Jundî Shâpûr as still a populous town, famous for its sugar-cane, though at the present day an almost uninhabited ruin alone marks the site.

Dizfûl, 'the Diz Bridge' or 'the Castle Bridge,' lies on the Diz river to the west of Junday Sâbûr. The city took its name from a famous bridge, said to have been built by Sapor II, and called Ḳanṭarah Andâmish by Iṣṭakhrî. The remains of it still exist. The city was in the 4th (10th) century also known as Ḳaṣr (the Castle of) Ar-Rûnâsh; Muḳaddasî, however, sometimes refers to it merely as the town of Al-Ḳanṭarah, 'the Bridge.' The place and its famous bridge had various other names. Thus Ibn Serapion calls it Ḳanṭarah-ar-Rûm, 'the Roman Bridge,' and the Diz he names the river of Junday Sâbûr. Again, Ibn Rustah writes of Ḳanṭarah-ar-Rûdh, 'the River Bridge,' and in Ibn Khurdâdbih we find Ḳanṭarah-az-Zâb, Zâb being according to him the name of the Diz river. In the 8th (14th) century

[1] I. S. 32. Ist. 90, 92. I. H. 172, 173, 175. Muk. 409, 411. A. Y. i. 588, 591, 599. Hfz. 82*a*. Mst. 169, 170. Yak. i. 411, 412; ii. 676. Hamzah, 47.

Mustawfî describes the bridge as built of 42 arches, being 320 paces in length, and the roadway 15 paces wide; he says it was then called the Andâlmishk (or Andamish) Bridge.

The town of Dizfûl occupied both banks of the river, and above the town a canal, cut through the rock on the east side, turned a great waterwheel working a mechanism which raised the water 50 ells and thus supplied all the houses of the town. The pasture lands round Dizfûl were famous, and the narcissus grew here abundantly. 'Alî of Yazd gives the name of Zâl to the river, and he describes the bridge at Dizfûl (a name which he writes Dizpul, in the Persian fashion) as built on 28 great arches, with 27 smaller ones between each two, making a total of 55. A reference to the modern map shows that at the present day the Dizpul river joins the Kârûn opposite Band-i-Ḳîr ('Askar Mukram), but in earlier times it must have come into the Dujayl somewhat lower down, and probably in its upper course the stream passed nearer to Junday Sâbûr than is now the case. At its junction, in the middle-ages, with the Dujayl, and probably to the north of Ahwâz, lay the two fertile districts, with their chief towns, called Great and Little Manâdhir, which Ibn Ḥawḳal in the 4th (10th) century describes as surrounded by palm-groves and growing much corn[1].

The country to the north and east of Dizfûl and Tustar, was, in the earlier middle-ages, known as the Lur Plain (Ṣaḥrâ Lur), being occupied by the Lur tribes who in later times migrated into Lesser and Greater Lur, the mountain districts, of which the first-named was included in the Jibâl province, as already noticed in Chapter XIV. In the 4th (10th) century, when Ibn Ḥawḳal wrote, the Lurs had evidently already begun to migrate, for he describes the neighbourhood as inhabited by the Kurds, and says of the Lur country that it was a most fertile though exceedingly mountainous district[2].

[1] I. R. 90. I. K. 176. I. S. 32. Ist. 93, 95, 197. I. H. 176, 177, 259. Muk. 384, 405, 408. Ykb. 361. Yak. ii. 130; iv. 111. Mst. 169. A. Y. i. 588, 591. For the various physicians of the name of Bukht-Yishû' who, though Christians, served the Abbasid Caliphs from Manṣûr to Hârûn-ar-Rashîd as court physicians, see Ibn Abi Usaybî'ah (edited by A. Müller), i. 125—143, 202.

[2] Ist. 88, 94. I. H. 171, 176. Muk. 409.

To the south-west of Dizfûl lie the ruins of Sûs, the ancient Susa, near the bank of the Karkhah river. This was a populous town in the middle-ages, being the centre of a district with many cities, and it was famous for its raw silk, as well as for oranges, while the sugar-cane grew here abundantly. The city was protected by an ancient fortress, and there were fine markets in the town, where stood a Friday Mosque built on round columns. Tradition asserted that the tomb of the prophet Daniel had been made in the bed of the Karkhah river which ran on the further side of Sûs, and a fine mosque marked the place on the bank which lay nearest to his supposed grave. Mustawfî, who describes the city as a flourishing place in the 8th (14th) century, speaks of the tomb of the prophet Daniel as standing (apparently on dry ground) to the west of it, adding that in his honour none of the fish in the river were ever molested by man. The neighbouring city of Karkhâ, or Karkhah, which now gives its name to the river flowing by the mounds of Sûs, lies some distance above these, and on the right or western bank. Muḳaddasî describes it as a small but populous town, holding its market weekly, on the Sunday. It was protected by a castle, and was surrounded by gardens[1].

A number of places are mentioned by the early geographers as lying on or near the Karkhah river, some to the westward, some below Sûs, which were important towns during the middle-ages, but of which no trace now remains on the modern map. Their positions are, however, approximately given by the Itineraries. Of these the most important was Baṣinnâ, which lay a short day's journey south of Sûs, on a canal (or possibly a minor affluent of the Karkhah river), which was known as the Dujayl or 'Little Tigris' of Baṣinnâ. It was a great place for trade, and the veils of Baṣinnâ were celebrated all over the Moslem world; beautiful carpets of felt also were made here, and wool-spinning was a chief industry. The city was defended by two castles, and the Friday Mosque, a bow-shot from the river bank, stood at the town gate; seven mills built in barges floated on the 'Little Tigris' according to Muḳaddasî. Near Baṣinnâ, and also about

[1] Ist. 88, 92, 93. I. H. 174. Muk. 405, 407, 408. Mst. 269. A. F. 311. Yak. iv. 252 (where *Karajah* is printed in error for Karkhah).

a day's journey from Sûs, but probably to the west of the Karkhah river, was the town of Bayrût or Bîrûdh, which Yâḳût visited in the 7th (13th) century. Muḳaddasî speaks of it as a large place, surrounded by date-groves, and on account of its flourishing commerce it was known as 'the Little Baṣrah.'

Mattût or Mattûth, where there was a strong castle, was also of this neighbourhood; it lay nine leagues to the south of Sûs, and on the road between Ahwâz and Ḳurḳûb. This last—where were made the celebrated Sûsanjird embroideries—was a town of some importance, lying half-way between Sûs and Ṭîb in 'Irâḳ, being one march from Sûs and two from Baṣinnâ. Another town of this district, the site of which has not been found, though probably it stood to the north of Ḳarḳûb, was Dûr-ar-Râsibî, which Yâḳût describes as situated between Ṭîb and Junday Sâbûr. This Dûr was famous as the birth-place and residence of Ar-Râsibî, who died in 301 (913), having been for many years the semi-independent governor of all the districts from Wâsiṭ to Shahrazûr, during the Caliphate of Muḳtadir. He was celebrated for his immense wealth, and of the goods and furniture that he left at his death Yâḳût gives a long and curious inventory[1].

The Karkhah river is joined at about the latitude of Ahwâz by streams coming down from Ḥawîzah (or Ḥuwayzah, the diminutive form of Ḥûz or Hûz, as already said, the name of the people of this province), which Mustawfî describes in the 8th (14th) century as one of the most flourishing cities of Khûzistân. Corn, cotton, and sugar-cane grew here abundantly, and the town had at that time a population of Sabæans or Sâbians. The town of Nahr Tîrâ or Nahr Tîrîn, on the canal or river of this name, which appears to have been a right bank affluent of the lower Karkhah, must also have been of the Ḥawîzah district. It lay a day's journey west of Ahwâz on the road to Wâsiṭ, and it was famous for the stuffs made there, which resembled those of Baghdâd.

The Karkhah river flows from the west into the Dujayl below Ahwâz, probably in the broad reach, already referred to, known as

[1] Ist. 171, 175. I. H. 93. Muk. 405, 408. Yak. i. 656, 786; ii. 616; iv. 65, 412. Hfz. 82 *b*. A. F. 313.

the Lotus river (Nahr-as-Sidrah). From the east, but lower down, is the junction of the Dawraḳ river, or canal, on which lay the city of this name, the capital of the Surraḳ district. The town was called Dawraḳ-al-Furs, 'of the Persians'; it was very spacious, with fine markets where goods of all sorts were warehoused, and the pilgrims from Fârs and Kirmân mostly passed through here on their road to Mecca. It was famous for its veils. Its Friday Mosque stood in the market-place, and on the river bank were many hamlets. Yellow sulphur was found here, near the hot sulphur springs, where the sick bathed and were healed. These, which were especially beneficial in skin diseases, gushed out from a hill side, the waters filling two tanks. In the 4th (10th) century wonderful Sassanian buildings were still to be seen at Dawraḳ, also a fire-temple, according to Ibn Muhalhal.

In the district near Dawraḳ were the two cities of Mîrâkiyân and Mîrâthiyân, which Muḳaddasî describes. The first lay on a tidal canal, and was surrounded by excellent lands; while Mîrâthiyân consisted of two quarters, with a Friday Mosque in each of them and markets that were much frequented. In the 4th (10th) century much of the water of the southern swampy lands of the Khûzistân district drained out to the Persian Gulf by channels running south from Dawraḳ, and these entered the sea at Bâsiyân. Near this town must have been the creek and island of Dawraḳistân, mentioned by Yâḳût and Ḳazwînî, where ships coming from India cast anchor. The town here was protected by a fortress, to which political prisoners were sent by the Caliph to be kept out of the way; and as late as the 7th (13th) century boats could pass up from here northwards, to 'Askar Mukram, by a series of canals or rivers that flowed to the eastward of the Dujayl[1].

The Dujayl below Ahwâz soon broadened out to become the tidal estuary, which was the lower part of the Lotus river or Nahr-as-Sidrah. On this estuary stood Sûḳ Baḥr, a town where, until the time of the Caliph Muḳtadir in the middle of the 4th (10th)

[1] Ist. 93. I. H. 176. Muk. 407, 412. Yak. i. 411; ii. 371, 618, 620. Mst. 169. Kaz. ii. 130, 246. Both Nahr Tîrâ and Manâdhir must have been important places in Omayyad days, for between the years 90 and 97 (709—716) both were mint cities.

century, there had been toll-barriers, vexatious and unlawful dues being here exacted. The town of Sûḳ-al-Arba'â (the Wednesday market) was in this neighbourhood, lying to the east of the Dujayl, and on a canal which divided the town into two quarters that were connected by a wooden bridge. The eastern quarter of Sûḳ-al-Arba'â was the more populous, and here was the mosque. The neighbouring town of Jubbâ was noted for its sugar-canes, and the lands near were occupied by many villages.

At the head of the broad waters of the great tidal estuary of the Dujayl was the fortress called Ḥiṣn Mahdî, with a mosque standing in the midst of its guard-houses (Rubâṭ), said to have been built by the Caliph Mahdî, father of Hârûn-ar-Rashîd. Ḥiṣn Mahdî stood a few miles above the point where the Aḍudî canal branched off to the westward, joining the head of the Dujayl estuary with the Blind Tigris at Bayân, and round it lay the district of the Sabkhah, or salt marshes (see Chapter III, p. 48). The estuary, or Fayḍ of the Dujayl went into the Persian Gulf at Sulaymânân, and this was a dangerous passage for ships, which appear to have reached Ahwâz more safely by threading the various canals and rivers going up by Bâsiyân to Dawraḳ and thence into the Lotus river. The fortress of Ḥiṣn Mahdî, the exact site of which is unknown, stood, we are told, at the junction of many roads, and commanded the upper reach of the Dujayl estuary, where it was nearly a league across, being immediately below where many streams from the Ḥawîzah country and the Dawraḳ river flowed in from the north-west and the east. Above this point began the Lotus channel, going up to Ahwâz, from which city Ḥiṣn Mahdî was 20 leagues distant[1].

Three days' march east of Ahwâz is the city of Râmhurmuz, still known by the name which it received from King Hurmuz, grandson of Ardashîr Bâbgân. In the 4th (10th) century it was famous for the silkworms reared here, and raw silk was largely exported. In Râmhurmuz there was a fine Friday Mosque, and excellent markets which had been built by 'Aḍud-ad-Dawlah, the Buyid prince. Muḳaddasî relates that every night the gates of the various wards occupied by the shops of the cloth-

[1] I. S. 30. Kud. 194. Ist. 93, 95. I. H. 172, 176. Muk. 412, 419. Yak. i. 185; ii. 12; iii. 193.

merchants, perfumers, and mat-weavers, were securely locked. There was, he adds, a celebrated library here, where lectures were delivered, and this had been built and endowed by a certain Ibn Sawwâr, who had also founded a similar institution at Baṣrah. Râmhurmuz got its water by a canal from the Ṭâb river, but this in summer-time often ran dry, and the town was everywhere so infested by gnats that according to Muḳaddasî mosquito curtains were a necessity. Mustawfî, in the 8th (14th) century, says that the name Râmhurmuz was then commonly shortened to Râmuz; the town was still a flourishing centre, much corn, cotton, and sugar-cane being grown in its districts.

Six leagues south-east of Râmhurmuz, on the road to Arrajân and not far from the river Ṭâb, which here marked the boundary of Fârs, was the Ḥawmah or district of the Zuṭṭ, otherwise known as the Jât tribes from India (identical it is said with the Gipsies). This district was watered from the Ṭâb river, and here stood the two populous villages called Az-Zuṭṭ and Al-Khâbarân. Beyond this, and two marches short of Arrajân, close to the Fârs frontier on the road coming from Arrajân to Dawraḳ, was the little town of Asak, where, according to Iṣṭakhrî, there was a small volcano. The place stood in the midst of palm-groves, and much *dûshâb*, or syrup of raisins, was made here and exported. Near Asak also were Sassanian remains, namely, a great Aywân or domed hall, a hundred ells in height, built by King Ḳubâdh over a spring. East of Asak, and a few miles short of Arrajân, but to the west of the bridges over the Ṭâb river, was the market town of Sanbîl in the midst of its district, which thus lay along the borders of Fârs[1].

The Lur districts lay east and north of Tustar along the upper course of the Dujayl (Kârûn river) and its numerous affluents. The country to the east and south of the upper Kârûn (which here makes a great bend and doubles back, between its source in the mountains west of Isfahân, and the point north of

[1] Ist. 92, 93, 94. I. H. 175, 176. Muk. 407, 413. Yak. i. 61. Mst. 169. By a strange error Yâḳût (ii. 791) mentions the village of Az-Zuṭṭ under the form Ar-Ruṭṭ, though he was perfectly well acquainted with the Zuṭṭ or Gipsies, and mentions a canal (ii. 930) called after them.

Tustar, where it finally turns south and flows down towards the Persian Gulf) Mustawfî describes as the Great Lur district, and this lay contiguous to the Shûlistân district over the border in Fârs. The chief town of Great Lur was Îdhaj, otherwise called Mâl-Amîr. Mukaddasî describes it in the 4th (10th) century as one of the finest towns of Khûzistân; and it stood near the hills, where, at a place called Asadâbâd, was the palace of the governor. In winter snow fell here abundantly, and was stored to be carried to Ahwâz for sale during the summer. The fields being irrigated by the rains the pistachio-trees produced fine crops of nuts. Ibn Baṭûṭah, who visited the place in the beginning of the 8th (14th) century, says that Îdhaj was already then more commonly known as Mâl-al-Amîr, 'the Amir's property,' a name which it still bears, Îdhaj having now become obsolete.

Îdhaj was further famous for its great stone bridge over the Dujayl, which Yâkût describes as one of the wonders of the world. This, the ruins of which still exist, was known as the Kanṭarah Khurrah Zâd, being so named after the mother of King Ardashîr, and it spanned the ravine by a single arch, rising 150 ells above the water level. In the gorge two leagues below the town was a mighty and dangerous whirlpool, known as Fam-al-Bawwâb, 'the Porter's Mouth.' The great bridge was repaired in the 4th (10th) century by the Wazîr of Rukn-ad-Dawlah, the Buyid prince, and it took two years' labour to bring this to completion. Its stones were joined by lead with iron clamps, and it is said that 150,000 dînârs (£75,000) were spent upon the work. Yâkût says that earthquakes were frequent in the neighbourhood of Îdhaj; also there were many mines, a certain alkali being found here, called Kûkalî, which was a sovereign remedy for the gout. He adds that an ancient fire-temple was to be seen at Îdhaj, which until the reign of Hârûn-ar-Rashîd had been constantly in use.

Occupying both banks of the river, and four leagues to the north-west of Îdhaj, was the small town called Sûsan, otherwise known as 'Arûj (or 'Arûḥ). Round this place stretched extensive gardens, producing grapes, citrons, and oranges, and Mustawfî says that the mountains, on which snow still lay in summer, were only four leagues distant. 'Arûj, or Sûsan, was also known as Jâbalak, and this place according to some authorities is to be identified with 'Shushan the palace' of the Book of Daniel. About 150 miles east of Mâl-Amîr, on the frontier of Fârs and near the easternmost of the affluents of the Kârûn river, is Lurjân (otherwise Lurdagân or Lurkân, all forms of the name of Lur), which Iṣṭakhrî describes as the capital of the Sardân (or Sardan) district,—a spacious town embowered in trees. Mustawfî praises it for its abundant grapes, and it was often held to be of the province of Fârs, on the borders of which it lay[1].

The main produce of Khûzistân was sugar, for the sugar-cane grew in almost all parts of it, and Muḳaddasî states that in the 4th (10th) century, throughout Persia, Mesopotamia, and Arabia, no sugar but that exported from Khûzistân was to be found. He says that Ahwâz, the capital, was renowned for a special kind of kerchief, such as women mostly wear; and Tustar produced the brocades (Dîbâj) that were famous all the world over, as well as rugs and fine cloth. Much fruit also was grown in Tustar for export, particularly melons. The district of Sûs was especially the home of the sugar-cane, and the city exported enormous quantities of this commodity; silk too was woven here and cloth stuffs. In 'Askar Mukram they made veils of raw silk, and napkins, also cloth. Baṣinnâ was famous for its curtains; Ḳurḳûb for felt rugs; and Nahr Tîrâ for long face-veils[2].

In Khûzistân all the rivers and canals were navigable for boats, and much of the traffic between the towns passed along the waterways. The high roads centred in Ahwâz, to which the traveller from Baṣrah journeyed either by water along the 'Aḍudî canal, or by land across the salt marsh (Sabkhah) from 'Askar Abu Ja'far, opposite Ubullah, to Ḥiṣn Mahdî; and thence through Sûḳ-al-Arba'â to Ahwâz[3].

The distances between the various cities of Khûzistân are given by Iṣṭakhrî and Muḳaddasî in much detail. From Ahwâz a road went west to Nahr Tîrâ, and on thence to Wâsiṭ in 'Irâḳ. The northern road from the capital passed through 'Askar Mukram

[1] Ist. 103, 126. I. H. 182, 197. Muk. 414. Kaz. ii. 201. Yak. i. 416; iv. 189. Mst. 151. I. B. ii. 29. For Sûsan compare Sir H. Layard and Sir H. C. Rawlinson in *J. R. G. S.* 1839, p. 83; and 1842, p. 103.

[2] Muk. 416.

[3] Kud. 194. Muk. 135.

to Tustar, whence by Junday Sâbûr and Sûs it struck westward to Ṭîb, whence again there was a high road to Wâsiṭ.

From Junday Sâbûr Muḳaddasî gives the route through the Lur mountains to Gulpaygân in the Jibâl province, north-west of Isfahân; and from 'Askar Mukram another road (given by Ḳudâmah and others) went east to Îdhaj, whence across the mountains this likewise reached Isfahân[1].

From 'Askar Mukram, and from Ahwâz, two roads converged on Râmhurmuz, whence continuing eastwards the frontier of Fârs was reached on the Ṭâb river over against Arrajân. These roads are given by Ḳudâmah and most of the other authorities, being a part of the high road from Baṣrah to Shîrâz. Iṣṭakhrî also gives another route, chiefly by water, from Ḥiṣn Mahdî to Arrajân, which passed by Bâsiyân on the coast to Dawraḳ, and thence by Asak to Arrajân. The stages north from Râmhurmuz to Îdhaj are recorded by Muḳaddasî, who also describes a route from Râmhurmuz across the Lur mountains to Isfahân. A second route passed from the Lur plains (north of Dizfûl) by Sâbûrkhwâst to Karaj of Abu Dulaf—the distances here, however, are only given in marches, and the stages are difficult if not impossible now to identify. A third route north, given by Muḳaddasî, went across the mountains from Arrajân in seven days' march to Sumayram (in Fârs), south of Isfahân, keeping along the frontier of Khûzistân and Fârs[2].

[1] Ist. 96. I. H. 178. Muk. 418—420. I. R. 187, 188. Kud. 197.

[2] Kud. 194. I. R. 188. Ist. 95. I. H. 177. Muk. 401, 420, 453, 459.

CHAPTER XVII.

FÂRS.

Division of province into five districts or Kûrahs. The district of Ardashîr Khurrah. Shîrâz. Lake Mâhalûyah. The Sakkân river. Juwaym. Dasht Arzin lake. Kuvâr. Khabr and Ṣimkân. Kârzîn and the Ḳubâd Khurrah district. Jahram. Juwaym of Abu Aḥmad. Mândistân. Îrâhistân. Jûr or Fîrûzâbâd. The coast districts of Fârs. Ḳays island. Sîrâf. Najîram. Tawwaj. Ghundîjân. Khârik and other islands of the Persian Gulf.

The province of Fârs had been the home of the Achæmenian dynasty, and the centre of their government. To the Greeks this district was known as Persis, and they, in error, used the name of this, the central province, to connote the whole kingdom. And their misuse of the name is perpetuated throughout Europe to the present day, for with us Persia—from the Greek Persis—has become the common term for the whole empire of the Shâh, whereas the native Persians call their country the kingdom of Îrân, of which Fârs, the ancient Persis, is but one of the southern provinces. The Arabs had inherited from the Sassanian monarchy the division of Fârs into five great districts, each called a Kûrah; and this division, which it will be convenient to retain in describing the province, continued in use down to the time of the Mongols. The five Kûrahs were:—(1) Ardashîr Khurrah, with Shîrâz, the provincial capital, for its chief town; (2) Sâbûr or Shâpûr Khurrah, with Shâpûr city for its chief town; (3) Arrajân, with the chief town of the same name; (4) Iṣṭakhr, with the ancient city of this name (Persepolis), the Sassanian capital of Fârs; and lastly (5) Dârâbjird, also with the chief town of the same name.

Further it must be noted that, during the Caliphate, Fârs

PROVINCES OF

MAP VI

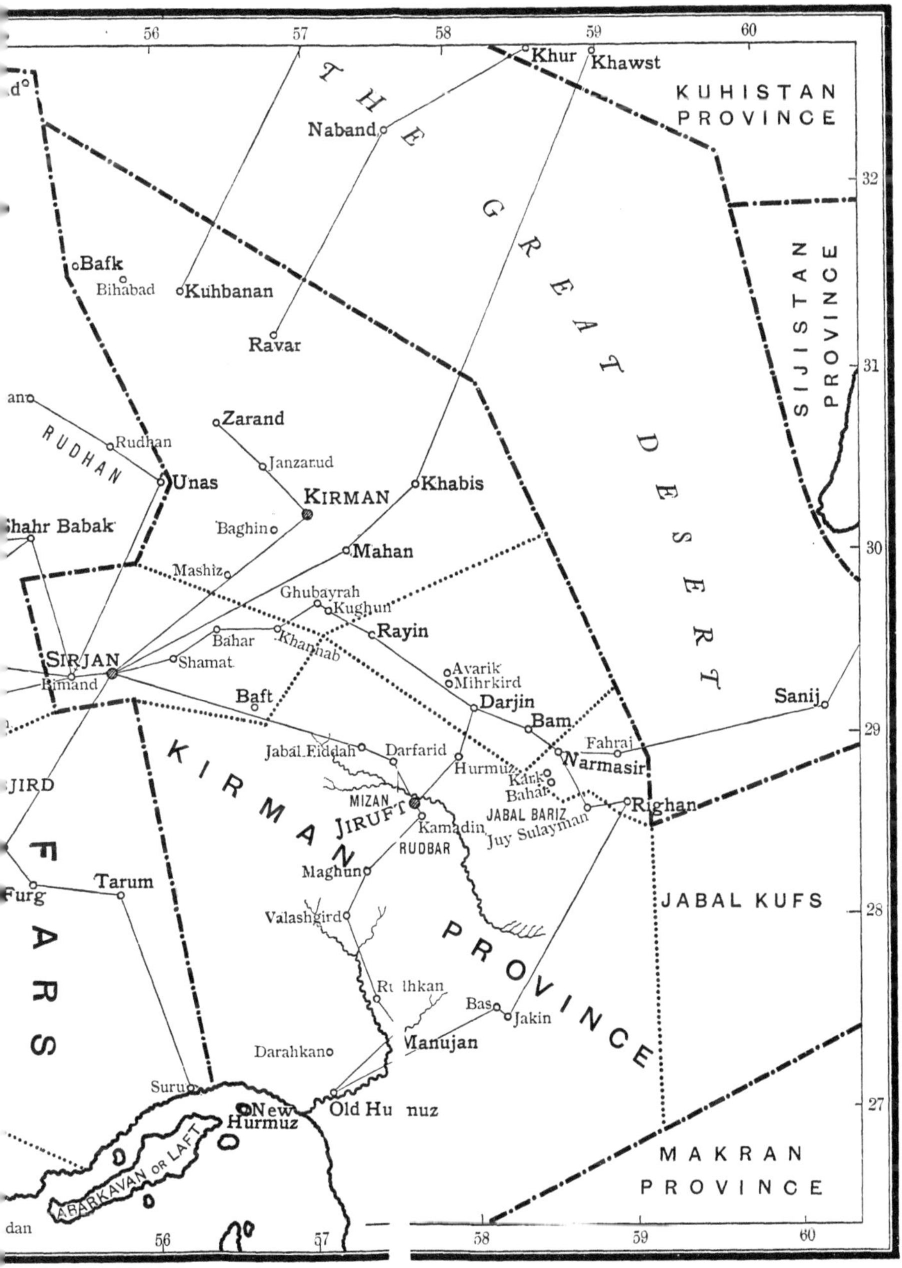

RS AND KIRMAN

included Yazd with its district, also the district of Rûdhân (between modern Anâr and Bahramâbâd), both of these having formed part of the Iṣṭakhr Kûrah. After the Mongol conquest, however, Yazd was of the Jibâl province, while at the present day it is counted as forming part of Kirmân, as is also the case with the former district of Rûdhân. In old Persian *Khurrah* has the meaning of 'Glory'; Ardashîr Khurrah and Shâpûr Khurrah, therefore, signify the districts which commemorate the glory of the founder of the Sassanian kingdom, Ardashîr, and of his famous son, Sâbûr or Shâpûr, the Greek Sapor. Lastly, the Arab geographers commonly divide Fârs between two regions, namely, the Hot Lands and the Cold Lands (*Jurûm* and *Sarûd*), by a line running east and west; and at the present day we find that this division of the lowlands near the coast from the highlands beyond the passes is still current under the names, respectively, of the *Garmsîr* and the *Sardsîr*, 'the hot' and 'the cold region,' which are also the terms employed by Mustawfî[1].

Shîrâz, the capital of Fârs, is an Arab foundation, and at the time of the Moslem conquest in the days of the Caliph 'Omar its site was the camping ground of the army sent to besiege Iṣṭakhr. As Muḳaddasî points out, Shîrâz probably owes its pre-eminence as a town to its central position, being supposed to lie 60 leagues from the frontiers at the four cardinal points of the compass, and 80 leagues from each of the four corners of the province. The chronicles state that Shîrâz was founded in the year 64 (684) by a certain Muḥammad, brother or cousin of Ḥajjâj, the famous governor of 'Irâḳ under the Omayyads; and it grew to be a large city in the latter half of the 3rd (9th) century when the Ṣaffârids had made it the capital of their semi-independent principality. In the 4th (10th) century Shîrâz is described as being nearly a league across, with narrow, but crowded markets. The city had then eight gates, the Gates of Iṣṭakhr, Tustar, Bandâstânah, Ghassân, Sallam, Kuvâr, Mandar, and Mahandar. Its water was from an underground channel carried down from Juwaym, a village five leagues to the north-west; and there was

[1] Muḳaddasî (p. 421) alone divides Fârs into six (in the place of five) Kûrahs; making a separate district of the country round Shîrâz. Ist. 97, 135. Baladhuri, 386. Muk. 447.

a Bîmaristân, or hospital, also the palace built by 'Aḍud-ad-Dawlah, the Buyid, who according to the *Fârs Nâmah* established a library here.

Half a league south of Shîrâz, this same Buyid prince, 'Aḍud-ad-Dawlah, surnamed Fanâ Khusraw, had built himself another palace and surrounded it by a new town, named after himself, Kard Fanâ Khusraw. Immense sums were spent on the gardens, which extended a league across; and the houses round this were occupied by wool-weavers, brocade-makers, and others, being all craftsmen whom the Buyids had brought to settle in Fârs from many distant lands. A yearly festival was held at Kard Fanâ Khusraw, which also became for a short time a mint city; but its glories did not survive its founder, and before the close of the 4th (10th) century it had fallen to ruin. As a suburb it came to be known as Sûḳ-al-Amîr (the Amir's Market), and the rents on shops are said to have produced 20,000 dînârs (£10,000) yearly.

The walls of Shîrâz were first built by Ṣamṣâm-ad-Dawlah or by Sulṭân-ad-Dawlah (son and grandson of 'Aḍud aforesaid), being originally eight ells thick, with a circuit of 12,000 ells, and no less than eleven gates. In the middle of the 8th (14th) century, these walls having fallen to ruin, Maḥmûd Shâh Injû, the rival of the Muẓaffarids, repaired them, building also bastions of burnt brick. When Mustawfî knew Shîrâz the city was divided into seventeen quarters, and had nine gates. These were the Gates of Iṣṭakhr; of Dârak (or Darâk Mûsâ), called after the mountain of this name, two leagues distant from Shîrâz, where the winter snow was stored in pits for use in summer-time; then the Gate of Bayḍâ; of Kâzirûn; of Sallam; of Ḳubâ (for which some MSS. give Fanâ or Ḳanâ); next Bâb-i-Naw (the New Gate); and lastly, Bâb-i-Dawlah and Bâb-i-Sa'âdah, 'the Gate of Government' and 'the Gate of Felicity.' Mustawfî, who gives the list, further remarks that Shîrâz is a very fine town, the market streets never being empty, but he admits that these last were inconceivably filthy. The water-supply was from the famous channel of Ruknâbâd, which had been dug by Rukn-ad-Dawlah the Buyid, father of 'Aḍud mentioned before, and from the canal of the Sa'dî orchard. In spring, torrents flowed down through the city from Mount Dârak; and thence drained into Lake Mâhalûyah.

There were three chief mosques: first the Old Mosque—Jâmi 'Atîḳ—built by the Ṣaffârid 'Amr, son of Layth, in the latter half of the 3rd (9th) century, and this mosque, Mustawfî states, was never empty; next, and dating from the latter half of the 6th (12th) century, was the New Mosque, built by the Salghârî Atabeg Sa'd ibn Zangî; and lastly there was the Masjid Sunḳur, in the Barbers' Square, built by the first Atabeg of the Salghârids. The hospital of 'Aḍud-ad-Dawlah still existed, and Shî'ahs visited the shrine of Muḥammad and Aḥmad, sons of the seventh Imâm Mûsâ-al-Kâẓim. The account which Ibn Baṭûṭah, the contemporary of Mustawfî, gives of Shîrâz bears out the preceding. He, too, speaks of the Old Mosque, the north door of which was known as the Bâb Ḥasan, 'the Gate Beautiful,' and of the shrine of Aḥmad, where there was a college. Further, he eulogises the five streams that flowed through the city; one, that of Ruknâbâd, rising at Al-Ḳulay'ah, 'the Little Castle,' in the hills, near to which was the fine orchard surrounding the tomb of the poet Sa'dî, who had died in 691 (1292), about half a century before the time of Ibn Baṭûṭah's visit. Sa'dî had flourished at the court of the Atabeg Abu Bakr, son of Atabeg Sa'd who had built the New Mosque, and in the orchard round his tomb, which was much visited, were magnificent marble tanks for clothes-washing, which Sa'dî had built on the Ruknâbâd stream.

At the close of the 8th (14th) century, Shîrâz had the good fortune to escape a siege by Tîmûr, who defeated the Muẓaffarid princes at the battle of Pâtîlah in the plain outside. The city suffered little damage, according to 'Alî of Yazd, for Tîmûr camped at the garden called Takht-i-Ḳarâchah, outside the gates of Sallam and Sa'âdah, opening towards Yazd. The same authority states that the other eight gates were then closed, and he also mentions the Red Castle Hill (Kûh Ḳal'at Surkh) near Shîrâz, the position of which is unknown. Of famous castles near Shîrâz Mustawfî mentions Ḳal'ah Tîz, standing on a solitary hill three leagues to the south-east of the city. There was a spring of water here, on the hill-top, and another in the plain below, which for a day's journey beyond was all waterless desert[1].

[1] The reading of the name is uncertain. Tîr, Tabr, Babr, Bîr and Tasîr or Tashîr, with many other variants occur in the MSS. of Mustawfî. Ist. 124.

Shîrâz stands on no great river, but its streams, as already said, drain eastward, flowing into the lake which occupies a depression in the plain a few leagues distant from the city. This lake is called Jankân by Iṣṭakhrî: Abu-l-Fidâ and Ibn Baṭûṭah refer to it as Jamkân; in the *Fârs Nâmah* and in Mustawfî it has the name of Mâhalûyah, and at the present day it is known as the Lake of Mâhalû. The water is salt, and from the salt-pans along its shore Shîrâz was supplied with this necessary commodity, as also with fish, which were abundant in its waters. The lake was 12 leagues round, the district of Kahrjân lying along its southern borders, while to the south-east was the city of Khawristân, otherwise called Sarvistân, where the date palm flourished and corn was grown, also other produce of both the hot and the cold regions. Kûbanjân, according to the *Fârs Nâmah* and Mustawfî, was a small town near Sarvistân[1].

The longest river in Fârs is the Nahr Sakkân, which rising some 30 miles to the north-westward of Shîrâz follows a devious course, going south-east for over 150 miles; then after making a great bend it runs due west for another 150 miles, but with many windings, and finally, after receiving the waters of the Fîrûzâbâd river from the north, discharges itself into the sea a little to the south of Najîram[2]. The name Sakkân is said by Iṣṭakhrî to be derived from the village of Sakk, which stands near the great bend westward; other authorities, however, spell the name variously: thus we find Sittajân, Thakkân, and Sîkân, while Mustawfî generally has Zakkân or Zhakkân. In the *Fârs Nâmah* and later

Muk. 429, 430, 456. F. N. 71 *a*, *b*. Yak. iii. 349; iv. 258. Mst. 170, 171, 179, 203. I. B. ii. 53, 77, 87. A. Y. i. 437, 594, 609, 613. The garden of Takht-i-Ḳarâchah, 'the Throne of Ḳarâchah,' was so named after the Atabeg Ḳarâchah, who became governor of Fârs on the death of Atabeg Châûlî in 510 (1116). It is said to be identical with the garden now known as Takht-i-Ḳajar.

[1] I. K. 52. Ist. 122, 131. Muk. 422, 455. F. N. 73 *a*, 80 *b*. Mst. 172, 226. A. F. 43. I. B. ii. 61. Yak. ii. 193, where *Jîkân* (for Jankân) is a clerical error.

[2] Its upper course is now known as the Ḳâra Aghâch, Black-tree river (in Turkish); its lower course is called the Mând river. The Sakkân is probably identical with the river Sitakus of Nearchus. See Colonel Ross, *P.R.G.S.* 1883, p. 712.

writers, the district where the river had its source is named Mâṣaram; according to Iṣṭakhrî it rose in Rustâḳ-ar-Ruwayḥân, which is the plain south of Juwaym and Khullâr. These are two important villages, lying 5 and 9 leagues distant respectively from Shîrâz, on the road to Nawbanjân, to the north of the Dasht Arzin plain. Near Juwaym, as already said, one of the Shîrâz streams took its rise. According to Mustawfî, Khullâr was famed for its millstones, though the people themselves possessed no mills, and had to send elsewhere to grind their corn. Its honey also was largely exported. Dasht Arzin (the Plain of the Bitter-almond) was famous for its magnificent pasture lands (Marghzâr), and the Lake of Dasht 'Arzin, which in the season of rains was 10 leagues across, was of sweet water; this, however, as often as not, dried up in summer. According to Iṣṭakhrî, the lake produced much fish, and Mustawfî adds that the forest near here abounded with lions[1].

The Sakkân river, 10 leagues south of Shîrâz, passed the town of Kavâr or Kuvâr, lying near its left bank. According to Mustawfî a dam had here been thrown across the stream to raise its water for irrigation, and the neighbouring pasture lands were famous. Both the sour cherry and the almond grew here plentifully, also large pomegranates. Beyond Kuvâr, also on the left bank of the river Sakkân, is the town of Khabr, noted for the tomb of Sa'îd, brother of Ḥasan-al-Baṣrî, the theologian. Mustawfî states that Khabr was larger than Kuvâr, and that near by was the famous castle of Tîr-i-Khudâ, 'God's Arrow,' so called from its inaccessibility, for it stood on a hill-top, so that no human arrow could attain it. Below Khabr the Sakkân river turned south, following a sinuous course through the district of Ṣimkân, the town of Ṣimkân being near its left bank at the junction of a great affluent coming from Dârâbjird on the east[2].

According to Mustawfî, Ṣimkân was a fine town standing on

[1] Juwaym, sometimes written Juwayn, is the present village of Goyun. Ist. 120, 122. I. K. 44. F. N. 77 *b*, 79 *b*, 80 *b*, 81 *a*. Yak. ii. 457. Mst. 177, 179, 214, 226.

[2] Ist. 105, 120. F. N. 71 *b*, 72 *a*, 81 *a*, 83 *a*, 86 *a*. Yak. ii. 399. Mst. 172, 173, 179. This district is now called Sîmâkûn, and often by mistake written *Akun* on the maps. See E. Stack, *Six Months in Persia*, ii. 232.

the stream where this was crossed by a bridge; and it was remarkable that all the lands above the bridge produced trees of the cold region only, such as the plane (Chinâr) and the nut; while below the bridge grew oranges and lemons with other fruits of the hot region. The wine made here was so strong that, before drinking, it had to be mixed with twice or thrice its weight of water. Not far distant was Hîrak, a large village of the dependencies of Ṣimkân. Near the right bank of the Sakkân river, and south of the Ṣimkân district, were the three towns of Kârzîn, Ḳîr and Abzar, the surrounding district being known as Ḳubâd Khurrah, 'the Glory of Ḳubâd,' in memory of one of the Sassanian kings. Iṣṭakhrî speaks of Kârzîn as being one-third the size of Iṣṭakhr (Persepolis); it had a strong castle up to which water could be drawn from the Sakkân river, and being on a great height many distant castles could be seen from it[1].

The town of Jahram (or Jahrum), which is sometimes counted as of the Dârâbjird district, lies south of Ṣimkân, and east of Kârzîn, surrounded by a fertile plain. It was famous for its great castle, lying five leagues distant from the town, called Ḳal'ah Khûrshah, which Niẓâm-al-Mulk, the great Wazîr of the Saljûḳs, had re-fortified, it having been originally built by Khûrshah, who was governor of Jahram under the Omayyad Caliphs[2]. To the south-east of Jahram is the town of Juwaym of Abu Aḥmad (so called to distinguish it from Juwaym near the head-waters of the Sakkân, see above, p. 253), which Muḳaddasî describes as lying on a small river, surrounded by palm-gardens, having a fine mosque which stood in a long market street. The district to the south-west was called Îrâhistân, and near the town stood the strong castle called Samîrân (or Shamîrân), which Mustawfî characterises as 'a nest of robbers and highwaymen.' The surrounding districts were famous pasture grounds, especially those lying between Juwaym

[1] Ist. 125. Muk. 422. F. N. 72 *a*, 73 *a*, 82 *b*, 83 *a*. Mst. 172, 179. According to the *Fârs Nâmah* (*folio* 78 *a*) and Mustawfî (p. 177) there would appear to have been another district called Kûrah Ḳubâd Khurrah on the banks of the Ṭâb river above Arrajân.

[2] Ist. 107. F. N. 69 *a*, 82 *b*. Mst. 175, 179. The name of the castle is written Khurûshah, Khûrshah, and Kharashah, in the various MSS., also Kharshad and Kharshar, but no mention of it occurs in the older Arab geographers.

and the bank of the Sakkân river, where were many stagnant pools and lion-haunted forests.

The town of Kâriyân, commanded by a strong fortress, lay one march west of Juwaym, and was celebrated for its fire-temple, from which the sacred fire anciently preserved here was distributed far and wide by the Zoroastrian priests. The fortress, which crowned a hill-top, was deemed impregnable. To the west of Kâriyân, and at the great westward bend of the Sakkân river, stood Lâghir, a place of some importance in the 8th (14th) century, when Mustawfî wrote, for it was a stage on the caravan road down from Shîrâz to Ḳays island. Lâghir also is mentioned in connection with Kaharjân (or Makarjân), a place no longer to be found on the map. Between Lâghir and the coast, but along the right bank and north of the Sakkân river, lay the desert of Mândistân, midway between Najîram and Bûshkânât; here were found neither permanent villages nor streams, but, none the less, as Mustawfî writes, on the rare occasions of sufficient rainfall, the whole desert might be made to grow crops of cotton and corn that at the close of the winter season would give profit of a thousand-fold[1].

Mândistân, the medieval name of this desert—meaning 'the Mând country'—is doubtless retained in the name of the Mând river, which, as already noted, is now used for the lower course of the Sakkân. About half-way between Lâghir and the sea the main stream receives an important affluent from the north, namely the river of Fîrûzâbâd. The city of Fîrûzâbâd was anciently called Jûr, and in Sassanian times this (in place of the later Shîrâz) had been the chief town of the district of Ardashîr Khurrah. Iṣṭakhrî reports that the plain here had originally been a lake, this having been drained by King Ardashîr, who built the city round an artificial mound—still existing here in the 4th (10th) century and later—called Aṭ-Ṭirbâl, 'the Look-out,' with a building named in Persian the Aywân (Archway), standing upon a great platform. At this time Jûr was as large as Iṣṭakhr, and the city was surrounded by a wall and ditch, with four gates, namely Bâb Mihr to the east,

[1] Ist. 117. Muk. 427, 428. F. N. 69 *b*, 73 *b*, 82 *b*, 86 *a*. Mst. 172, 173, 175, 179, 180. J. N. 268. Kaz. ii. 162.

Bâb Bahrâm opposite, Bâb Hurmuz to the north, and Bâb Ardashîr to the south.

The name Jûr, in Persian pronounced *Gûr*, means 'a grave,' and it was held inauspicious by the courtiers of 'Aḍud-ad-Dawlah the Buyid, who was fond of coming here, that the Amîr should be said to be residing in Gûr, 'the grave.' Hence Jûr was renamed Fîrûzâbâd—'the Abode of Luck'—and so it is called at the present day. Muḳaddasî, who gives the story, speaks of the great town square (*Raḥbah*), and the beautiful rose gardens of Fîrûzâbâd, also of the well-cultivated country round, stretching a day's march across. Water for the town was brought from a neighbouring hill by means of a syphon tube, and according to the Persian geographers there was a great castle four leagues from the town, called Ḳal'ah Sahârah (or Shahârah). The Fîrûzâbâd river is named by Iṣṭakhrî the Tîrzah; the *Fârs Nâmah* and Mustawfî call it the Burâzah (or Barârah) river. It rose in the Khunayfghân district, and was said to have been turned from its original course by Alexander the Great, who, when besieging Jûr, flooded the country round and made the lake, which was subsequently drained by Burâzah the Sage, in the reign of King Ardashîr. He afterwards built an aqueduct that conveniently brought the waters of the stream into the town, and from him the river took its name of the Nahr Burâzah. Ḳazvînî says there was a celebrated fire-temple in Fîrûzâbâd, and refers to a wonderful spring of water that gushed out at the city gate; the red roses of Jûr, too, he adds, were famous the world over. The country to the north was, as already said, the district of Khunayfghân, or Khunayfḳân, which the Persians pronounced Khunâfgân; and among the hills there was a large village of this name, whence a difficult and stony road led down to Fîrûzâbâd[1].

The coast of the Ardashîr Khurrah district was known as the Sîf (or shore), and there were three Sîfs, all of the hot region, or Garmsîr, lying along the Persian Gulf. These were named respectively the Sîf 'Umârah to the eastward of Ḳays island; the Sîf Zuhayr on the coast south of Îrâhistân and round Sîrâf; and lastly the Sîf Muẓaffar to the north of Najîram; the 'Umârah,

[1] Ist. 105, 121, 123. Muk. 432. F. N. 70 *a*—72 *b*, 79 *b*, 82 *a*. Mst. 172, 179, 219. Kaz. ii. 121.

Zuhayr, and Muzaffar being the names of three Arab tribes who, having crossed to the northern coasts from the other side of the Persian Gulf, had here settled in Fârs. In the 4th (10th) century Sîf 'Umârah was famous for an impregnable castle on the sea, called Ḳal'ah-ad-Dîkdân (or Dîkbâyah), also known as Ḥiṣn Ibn 'Umârah, where twenty ships could find safe harbourage, and the only entrance into the castle was by working a crane set on the walls. A short distance to the west of this lay the island of Ḳays, or as the Persians wrote the name, Kîsh, which in the course of the 6th (12th) century became the trade centre of the Persian Gulf after the ruin of Sîrâf, which will be described presently. A great walled city was built in Ḳays island, where water tanks had been constructed, and on the neighbouring sea-banks was the famous pearl fishery. Ships from India and Arabia crowded the port, and all the island was full of palm gardens. In summer, says Ḳazvînî, the heat was greater than the hottest room in the bath (Ḥammâm): none the less Ḳays was a very populous town. The island lay about four leagues from the coast, where the port of embarkation was Huzû, to which, in the 7th (13th) century, a caravan road came down from Shîrâz through Lâghir. Huzû, though much ruined when Yâḳût wrote, had been a strong fortress in the 4th (10th) century under the Buyids, who made it their state prison. Close to the town was the village called Sâviyah (with variants in the MSS. Tâbah or Tânah and the true reading is unknown)[1].

[1] Ist. 116, 140. I. H. 188. Yak. ii. 711; iv. 333, 974. F. N. 74 *b*. Mst. 171, 173, 180. Kaz. ii. 161. The name of the island is spelt Ḳays, Ḳaysh, and Kîsh (with dotted ḳ or undotted k).

The stages on the road down from Lâghir to Huzû are given by Mustawfî (p. 200), but as no modern traveller has followed this route the names are not to be found on the map, and are most uncertain; the distances are in farsakhs (leagues). 'From Lâghir 6 to Fâryâb district, thence 6 to the city of Ṣaj (Ṣaḥ, Ḥaj, Ḍaḥ, with many other variants), thence 5 to Âb-Anbâr-i-Kinâr, thence 5 to Haram (Sîram or Marmaz), thence 6 down many steep passes to the village of Dârûk (Dârzak, Ûrak or Dâvrak), thence 6 to Mâhân (Hâmân or Mâyân), and thence 6 by the pass of Lardak to Huzû on the sea-shore.' The district Mustawfî calls Fâryâb is evidently identical with Bârâb, half-way between Kâriyân and Kurân, as given by Muḳaddasî (p. 454). The city of Ṣaj is a puzzle, none being known in this region, but possibly we should read

To the westward of Sîf 'Umârah along the sea-shore was the Zuhayr coast, of which Kurân, inland, was the chief town, Sîrâf, and Nâband being its famous harbours; and the region went as far as Najîram beyond the mouth of the Sakkân river. Inland of this was the Îrâhistân district. According to Iṣṭakhrî, Kurân produced an edible clay, green in colour, that tasted like beetroot. Mustawfî counts Kurân as of Îrâhistân, and says its lands only produced dates. Due south of it was the district and town of Mîmand, not far from the port of Nâband, which last stood at the head of a creek known as the Khawr or Khalîj of Nâband. Mîmand, according to Mustawfî, produced quantities of grapes, also other fruits of the hot region, and it was famous for its clever craftsmen[1].

Further up the coast, to the north-west of Nâband, was the port of Sîrâf, the chief emporium of the Persian Gulf in the 4th (10th) century, prior to the rise of Ḳays island into pre-eminence. Sîrâf, Iṣṭakhrî says, nearly equalled Shîrâz in size and splendour; the houses were built of teak-wood brought from the Zanj country (now Zanzibar), and were several storeys high, built to overlook the sea. This author writes that a merchant of his acquaintance here had spent 30,000 dînârs (£15,000) on his house, and the Sîrâf merchants were accounted the richest in all Fârs, a fortune of sixty million dirhams (about two millions sterling) having been gained here by commerce. There were no gardens round the city, fruit and other produce being brought in from the mountains of Jamm, where there was a great castle called Samîrân. Muḳaddasî speaks of Sîrâf as commercially the rival of Baṣrah; its houses were the finest he had ever seen, but it had been in part ruined by an earthquake, lasting seven days, which had occurred in 366 or 367 (977), and with the fall of the Buyid dynasty the place began to decay. The *Fârs Nâmah* states that its final ruin was the work of Rukn-ad-Dawlah Khumârtagîn, the

Jamm (Ist. 106). This route, unfortunately, is not reproduced in the *Jahân Numâ*, nor is it given by any Arab geographer. The coast of the Banî-aṣ-Ṣaffâr would appear to have been identical with the 'Umârah coast, to judge by what Iṣṭakhrî (p. 141) and Yâḳût (iii. 217) write.

[1] Ist. 104, 141, 152. Yak. i. 419; ii. 489; iii. 212, 217. Mst. 172, 173. A. F. 322.

Amîr of Kays island, who made the latter the port of call, though he had his war-ships still built at Sîrâf; but when Yâkût visited the place at the beginning of the 7th (13th) century, only the mosque, with its columns of teak-wood, remained standing, though the ruins of the town could be traced up the neighbouring gorge from the sea-side. Ships then went on to Nâband for shelter, as the harbour of Sîrâf was already silted up. Yâkût adds that the name of Sîrâf was in his time pronounced Shîlâv by the natives.

Najîram, a port of some importance to the westward of Sîrâf, beyond the mouth of the Sakkân river, was at the beginning of the Muzaffar coast, which stretched thence as far as Jannâbah in the Kûrah (district) of Arrajân. Najîram possessed two mosques when Mukaddasî wrote, with good markets, and cisterns for storing rain-water. The Dastakân district was also of the Sîf Muzaffar, and in the 4th (10th) century its chief town was called Saffârah. The district itself appears to have been in the neighbourhood of Jannâbah, but the exact position of the town of Saffârah is unknown[1].

Near the frontier of the Arrajân district, the river of Shâpûr debouches, and some distance from its mouth, probably above the junction of the Jirrah river, to be mentioned later, must have stood the important commercial town of Tawwaj or Tavvaz. In the 4th (10th) century Iştakhrî speaks of this place as about the size of Arrajân; it was very hot, and stood in a gorge of the lowlands, palm-trees growing here abundantly. Tawwaj, which was a place of great trade, was famous for its linen stuffs, woven in divers colours, with a gold-thread ornament. The Shâpûr river, which flowed near the city, was often called the Tawwaj river; and the town is said to have been peopled with Syrian Arabs, brought hither by 'Adud-ad-Dawlah the Buyid. At the beginning of the 6th (12th) century Tawwaj had already much fallen to ruin. Its site has never been identified, but the position of the town is given as on or near the Shâpûr river, in a gorge, being 12 leagues from Jannâbah on the coast, and four from

[1] Possibly this Dastakân district is identical with the coast of the Banî-aş-Şaffâr, already mentioned. Ist. 34, 106, 116, 127, 141, 154. Muk. 422, 426, 427. F. N. 73 *b*, 74 *a*. Yak. iii. 211, 217. Mst. 172. The ruins of Sîrâf are described by Captain Stiffe in the *J. R. G. S.* 1895, p. 166.

the pass that leads down from Darîz. Tawwaj was a famous place at the time of the first Moslem conquest, and its mosque dated from those early days; but when Mustawfî wrote, it had become a complete ruin.

The important town of Ghundîjân, in the district of Dasht Bârîn, was of this neighbourhood. The position of Ghundîjân, of which apparently no ruins now exist, is given in the *Fârs Nâmah* as standing four leagues from Jirrah and 12 from Tawwaj; and the author speaks of the Jirrah river as flowing by 'a part of Ghundîjân.' In the 4th (10th) century the town is said to have equalled Iṣṭakhr (Persepolis) or Jannâbah in size; carpets and veils were made here, and the district was counted as of the hot region. Muḳaddasî describes a stream among the Ghundîjân hills as producing a poisonous hot vapour, so that none could approach it, and birds flying over the stream fell down suffocated; but there were also hot mineral springs here that healed the sick. The population of Ghundîjân, according to Mustawfî, consisted mostly of shoemakers and weavers, and in his day the name Ghundíjân had taken the place of Dasht Bârín in the common speech for the district. In the neighbourhood was a strong castle, called Ḳal'ah Ram Zavân (or Dam Darân, with many other variants), where great cisterns had been dug for storing water. The district of Bûshkânât lay half-way between Ghundîjân and the Mândistân desert (see p. 255) to the north of Najîram. According to Mustawfî there were no towns here, but dates grew and were the chief crop, for Bûshkânât was of the hot region of the Gulf[1].

[1] Muḳaddasî and Yâḳût with many of the older authorities state that Dasht Bârîn was the name of the *town*, Ghundîjân being that of the *district*. Originally, however, this can hardly have been the case, since the name Dasht Bârîn, meaning 'the Plain' of Bârîn, is not applicable to a town. The name of a district or province in the East is very frequently taken over by the chief town, and following this rule when Ghundîjân fell out of use, the name Dasht Bârîn may have taken its place, being used then for town or district indifferently, as Mustawfî remarks later, but contrariwise of the name Ghundîjân. Ist. 106, 128, 130, 152, 153. Muk. 422, 423, 432, 435, 445, 448. F. N. 73 *a*, 76 *a*, 79 *b*, 82 *b*, 86 *a*. Mst. 171, 177, 179, 218. Yak. i. 199, 890; ii. 576; iii. 5, 820. Tawwaj is often included in the Shâpûr Khurrah district by the earlier geographers.

The island of Khârik, which lay off the mouth of the Shâpûr river. was included in the Ardashîr Khurrah district, and was a port of call for ships sailing from Baṣrah to Ḳays island and India. Yâḳût had visited the island, and says that from its hills Jannâbah and Mahrubân, both on the coast of the Arrajân district, were visible. The soil of the island was fertile, producing many fruits, and the date palm grew well here. In the neighbouring sea was one of the best pearl fisheries. Many of the other islands in the Persian Gulf are described by our authorities as of the Ardashîr Khurrah district; but Khârik and Ḳays were commercially the two most important, and of the others named some are not easy to identify. Uwâl was the chief of the Baḥrayn islands, on the Arabian coast, and it is mentioned in the annals of the first Moslem conquest. Bûshahr (Bushire of the present day) first appears in the pages of Yâḳût, and opposite to it on the mainland, as stated by Balâdhurî, was Rîshahr or Râshahr of Tawwaj. The island called Lâwân (Allân, Lân, or Lâr are all variants), by the distances given, must be the present island of Shaykh Shu'ayb lying to the west of Ḳays, and Abrûn island is doubtless the modern Hindarâbî which with Chîn (or Khayn) lies near Ḳays.

The great island at the narrows of the Gulf now called Kishm, also the Long Island (Jazîrah Ṭawîlah), is probably that referred to in our medieval authorities under the various names—possibly merely manuscript variants—of Banî (or Ibn) Kawân, Abarkâfân, and Abarkumân. Yâḳût states that it was also known as Lâft. The island of Khâsik or Jâsik was one of its neighbours, or was possibly merely another name for Kishm (the Long Island). Its population were hardy boatmen, and according to Ḳazvînî they were much given to piracy and raiding. Near each of these islands were pearl fishery banks, but most of them were uninhabited, except during the fishing season. Beyond and east of Kishm was the island of Hurmuz (Ormuz), which being in Kirmân will be spoken of in the chapter treating of that province[1].

[1] Ist. 32. I. K. 61. Baladhuri, 386, 387. Yak. i. 395, 503; ii. 387, 537; iv. 341, 342. Mst. 181, 222. Kaz. ii. 117.

CHAPTER XVIII.

FÂRS (*continued*).

The district of Shâpûr Khurrah. Shâpûr city and cave. The Ratîn river. Nawbanjân. The White Castle and Sha'b Bavvân. The Zamms of the Kurds. Kâzirûn and its lake. The rivers Ikhshîn and Jarshîk. Jirrah and the Sabûk bridge. The Arrajân district and Arrajân city. The Ṭâb river. Bihbahân. The river Shîrîn. Gunbadh Mallaghân. Mahrubân. Sînîz and Jannâbah. The river Shâdhkân.

The district of Sâbûr Khurrah, 'the Glory of Shâpûr' (Sâbûr, as already said, being the Arabic form of the Persian name), was the smallest of the five Kûrahs or districts of Fârs, and its limits were comprised within the basin of the upper Shâpûr river and its affluents.

The chief town of the district in early days was the city of Shâpûr, the name of which had originally been Bishâpûr[1], more commonly known as Shahristân, 'the town-place' or 'the capital.' Ibn Ḥawkal states that Shâpûr city was in his day as large as Iṣṭakhr and more populous, but Mukaddasî in the latter part of the 4th (10th) century speaks of the town as already for the most part gone to ruin, its population having migrated to the neighbouring and rising city of Kâzirûn. Shâpûr, however, was then still a rich place, for its lands produced sugar-cane, olives, and grapes abundantly, and fruits and flowers, such as the fig, the jasmine, and the carob, were seen on every hand. The castle was

[1] In the MSS. the name is generally (but probably incorrectly) written Nashâpûr or Nishâpûr. Bishâpûr stands for Bih-Shâpûr, the older form being Wih-Shâpûr, meaning 'the good Sapor' or 'the excellence of Sapor.' This prefix Bih occurs in other place-names; cf. Bih Ardashîr, or Guwâshir, in Chapter XXI, p. 303.

called Dunbulâ, and the town wall had four gates, namely those of Hurmuz, Mihr, Bahrâm, and lastly the City gate (Bâb-ash-Shahr). Outside the town was a Friday Mosque, and another called Masjid-al-Khiḍr, or the mosque of Elias. In the beginning of the 6th (12th) century the author of the *Fârs Nâmah* describes Shâpûr as having completely fallen to ruin; and when Mustawfî wrote a couple of centuries later the name of Shâpûr or Bishâpûr had been transferred to the neighbouring Kâzirûn district.

Mustawfî apparently knew the Shâpûr river under the name of the Shahriyâr Rûd, and the city, he says, had been named Dîndâr by its first founder, the mythical King Tahmurath, the 'Devil-binder.' Afterwards Alexander the Great laid it in ruins, and King Shâpûr rebuilt it, when it was known, according to Mustawfî, as Banâ Shâpûr, and later as Nâshâpûr or Bishâpûr. Its crops were famous in the 8th (14th) century: the iris, violet, jasmine, and narcissus grew abundantly, and much silk was woven here. Mustawfî further refers to the well-known colossal statue of King Shâpûr in the cave near the ruins. This he describes as 'a black statue of a man, larger than life, standing in a temple (*Haykâl*); some say it is a talisman, others that it is merely a real man whom God has turned to stone. The kings of that country were used to visit it, and to pay it honour anointed the statue with oil.' Already in the 4th (10th) century Muḳaddasî refers to the cave, which, he says, lay one league distant from the city of Nawbandajân. The colossal figure of King Sapor he describes as crowned and standing at the mouth of the cave, in which water fell continually, and a violent wind blew. At the base of the statue were the semblances, sculptured, of 'three green leaves.' The foot of the image measured ten spans in length, while the total height was eleven ells[1].

The upper course of the Shâpûr river was called the Nahr Ratîn by the Arab geographers, and it came from the Upper Khumâyijân or Khumâyigân district, of which one of the principal villages, according to Mustawfî, was Dîh 'Alî. Lower Khumâyijân was counted

[1] I. H. 194. Muk. 432, 444. F. N. 74 *b*, 75 *a*, where the name is spelt Bishâvûr and Bîshâpûr. Mst. 175, 176. C. A. De Bode, *Travels in Luristan* (London, 1845), i. 214.

as of the Iṣṭakhr Kûrah (the Persepolis district, to be described in the next chapter) lying round Bayḍâ on an affluent of the Kur river, and both these Khumâyijân regions were famous for the products of the colder hill country, such as nuts and pomegranates, while much excellent honey was exported. The people were mostly muleteers, who travelled with caravans. To the westward of Khumâyijân was the district of Anburân with the city of An-Nawbandajân, otherwise called Nûbandagân or Nawbanjân. This place, when Iṣṭakhrî wrote, was larger than Kâzirûn, the climate was hot and the date palm grew here. Muḳaddaṣî speaks of its fine markets, of the gardens with their abundant water-supply, also of its mosque. In Saljûḳ times Nawbandajân had fallen to ruin, but in the 5th (11th) century the town was rebuilt by the celebrated Atabeg, the Amîr Châûlî[1].

Two leagues distant from Nawbanjân began the famous valley, one of the four earthly paradises of the Moslems, called Sha'b Bavvân, the waters of which drained to the Kur river in the Iṣṭakhr Kûrah. The valley was three and a half leagues in length by one and a half across, and its fertility was beyond compare; being due, according to Mustawfî, to the nature of the hills on either side of the valley, which stored the winter snows and thus afforded water throughout the summer droughts. A couple of leagues to the north-east of Nawbanjân is the great mountain fastness called the White Castle—Ḳal'ah Safîd, and Isfîd Diz—or the Castle of Isfandiyâr, occupying a flat-topped table-mountain, many miles in circuit, bounded by precipitous sides. Muḳaddasî possibly mentions it under the name of the Ḳaṣr Abu Ṭâlib, which, he says, was called 'Ayân. The *Fârs Nâmah* states that Ḳal'ah Safîd had been rebuilt by a certain Abu Naṣr of Tîr Murdân in the earlier years of the Saljûḳs, and that at the beginning of the 6th (12th) century it was in the

[1] The Amîr Châûlî (often written Jâûlî), whose name so frequently occurs in the *Fârs Nâmah* and Mustawfî in connection with the rebuilding of towns or castles in Fârs, and the reconstruction of river dams, was governor of the province for Sulṭân Muḥammad the Saljûḳ. Atabeg Châûlî Saḳâuh (meaning 'the Falcon') received the surname of Fakhr-ad-Dawlah, and died in 510 (1116) after having been the semi-independent governor of both Kirmân and Fârs for nearly a score of years.

hands of their governor. The mountain summit, which was 20 leagues in circuit, had only one road leading to the top, and this was guarded below by the castle called Dizak Nishnâk. The summit was a level plain, with many springs and gardens, and fruit grew here abundantly. The siege of Ḳal'ah Safîd by Tîmûr, at the close of the 8th (14th) century, made it famous in history. He was marching from Bihbahân to Shîrâz, and took the place by storm, after a two days' investment, in the spring of 795 (1393)[1].

One march east of Nawbanjân, on the road to Shîrâz, lay Tîr Murdân, a small town surrounded by six villages, of which the most important was called Karjan (or Jarkan), lying five leagues from Nawbanjân. The surrounding region was well watered, very fertile, and much honey was exported. To the west of Nawbanjân, on the road to Arrajân, was the town of Anburân, in this district; also the Bâsht Ḳûṭâ district, with the town of Bâsht, which still exists. Two rivers, the Darkhîd and the Khûbdhân, traversed this region. The Nahr Khawrâwâdhân, otherwise the Khûbdhân river, had on its banks the town of the same name, distant four leagues from Nawbanjân, and Khûbdhân town in the 4th (10th) century was a populous place, with a mosque and good markets. Four or six leagues west of this river, and two stages distant from Nawbanjân, was the small town of Darkhîd, on the river of the same name, which last came from, or some authorities say flowed into, a small lake. It is mentioned that the Darkhîd river was a sufficiently large stream to be unfordable. The Khûbdhân river was an affluent of the river Shîrîn, which will be noticed when describing the Arrajân district, and either the Khûbdhân river or the Darkhîd was crossed by a great bridge, built by a certain Abu Ṭâlib of Nawbanjân, who had erected the castle of 'Ayân mentioned in the previous paragraph. Iṣṭakhrî and Muḳaddasî are at variance as to which of the rivers this celebrated bridge traversed. Later authorities add to the confusion by giving different names to these rivers, which it is difficult

[1] Ist. 110, 111, 120, 127. Muk. 434, 437, 447. F. N. 76 *b*, 78 *a*, 81 *b*. Mst. 177, 178, 219. A. Y. i. 600. Dizakî Nishkuman and Astâk are variants of the name of the lower castle in the MSS. Ḳal'ah Safîd is well described by Macdonald Kinneir, *Persian Empire*, p. 73.

or impossible now to identify with any of the existing streams shown on our maps. The bridge is described by Muḳaddasî as having been built in his day, 'and there is none to equal it in all Syria and Mesopotamia.' This was in the latter half of the 4th (10th) century, and Yâḳût in the 7th (13th) century apparently refers to it as still existing. Many of these places are also mentioned by 'Alî of Yazd, in describing the march of Tîmûr from Bihbahân to Shîrâz[1].

In this mountain region of Fârs, known later as the Jabal Jilûyah, the five Kurdish tribes, called collectively the Zamm-al-Akrâd, had in the 4th (10th) century their pastures and camping grounds. Muḳaddasî speaks of a castle in the mountain near here that belonged to them, standing in a wide district with many gardens stocked with fruit trees and date palms[2].

The city of Kâzirûn, from the latter half of the 4th (10th) century when Shâpûr fell to ruin, became the most important town of the Shâpûr district. Ibn Ḥawḳal describes it as in his time smaller than Nawbandajân, but well-built, the houses being of stone set in mortar. Muḳaddasî, a little later, refers to it as 'the Damietta of Persia,' already commercially important as the centre of the linen trade, and 'Aḍud-ad-Dawlah the Buyid had recently built in the town a great house (*Dâr*) for the merchants, the rooms in which produced a yearly rent of 10,000 dirhams (£400). The houses of the town, Muḳaddasî tells us, were all like palaces, each with a garden; the mosque crowned a hillock. According to Mustawfî Kâzirûn had originally consisted of three neighbouring villages, named Nûrd, Darbast, and Râhshân, built on the water conduits of these names, which, it is stated, were still preserved in the town quarters. The dates of Kâzirûn were

[1] The spelling of the names varies greatly. Khawrâwâdhân is contracted to Khûbdhân, also written Khwâbdhân, Khabâdhân and Khâvdân, or Khâvarân in 'Alî of Yazd. Darkhîd is also written Darkhuwîd, but Dakhûnad (as given by Muḳaddasî) is probably only a clerical error. Ist. 110, 120. Muk. 435, 440. F. N. 76 *a*, *b*, 79 *a*, 80 *b*. Mst. 176, 218. Yak. i. 905; ii. 487; iii. 838. Ibn-al-Athîr, viii. 122, 202. A. Y. i. 600.

[2] Ist. 98, 113. Muk. 435. Yak. ii. 821. Mst. 176, 206. *Zamm* means in Kurdish 'a tribe' (more correctly written *Zûmah*), and by mistake the word has often been given as *Ramm*. See the translation by Prof. De Goeje of Ibn Khurdâdbih, p. 33, footnote.

excellent, especially of a kind called Jîlân, and a cotton stuff, known as Kirbâs, was exported largely. The neighbouring pastures, called Marghzâr Narkis, 'the narcissus meads,' were famous. The district round Kâzirûn was known as the Shûl country, according to Ibn Baṭûṭah, who passed through here in the year 730 (1330), and at the present day this region is called Shûlistân. In the plain, a short distance to the east of the city, lies the Kâzirûn lake, which in the 4th (10th) century was known as the Buḥayrah Mûz, or Mûrak (for the reading of the name is uncertain). It was 10 leagues in length, very salt, and contained much fish. The two famous passes on the road above the lake going up to Shîrâz, which are now known to travellers as the Old Woman's Pass and the Maiden's Pass (Kutâl Pîr-i-Zan, and Kutâl-i-Dukhtar), are named by Mustawfî, the Hûshang Pass, which lies three leagues from Kâzirûn, and the Mâlân Pass, which is above it and is likewise very steep[1].

The roads down to the coast from Kâzirûn lead by Darîz to Kumârij, and thence by Khisht on the Shâpûr river to Tawwaj, which has been described in the previous chapter (p. 259). Darîz was a small town, and already in the 4th (10th) century famous for its linen weavers; Khisht, lying beyond it, had a strong castle, according to Muḳaddasî, and was surrounded by broad lands. The *Fârs Nâmah* mentions Khisht and Kumârij together, and Mustawfî gives the people of both places a bad character as being inveterate robbers.

A short distance below Khisht the river Shâpûr received on its left bank the waters of the Jirrah river, which was known as the Nahr Jarshîḳ to the Arab geographers, and the latter, a few miles before it fell into the Shâpûr river, was joined on its left bank by the tributary stream called by them the Nahr Ikhshîn. The Ikhshîn river took its rise among the valleys of the Dâdhîn country, and according to Iṣṭakhrî, its waters, which were sweet and drinkable, had the property of dyeing to a green colour any cloth that was steeped therein. The Jarshîḳ river rose in the

[1] Ist. 122. I. H. 197. Muk. 433. Mst. 176, 180, 200, 226. Of the three town quarters of Kâzirûn variants in the MSS. are Nûr, Darîst, and Rahibân or Rahiyân. I. B. ii. 89. The *Fârs Nâmah* (f. 80*b*) writes the name of the lake *Mûr* very clearly. It is sometimes called Daryachih Shûr, 'the Salt Lake.'

hills to the south of Jirrah, in the Mâṣaram country (which according to Mustawfî was a district stretching from this river to as far north as the head-waters of the Sakkân river), and before reaching the town of Jirrah it was crossed by an ancient stone bridge called the Ḳanṭarah Sabûk. The river next watered part of the Dâdhîn district, and finally, after receiving the Ikhshîn river, fell into the Shâpûr river some distance above Tawwaj. The *Fârs Nâmah* and Mustawfî state that the country at the head-waters of the Jirrah river, near the town of Jirrah, formed part of the Ghundîjân district, and this gives a clue to the position of Dasht Bârîn, which, as we have seen on a previous page, belonged to the Ardashîr Khurrah district. The city of Jirrah is described by Muḳaddasî as crowning a hill-top, and possessing many palm gardens. Yâḳût states that the common people in his day pronounced the name Girrah, which is confirmed by the *Fârs Nâmah* and Mustawfî; they also refer to its corn crops and dates, for all the lands round the city were extremely fertile[1].

The Arrajân district is the westernmost of the five Kûrahs of Fârs, and Arrajân, its chief town, lay at its westernmost border, on the Ṭâb river, which on this side forms the boundary between Fârs and Khûzistân. The ruins of Arrajân lie a few miles to the north of the present town of Bihbahân, which has taken its population and become the chief town of the district since the close of the 6th (12th) century.

In the 4th (10th) century Arrajân was a fine town, surrounded by date-gardens and olive-groves. It had six gates, which were by order closed at night, and were named, respectively, the Ahwâz, Rîshahr, and Shîrâz gates, then the gate of Ar-Ruṣâfah, the gate of the Maydân (or Square), and lastly Bâb-al-Kayyâlîn or the 'Gate of the Weighers.' The mosque and market streets were magnificent. Soap was largely manufactured in the town. Near Arrajân, and crossing the Ṭâb river on the high roads into Khûzistân, were two famous bridges, the remains of which still exist. One was said to have been built by a certain Daylamite physician of Ḥajjâj, governor of 'Irâḳ under the Omayyad

[1] Ist. 120, 127, 152. Muk. 433, 434, 435. F. N. 75 *b*, 76 *a*, 79 *b*. Mst. 176, 177, 218, 219. Yak. ii. 36, 67.

Caliphs, and is described by Iṣṭakhrî as having but a single arch, 80 paces across in the span, and sufficiently high for a man, mounted on a camel and bearing a banner, to pass freely under the key-stone. This bridge, which was known as the Ḳanṭarah Thakân, stood a bow-shot from the city of Arrajân on the road to Sanbîl. The second stone bridge was more than 3000 ells in length, and dated from the times of the Sassanian kings, being known as the Ḳanṭarah-al Kisrawîyah or 'the Bridge of the Chosroes.' It was on the road leading to the village of Dahlizân. In a hill near Arrajân, according to Ḳazvînî, was a cave whence bitumen (*Mûmiyâ*) was taken from a spring, and this was celebrated all the world over for its medicinal properties, while in the town of Arrajân itself a fathomless well called the Bîr Ṣâhik existed, the water of which was never known to fail, even in the driest summer season.

Mustawfî, in the beginning of the 8th (14th) century, states that Arrajân was then called Arkhân or Arghân by the common people, and at the end of this century 'Alî of Yazd refers to the river Ṭâb as the Âb-i-Arghûn. Arrajân had suffered much, according to Mustawfî, on its capture in the 7th (13th) century by the Ismailian heretics (the Assassins, subjects of the Old Man of the Mountain), and the town had never recovered its former prosperity. There had been Ismailian strongholds on the hill-tops in the neighbourhood, one called Ḳal'ah Ṭîghûr, and another Diz Kilât, and the garrisons of these places had frequently plundered the city and its districts. By the latter half of the 8th (14th) century, Arrajân had fallen completely to ruin, and it was replaced shortly after this by the town of Bihbahân, situated some half-dozen miles lower down the Ṭâb river. Bihbahân, the name of which occurs in none of the Arab geographers, is first mentioned by 'Alî of Yazd, in his description of the march of Tîmûr from Ahwâz to Shîrâz in the spring of 795 (1393), and from this date onward Bihbahân has been the chief town of the region formerly known as the district of Arrajân[1].

[1] Ist. 128, 134, 152. I. R. 189. I. K. 43. Muk. 425. Kaz. ii. 94, 160. Mst. 177, 178. A. Y. i. 600. In his *Mirât-al-Buldân* (Tihrân lithograph, A.H. 1294, vol. I. p. 306) the Ṣanî'-ad-Dawlah says that Bihbahân was first settled by the Kûhgilû nomads, by order of Tîmûr, these having migrated from

The river Ṭâb of the Arab geographers is now known as the Jarâḥîyah, Jarâḥî, or Kurdistân river, for by some confusion the name of Ṭâb has, at the present day, been transferred to the Khayrâbâd affluents of the Hindiyân or Zuhrah river, a different stream which flows out to the Persian Gulf at Hindiyân. The Ṭâb river of the middle-ages had its source, if we may accept the combined authority of Iṣṭakhrî and Muḳaddasî, in the mountains to the south-west of Isfahân, at Al-Burj over against Sumayram in the Iṣṭakhr district. Thence coming down to the district called As-Sardan, in Khûzistân, the Ṭâb was joined on its left bank by the river Masîn, the village of Masîn lying near the point of junction, and the combined streams flowed on to Arrajân. Below this city the Ṭâb watered the Rîshahr district, and then curving round abruptly to the south reached the sea to the west of Mahrubân. The Masîn river above-mentioned also rose in the mountains near Sumayram, and flowed past a place called Sîsḥat, according to the *Fârs Nâmah* and Mustawfî, before it joined the Ṭâb. It is said to have been 40 leagues in length, and was a sufficiently broad river not to be easily fordable. Near the upper course of the Ṭâb was the district of Bilâd Shâpûr, or Balâ Sâbûr, of which the chief town was called Jûmah, which stood on the frontier between Fârs and Khûzistân. The district had been very fertile, but when Mustawfî wrote the lands had already gone out of cultivation. Along the course of the Ṭâb river, according to the *Fârs Nâmah*, was also the region called Kûrah Ḳubâd Khurrah, but all earlier authorities give this as the name of the district round Kârzîn, as has been already described on p. 254[1].

Kûfah. For the ruins of Arrajân, and of the two bridges now known as the Pul-i-Bigam and the Pul-i-Dukhtar (the Lady's and the Maiden's bridge), see De Bode, *Luristan*, i. 295, 297. The name of the first bridge is often given as Ḳanṭarah Rakân or Takân in the MSS. Ibn Ḥawḳal (p. 170) further mentions a wooden bridge as crossing the Ṭâb river, passing at a height of ten ells above the the water level.

[1] Ist. 119. Muk. 24, 425. F. N. 77 *b*, 78 *a*, 79 *a*. Mst. 176, 177, 218. The Arab geographers evidently confounded the upper course of the Arrajân river (the Ṭâb) and its affluent (the Masîn) with the streams which we know to be the upper branches of the Kârûn. It is to be further noted that the Arrajân river, in its lower course near the Persian Gulf, has evidently changed its

Below Arrajân the Ṭâb river, as already said, curved round the Rîshahr district (not to be confounded with Rîshahr of Bushire mentioned above, p. 261); and here, besides the town of Rîshahr, lying half-way between Arrajân and Mahrubân, there was a town called Daryân (otherwise Dayrjân or Darjân) which in the 4th (10th) century had fine markets and lay in a fertile district. Rîshahr continued to be an important place in Saljûḳ times, and the *Fârs Nâmah* speaks of its castle, and states that ships were built here. According to Mustawfî the Persians called the place Barbiyân, and the original name, he says, had been Rîṣahr. Linen stuffs were manufactured here, and the population traded largely with the Gulf ports. The summer heat was terrific, and people went up to Diz Kilât, one league away, which as just mentioned had formerly been a castle of the Ismailians. Near Rîshahr was Hindîjân, a small town and district on the lower course of the Arrajân river, and Muḳaddasî relates that this Hindîjân or Hinduwân town was a great market for sea fish and possessed a fine mosque. In the Hindîjân district were the remains of fire-temples, and some waterwheels of ancient construction. Further, there were supposed to be hidden treasures, 'as in Egypt,' and Ḳazvînî speaks of a well, from which arose a poisonous vapour, so that birds flying above fell dead into it. Lastly, at Ḥabs, a town in this district on the road to Shîrâz, there had been a toll-house in Saljuḳ times[1].

Jallâdgân, otherwise pronounced Jallâdjân, was a neighbouring district lying between the lower courses of the rivers Ṭâb and Shîrîn. The river Shîrîn—'the Sweet Water'—rose in the hills called Jabal Dînâr of the Bâzranj or Bâzrang district, and passed through the district of Furzuk, lying four leagues south-east of Arrajân. According to 'Alî of Yazd, Tîmûr, marching from

bed since the 4th (10th) century. Muḳaddasî speaks of it as debouching near Sînîz, but this is possibly only a clerical error for 'near the Tustar' river, in other words the estuary of the Dujayl.

[1] Ist. 112, 113, 119, 121. Muk. 422, 426, 453. F. N. 78 *a*, *b*. Mst. 177, 178. Yak. iv. 963, 993. Kaz. ii. 186. Hindîjân, Hinduwân, and Hindiyân appear to be all intended for the same place. For Ḥabs the MSS. give Khabs, Jîs, Jins and every possible variation; it was a post-stage, as mentioned in the Itineraries.

Bihbahân to Shîrâz, crossed the Shîrîn river on the day after leaving Bihbahân; four days later he reached the Khâvdân river (already noticed, p. 265, under the name of Khûbdhân), and thence marched to Nawbanjân. We have seen that the Khûbdhân river was a tributary of the Shîrîn, and this last appears to be identical with the stream now known in its upper course as the Khayrâbâd river (with many affluents), and lower down as the Zuhrah river, which is the river marked on modern maps as the Ṭâb, or Hindiyân. On one of the tributaries of the river Shîrîn was situated Gunbadh Mallaghân, an important place lying on the road from Nawbanjân to Arrajân which is now called Dû Gunbadân, 'the Two Domes,' and still shows extensive ruins. Of this neighbourhood were the Dînâr hills, and the district of Bâzrang already mentioned; also Ṣarâm, where the climate in winter was extremely cold, and the mountain summits near by never entirely free from snow even in summer. The town of Gunbadh Mallaghân, however, was of the hot region, and famous for its date palms. The name is also spelt Gunbad Mallajân or Malaḳân, and Muḳaddasî in the 4th (10th) century speaks of the village here as in ruins. According to the *Fârs Nâmah* in the beginning of the 6th (12th) century the small town here was protected by a castle, in which rations of corn, to last the garrison for three or four years, were kept in store. Many other like castles crowned the adjacent hills, among the rest one named Ḳal'ah Khing being especially mentioned. Mustawfî states that the neighbouring district was known as Pûl Bûlû (some MSS. give Pûl Lûlû) and was very fertile, producing famous apricots; and he declares the castle of Gunbad Mallaghân was so strong that one man might hold it against an army[1].

Not far from the mouth of the river Shîrîn—which, as already said, is the modern Ṭâb or Zuhrah river—lay the port of Mahrubân, close to the western frontier of Fârs, and this was the first harbour reached by ships bound to India after leaving Baṣrah and the

[1] Ist. 111, 112, 113, 119, 120. Muk. 435. F. N. 76 *b*, 77 *a*, 78 *b*, 79 *a*, 83 *b*, 85 *b*. Mst. 176, 177, 178, 179, 218. Yak. iii. 5; iv. 630. A. Y. i. 600. Hfz. 31 *b*. De Bode, *Luristan*, i. 258. To the north of Dû Gunbadân is the castle now known as Ḳal'ah Arû; possibly this is the place named Khing in the *Fârs Nâmah*.

Tigris estuary. Mahrubân was accounted the port of Arrajân, and in the 4th (10th) century was very populous, and had a fine mosque and good markets. According to Mustawfî the Persians called it Mâyruyân, or Mahruyân; linen was made here, and dates were exported, but the shipping was always the chief source of income. Nâṣir-i-Khusraw touched at Mahrubân in 443 (1052), and describes the town as lying along the sea-shore on the eastern side of the bay. The markets were excellent, and the mosque bore the name of Ya'ḳûb, son of Layth the Ṣaffârid. Water was stored in cisterns, there were three great caravanserais for travellers who landed here for Arrajân, and the commerce of the place was considerable. The next port down the Gulf, east of Mahrubân, was Sînîz or Shînîz, whose ruins lie on the creek now called Bandar Daylam. Iṣṭakhrî describes the place as larger than Mahrubân in the 4th (10th) century. There was a small bay (*Khawr*), and the town lay half a league from the open sea; the climate here was very hot, and date palms grew abundantly. Muḳaddasî speaks of the mosque and the palace of the governor, and of the markets as being well provided with wares. According to Yâḳût, Sînîz was half ruined by the Carmathians, who sacked the port in 321 (933). The *Fârs Nâmah* however, and Mustawfî, in the 6th and 8th (12th and 14th) centuries, speak of it still as a flourishing place, where flax was grown and much linen made. The port was defended by a fortress (Ḥiṣâr), and the oil for lamps that came from its district was exported far and wide[1].

South of Sînîz was Jannâbah (or Jannâbâ), the ruins of which still exist, lying near the mouth of the river which the Arab geographers called the Nahr-ash-Shâdhkân. Jannâbah according to Iṣṭakhrî was extremely hot, and its creek (*Khawr*) was not a safe anchorage. The town was larger than Mahrubân and had excellent markets; further, it was celebrated as the birth-place of Abu Ṭâhir the Carmathian. The Persians called the place Ganfah, or Âb-i-Gandah, from its 'foul water,' and four neighbouring villages lying on the sea-coast were counted as of its

[1] Ist. 34, 128. Muk. 426. N. K. 90. Yak. i. 502; iii. 221. F. N. 78*b*, 79*a*. Mst. 178.

dependencies. The river Shâdhkân rose in the Bâzrang district, and, passing through the Dastaḳân plain, flowed thence out to the sea. Which stream on the present map it corresponds with is not quite clear, but it must undoubtedly be one of the two short rivers which enter the Persian Gulf near Jannâbah. In point of fact, however, no large stream now exists here, though Mustawfî especially states that this was a 'large river and not easily fordable, being nine leagues in length'; he therefore had in mind a stream of some considerable size[1].

[1] Ist. 32, 34, 119, 128. Muk. 426. F. N. 78 *b*. Mst. 178, 218.

CHAPTER XIX.

FÂRS (*continued*).

The Iṣṭakhr district, and Iṣṭakhr city or Persepolis. Rivers Kur and Pulvâr. Lake Bakhtigân and the cities round it. The Marvdasht plain. Bayḍâ and Mâyin. Kûshk-i-Zard. Sarmaḳ and Yazdikhwâst. The three roads from Shîrâz to Isfahân. Abarḳûh. Yazd city, district and towns. The Rûdhân district and its towns. Shahr-i-Bâbak and Harât.

The Kûrah or district of Iṣṭakhr occupied the whole of the northern part of Fârs, and this, as already said, in the middle-ages included Yazd, with the neighbouring towns and lands lying along the border of the Great Desert. The capital of the district was Iṣṭakhr, as the Arabs named the Sassanian town which the Greeks had called Persepolis.

The city of Iṣṭakhr lay on the river Pulvâr, a few miles above its junction with the Kur river, and some distance to the westward of the remains of the great Achæmenian platform and palaces. At the time of the Moslem conquest Iṣṭakhr was one of the largest, if not the most important of the Sassanian cities of Fârs, and it was taken by capitulation. In the 4th (10th) century, Ibn Ḥawḳal describes the town as a mile broad, and as having formerly been surrounded by a wall which, he says, had recently been destroyed. At the city gate, crossing the river, was the Khurâsân bridge (why so called is not stated), a very fine structure, and the houses stretched far beyond this into the country, being surrounded by gardens which produced rice and pomegranates. The other Arab geographers add nothing to this account, and the Moslem writers give no information of interest about the celebrated Achæmenian buildings and tombs, which they generally ascribe to Jamshîd and King Solomon. Mustawfî states that the ruin of Iṣṭakhr (and hardly any trace of the Moslem city now remains) was due to the turbulent outbreaks of its inhabitants. Finally in the latter half of the 4th (10th) century Ṣamṣâm-ad-Dawlah, son of 'Aḍud-ad-Dawlah the Buyid, was forced to send an army against Iṣṭakhr under the Amîr Ḳutlumish; as a result the town was laid in ruins, and from that time onward Iṣṭakhr was reduced to the size of a village, containing perhaps a hundred men, as described in the *Fârs Nâmah* at the beginning of the 6th (12th) century.

On the hills to the north-west of the city were three great fortresses, known as the Castle of Iṣṭakhr Yâr, 'the Friend of Iṣṭakhr,' the Ḳal'ah Shikastah, 'the Broken Castle,' and the Castle of Shankavân. Collectively these castles were called Sih Gunbadhân, 'the Three Domes'; and from a deep gorge in the mountains, where a dam had been built, water was brought to the first of these castles, in which 'Aḍud-ad-Dawlah the Buyid had constructed great tanks, carefully roofed over on twenty columns, so as to be capable of supplying the needs of a thousand men during a year's siege. There was here an exercising-ground, or Maydân, on the hill-top, which had also been planned and constructed by 'Aḍud-ad-Dawlah[1].

The Pulvâr river—which the Arab geographers call the Furwâb, and which in Persian is written Purvâb—rises to the north of Ûjân or Uzjân at Furvâb village in Jawbarḳân. Flowing at first eastward, it turns to the south-west above Pasargadæ at the tomb of Cyrus, which the Moslems call the Shrine of the Mother of King Solomon (Mashhad-i-Mâdar-i-Sulaymân), and, running through the Iṣṭakhr gorge, passes this city and enters the plain of Marvdasht, where it falls into the river Kur a short distance above the great dam called Band-i-Amîr. The river Kur rises in the district of Kurvân, a little to the south of Ûjân, and not far therefore from the source of the Pulvâr river, but it takes at first the opposite direction. Flowing towards the north-west it makes a great circular sweep, passing under the Shahriyâr bridge, on the summer road

[1] Baladhuri, 388. I. H. 194. Muk. 435. F. N. 67 *b*, 81 *b*, 83 *a*. Mst. 173, 174, 178, 179. Hfz. 85 *b*. The ruins of the three castles still exist, and one of them was visited by J. Morier, *Second Journey through Persia* (London, 1818), pp. 83, 86. De Bode, *Luristan*, i. 117.

from Shîrâz to Isfahân, which stands in the Ûrd district. Passing southward the Kur next flows near the villages of Kûrad and Kallâr, turning then to the south-east, when it receives an affluent from the Sha'b Bavvân valley (see above, p. 264), and traverses in turn the districts of Râmjird and Kâmfîrûz. Passing into the Marvdasht plain it here receives on its left bank the Pulvâr river, then waters the districts of Upper and Lower Kirbâl, and flowing near the large village of Khurramah falls into Lake Bakhtigân, between the Jafûz district to the south, and the Kâskân district on the left bank.

The *Fârs Nâmah*, and other Persian authorities, state that the Kur was known in its upper reach as the Rûd 'Âṣî, 'the Rebel River,' because till it was hemmed back by a dam (*band*) its waters could not be used for purposes of irrigation. The first of these dams on the Kur was called the Band-i-Mujarrad, 'the Bare Dyke.' This was of very ancient construction, and having fallen to decay had been restored by the Atabeg Fakhr-ad-Dawlah Châûlî in the beginning of the 6th (12th) century, after whom the dyke was called the Fakhristân, a name it still bore in the time of Ḥâfiẓ Abrû. Below the junction of the Pulvâr the Kur was dammed back by the celebrated Band-i-Amîr[1] or Band-i-'Aḍudî, part of the works being also known as the Sikr (Weir) of Fanâ Khusraw Khurrah. All these names came from 'Aḍud-ad-Dawlah the Buyid, who had constructed this dam to water the district of Upper Kirbâl. According to the contemporary account of Muḳaddasî, this dam was 'one of the wonders of Fârs.' The foundations of the dam were laid in masonry, with lead joints, and it threw back the waters of the Kur river, forming an extensive reservoir. Along this 'Aḍud-ad-Dawlah had erected ten great waterwheels, which raised the water to a still higher level, thus to irrigate 300 villages, and at each waterwheel was a mill for grinding corn. Soon afterwards a great town was founded near the dam. The lowest of the dams upon the Kur river was called the Band-i-Ḳaṣṣâr—'the Fuller's Dam'—and served to raise the waters to irrigate the district of Lower Kirbâl. This dam was an ancient structure, but having fallen to ruin in the beginning of the 6th (12th) century it was repaired by the Atabeg Châûlî

[1] Hence 'Bendemeer's stream' of Moore's *Lallah Rookh*.

aforesaid, who also effected a much needed restoration of the Band-i-Amîr[1].

The great lake of Bakhtigân into which the Kur flows, though at the present day surrounded by desert lands, was in the middle-ages bordered by many villages and towns situated in richly cultivated territories. The waters of the lake form two great bays, of which the southern one in medieval times was known as Bakhtigân, the northern part of the lake being called the Buḥayrah Bâsafûyah or Jûbânân. The waters were salt, and abounded in fish, which supplied the Shîrâz market, and the lake shore was covered with reeds that, when cut, served as fuel. The Jafûz district was at the western end of the lake, with the town of Khurramah (still existing as an important village) lying 14 leagues distant from Shîrâz, on the road to Kirmân which went along the southern shore of Bakhtigân. Muḳaddasî speaks of Khurramah in the 4th (10th) century as a town with broad lands and a castle crowning a hill-top; this last was very strong and well built, according to Mustawfî writing in Mongol times, and the *Fârs Nâmah* refers to its cisterns[2].

The south-eastern end of Lake Bakhtigân was of the Dârâbjird district, and here lay Khayrah and Nîrîz, which will be spoken of in the next chapter. Near the eastern end, in what is now a waterless desert, stood in the 4th (10th) century the two important towns of Great and Little Ṣâhak or Ṣâhik, a name which the Persians wrote Châhik (meaning 'a small pit' or 'well'). At Great Ṣâhik the two roads—one along the north side of Bakhtigân lake, from Iṣṭakhr; the other by the southern shore, from Shîrâz—came together, and from Great Ṣâhik one single road went on to Kirmân. Muḳaddasî describes Great Ṣâhik as a small town, famed for its calligraphists, who wrote fine copies of the Ḳurân. In the neighbourhood, according to Mustawfî, were steel and iron mines, and the *Fârs Nâmah* speaks of the excellent swords made here.

On the road from Great Ṣâhik to Iṣṭakhr, and lying on the

[1] Ist. 121. Muk. 444. F. N. 79 *b*. Mst. 216, 218. Hfz. 32 *a*. Yak. iii. 107.

[2] Ist. 122, 135. Muk. 437. F. N. 80 *a*, 82 *b*, 87 *b*. Mst. 174, 179, 225, 226.

northern shore of that part of Lake Bakhtigân which was called Bâsafûyah or Jûbânân, were two towns of importance during the middle-ages, all traces of which seem to have disappeared from the map. The easternmost, lying six or eight leagues from Great Ṣâhik, was the city of Bûdanjân, known as Ḳariyat-al-Âs, 'the Myrtle Village,' which Mustawfî gives under the Persian form of Dîh Mûrd. The country round produced plentiful corn crops, and the myrtle, after which the town was called, flourished here. To the westward of Ḳariyat-al-Âs, and six or seven leagues further on the road towards Iṣṭakhr, was Ḳariyat 'Abd-ar-Raḥmân, otherwise called Abâdah, a city standing in the district of Barm. The town possessed fine houses and palaces, and Ḳazvînî relates that the water in its wells was intermittent, sometimes rising up and overflowing the surface of the ground, and at other times being so deep down in the pits as almost to disappear from view. In Saljûḳ times Abâdah had a strong castle, with engines of war, and great water cisterns[1].

The broad plain of Marvdasht is traversed by the lower reaches of the Kur river after it has received the waters of the Pulvâr; it is overlooked from the north by Iṣṭakhr with its three castles, and was divided further into various districts. Lower and Upper Kirbâl lay near the western end of the Bakhtigân lake; Ḥafrak and Ḳâlî came higher up the Kur river, and the meadow lands of Ḳâlî bordered the banks of the Pulvâr. In the Ḥafrak district (spelt Ḥabrak in the older MSS.) was the strong castle of Khuvâr, near the village of the same name. The place is mentioned by Iṣṭakhrî, and several times in the *Fârs Nâmah*, where its position is given as half-way between the 'Aḍudî dam on the Kur, and Abâdah on Lake Bakhtigan, being 10 leagues from either place. Khuvâr is referred to also twice by Yâḳût, who, however, evidently did not know its position. Its water was taken from wells, and the fortifications of the castle were very strong. The plain of Marvdasht was famous for its corn lands, being well irrigated from the dams on the Kur. According to

[1] I. K. 48, 53. Kud. 195. Ist. 101, 131. Muk. 437. F. N. 66*a*, 68*a*, *b*, 83*a*. Mst. 175, 179. Kaz. ii. 160. Besides the *city* of Abâdah (or Abâdhah) there was the village of the same name, on the road from Iṣṭakhr to Isfahân, which will be mentioned later.

the *Fârs Nâmah* it took its name from the hamlet of Marv, which originally had been one of the quarters of Iṣṭakhr city, where later were the gardens of Jamshîd, below the Achæmenian ruins[1].

Above Marvdasht came the Kâmfîrûz district, for the most part on the right bank of the Kur, of which the chief town was, and is, Bayḍâ. Al-Bayḍâ means in Arabic 'the White' (town); and this is one of the few instances in which an Arabic name has been adopted by the Persians (who pronounced it *Bayzâ*), and kept in use down to the present day. Bayḍâ was so called because it 'glistened from afar,' and Ibn Ḥawḳal adds that its name among the Persians had been Nasâtak, meaning, according to Yâḳût, Dâr-i-Isfîd or 'White Palace.' Part of the Moslem army had camped here, when besieging Iṣṭakhr city; and Bayḍâ was as large a place as this last in the 4th (10th) century, Muḳaddasî referring to it as a fine town, with a large mosque, and a much-visited shrine. The pasture lands around it were famous, and the light-coloured soil made the city stand out 'glistening white' among its green corn-lands. The Kâmfîrûz district comprised many villages, which Iṣṭakhrî names, and its oak (Balûṭ) forests were in his day haunted by fierce lions, which were the terror of the cattle on its pasture lands.

North and east of the Kâmfîrûz district was the district of Râmjird, of which the chief city was Mâyin. Half-way between Shîrâz and this place was the town called Hazâr or Âzâr Sâbûr, otherwise Naysâbûr, which is often mentioned in the 4th (10th) century. Muḳaddasî describes it as a small town, possessing broad lands, irrigated by underground channels; and it was the first stage out from Shîrâz going to Mâyin, on the summer or mountain road from Shîrâz to Isfahân. Mâyin, the capital of Râmjird, is described by Muḳaddasî as a populous city with fruitful lands. Mustawfî reports that under the Mongol dynasty its revenues amounted to 52,500 dînârs (about £17,500 in the Îl-Khânid currency). There was in the town a famous shrine of a certain Shaykh Gul Andâm; and at the foot of the pass, on the road north, was the Mashhad of Ismâîl, son of the seventh Imâm Mûsâ-al-Kâẓim. The district of Râmjird owed its great productiveness to the irrigation canals

[1] Ist. 104. F. N. 66 *b*, 67 *b*, 83 *a*, 84 *b*, 86 *a*, *b*. Mst. 174, 175, 179, 180. Yak. i. 199; ii. 480.

taken from above the dam on the Kur at Band-i-Mujarrad, which, as already stated, the Atabeg Châûlî had restored. In Râmjird also was the castle called Sa'îdâbâd, crowning the summit of a steep hill, the road up being one league in length. In old days it was called Isfîdbâdh (the White Place), and in the times of the Omayyad Caliphs it had frequently been held against their armies by rebel chieftains. Finally Ya'ḳûb, son of Layth the Ṣaffârid, at the close of the 3rd (9th) century took possession of it, and, after strengthening the fortifications, used it as a state prison 'for those who opposed him.' The name Isfîdbâdh is possibly a misreading, being sometimes written Isfandyâr, and it is apparently identical with the Isfîdân of the *Fârs Nâmah* and Mustawfî, near which was the village of Ḳumistân, with a great cavern in the adjacent hill[1].

Near the left bank of the Kur river, not far from Mâyin, stood the town and castle of Abraj (often miswritten *Îraj*), which is mentioned by Iṣṭakhrî as of this district, and the place is still to be found on the map. The *Fârs Nâmah* and Mustawfî describe Abraj as a large village at the foot of a mountain, on whose slope its houses were partly built. Its castle, the Diz Abraj, was in part fortified by art, part being already impregnable by the precipices of the hill summit on which it stood; it had gardens too, and was well supplied with water. The town of Ûjân, or Uzjân, which lies one march north of Mâyin, is mentioned by Mustawfî, but no details are given. Ûjân is probably identical with the place named Ḥûsgân (for Ḥûsjân) by Ḳudâmah, where the name is printed in error Khûskân, and in the text of Muḳaddasî, again, it is misprinted Ḥarskân[2].

[1] Kud. 196. Ist. 111, 117, 126, 132. I. H. 197. Muk. 432, 437, 458. F. N. 66*a*, 68*a*, 81*b*. Mst. 174, 175, 180. Yak. ii. 561; iii. 93, 838. The fortress of Sa'îdâbâd is probably the modern Manṣûrâbâd, as described by H. Schindler, *P. R. G. S.* 1891, p. 290.

[2] Kud. 196. Ist. 102, 136. Muk. 457, 458. F. N. 66*b*, 83*a*. Mst. 174, 179. Abraj, as given in the *Fârs Nâmah*, is undoubtedly the true pronunciation, Îraj (as printed in the texts of Iṣṭakhrî and Muḳaddasî) being due to a clerical error in the MSS., and this has been adopted by Yâḳût (i. 419). The old castle exists above Abraj, and is now known as Ishkanvân, which recalls the name of Shankavân mentioned above (p. 276) as one of the three castles of Iṣṭakhr. See Schindler, *P. R. G. S.* 1891, p. 290.

The most direct road from Shîrâz to Isfahân went by way of Mâyin, and thence by Kûshk-i-Zard through Dîh Girdû and Yazdikhwâst to Ḳûmishah on the frontier of Fârs. From Mâyin the road went up the pass, going north to the crossing of the Kur river at the Shahriyâr bridge, near which was the guard-house of Salâḥ-ad-Dîn in the plain called Dasht Rûn or Dasht Rûm. North of this, again, according to Mustawfî, came the Mother and Daughter Pass (Garîvah-i-Mâdar-wa-Dukhtar), and then Kûshk-i-Zard, 'the Yellow Kiosque,' which is probably identical with the Ḳaṣr Ayin, or A'în, of Iṣṭakhrî and Muḳaddasî. The plains of greater and lesser Dasht Rûn were famous as pasture grounds, and the arable lands gave four crops a year, these being watered by the Kur river and its affluents. Kûshk-i-Zard is first mentioned in the *Fârs Nâmah*, where the name is more generally written Kushk-i-Zar, or 'the Golden Kiosque.' To the north again, between Kûshk-i-Zard and Dîh Girdû, stretched the even more fertile pasture lands of the Ûrd or Âvard district, the chief towns of which, according to Iṣṭakhrî, were Bajjah and Taymaristân (written Ṭaymarjân in the *Fârs Nâmah*). Mustawfî mentions Dîh Girdû, and it appears in the *Fârs Nâmah* as Dîh Gawz (for *Jawz*), both names signifying 'Nut Village.' The earlier Arab geographers do not mention this name (which is Persian in the forms given above), but by its position in the Itineraries, modern Dîh Girdû must be equivalent to Iṣṭakhrân of Ḳudâmah and Iṣṭakhrî.

Along the eastern borders of the Dasht Ûrd plain lie Iklîd, Sarmaḳ, and Abâdah village, then Shûristân and Sarvistân village, half-way between Dîh Girdû and Yazdikhwâst. Iklîd had a fine castle according to the *Fârs Nâmah*, and like Sarmaḳ was famous for its corn lands. The name of Sarmaḳ is spelt Jarmaḳ by Muḳaddasî; it was a well-built town surrounded by trees, among which those bearing the yellow plum were notable, this fruit being dried and largely exported to other places. The village of Abâdah, a stage on the present post-road from Shîrâz to Isfahân, is first mentioned in the *Fârs Nâmah*, and later by Mustawfî; the same also is to be said of Shûristân which lies on a salt river flowing east to the desert. The village of Sarvistân, Muḳaddasî states, had a mosque in the 4th (10th) century, and the place was well supplied with water from the neighbouring hills. The name of Yazdikhwâst, the town lying to the north of this, first occurs in the *Fârs Nâmah*, but it is doubtless the same place mentioned by Muḳaddasî under the curtailed form of Azkâs. Mustawfî gives Yazdikhwâst with Dîh Girdû, but adds no particulars. The name is often spelt Yazdikhâs[1].

Ḳûmishah, which Muḳaddasî spells Ḳûmisah, was, as already said, on the northern frontier of Fârs, and it was often counted as belonging to Isfahân. Mustawfî mentions the clay-built castle of Ḳûlanjân which defended it, and tells us that it was surrounded by fruitful districts. To the westward of Yazdikhwâst is situated the town of Sumayram near the head-waters of the Tâb river, and through it passed the western road from Shîrâz to Isfahân. Muḳaddasî describes Sumayram as having a well-built mosque standing in the market street. Nuts and other fruits of the cold region abounded here, and the town was protected by a strong castle, with a plentiful spring of water within the fortifications. Yâḳût states that the name of this castle was Wahânzâd. The western road from Shîrâz to Isfahân passed through Bayḍâ in the Marvdasht plain, and thence went on to Mihrajânâvâdh (or Mîhrajânâbâd), which Muḳaddasî describes as a town with broad lands, apparently lying on the banks of the river Kur, or on one of its western affluents. Between this and Sumayram the only important places were Kûrad and Kallâr (already mentioned as on the Kur), two neighbouring towns, famous according to Muḳaddasî and Mustawfî for their corn lands and the fruit trees of the cold region. Iṣṭakhrî refers to their well-built houses, but apparently all trace of these two places has disappeared[2].

The shortest of the three roads from Shîrâz to Isfahân is that already described, by Mâyin and the Dasht Rûn plain, and this is called the Winter Road in the *Fârs Nâmah*. The Summer

[1] I. K. 58. Kud. 196. Ist. 103, 132. Muk. 437, 458. F. N. 65 *b*, 66 *a*, 80 *b*, 81 *a*, 83 *a*, *b*, 84 *a*, *b*. Mst. 174, 175, 179, 200. Yak. i. 197. I. B. ii. 52.

[2] Ist. 126. F. N. 66 *a*, 84 *a*, *b*. Muk. 389, 437, 457, 458. Mst. 175. Yak. iii. 151; iv. 942. It is to be remarked that while Muḳaddasî (p. 458) in his itinerary refers to Kûrad and Kallâr as though these two villages stood close one beside the other, in the *Fârs Nâmah* itinerary (f. 84 *b*) Kallâr is placed five leagues north of Kûrad.

Road was much longer, and was the easternmost of the three, going by Iṣṭakhr through Kamîn and past the tomb of Cyrus to Dîh Bîd, where, to the right, a road branched off to Yazd. The Isfahân road continued westward through Sarmaḳ and Abâdah village to Yazdikhwâst and Ḳûmishah. Kamîn, not far from the eastern bank of the Pulvâr river, was according to Mustawfî a town of considerable importance in the 8th (14th) century, standing in a corn-producing district, and its fine pasture lands lying along the river are specially mentioned. Higher up, at the bend of the Pulvâr, is Pasargadæ and the tomb of Cyrus, which, it may be remembered, the Moslems identify as the tomb of the mother of Solomon. The four-sided stone mausoleum, still to be seen here, was held to be protected by a talisman, and according to the *Fârs Nâmah* anyone attempting to take up his abode within its walls suddenly became blind. The surrounding pasture lands were called the Marghzâr of Kâlân. Dîh Bîd, 'Willow Village,' the next stage north of this, where the road forked, is given by Muḳaddasî and the other Arab geographers as Ḳariyat-al-Bîdh, and to the north again, about half-way between Iṣṭakhr and Yazd, stood the city of Abarḳûh.

Abarḳûh or Abarḳûyah—sometimes shortened to Barḳûh—is said by Ibn Ḥawḳal to have been a fortified town one-third the size of Iṣṭakhr, with great markets, and Muḳaddasî refers to its fine mosque. Mustawfî says the population were all craftsmen, and the lands round produced much corn and cotton; he further adds that the climate of the city had this remarkable peculiarity—that no Jew could remain alive here above forty days, hence 'among the population of Abarḳûh were no Jews.' In the town itself Mustawfî describes the tomb of the famous saint called Ṭâûs-al-Ḥaramayn, 'Peacock of the Two Sanctuaries' (Mecca and Medina); and it was an acknowledged fact that such was the saint's humility, that the shrine over his grave would never suffer itself to be covered by a roof. However often a roof was erected over the tomb, says Mustawfî, it was invariably destroyed by a supernatural power, lest the saint's bones should become the object of idolatrous worship. In the neighbourhood of Abarḳûh was the village of Marâghah (or Farâghah), where there were magnificent cypress trees, celebrated all the world over as larger and finer than those even of Balkh, or of Kishmar in Ḳuhistân[1].

Yazd in early times had been known as Kathah, and this name, when the town came to be called more particularly Yazd, had passed to its district, otherwise known as the Ḥawmah, or Jûmah (of Yazd). Ibn Ḥawḳal in the 4th (10th) century describes the place as a well-built and well-fortified city, with two iron gates—Bâb Izad and Bâb-al-Masjid—the latter near the mosque which stood in the extensive suburb. A small stream flowed out of the castle hill, the lands round were extremely fertile, although so near the Great Desert, and fruit was largely exported to Isfahân. In the neighbourhood a lead mine was productively worked. Ḳazvînî and others speak of the heavy silk stuffs that were woven in Yazd, all of most beautiful patterns. Mustawfî adds that the town was built of sun-dried bricks, which here lasted as burnt bricks elsewhere, for hardly any rain ever fell, though water was plentiful, being brought in by channels from the hills, and each house had its own storage tank.

One stage to the north of Yazd was Anjîrah, 'Fig Village,' then at the second stage Khazânah (often incorrectly printed Kharânah), a large village with farms and gardens, defended by a fortress on a neighbouring hill; and at the third stage, on the desert border, lay Sâghand. This last, according to Ibn Ḥawḳal, was a village with a population of 400 men, defended by a castle, and its lands were well irrigated by underground water channels. The three towns of Maybud, 'Uḳdah, and Nâyîn lie to the north-west of Yazd, one beyond the other along the desert border, and are generally accounted dependencies of Yazd, though many authorities give Nâyîn to Isfahân. Nâyîn according to Mustawfî was defended by a castle, and the circuit of its walls was 4000 paces. Our authorities, however, give no details about any of these places, merely mentioning their names[2].

1 Ist. 129. I. H. 196. Muk. 437, 457. F. N. 81 *b*, 84 *b*. Mst. 174, 175, 180, 200. J. N. 266. The phenomenon of the roofless tomb is also described by Ibn Baṭûṭah (ii. 113) as a characteristic of the shrine of Ibn Ḥanbal in Baghdâd, and Professor Goldziher has some interesting remarks on this curious superstition in his *Muhammedanische Studien* (i. 257).

2 Ist. 100. I. H. 196, 294, 295. Muk. 424, 437, 493. Kaz. ii. 187. Mst. 153. Yak. iii. 694; iv. 711, 734.

About 75 miles south of Yazd, and half-way between that city and Shahr-i-Bâbak, is the town of Anâr, from which Bahrâmâbâd is 60 miles distant in a south-easterly direction, and both towns are now included in the Kirmân province. During the middle-ages, however, the whole of this district formed part of Fârs and was known as Ar-Rûdhân, of which the three chief towns were Abân (now Anâr), Adhkân, and Unâs (near Bahrâmâbâd)[1].

Unâs, the chief town of the district, was, according to Iṣṭakhrî, of the size of Abarḳûh, and Muḳaddasî speaks of a fine mosque here, approached by steps from the market street, also baths, and well-irrigated gardens, though all round the town lay the sands of the desert. The fortress of Unâs was very strong, and had eight gates, which Muḳaddasî enumerates, for he had visited the place. The place, too, was famous for its fullers, who lived within the town, for there were no suburbs. The Rûdhân district is said to have extended over 60 leagues square. Originally, as at the present day, it had been included in Kirmân, but in the 4th (10th) century it was added to Fârs, and according to the *Fârs Nâmah* this arrangement continued down to the time of Alp Arslân the Saljûḳ, who, after conquering all these regions in the middle of the 5th (11th) century, finally re-annexed Rûdhân to Kirmân[2].

Between Rûdhân and Shahr-i-Bâbak is the small town of Dîh

[1] Our authorities state that Abân was 25 leagues from Fahraj (which is five leagues S.E. of Yazd); the town of Ar-Rûdhân lay 18 leagues beyond Abân, and Unâs was one short march or two post-stages (*Barîd*) from Ar-Rûdhân. Further, Unâs lay one long march and two leagues (or one Barîd) from Bîmand, which last was four leagues west of Sîrjân; and from Ar-Rûdhân to Shahr-i-Bâbak was three days' march, the first march being to Ḳariyat-al-Jamâl, 'Camels' Village.' These distances, plotted out, show that the positions of modern Anâr and Bahrâmâbâd respectively coincide with medieval Abân and Unâs; while the town of Ar-Rûdhân, which is presumably the place elsewhere called Adhkân, must have stood between the two, near the modern village of Gulnâbâd. Ist. 135, 168. I. K. 48. Muk. 457, 473. Yâḳût confuses the matter: he mentions (iii. 925) the town of Anâr as though it were identical with Unâs, which from the distances given above is impossible; Anâr is here probably merely a clerical error for Unâs, which in another passage (i. 367) he counts as of Kirmân.

[2] Ist. 100, 126. Muk. 437, 438, 462. F. N. 64 *b*. Yak. ii. 830. Anâr is still most fertile and produces a considerable surplus of grain, which is exported.

Ushturân or in Arabic Ḳariyat-al-Jamâl, 'Camels' Village,' where, Muḳaddasî relates, there was a tall minaret to the mosque, and fine gardens lying on a stream below the town. Shahr-i-Bâbak, the city of Bâbak or Pâpak, father of Ardashîr, the first Sassanian monarch, was a town often counted as of Kirmân. The place still exists, and it is mentioned by Iṣṭakhrî, Muḳaddasî, and others, who however give us no details. Mustawfî includes it in Kirmân, and says that corn, cotton, and dates grew here abundantly. Two stages west of Shahr-i-Bâbak, on the road to Iṣṭakhr, is the small town of Harât, which the *Fârs Nâmah* couples with Ṣâhik (already mentioned, p. 278). Iṣṭakhrî speaks of Harât as being, in the 4th (10th) century, larger than Abarḳûh; it exported much fruit, according to Muḳaddasî, chiefly apples and olives, and had excellent markets, with streets round its mosque, and a fine stream of water traversed its gardens. Harât had but one gate; and Muḳaddasî names the little town of Far'â as of its neighbourhood. Writing in the 7th (13th) century Ḳazvînî states that the Ghubayrâ plant (possibly the penny-royal) grows abundantly in the gardens of Harât, and when the flowers are in bloom the women of this town were wont, he says, to become wildly excited. To the south-east of Ṣâhik, on the borders of the Dârâbjird district, is the town of Ḳuṭruh, still a place of some importance, where, according to the *Fârs Nâmah* and Mustawfî (who spells the name Kadrû), there were excellent iron mines[1].

[1] Major Sykes (*Ten thousand Miles in Persia*, p. 78) found the ruins of a fire-temple near Shahr-i-Bâbak. Ist. 102. I. H. 182. Muk. 52, 423, 424, 425, 436, 437, 455. F. N. 66 *a*, 68 *a*. Yak. i. 75, 178. Mst. 175, 182. Kaz. ii. 186. The name of Harât village is identical in spelling with Herât, the famous city of Khurâsân.

CHAPTER XX.

FÂRS (*continued*).

The Dârâbjird Kûrah or Shabânkârah district. Dârâbjird city. Darkân and Îg. Nîrîz and Iṣṭâhbânât. Fasâ, Rûnîz, and Khasû. Lâr and Furg. Ṭârum. Sûrû. The trade and manufactures of Fârs. The high roads across Fârs.

The Dârâbjird Kûrah was the easternmost of the five districts of Fârs, and it almost exactly corresponded with the province of Shabânkârah, which, under the Mongol dominion, was divided off from Fârs and formed a separate government. The Shabânkârah according to the author of the *Fârs Nâmah* (who, however, does not apply this name to the Dârâbjird district) were a tribe descended from the Faḍlûyah, a family of Daylamite origin, and they had been of the Ismailian sect of the Shî'ahs. In Saljûḳ times they and the Kurds had waged successful war against the Atabeg Châûlî, and after the decay of Saljûḳ power the Shabânkârah took possession of the eastern region of Fârs, to which they gave their name. The Shabânkârah province is mentioned by Marco Polo, under the form of Soncara, as the seventh out of the eight 'kingdoms' into which he divides Persia; the name, however, has again fallen out of use, and this territory is now known as Dârâbjird[1].

[1] *The Book of Ser Marco Polo*, Sir H. Yule, London, 1874, 2nd ed., i. 84. Shabânkârah appears in the chronicle of Ibn-al-Athîr (x. 362) spelt Ash-Shawânkârah. The chiefs of the tribe who opposed Atabeg Châûlî in the beginning of the 6th (12th) century were Faḍlûh and his brother Khasrû. This last name is probably that more correctly written Ḥasûyah (possibly for Ḥasanûyah) in the *Fârs Nâmah*.

The capital of the district, under the Caliphate, was the city of Dârâbjird or Dârâbgird, which Iṣṭakhrî describes as a walled town with a water-ditch, having four gates, and in the midst of the town stood a rocky hill. Muḳaddasî states that the city was circular and measured a league across in every direction, its gardens were very fruitful, its markets well supplied, and water ran in channels through the town. Near Dârâbjird was the celebrated Ḳubbat-al-Mûmiyâ, 'the Bitumen or Naphtha Dome,' closed by an iron door and only opened once a year, when an officer of the Sulṭân went in and gathered in a box the twelve months' accumulation of the precious Mûmiyâ, which was then sealed up and despatched to Shîrâz for the royal use. At the beginning of the 6th (12th) century, according to the *Fârs Nâmah*, Dârâbjird city was then mostly in ruins, though there was a strong fortress in its midst. Round about extended the famous meadow lands (Marghzâr) of Dârâbjird, and in the neighbourhood was a hill where rock salt, of seven colours, was dug out. According to Mustawfî there was a strongly fortified pass near Dârâbjird, commanded by a great castle, known as Tang-i-Zînah[1].

Under the Shabânkârah, the capital of the Dârâbjird province was removed to Dârkan (or Zarkân), to the north of which stood the fortress of Îg (or Avîg). The Arab geographers of the 4th (10th) century mention these, writing the names Ad-Dârkân or Ad-Dârâkân and Îj, and Iṣṭakhrî says there was a mosque in his day in both these places. Mustawfî, who generally spells the name Zarkân, and refers to the castle as the Ḳal'ah Avîg, says that the surrounding district was very fertile, growing cotton, corn, dates, and other fruits. According to him the castle of Avîg had been first fortified in Saljûḳ times by the Khasûyah tribe, and Yâḳût adds that fruit from here was exported even as far as to the island of Kîsh (Ḳays).

To the north-east of Îg are the town and district of Nayrîz (or Nîrîz) at the eastern end of Bakhtigân; to which lake, at times, it has given its name. Muḳaddasî speaks of the Great Mosque of Nayrîz in the market street, and the ruins of this building, bearing

[1] Ist. 123, 155. Muk. 428. F. N. 68 *b*, 81 *a*, 86 *b*. Mst. 181. The Bitumen Dome, or one similar, is stated by Ibn-al-Faḳîh (p. 199) to have been near Arrajân; see p. 269.

the date 340 (951) still exist. Close to the shore of the lake stands the town of Khîr (spelt also Khayâr and Al-Khayrah), which is mentioned, from the 4th (10th) century onwards, as a stage on the road along the south side of Lake Bakhtigân going from Shîrâz to Kirmân. Mustawfî and the *Fârs Nâmah* name the district round Khayrah Mîshkânât; it was famous for its raisins (*kishmish*), and both Nîrîz and Khayrah were protected by strong castles[1].

Half-way between Khayrah and Îg lies the town of Iṣṭahbânât, a name which the Arab geographers also spell Al-Iṣṭahbânân or sometimes Al-Iṣbahânât, which is shortened by the Persians into Iṣṭahbân. Mustawfî describes it as a town buried in trees, with a strong castle in its vicinity. It had been laid in ruins by the Atabeg Châûlî, who had, however, subsequently caused it to be rebuilt; and the castle in the 8th (14th) century was occupied by the Khasûyah tribe.

The town of Fasâ, pronounced Pasâ by the Persians, was in the 4th (10th) century the second city of the Dârâbjird district, being almost of the size of Shîrâz. It was well built, much cypress-wood being used in the construction of the houses, and was very healthy. The markets were excellent, there was a ditch round the town, which was further defended by a castle, and large suburbs stretched beyond the city gates. Dates, nuts, and oranges in abundance came from its gardens. Muḳaddasî states that the Great Mosque, built of burnt brick and with two courts, rivalled that of Medina for splendour. The *Fârs Nâmah* speaks of Fasâ as being almost of the size of Isfahân. The Shabânkârah had ruined it, but the city had been rebuilt by the Atabeg Châûlî. Mustawfî adds that anciently the city was called Sâsân, and it had been built triangular in plan. Its water-supply, which was abundant, was taken from underground channels, for there were no wells. Shaḳḳ Mîskâhân and Shaḳḳ Rûdbâl (or Rûdbâr) were of its dependencies, and in the neighbourhood stood the strong castle of Khwâdân, where there were great cisterns for storing water[2].

[1] Ist. 107, 108, 132, 136, 200. Muk. 423, 429, 446, 455. F. N. 68*b*, 69*a*, *b*. Mst. 181. Yak. i. 415; ii. 560. Captain Lovett, *J. R. G. S.* 1872, p. 203.

[2] Ist. 108, 127, 136. Muk. 423, 431, 448. F. N. 69*a*, 70*a*, 82*b*, 83*a*. Mst. 175, 179, 181. J. N. 272.

The town of Kurm lies some miles north of Fasâ, on the road to Sarvistân, and is given thus in the Itineraries. According to the *Fârs Nâmah* its district and that of Rûnîz (or Rûbanz) belonged to Fasâ; the latter district forming part of the Khasû territory, which Muḳaddasî marks as lying one march south-west from Dârâbjird on the road to Juwaym of Abu Aḥmad (see above, p. 254). The earlier geographers give the form of the name as Rûnîj (or Rûbanj), and it is probable that this town is identical with the present Khasû (or Kusû). Mustawfî speaks of Kurm and Rûnîz as two towns enjoying a warm climate with an abundant water-supply; and according to Muḳaddasî the Khasû (or Khashû) territory extended far to the eastward, for besides Rûnîj it included the towns of Rustâḳ-ar-Rustâḳ, Furg, and Ṭârum. Mustawfî counts Khasû as belonging to Dârâbjird[1].

Due south of Rûnîz is the small town of Yazdikhwâst, which is mentioned by Muḳaddasî and Yâḳût as of the Dârâbjird dependencies, and south of this again is the city of Lâr. Lâr is not mentioned by any of the earlier Arab geographers, nor does the name occur in the *Fârs Nâmah*, which dates from the beginning of the 6th (12th) century. Mustawfî, in the earlier part of the 8th (14th) century, is our first authority to speak of Lâr, as the name of a district (*vilâyat*) by the sea, most of its population, he adds, being merchants who were given to sea voyages. Corn, cotton, and dates were grown here. His contemporary Ibn Baṭûṭah visited Lâr city about the year 730 (1330), and describes it as a large place, with many gardens and fine markets. Under Shâh Shujâ' of the Muẓaffarid dynasty at the close of the 8th (14th) century, and later under the Tîmûrid princes, Lâr became a mint city, which proves it to have been in those days a place of some size and importance.

[1] I. K. 52. Ist. 108, 116, 132. Muk. 422, 423, 454, 455. F. N. 69*b*. Mst. 181. The pronunciation Rûbanj, adopted in the text of Muḳaddasî, is apparently on the authority of Yâḳût (ii. 828), who carefully spells the word letter for letter. The MSS. of the *Fârs Nâmah* and Mustawfî almost invariably give Rûnîz (for an older form Rûnîj), which is still the name of a district in these parts. It seems probable therefore that *Rûbanj*, as printed in Iṣṭakhrî and Muḳaddasî, is a clerical error, and that by a shifting of the diacritical points we should everywhere read Rûnîj, or Rûnîz, in the place of Rûbanz and Rûbanj.

Furg, which lies three marches south-east of Dârâbjird, is still a considerable town. Muḳaddasî, who spells the name Furj, states that beside it lay the twin city of Burk, but the two names would appear merely to be variants of the original Persian place-name. The city called Burk stood on a hillock, 'like a camel-hump,' two leagues from the mountains; it possessed a mosque in the market street, was a fine place and an agreeable residence. Its neighbour, Furg, had a castle on a hill, was not in the 4th (10th) century a large town, but had its own mosque and many baths, water being plentiful in both cities. Very naturally the names of the two cities were often confounded, one replacing the other. The *Fârs Nâmah* writes the name Purk or Purg, and says that its castle was impregnable, being built of stone and very large. Mustawfî adds that both corn and dates were grown in Burk (as he writes the name) most abundantly. Rustâḳ-ar-Rustâḳ is described by Muḳaddasî as a small town with good markets, lying in the midst of a fertile district measuring four leagues across in every direction. It lies one march to the north-west of Furg, on the road to Dârâbjird[1].

The town of Târum, also spelt Ṭârum, like the district of this name in the Jibâl province (see above, p. 225), lies two marches east of Furg, on the road to the coast. Muḳaddasî refers to its mosque, and praises the markets, gardens, and palm-trees, for a stream ran through the town. Much honey was produced here, and according to the *Fârs Nâmah* it was nearly the size of Furg, and had a strong castle well supplied with cisterns. From Ṭârum the caravan road went almost due south to the coast, where lay the port of Sûrû, or Shahrû, over against the island of Hurmuz. Mustawfî names the port Tûsar, but the reading is uncertain. The Arab geographers speak of Sûrû as a village of fishermen, having no mosque, and dependent for the water-supply on wells dug in the neighbouring hills. There was, Muḳaddasî adds, much trade with 'Omân across the gulf, and the place, which he speaks of as a small town, lay exactly on the Kirmân frontier[2].

[1] Muk. 428, 454, note *n*. F. N. 69 *a*, 83 *a*. Mst. 181. Yak. ii. 560. I. B. ii. 240. The town of Burk appears to be identical with the old fort of Bahman, with a triple wall and ditch, which lies about a mile south of the present town of Furg. Stack, *Persia*, i. 756.

[2] Ist. 167. I. H. 224. F. N. 69 *a*. Muk. 427, 429. Mst. 181, 201.

The trade and manufactures of the province of Fârs, in the 4th (10th) century, are carefully described by both Iṣṭakhrî and Muḳaddasî. At this time, as already stated, the chief port of Persia, on the gulf, was Sîrâf. This place distributed all imports by sea, and to it were brought rare and precious Indian goods, such as were known collectively in Arabic under the name of *Barbahâr*. Iṣṭakhrî gives the imports of Sîrâf as follows:—aloes-wood (for burning), amber, camphor, precious gems, bamboos, ivory, ebony, paper, sandal-wood, and all kinds of Indian perfumes, drugs, and condiments. In the town itself excellent napkins were made, also linen veils, and it was a great market for pearls.

At all times Fârs has been celebrated for the so-called attar of roses (*'Aṭar* or *'Iṭr* in Arabic signifies 'a perfume' or 'essence'), which, of divers qualities, was more especially made from the red roses that grew in the plain round Jûr or Fîrûzâbâd. The rose-water was exported, Ibn Ḥawḳal writes, to all parts of the world, namely, to India, China, and Khurâsân, also to Maghrib or North-west Africa, Syria, and Egypt. Besides the essence of roses, Jûr also produced palm-flower water, and special perfumes distilled from southernwood (in Arabic *kaysûm*, the *Artemisia abrotanum*), saffron, lily, and willow flowers. The city of Shâpûr and its valley produced, according to Muḳaddasî, ten different kinds of perfumed oils, or unguents, made from the violet, water-lily (*Nînûfar*), narcissus, palm-flower, common lily, jasmine, myrtle, sweet-marjoram, lemon, and orange flowers, and these oils were exported far and wide over the eastern world.

The carpets and embroideries of Fârs have in all times been celebrated, and in the East, where robes of honour have always been the mark of distinction, specially brocaded stuffs were manufactured for the sole use of the Sulṭân, on which his name or cypher was embroidered. These were known as *Ṭarâz*, and the town of Tawwaj was famous for their manufacture, as was also Fasâ, where peacock-blue and green stuffs, shot with gold thread, were embroidered for the royal use.

The remaining products of Fârs may best be grouped under the cities producing them. The looms of Shîrâz produced a variety of fine cloths for making cloaks, also gauzes and brocades, and stuffs woven of raw silk (*ḳazz*). Jahram was famous for long carpets and woollen rugs, hangings for curtains, and small prayer-carpets, such as were carried to and from the mosque. Besides the scented oils already mentioned, Shâpûr exported various medicaments, as well as sugar-canes, shaddocks, nuts, olives, and other kinds of fruit, and osiers. Kâzirûn and Darîz produced linen stuffs and fine gauzes, an imitation of the Egyptian brocades known by the name of *dabîḳ*, and fringed towels. Ghundîjân, the capital of Dasht Bârîn, produced carpets, curtains, cushions, and the Ṭarâz embroideries for the Sulṭân's use. Arrajân was famous for a kind of syrup, made from raisins, which was called *dibs*, or *dûshâb*. Good soap was also manufactured here, also thicker woollens and napkins, and the town was an emporium for Indian goods (*Barbahâr*). The neighbouring port of Mahrubân exported fish, dates, and excellent water-skins. At Sînîz the special kind of gauze known as *ḳaṣṣâb* was made, also linen stuffs, for which Jannâbah was also famed.

Iṣṭakhr manufactured stuffs for veils, while the towns of the Rûdhân district produced excellent cloth, a particular kind of sandal called *Shimshik*, water-skins, and divers condiments. Yazd and Abarḳûh yielded cotton stuffs.

In Dârâbjird were manufactured all kinds of cloths, fine, medium, and coarse in texture, also embroideries, fine carpets, and matting. Jasmine-oil and perfumes and the aromatic grains found wild here were exported. The *Mûmiyâ* or bitumen, from Arrajân and Dârâbjird, has already been mentioned. Iṣṭakhrî describes a boneless fish, said to be excellent eating, which lived in the moat of Dârâbjird. Furg produced much the same commodities, together with dibs-syrup; and the like came from Ṭârum, where various kinds of water-skins were manufactured and very serviceable buckets. Fasâ was especially known for its goat-hair, and raw-silk stuffs, also carpets, rugs, towels, napkins, and silk embroidered hangings, particularly of the famous peacock-blue and green colour, shot with gold thread. Cardamums and dye-stuffs came also from Fasâ, and much felt was made, the tents of this material known as *khargâh* being largely exported. Lastly in Fârs, according to Ibn Ḥawḳal, there were silver mines at Nâyin; iron and quicksilver were found in the hills of

Iṣṭakhr, besides lead, copper, sulphur, and naphtha in divers regions. No gold-mine was known. Dye-stuffs of various kinds were common throughout Fârs, so that the land, he says, was full of dyers and their dye works[1].

The high roads of Fârs are described in detail by a long list of authorities, both Arab and Persian, and the distances in these itineraries are generally given in leagues (*farsakh*). Unfortunately Ya'ḳûbî, one of our best authorities for the Road Books, is entirely wanting for Fârs, and Ibn Rustah also for the most part fails us, but beginning with Ibn Khurdâdbih and Ḳudâmah in the 3rd (9th) century, we have Iṣṭakhrî and Muḳaddasî in the 4th (10th) century, and in the first years of the 6th (12th) century the roads of this province are all minutely given by the Persian author of the *Fârs Nâmah*, whose description is for the geography of this period an immense gain which unfortunately is lacking to us for the rest of Persia. Mustawfî, also a Persian authority, registers in the 8th (14th) century the changes effected by the Mongol conquest, and at the close of this century 'Alî of Yazd describes in detail the march of Tîmûr from Ahwâz to Shîrâz, which lay along one of the trunk roads.

In this province the roads all radiated from Shîrâz, and it will be convenient first to describe those leading down to the coast. Sîrâf, Ḳays island, and lastly, Hurmuz island, each in turn became the chief port of the Persian Gulf, and the high roads went down to these, just as at the present day the caravan and post road goes down to Bushire which has now succeeded to the supremacy of Hurmuz. The easternmost of the roads to the coast leads to the port over against the island of Hurmuz, whence also by coasting Hurmuz city on the mainland was reached. Both of these places will be described in Chapter XXII. Leaving Shîrâz this road went by Sarvistân and Fasâ to Dârâbjird, Furg, and Ṭârum, whence turning due south it struck the coast, in early times at Sûrû, or Shahrû, or, as Mustawfî calls it, Tûsar. Not far from here, in Ṣafavid days, the port of Bandar 'Abbâs which still exists was founded, as will be noted later. Of this road we have five separate accounts[2].

[1] Ist. 152—155. I. H. 213—215. Muk. 442, 443.

[2] I. K. 52, 53. Ist. 131, 132, 170. Muk. 454, 455. F. N. 85 *a*. Mst. 200.

The next road, running almost due south from Shîrâz, went in early times to Sîrâf. After the ruin of this port caravans followed a branch to the south-east at a point half-way down to the coast, the new road leading to the port opposite the island of Ḳays, and this is the route described by Mustawfî. Muḳaddasî also gives an important by-road, going south-west from Dârâbjird, on the Hurmuz route, to Sîrâf, and this cuts across the road from Shîrâz to Ḳays island given at a later date by Mustawfî. Starting from Shîrâz all these routes went by Kavâr to Jûr or Fîrûzâbâd. Here the older road branched to the right, going down to Sîrâf. The road given in the *Fârs Nâmah* turned to the left at Fîrûzâbâd, going by Kârzîn to Lâghir, whence, through Kurân, Sîrâf was reached. The route given in Mustawfî leaves the city of Fîrûzâbâd a few leagues to the eastward, and goes down like the *Fârs Nâmah* road to Lâghir, where, branching south-east and to the left, it passed through Fâryâb and the desert to Huzû, the port opposite Ḳays island. Unfortunately this road from Lâghir to Huzû is only found in Mustawfî, and the MSS. give most uncertain readings for the names of the various stages. Apparently, too, no modern traveller has gone by this road, so that we are at a loss for corrections, our maps being here a blank. The cross-road from Dârâbjird, given by Muḳaddasî, goes by Juwaym of Abu Aḥmad to Fâryâb or Bârâb, a stage on Mustawfî's route, and then to Kurân, on the *Fârs Nâmah* route, whence it led direct to Sîrâf[1].

The western road to the coast followed in its upper section the present track from Shîrâz to Bushire, for it passed by Kâzirûn and Darîz to Tawwaj, the important commercial town of the 4th (10th) century, and thence to the port of Jannâbah. The *Fârs Nâmah* gives an important variant to this route, going by the Mâṣaram country to Jirrah, and thence by Ghundîjân to Tawwaj; at Ghundîjân, however, a branch turning off south went down to the port of Najîram, which lies some distance to the west of Sîrâf. Mustawfî only gives the road westwards from Shîrâz as far as Kâzirûn, in his day Tawwaj was in ruins, and at that time the chief port on the Persian Gulf was Ḳays island[2].

[1] Ist. 128, 129. Muk. 454, 455. F. N. 86 *a*, *b*. Mst. 200, also v. supra, p. 257, note 1.

[2] Ist. 130. Muk. 453, 454, 456. F. N. 86 *a*. Mst. 200.

The most fully detailed of all the roads in Fârs is that going from Shîrâz, north-west, to Arrajân and Khûzistân, for we have no less than eight separate accounts of it, though they vary as to some of the stages; the last being that given by 'Alî of Yazd describing in the reverse direction the march of Tîmûr in 795 (1393) from Ahwâz through Bihbahân to Shîrâz, when, on his way, he stormed the great White Fortress of Ḳal'ah Safîd. Leaving Shîrâz, the high road to Khûzistân, as described in the Road Books, goes north-west by Juwaym (Goyun) to Nawbanjân, and thence through Gunbadh Mallaghân to Arrajân, whence by the great bridge over the Ṭâb river it reached Bustânak on the frontier of Fârs. Muḳaddasî and the earlier geographers add the distances from Arrajân to the port of Mahrubân, and thence south-east along the coast to the port of Sînîz and on to Jannâbah[1].

From Shîrâz to Isfahân there were three separate routes in use during the middle-ages. The westernmost turned off to the right, at Juwaym, from the Arrajân road, going to Bayḍâ in the Marvdasht plain, and thence by Kûrad and Kallâr to Sumayram and Isfahân. This route is described by Ibn Khurdâdbih and Muḳaddasî. The middle route is the summer road through the hill country, which goes from Shîrâz to Mâyin, and thence by Kûshk-i-Zard and Dîh Girdû through Yazdikhwâst to Isfahân. This road, with some variants in the names of the stages, is given by the earlier Arab geographers and also by the later Persian authorities. The easternmost of the three roads (the winter or caravan road, through the plains) went from Shîrâz north-eastward to Iṣṭakhr and thence to Dîh Bîd. Here a main route went off to the right going by Abarḳûh to Yazd, while the road to Isfahân turned to the left, and passing through Surmaḳ and Abâdah village joined the summer road at Yazdikhwâst, whence by Ḳûmishah Isfahân was reached. This winter road, which at the present time is the usual post-road from Shîrâz to Isfahân, is given by Muḳaddasî and the *Fârs Nâmah*: the stages to Yazd are enumerated by nearly all our authorities[2].

[1] I. Ḳ. 43, 44. Kud. 195. I. R. 189, 190. Ist. 133, 134. Muk. 453, 455. F. N. 85 *b*. Mst. 201. A. Y. i. 600.

[2] By the *Western Road*: I. K. 58. Muk. 457, 458. By the *Summer Road* or *Hill Road*: Kud. 196, 197. Ist. 132, 133. Muk. 458. F. N. 83 *b*. Mst.

The roads from Shîrâz to Shahr-i-Bâbak and thence on to Sîrjân, one of the capitals of Kirmân, followed two routes, one to the north of Lake Bakhtigân, the other passing along the southern shore of the lake. The northern route went first from Shîrâz to Iṣṭakhr (Persepolis), and from here to Shahr-i-Bâbak we have two roads, one direct by Harât village, the other by Abâdah city to Ṣâhik, where it joined the road along the southern shore of the lake. This last left Shîrâz, going eastward by the northern side of Lake Mâhalû to Khurramah, whence by the southern shore of Bakhtigân it reached Khayrah. From here the *Fârs Nâmah* gives the distances of a branch road to Nîrîz and Ḳuṭruh. The main road went from Khayrah to Great Ṣâhik, where, as already said, it was joined by the route from Iṣṭakhr along the northern lake shore, and from Great Ṣâhik it crossed a desert tract, going north-east to Shahr-i-Bâbak. Both by the northern and the southern shore of Lake Bakhtigân full itineraries exist in the Arab and Persian authorities, but the names of some of the intermediate stages are uncertain, namely of villages that no longer exist at the present day, for the whole of this country has gone out of cultivation and become depopulated since the close of the middle-ages[1].

200. By the *Winter Road*: Muk. 458. F. N. 84*b*. By the *Yazd Road*: I. K. 51. Ist. 129. Muk. 457. F. N. 86*b*. Mst. 201.

[1] The road *viâ* Harât: Muk. 455, 456, 457. The road *viâ* Abâdah and north lake shore: I. K. 53. Kud. 195. Ist. 130, 131. F. N. 84*b*. The road *viâ* Khayrah and south lake shore: I. K. 48. Muk. 455. F. N. 85*a*. Mst. 201. For the roads which centred in Sîrjân, coming up from Fârs, see the next chapter, note 1, p. 302, and Chapter XXII, p. 320.

CHAPTER XXI.

KIRMÂN.

The five districts of Kirmân. The two capitals. Sîrjân, the first capital, its position and history. Bardasîr, the second capital, now Kirmân city. Mâhân and its saint. Khabîṣ. Zarand and Kûhbinân, Cobinan of Marco Polo.

The province of Kirmân, as Iṣṭakhrî writes, is for the most part of the hot region, only a quarter of the country being mountainous and producing the crops of a cold climate, for the larger part of the province belongs to the Desert, the towns lying singly, and separated one from another by broad stretches of uncultivated land, and not standing clustered in groups as was the case in Fârs. Yâḳût states that under the Saljûḳs Kirmân had been most populous and flourishing, but already in the 7th (13th) century, when he wrote, ruin had set in, lands going out of cultivation. Finally this evil state was rendered permanent by the devastation which resulted from the invasion of Tîmûr at the close of the 8th (14th) century.

Muḳaddasî in the 4th (10th) century divides the province of Kirmân into five Kûrahs or districts, called after their chief towns; namely (i) Bardasîr, with the sub-district of Khabîṣ to the north; next (ii) Sîrjân, on the Fârs frontier; then (iii) Bam and (iv) Narmâsîr on the desert border to the east; and lastly (v) Jîruft to the south, running down to the sea-coast of Hurmuz. On the north and east the frontier was the Great Desert, on the south-west the sea-coast, while on the west the Kirmân frontier, round about Sîrjân, ran out 'like a sleeve' into the lands of the Fârs province, as

Iṣṭakhrî puts it, and according to some early accounts Shahr-i-Bâbak was herein included as of the Kirmân province[1].

The present capital of the province is the city of Kirmân, the province and its chief town being of the same name, as is so often the case in the East. During the middle-ages, however, the Kirmân province had two capitals, namely Sîrjân and Bardasîr, of which the latter town is identical with the modern city of Kirmân, standing near what is still known as the Bardasîr district.

Sîrjân, the older Moslem capital of Kirmân, was already the chief city under the Sassanians. The Arab geographers always write the name As-Sîrjân or Ash-Shîrajân (with the article), and though no town of the name now exists, the district of Sîrjân still occupies the western part of the Kirmân province, with Sa'îdâbâd for its chief town. The recently discovered ruins at Ḳal'ah-i-Sang[2], on a hill spur some 5 miles to the east of Sa'îdâbâd, on the Bâft road, are evidently the site of Sîrjân, the ancient capital, for they are those of a great city, and the distances given in the medieval itineraries show that these ruins exactly occupy the position of Sîrjân city; and though the modern Sîrjân district covers but a portion of the older Kûrah, it has preserved for us the ancient name. After the Arab conquest Sîrjân continued to be the capital of the Moslem province until the middle of the 4th (10th) century, when all southern Persia came under the power of the Buyids. The governor they sent to Kirmân was a certain Ibn Ilyâs, and he for an unknown reason changed his residence to Bardasîr (the modern Kirmân city), and later, with the transference of all the government offices thither from Sîrjân, this last fell to be a place of secondary importance. When Iṣṭakhrî wrote, however, Sîrjân was still the largest city of Kirmân. He states that there was little wood used

[1] Ist. 158, 163, 165. Muk. 460, 461. Yak. iv. 263.

[2] Ḳal'ah-i-Sang, otherwise known as Ḳal'ah-i-Bayḍâ (the Stone or the White Fort), occupies a limestone hill rising some 300 feet above the plain, and egg-shaped, being about 400 yards in length. The ruins, still surrounded by a low wall of sun-dried brick, built on older foundations, were discovered and first visited by Major Sykes, in 1900, who has described them in detail, p. 431 of *Ten Thousand Miles in Persia* (London, 1902).

in its houses, since these were all built with vaulted roofs. Muḳaddasî describes the place under the Buyid rule as larger and more populous than Shîrâz. It had two chief markets, the old and the new, and both were full of goods, especially clothes and stuffs for making them, for which it was famous. The streets were well built, and most of the houses had gardens. The city was closed by eight gates (Muḳaddasî cites their names, some of which however are uncertainly written in the MSS.), and near that called Bâb Ḥakîm, 'the Physician's Gate,' 'Aḍud-ad-Dawlah, the Buyid, had built a great palace. The Friday Mosque stood between the old and the new market, its minaret had been erected by 'Aḍud-ad-Dawlah, and the water of the town was derived from two underground channels that had been dug in the 3rd (9th) century by 'Amr and Ṭâhir, sons of Layth the Ṣaffârid.

Yâḳût, who states that when he wrote—7th (13th) century—Sîrjân was the second city of Kirmân and contained forty-five mosques, large and small, asserts that the town in his day was known under the name of Al-Ḳaṣrân, 'the Two Palaces,' but he gives no explanation. The name of Sîrjân frequently occurs in the chronicles of Ibn-al-Athîr and Mîrkhwând, when relating the history of the Buyids and Saljûḳs. Mustawfî, after the Mongol conquest, described it as having a strong castle and its lands grew both cotton and corn. Sîrjân afterwards passed into the possession of the Muẓaffarid princes, who reigned in Fârs at Shîrâz, but conquered all Kirmân from the Ḳarakhitay dynasty at the beginning of the 8th (14th) century. In the year 789 (1387) Tîmûr marched into Fârs, appeared in force before Shîrâz, received the submission of the Muẓaffarid princes, and was induced when he left Fârs to conquer Irâḳ, to reinstate some of them as tributaries. Left to themselves, however, they fomented rebellion, and in 795 (1393) Tîmûr again entered Fârs, overthrew the Muẓaffarid forces in a pitched battle, and appointed his own son Prince 'Omar Shaykh governor of Fârs and Kirmân.

Many districts, however, especially in Kirmân, refused to submit to Tîmûr, and Gûdarz, the governor of Sîrjân, held out in the name of the Muẓaffarids, so that Prince 'Omar Shaykh at last had to send troops to lay formal siege to that stronghold. According to the account given by 'Alî of Yazd, the Ḳal'ah (castle) of

Sîrjân had been recently repaired, so that the place was very strong, and after the lapse of a year, as the siege operations were making no progress, 'Omar Shaykh set out for Sîrjân in person, to bring matters to a crisis. He was however at this moment recalled by his father, and met his death by mischance while travelling through Kurdistân to join Tîmûr at the royal camp before Âmid in Upper Mesopotamia. This was in 796 (1394) and for another two years Sîrjân still held out, the garrison ultimately yielding to famine rather than to force of arms; and by order of Tîmûr, when Gûdarz at length did surrender, he and his few remaining soldiers were all massacred in cold blood, as a warning to the disaffected throughout the province. Sîrjân was left a ruin, and though Ḥâfiẓ Abrû, writing in the reign of the successor of Tîmûr, still speaks of Sîrjân as the second city of Kirmân (second to Bardasîr), with a strong castle crowning a high rock, the name of Sîrjân after this date disappears from history, and its exact site has only quite recently been discovered in the ruins of Ḳal'ah-i-Sang, as already said[1].

As mentioned above, the modern capital of the province is Kirmân city, and this, though not the first Moslem capital, appears to have been an important town from early Sassanian

[1] Ist. 166. Muk. 464, 470. Yak. iv. 106, 265. Mst. 182. Hfz. 140 *a*. A. Y. i. 618, 667, 784. Mîrkhwând, pt. iv. 170; pt. vi. 48, 69. The position of Sîrjân is given by the Arab geographers in marches from various known places, often with an equivalent total in farsakhs or leagues. Unfortunately in the Kirmân province the stage-by-stage itineraries, with details of places passed (as we have for the Jibâl province, and the whole of Fârs), are lacking. The following, however, is a summary of the distances recorded, and they agree with the position of Ḳal'ah-i-Sang for Sîrjân city. From Shahr-i-Bâbak on the north-west, where the high roads coming up from Shîrâz and Iṣṭakhr united, Sîrjân was distant 24 and 32 leagues by different roads, and it was 38 to 46 leagues, or three long marches, from Great Ṣâhik. From Rustâḳ-ar-Rustâḳ (one short day's march north-west of Furg) Sîrjân was four marches, and from Nîrîz five and a half marches distant. Going east and south-east, the road from Sîrjân to Jîruft measures six marches or 54 leagues; while to Râyin it was five marches, and to Sarvistân (to the south-east of Râyin) 45 or 47 leagues. Finally, from Sîrjân to Mâhân was counted as three marches, and to Bardasîr (Kirmân city) two marches. The authorities for these distances are as follows:—I. K. 48, 49, 53, 54. Kud. 195, 196. I. F. 206, 208. Ist. 131, 135, 168, 169. Muk. 455, 464, 473.

times. In regard to its origin, we have it stated by Ḥamzah of Isfahân, an historian of the 4th (10th) century, that King Ardashîr, the founder of the Sassanian dynasty, built a city called Bîh-Ardashîr, meaning 'the good place of Ardashîr'; this name the Arabs corrupted in their pronunciation to Bihrasîr (or Bihdasîr) and Bardasîr (or Bardashîr); while the Persians, as Muḳaddasî informs us, pronounced it Guwâshîr, from Wîh-Artakhshîr the more archaic form of Bîh-Ardashîr. Yâḳût adds that the name was in his day spelt Juwâsîr, Juwâshîr, or Gawâshîr, these being all equivalent to, and used indifferently with, the Arabic form Bardasîr[1].

This city of Bardasîr, which became the new capital of the Kirmân province under the Buyids, is without doubt identical in every respect with the modern city of Kirmân, as is proved by its position as given in the Itineraries, and from the description by the Arab geographers of various buildings in Bardasîr, and natural features, all of which still exist, and are to be recognised in Kirmân city. The Arab and Persian chronicles, it will be seen, fully bear out the identification, for after the 4th (10th) century Bardasîr, indifferently called Guwâshîr, becomes in their narratives the capital of Kirmân, and these names are in time replaced by 'the city of Kirmân,' or briefly Kirmân, the province—as is so often the case—giving its name to the capital.

Muḳaddasî, writing at some length upon Bardasîr, describes it, at the time when the Buyid governor had made it the new capital, as a well-fortified though not a very large city. Outside the town was a great castle (Ḳal'ah) standing high up on a hill with gardens, where there was a deep well, dug by the governor Ibn Ilyâs, and hither the aforesaid Ibn Ilyâs was accustomed to ride up every night to sleep on the height. At the town gate was a second fortress (Ḥiṣn) surrounded by a ditch, which was crossed by a bridge; and in the centre of the town was a third castle (Ḳal'ah) overlooking the houses, alongside of which

[1] Ḥamzah, 46. Muk. 460, 461. Yak. i. 555; ii. 927; iv. 265. The pronunciation *Yazdashîr* sometimes given is merely a clerical error, from a mis-setting of the diacritical points in the Arabic writing. At the present day Bardasîr is the name of the small district lying to the south-west of modern Kirmân city, of which the chief town is Mâshîz. As the name of a *town* Bardasîr is unknown. For another instance of *Bih* or *Wih* in Persian place-names, see above p. 262, note.

stood the Great Mosque, a magnificent building. The city had four gates, the first three being called after the towns whither their roads led, namely, Bâb Mâhân, Bâb Khabîṣ, and Bâb Zarand; the fourth was the Bâb Mubârik, 'the Blessed Gate,' or possibly so called after somebody of the name of Mubârik, or Mubârak. Muḳaddasî adds that the place was full of gardens, wells were common, and underground channels gave an abundant water-supply[1].

From the time when Ibn Ilyâs in the reign of 'Aḍud-ad-Dawlah removed the government offices (Dîvân) to Bardasîr, this town, as already said, remained the chief capital of Kirmân, and followed the fortunes of the province, which, as a rule, was annexed by whoever was the ruler of Fârs. In the early part of the 5th (11th) century, the Buyids fell before the rising power of the Saljûḳs, who were masters of the Kirmân province from 433 to 583 (1041 to 1187). Under them, though Sîrjân is one of their chief cities, Bardasîr continued as the 'Dâr-al-Mulk' or official capital of this governorship. In the Saljûḳ chronicle written by Ibn Ibrâhîm the name of the capital is given sometimes as Bardasîr, sometimes as Guwâshîr; while in the corresponding chapters of the *Rawḍat-as-Ṣafâ*, Mîrkhwând invariably refers to the Saljûḳ capital as 'the city of Kirmân,' or more briefly as Kirmân, and the name Bardasîr is nowhere mentioned by him. The two names, therefore—Bardasîr and Kirmân—were for a time used indifferently to denote one and the same place. Ibn-al-Athîr, for example, under the year 494 (1101), relates how Îrân Shâh the Saljûḳ was expelled 'from the city of Bardasîr, which same is the city of Kirmân[2].'

In 583 (1187) the province of Kirmân was overrun by the

[1] Muk. 461.

[2] Ibn-al-Athîr, x. 219. This passage has a fallacious appearance of being conclusive evidence that Bardasîr was later Kirmân city. But though the fact is beyond doubt from both history and topography this passage is no real proof of it, for 'the city of Kirmân' (*Madînah Kirmân*) merely means the capital (city) of Kirmân (province), and is ambiguous. In an earlier volume, Ibn-al-Athîr (iii. 100) relating how, under the Caliphate of 'Omar, Sîrjân was first taken by the Arab armies, adds the words 'which same is the (capital) city of Kirmân' (*Madînah Kirmân*), though Sîrjân certainly is *not* the modern city of Kirmân, as might be inferred at first sight from this passage.

Ghuzz Turkomans, who plundered and half-ruined Bardasîr, and temporarily made Zarand the capital of the province. The power of the Saljûḳs was then on the wane, and in 619 (1222) all Kirmân passed under the sway of the short-lived dynasty generally known as the Ḳarâkhitay. Ḳutluḳ Khân, the first prince of this line, is described by Mîrkhwând as taking possession of 'the city of Kirmân,' and later it is stated that he was buried in the Madrasah, or college, which he himself had caused to be built 'in the quarter called Turkâbâd, outside the city of Kirmân.' On the other hand, both Mustawfî in his *Guzîdah*, and Ibn Ibrâhîm in the Saljûḳ chronicle, state that Ḳutluḳ Khân, in the year 619 (1222), took possession of 'the city of Bardasîr' (or Guwâshîr as the *Guzîdah* has it), thus becoming ruler of all the Kirmân kingdom. Lastly the contemporary authority of Yâḳût gives Bardasîr as the name at this time (13th century A.D.) of the capital of Kirmân[1].

The Mongol conquest of Persia did not materially affect Kirmân, and the daughter of the last prince of the Ḳarâkhitay in the first years of the 8th (14th) century married the Muẓaffarid ruler of Fârs, who afterwards took over the province of Kirmân, under Mongol overlordship. Mustawfî, speaking of the capital Guwâshîr, otherwise Bardashîr, describes the Old Mosque as dating from the close of the 1st century of the Hijrah, and the reign of the Omayyad Caliph 'Omar II, who died in 720 A.D. He also speaks of the garden laid out by the Buyid governor Ibn Ilyâs, called Bâgh-i-Sîrjânî, namely, 'the garden of him who came from Sîrjân,' which when he wrote in 730 (1330) was still flourishing. Ibn Ilyâs, Mustawfî adds, had also built the castle on the hill, already recorded as having been described by Muḳaddasî, and within the town there was the mosque called the Jâmi'-i-Tabrîzî, founded by Tûrân Shâh, the Saljûḳ, and the celebrated shrine over the grave of the saint Shâh Shujâ' Kirmânî. A somewhat later authority, Ḥâfiẓ Abrû, states that Turkhân Khâtûn, daughter of Ḳutluḳ Khân of the Ḳarâkhitay, in the year 666 (1268), erected a magnificent Jâmi' (Friday Mosque) in

[1] Mst. *Guzîdah*, Chapter IV, section X, Reign of Burâḳ Hâjib. Ibn Ibrâhîm, 4, 54, 200, 201. Mîrkhwând, part iv. 104, 105, 128, 129. Yak. iv. 265.

Kirmân, besides other mosques and colleges, one of which will be noticed presently; and the same author, writing in 820 (1417), refers to the city indifferently under the two names of Bardasîr (or Guwâshîr) and Kirmân[1].

These descriptions of Bardasîr given by our various authorities, from Muḳaddasî in the 4th (10th) century down to Ḥâfiẓ Abrû in the early part of the 9th (15th) century, clearly refer to many of the buildings that still exist, mostly in ruin, in the present city of Kirmân. Thus, as we have seen, Muḳaddasî mentions the three fortresses or castles for which the city was famous, and in the Saljûḳ chronicle frequent reference is made by Ibn Ibrâhîm to the castle on the hill (Ḳal'ah-i-Kûh), to the old castle, and to the new castle,—which are evidently identical with the three places described by Muḳaddasî. In modern Kirmân we find that there is, in the first place, an ancient fortress crowning the hill near, and to the east of the city, now generally known as the Ḳal'at-i-Dukhtar or the 'Maiden's Fort,' which is attributed to King Ardashîr in the popular belief. Next, still further to the south-east, is a second hill, fortified of old with walls and towers, now crumbling to ruin, which is known as Ḳal'ah Ardashîr, and this must be the fortress 'outside the city gate'; while, lastly, the older fortress, within the town, doubtless stood on the site of the present governor's palace[2].

The mosque of Tûrân Shâh, mentioned by Mustawfî, still exists under the name of Masjid-i-Malik; while another building, connecting Kirmân city with the time when it was still called Bardasîr, is the magnificent green (or blue) dome, the Ḳubbat-i-Sabz, which, until quite recently, covered the tomb of Turkhân Khâtûn, the daughter of Ḳutluḳ Khân, already mentioned, of the Ḳârâkhitay. This princess, as history relates, some time after her father's death, ousted her brother from the throne, and then during twenty-five years remained virtual ruler of Kirmân, governing in the name of her husband—a nephew of Ḳutluḳ Khân—and of her two sons, whom in turn she allowed nominally to succeed to the throne. Mîrkhwând states that she died

[1] Mst. 182. Hfz. 139 *b*, 140 *a*.

[2] A plan of Kirmân city is given by Major Sykes (p. 188), also a view of these two ancient forts (p. 190), in *Ten Thousand Miles in Persia*.

in 681 (1282) and was buried under the dome of the Madrasah-i-Shahr, or city college. The green dome within which her tomb was placed bore an inscription on its walls, giving the names of the architects, with the date 640 (1242) when the building was completed, namely during the nominal reign of the son of Ḳutluḳ Khân, whom his sister Turkhân Khâtûn afterwards set aside[1].

Of other towns in the Bardasîr district the Arab geographers give on the whole but meagre accounts; groups of villages, so common in Fârs, did not exist, and generally in Kirmân each town was separated from its neighbour by a wide stretch of desert country. A score of miles to the south-west of Kirmân city lies Baghîn, and a like distance beyond this Mâshîz, both on the road from Kirmân to Sîrjân. At the present time these are the only towns in this quarter, and both are frequently mentioned by Ibn Ibrâhîm, in the Saljûḳ chronicle, when relating events of the latter half of the 4th (10th) century. It is curious therefore that neither Baghîn nor Mâshîz should be mentioned by any of the earlier Arab geographers, nor by Mustawfî, nor, apparently, by any of the Persian authorities who have described the campaigns of Tîmûr. Two short marches to the south-east of Kirmân city lies the town of Mâhân, at the present day celebrated for the shrine at the tomb of Ni'mat-Allah, the Ṣûfî saint and 'Nostradamus' of Persia, whose prophecies are still current throughout Moslem Asia. He died in 834 (1431) aged over a hundred years, and is said to have been a friend of the poet Ḥâfiẓ. In the 4th (10th) century Muḳaddasî describes Mâhân as a town chiefly inhabited by Arabs. The mosque was near the fortress, which, surrounded by a ditch, stood in the middle of the town; and for a day's march around the land was covered with gardens which were irrigated from a stream of running water.

[1] The Ḳubbat-i-Sabz was completely ruined by an earthquake in 1896. It is described by Major Sykes, who gives an illustration (*Persia*, p. 264) representing the building as he saw it before the earthquake. Major Sykes gives a description of it, p. 194, as also of the mosque of Tûrân Shâh, who reigned from 477 to 490 (1084 to 1097). Ibn Ibrâhîm, 28, 34, 177, 187, 189, 190, 194. Mîrkhwând, part iv. 129, 130. See also Stack, *Persia*, i. 202, 204. Schindler, 'Reise in Persien,' *Zeitschrift der Gesellschaft für Erdkunde* (Berlin), 1881, pp. 329, 330.

Ghubayrâ and Kûghûn, two towns lying one league apart, of which apparently no trace remains at the present day, were to the south of Mâhân, being one march west of Râyîn (which still exists). In the 4th (10th) century Muḳaddasî describes Ghubayrâ as a small town surrounded by villages, with a fortress in its midst, while outside was the market recently built by the Buyid governor Ibn Ilyâs, already many times mentioned. Both this place and Kûghûn had fine mosques and the water was from underground channels. Some fifty miles east of Kirmân, and on the borders of the Great Desert, lies Khabîṣ, which was counted as three marches distant from Mâhân. The level was low, for the desert is here far below the plateau of central Persia on which the city of Kirmân stands, and Khabîṣ, as Iṣṭakhrî remarks, is very hot, and the date palm was consequently much grown. Muḳaddasî adds that there was a fortress here, and the town had four gates. It was very populous, much silk was manufactured, for the gardens were celebrated for their mulberry-trees, being watered by a stream that passed through the town. Excellent dates, too, were exported[1].

Two marches to the north-west of Kirmân is the city of Zarand, and half-way between the two, during the middle-ages, lay the town of Janzarûdh, of which apparently no trace remains. Muḳaddasî describes Janzarûdh as possessing a mosque standing in the market, where abundance of fruit was sold, for the town was on a river, the Janz. Zarand still exists, and Muḳaddasî speaks of the castle near by, which Ibn Ilyâs, the governor, had recently built. Zarand was in the 4th (10th) century a place of considerable size, it had six town gates, and the mosque was in the Maydân or public square, which was surrounded by market streets. Here a kind of fine gauze, used for linings and called *biṭânah*, was made. These Zarandî gauzes were largely exported to Fârs and 'Irâḳ, and in the 4th (10th) century were in great repute.

[1] Ibn Ibrâhîm, 66, 108, 109, 121. Ist. 234. Muk. 462, 463. Col. C. E. Yate, *Khurasan and Sistan*, p. 11. Major Sykes (*Persia*, p. 41) found a grave-stone in Khabîṣ dated 173 (789), also the ruins of a building that appears to have been a Christian church, or some non-Moslem shrine. As of the Khabîṣ sub-district Muḳaddasî (p. 460) mentions the four towns of Nashk, Kashîd, Kûk, and Kathrawâ, but no details are given of position, and apparently all trace of them is now lost.

Fifty miles north of Zarand lies Râvar on the border of the Great Desert, and west of this is Kûbinân, which was visited by Marco Polo. Both towns are described by Muḳaddasî, who says that Râvar in the 4th (10th) century was larger than Kûbinân, and had a strong fortress, which served to protect the frontier. Kûbinân or Kûhbanân he speaks of as a small town with two gates, and a suburb where there were baths and caravanserais. The mosque was at one of the town gates, and was surrounded by gardens which stretched to the foot of the neighbouring mountains. In the vicinity is the town of Bihâbâd, a name which Muḳaddasî writes Bihâvadh, and he couples it with Ḳavâḳ, a populous hamlet, which lay three leagues distant, both places being of the cold region and possessing many gardens. Bihâbâd still exists, but Ḳavâḳ no longer appears on the map. Yâḳût in the 7th (13th) century states that both Kûhbanân and Bihâbâd were in his day celebrated for the *tûtiyâ* or tutty (an impure oxide of zinc), which was manufactured and exported hence to all countries. Mustawfî in the next century also refers to Kûhbinân, which Marco Polo, his contemporary, calls 'the city of Cobinan,' and the Venetian traveller carefully describes the manufacture here of the tutty, 'a thing very good for the eyes.' Already in the 4th (10th) century this was one of the notable exports of the Kirmân province, and Muḳaddasî states that because it came out of the crucible in finger-like pieces, it was commonly known as *Tûtiyâ Murâzîbîy*, 'cannular tutty.' These bunches of 'pipes,' he says, were separated one from another by water being poured over the hot mass, and it was purified by being roasted in long furnaces which he himself had seen built on the mountain side, near where the ore was extracted. The same was done also in the case of iron[1].

[1] Ist. 233. I. H. 224, 292. Muk. 462, 470, 493. Yak. i. 767; iv. 316. Mst. 183. See *The Book of Ser Marco Polo*, Yule, i. 127—130, for the description of the manufacture of tutty, which Major Sykes (*Persia*, p. 272) saw made in Kûhbanân at the present time, and in the identical manner above described. The name of Râvar is often miswritten *Zâvar* by a clerical error; and similarly Kûhbanân appears under the forms of *Kûbayân* and *Kûhbayân* from a misplacing of the diacritical points. *Banân* is the Persian name for the wild pistachio, Kûhbanân therefore signifying the mountain where this tree grows.

Some fifty miles west of Kûhbanân, and on the edge of the desert half-way between that town and Yazd, lies at the present day the hamlet of Bâfḳ. There are in the Kirmân province two towns of very similar names, Bâfḳ aforesaid, and Bâft or Bâfd, the latter lying 80 miles south of Kirmân city, and 200 miles distant from the northern Bâfḳ. The confusion is worse confounded by the fact that (northern) Bâfḳ is often now pronounced Bâfd, and hence is identical in name with the town south of Mâshîz, for dialectically the change of the dotted *ḳ* into *d* or *t* is common in Persian. A town of Bâfd is mentioned by Yâḳût as a small city of the Kirmân province, lying on the road to Shîrâz, and of the hot country. Ibn Ibrâhîm in the Saljûḳ chronicle mentions the names of both Bâft and Bâfḳ, but neither by him nor by Yâḳût are details afforded sufficient to identify the places[1].

[1] Yak. i. 474. A. F. 336. Ibn Ibrâhîm, 31, 43, 67, 90, 158, 159, 164, 172. Stack, *Persia*, ii. 13.

CHAPTER XXII.

KIRMÂN (*continued*).

The Sîrjân district. Bam and Narmâsîr districts. Rîgân. Jîruft and Ḳamadîn, Camadi of Marco Polo. Dilfarîd. The Bâriz and Ḳafṣ mountains. Rûdhkân and Manûjân. Hurmuz Old and New, Gombroon. The trade of the Kirmân province. The high roads.

The Sîrjân district—of which Sîrjân city, the older capital of the Kirmân province, which has already been described in the previous chapter, was the chief town—lay to the west of the Bardasîr district, and on the frontier of Fârs. Muḳaddasî mentions a number of towns in this district which now, unfortunately, no longer appear on the map, though their positions in relation to the site of Sîrjân city are known.

Four leagues west of Sîrjân, and close to the Fârs frontier, was Bîmand, described in the 4th (10th) century as an impregnable fortress, having iron gates. It was a place of importance too, as being the point of junction of the three high roads—from Shahr-i-Bâbak (north), from Rûdhân (north-east), and from Ṣâhik (west)—whence these all converged on Sîrjân. Muḳaddasî describes Bîmand as having a Great Mosque standing in the middle of its market street, and its water was from underground channels. Then one day's march to the east of Sîrjân, on the road to Râyîn, was a place called Shâmât, a town with many gardens and vineyards, exporting much fruit to outlying villages, and with a Friday Mosque standing in its midst. The town also bore the alternative name of Kûhistân. One march again east of Shâmât was Bahâr, and another day's march led to Khannâb, both places growing many dates. Beyond Khannâb lay Ghubayrâ, already described as of the Bardasîr district. Two days' march to the south-east of Sîrjân, on the road to Jîruft, stood a town the name of which is written either Vâjib or Nâjat (with some other variants). Muḳaddasî describes it as a very pleasant and populous place with many gardens, the water being supplied by underground channels, and the Great Mosque standing in the midst of its market streets[1].

The district of Bam (or Bamm, as the Arab geographers write it), surrounding the town of this name, lies to the south-east of Mâhân, at the border of the Great Desert, on the eastern frontier of Kirmân. Ibn Ḥawḳal describes Bam in the 4th (10th) century as larger and healthier than Jîruft, the town being surrounded by palm-groves. Near by stood the celebrated castle of Bam, held to be impregnable, and there were three mosques, the Masjid-al-Khawârij, the Mosque of the Clothiers (Al-Bazzâzîn), and the Castle Mosque. Cotton stuffs were largely manufactured here and exported; also napkins, the cloths for turbans, and the scarfs for head-wear known as *Ṭaylasân*. Muḳaddasî records that the city wall, which made a strong fortification, had four gates, namely, Bâb Narmâsîr, Bâb Kûskân, Bâb Asbîkân, and Bâb Kûrjîn. There were great markets both within the city and outside in the suburbs, while on the river which passed by the castle was the market of the Jarjân bridge. A celebrated bath-house stood in the Willow street (Zuḳâḳ-al-Bîdh). A league distant from Bam was the mountain called Jabal Kûd, where there were mills, surrounded by a large village, and where much cloth was manufactured. Mustawfî in the 8th (14th) century still refers to the strong castle of Bam, and speaks of its climate as rather hot[2].

Râyîn, lying due south of Mâhân, and about 70 miles north-west of Bam, is described by Muḳaddasî as a small town, with its

[1] I. K. 49, 54. Ist. 168, 169. Muk. 464, 465. For Nâjat Ibn Ḥawḳal reads *Nâjtah*, and *Bâkhtah*, *Fâkhtah*, or *Ḳâkhtah*, are the variants in Ibn Khurdâdbih; all of which may possibly be merely clerical errors for Bâft, the town mentioned in the last chapter (p. 310), which still exists approximately in the position indicated.

[2] The ancient fort of Bam, which stands at the present day, is described by Major Sykes (*Persia*, pp. 216, 218). The ruins of the medieval town are on the river bank at Guzârân, about a mile distant from the fort.

mosque standing in the market-place, and gardens extending all round the habitations. At one-third of the way from Râyîn to Bam stood the neighbouring towns of Avârik and Mihrkird (or Mihrîjird), of which the former still exists, the name being now pronounced Abârik. Between the two, in the 4th (10th) century, stood a castle built by the Buyid governor, Ibn Ilyâs. The water-supply was from a river, and the houses were clay-built. Between Abârik and Bam stands Daharzîn, which Muḳaddasî writes Dârzîn, other spellings being Dârjîn and Dayrûzîn. It had a fine Friday Mosque, and was a pleasant place, surrounded by gardens irrigated from a neighbouring stream[1].

The Narmâsîr district (in Persian Narmâshîr) lay south-east of Bam and on the desert border; its capital, the city of Narmâsîr, stood half-way between Bam and Fahraj. Fahraj still exists and in the 4th (10th) century, Narmâsîr was an important town; Muḳaddasî speaks of its many fine palaces, and of its numerous population. Merchants from Khurâsân trading with 'Omân lived here, for Narmâsîr stood on the Pilgrim road from Sîstân to Mecca and was a mart for Indian goods. Narmâsîr was then smaller than Sîrjân, but fortified, and it had four gates, Bâb Bam, Bâb Sûrḳân, the Gate of the Oratory (Muṣallâ), and lastly the Gate of the Kiosque (Kûshk). The Friday Mosque was in the midst of the markets. To its gate was an ascent of ten steps of burnt-brick stairway, and a fine minaret, famous in all the country round, towered above. The castle was known as the Ḳal'ah Kûsh-va-Rân (the name unexplained), and at the Bam gate were three forts called Al-Akhwât, 'the Sisters.' Palm-groves and gardens surrounded the town. At the present day no town of Narmâsîr appears on the map, but the ruins at the site called Chugukâbâd, 'Sparrow-town,' lying on the right bank of the sluggish river which winds through the Narmâsîr plain, must be the remains of the great medieval city. The place is now a complete wilderness, though as late as the 8th (14th) century Mustawfî still refers to Narmâsîr as a populous city.

Twenty miles due south of Fahraj is Rîkân (also spelt Rîḳân or Rîghân), the fortifications of which Muḳaddasî describes. The

[1] I. H. 223, 224. Muk. 465, 466, 470. Mst. 182. Yak. iv. 700. Abârik and Dârzîn are described by Major Sykes (*Persia*, p. 214).

Great Mosque stood near the town gate, and outside were palm gardens. Mustawfî refers to it as a very hot place, where dates and corn were grown abundantly. Between Rìghân and Bam stands Kurk, which Muḳaddasî couples with the neighbouring town of Bâhar (not to be confounded with the differently spelt Bahâr of Sîrjân, see p. 311), and both were populous towns in the 4th (10th) century, being surrounded by palm-groves. The town of Nisâ was also of the Narmâsîr district, but its position is unknown. It is stated that it had gardens in the plain, and a mosque in its market-place, and it was watered by a river[1].

The whole of the southern half of the Kirmân province, and down to the coast, was included in the district of Jîruft. Jîruft (or Jayruft) during the middle-ages was a city of much importance, and past it ran the only river which the Arab geographers mention by name in this province. The ruins of Jîruft (the name is now preserved in the Jîruft district only) are those now known as the Shahr-i-Daḳiyânûs, 'the City of the Emperor Decius,' who figures as a proverbial tyrant in the East, for in his reign the Seven Sleepers entered the Cave, as mentioned in the Ḳurân (chapter XVIII, v. 8, and see above, p. 155), the story being amplified of course in the popular legends. Near these ruins runs the stream now known as the Khalîl Rûd (or Ḥalîl Rûd), which the Arab and Persian geographers name the Dîv Rûd, 'the Demon Stream,' from its swiftness. It is an affluent of the Bampûr river, and drains east to the Hâmûn or swamp.

In the 4th (10th) century Ibn Ḥawḳal describes Jîruft as a great city, measuring two miles across, 'the mart of Khurâsân and Sijistân,' lying in a fruitful neighbourhood where the crops of both the hot and the cold regions were grown. The chief exports of the city were indigo, cardamoms, sugar-candy, and the *dûshâb* or raisin syrup. The surrounding district was called Al-Mîzân (Iṣṭakhrî writes Al-Mîjân), where the numerous gardens produced dates, nuts, and oranges. Snow came from the neighbouring hills, and water was supplied by the Dîv Rûd, which made a great

[1] I. K. 49. Muk. 463, 464. In Mustawfî (p. 182) for *Mâshîz* as given in the lithographed edition, 'Narmâsîr' must be read, according to all the best MSS., confirmed by the Turkish text of J. N. 257. For Chugukâbâd, see Sykes, *Persia*, p. 220.

noise flowing over the rocks. There was water-power here for turning from twenty to fifty mill-wheels. Provisions were also brought into the city from the neighbouring valley of Darfârid, and according to Muḳaddasî the sweet melons from here and the narcissus flowers, from which a perfume was made, were both celebrated. The town itself, which had a fortified wall, was closed by four gates, namely, Bâb Shâpûr, Bâb Bam, Bâb Sîrjân, and Bâb-al-Muṣallâ, 'the Oratory Gate.' The Great Mosque, built of burnt brick, was near the Bam gate, at some distance from the market streets. Muḳaddasî adds that Jîruft was in his time a larger city than Iṣṭakhr, and that its houses were mostly built of clay bricks on stone foundations.

Yâḳût states that the fertile district round Jîruft was called Jirdûs, and Mustawfî refers to the lion-haunted forests which had originally surrounded the town, but which in his day had given place to immense palm-groves. Ibn Ibrâhîm in the Saljûḳ chronicle during the 6th (12th) century frequently refers to Ḳamâdîn, 'a place at the gate of Jîruft where foreign merchants from Rûm (Greece) and Hind had their warehouses and where travellers by sea and land could store their goods'; and in another passage he mentions the 'precious goods from China, Transoxiana, and Khitây, from Hindustân and Khurâsân, from Zanzibâr, Abyssinia, and Egypt, also from Greece, Armenia, Mesopotamia, and Adharbâyjân,' which were all to be found for buying and selling in the storehouses of Ḳamâdîn. The Persian Ḳamâdîn is the place mentioned by Marco Polo under the name of Camadi, or the 'city of Camadi.' It had been formerly 'a great and noble place,' but when Marco Polo visited it 'was of little consequence, for the Tartars in their incursions have several times ravaged it.' This explains why both Jîruft and Ḳamâdîn, after the close of the 7th (13th) century, disappear from history, and the map no longer bears these names. Round Jîruft was the Rûdhbâr district, mentioned by the Arab geographers, which reappears in Marco Polo under the name of 'Reobarles[1].'

[1] For the ruins of Shahr-i-Daḳiyânûs, lying on the right bank of the Ḥalîl Rûd, a short distance to the west of modern Sarjâz, see Keith Abbott in *J.R.G.S.* 1855, p. 47; and Sykes, *Persia*, p. 267. Ist. 166. I. H. 222. Muk. 466,

One march to the north-east of Jîruft, and half-way to Dârjîn, lay the large hamlet of Hurmuz-al-Malik ('of the King,' so called to distinguish it from the port of Hurmuz), which was also known as Karîyat-al-Jawz, 'Nut Village.' According to Idrîsî—but it is not clear whence he got his account—this was an ancient city founded by the Sassanian king Hurmuz in the third century A.D., and it had been the chief town of the province of Kirmân, until, falling to ruin, the administration had been transferred to Sîrjân, which remained the capital of the province under the later Sassanians. The position of Hurmuz-al-Malik is indicated by Mukaddasî and other early geographers, but they give no details; Idrîsî adds that in his day (or more probably in the time of the unknown author from whom he takes his account) this Hurmuz was a handsome though small town, inhabited by a mixed population, having abundant water, and good markets with much merchandise. It lay, he says, one march distant from Bam[1].

A day's march to the north of the ruins of Jîruft lies Dilfarîd, which Mukaddasî calls Darfânî, and Ibn Hawkal Darfârid. It lay in a fruitful valley producing crops of both the hot and cold regions, and, as already stated, was the granary of Jîruft. One march to the north-west of this again was the Jabal-al-Ma'âdin—'Hill of Mines'—where silver was found, more especially in a gorge that ran up into the Jabal-al-Fuddah or 'Silver Hill[2].'

To the eastward of Jîruft was the hill country called Jabal Bâriz, described as clothed with great forests in the 4th (10th) century, and here at the time of the first Moslem conquest the hunted Magians had found safe refuge from the troops sent against them by the Omayyad Caliphs. This country was only brought under the Moslem yoke by the Saffârid princes; it was afterwards famous for its iron mines. Nearer the coast, and to

470. Yak. ii. 57. Mst. 182. Ibn Ibrâhîm, 48, 49, 83. Schindler, *J.R.A.S.* 1898, p. 43; and *The Book of Ser Marco Polo* (Yule), i. 98.

[1] Ist. 161, 189. I. H. 219, 225. Muk. 473. Idrisi, Jaubert, i. 423, and text in Paris MSS. *Arabes*, No. 2221, *folio* 157 *b*; No. 2222, *folio* 104 *a*. Yak. ii. 151. Major Sykes (*Persia*, p. 444) would identify Hurmuz-al-Malik (which no longer exists under this name) with *Carmana omnium mater* of Ammianus Marcellinus.

[2] Ist. 165, writes the name, probably merely by a clerical error, *Durbây*. I. H. 221, 222. Muk. 467, 471. A. F. 335.

the south-east of Jîruft, lay the hilly region known as Jabal-al-Kufṣ, the outlying regions of which, in the 4th (10th) century, were inhabited by mountain folk, while the Balûṣ (or Balûch) tribes wandered on their eastern borders, towards the lower limits of the Great Desert. Of the robber tribes of the Kufṣ mention will be made later when describing the Great Desert. Part of this outlying country was known as Al-Khawâsh, namely of the tribes called Al-Akhwâsh. These were camel-men, who lived in a valley where by reason of the heat much sugar-cane was grown for export to Sijistân and Khurâsân—this being the tract of mountainous country which intervenes between the southern end of the Great Desert and Makrân. In these highlands were seven separate mountains, each ruled, it was said, by its own chief, and 'Aḍud-ad-Dawlah the Buyid, in the 4th (10th) century, had made an expedition to conquer them. These people then had no horses, they were regarded as of the Kurds, for they owned flocks and herds, lived in hair-tents and possessed no cities. The date palm flourished abundantly in the lower regions of this country[1].

Some fifty miles south-west of Jîruft lies Gulâshkird, which Muḳaddasî writes Valâshgird, stating that it was a strongly fortified town protected by a castle known as Kûshah, and with its gardens irrigated by underground watercourses. Maghûn, a town with many gardens growing orange-trees and the indigo plant, lay one march north of Valâshgird towards Jîruft; its ruins are probably those now known as Fariyâb or Pariyâb[2]. Fifty miles south of Valâshgird was the important town of Manûḳân, now called Manûjân, which Muḳaddasî refers to as 'the Baṣrah of Kirmân' to mark its commercial importance. The town consisted of two opposite quarters, divided by the dry gorge called Kalân; one quarter was called Kûnîn, the other Zâmân, and a fort, which still exists, stood between the two, with the mosque known as the

1 Khwâsh is now the chief town of the Sarhad, a mountainous district described by Major Sykes (*Persia*, pp. 130, 353), which lies to the east of Narmâshîr. Ist. 163, 164, 168. I. H. 220, 221, 224. Muk. 471. Yak. iv. 148, where for *Al-Kârin* we should read Al-Bâriz.

2 Major Sykes (*Persia*, p. 269) refers to Fariyâb, which 'was once a great city, and was destroyed by a flood, according to local legend.'

Jâmi' Sayyân. One march from here, in the sandy plains nearer the coast, was the town of Darahkân; no trace of which, however, now appears to exist. There was a mosque in the town, and its gardens produced much indigo, water being procured by underground channels.

Between Valâshgird and Manûjân runs a river with many tributaries, now known as the Rûdkhânah-i-Duzdî: it is mentioned by Iṣṭakhrî as the Nahr-az-Zankân, and by Yâḳût as the Râghân river. Muḳaddasî refers to the populous town of Rûdhkân, which probably stood on its course, as surrounded by gardens growing date palms and orange-trees. To the north-east of Manûḳân, and on the road to Rîgân, being three marches from the port of Hurmuz, stood the twin cities of Bâs and Jakîn, each with its mosque and market. Nahr or Jûy-Sulaymân (Solomon's Brook), a populous town, one march west of Rîgân, is referred to by Muḳaddasî as of the Jîruft district. Its fertile lands were watered by a stream which ran through the town, in the centre of which stood a mosque and a castle. Lastly, in the northern part of the mountainous district of Jabal-al-Ḳufṣ, Muḳaddasî mentions the town of Ḳûhistân, for distinction called after a certain Abu Ghânim. It was very hot, and palm-groves grew all round the town, in the midst of which was a castle beside the mosque[1].

Old Hurmuz, or Hurmuz of the mainland, lay at a distance of two post-stages, or half a day's march, from the coast, at the head of a creek called Al-Jîr, according to Iṣṭakhrî, 'by which after one league ships come up thereto from the sea,' and the ruins of the town are still to be seen at the place now known as Minâb, vulgarly Minao. In the 4th (10th) century Old Hurmuz was already the seaport for Kirmân and Sijistân, and in later times, when New Hurmuz had been built on the island, this place supplanted Ḳays, just as Ḳays had previously supplanted Sîrâf, and became the chief emporium of the Persian Gulf. Iṣṭakhrî speaks of the mosque and the great warehouses of (Old) Hurmuz, many of the latter being in the outlying villages, two leagues from the town. Palm-groves were numerous and dhurrah was cultivated, also indigo, cummin, and the sugar-cane. Muḳaddasî praises the markets of Hurmuz, its water was from underground

[1] Ist. 169. Muk. 466, 467. Yak. iv. 330.

channels, and its houses were built of unburnt brick. On the sea-shore, half a day's march distant, was Al-'Arṣah, 'the Camp,' presumably at the entrance of the Hurmuz creek.

The adjacent island is mentioned by Ibn Khurdâdbih, in the middle of the 3rd (9th) century, under the name Urmûz (which Mustawfî spells Urmûṣ), and this is doubtless the later island of Jirûn. At the beginning of the 8th (14th) century—one authority gives the year 715 (1315)—the king of Hurmuz, because of the constant incursions of robber tribes, abandoned the city on the mainland, and founded New Hurmuz on the island aforesaid called Jirûn (or Zarûn), which lay one league distant from the shore. At this period New Hurmuz was visited by Ibn Baṭûṭah, and it is described by his contemporary Mustawfî, who notes the abundance of the date palms and sugar-cane growing here. Ibn Baṭûṭah states that Old Hurmuz in his day was known as Mûghistân, and the new town had taken the name of the island, being called Jirûn. It had a Friday Mosque, and fine markets, where goods from Sind and India were brought for sale.

At the close of the 8th (14th) century, Tîmûr ordered an expedition against the coast towns near Old Hurmuz, and seven castles in its neighbourhood were all taken and burnt, their garrisons escaping to the island of Jirûn. These seven castles, as enumerated by 'Alî of Yazd, were, Ḳal'ah-Mînâ, 'the Castle of the Creek,' at Old Hurmuz, Tang-Zandân, Kushkak, Ḥiṣâr-Shâmîl, Ḳal'ah-Manûjân (the town already mentioned), Tarzak, and Tâziyân. In 920 (1514) Hurmuz, more generally called Ormuz, was taken by the Portuguese under Albuquerque, and their port of landing on the mainland became celebrated under the name of Gombroon. This is the place which a century later Shâh 'Abbâs renamed Bandar 'Abbâs; it is the present harbour for Kirmân, and probably occupies the position of Sûrû or Shahrû mentioned above in the chapter on Fârs. The name Gombroon is said to be a corruption of Gumruk (from the Greek Κουμερκί), which became the common term for a 'custom-house' throughout the East. In the Turkish *Jahân Numâ* it is referred to as 'Gumrû, the port of Hurmuz, whence to the city of Lâr (in Fârs) it is four or five days' march[1].'

[1] I. K. 62. Ist. 163, 166, 167. I. H. 220, 222, 223. Muk. 466, 473.

Commercially Kirmân stood far behind Fârs, and the Arab geographers give us no detailed account of the trade of the province. Kirmân as a whole, Muḳaddasî states, grew dates and dhurrah as food-stuffs; dates were exported to Khurâsân, and indigo to Fârs, while the cereal crops raised in the Valâshgird district were taken down to Hurmuz, and thence shipped to more distant countries[1].

The geographers of the 3rd and 4th (9th and 10th) centuries give far less detail concerning the high roads of Kirmân than is the case when they are treating of the Fârs province. Further, as a rule, only the inexact measurement of the day's march (*marḥalah*) is given, and for most of the roads the reckoning from stage to stage in leagues (*farsakh*) is wanting.

The roads from Fârs into Kirmân converged on Bîmand, which, as already said, lay four leagues to the west of Sîrjân. From the north-east, one road from Unâs and the Rûdhân district came down to Bîmand (given by both Iṣṭakhrî and Muḳaddasî); while from Great Ṣâhik to Bîmand (and Sîrjân) we have two roads, both measured in farsakhs, one by Shahr-i-Bâbak (given by Ibn Khurdâdbih only), and another leading directly across the desert to Bîmand, to which there are two alternative routes, one (Ibn Khurdâdbih) by Ḳariyat-al-Milḥ, 'Salt Village,' the other by Rubâṭ-Pusht-Kham, 'Crook-back Guard-house' (Ḳudâmah and Iṣṭakhrî). Further, Muḳaddasî gives the road from Nîrîz (in marches) to Bîmand and Sîrjân; while both he and Iṣṭakhrî describe the route from the south-west which came up from Rustâḳ-ar-Rustâḳ in somewhat over four days' march, going direct to Sîrjân[2].

From Sîrjân to Bardasîr (Kirmân city) it was two days' march. Mustawfî says 20 leagues, but no halting-place or town is

Mst. 182, 222. I. B. ii. 230. A. F. 339. A. Y. i. 789, 809, 810. J. N. 258, 260. The name of the king who transferred the capital to the island is variously given as Shams-ad-Dîn, Ḳutb-ad-Dîn, or Fakhr-ad-Dîn. The island of Hurmuz was taken by the English in 1622; for its present state see Stiffe, *Geographical Magazine*, 1874, i. 12, and *J.R.G.S.* 1894, p. 160. The name is spelt indifferently Hurmuz, and Hûrmûz.

[1] Muk. 470.

[2] I. K. 48, 53. Kud. 195. Ist. 131, 168. Muk. 455, 473. Mst. 201.

mentioned in between, although, as already remarked, both Mâshîz and Baghîn must have been near the road followed, and both these places are frequently mentioned as existing in the 4th (10th) century by Ibn Ibrâhîm, who wrote in the 11th (17th) century. From Bardasîr (Kirmân) it was two marches to Zarand, Janzarûdh lying half-way between the two. From Sîrjân to Mâhân it was three days' march, and thence three more to Khabîṣ, but the intermediate stages cannot be identified[1].

From Sîrjân, eastward, the great caravan road towards Makrân went through a number of towns that no longer exist, coming to Râyîn, thence on by Darzîn, Bam, and Narmâsîr to Fahraj on the desert border. The stages along this road are given in farsakhs (leagues) by both Ibn Khurdâdbih and Ḳudâmah, besides the stations by the day's march (*marḥalah*) in two of our other authorities[2].

From Sîrjân south-east to Jîruft, in spite of the route being described in leagues by Ibn Khurdâdbih, and in marches by Iṣṭakhrî, none of the places mentioned, except Darfârid, can be surely identified; for, possibly with the exception of the southern Bâft, none of them are found on the map, and the true reading of the many variants in the MSS. is by no means certain. From Jîruft the road turned south, and passing through Valâshgird and Manûḳân, came to the coast at (Old) Hurmuz. According to Iṣṭakhrî, at Valâshgird a branch struck off westward to the frontier of Fârs, passing through a series of towns or villages that have now entirely disappeared, and unfortunately even the terminus of this road on the Fârs frontier cannot now be fixed[3].

From Old Hurmuz, up to Rîgân and Narmâsîr, Muḳaddasî gives the route in marches, passing through the towns of Bâs and Jakîn; while going south from Râyîn to Jîruft the distances through Darjîn and Hurmuz-al-Malik are given in marches by Iṣṭakhrî[4].

1 Ist. 169. Muk. 473. Mst. 201.
2 I. K. 49. Kud. 196. Ist. 168. Muk. 473.
3 I. K. 54. Ist. 169. 4 Ist. 169. Muk. 473.

CHAPTER XXIII.

THE GREAT DESERT AND MAKRÂN.

The extent and characteristics of the Great Desert. The three oases at Jarmaḳ, Nâband and Sanîj. The chief roads across the Desert. The Makrân province. Fannazbûr and the port of Tîz. Other towns. Sind and India. The port of Daybul. Manṣûrah and Multân. The river Indus. The Ṭûrân district and Ḳuṣdâr. The Budahah district and Ḳandâbîl.

The Great Desert of Persia stretches right across the high plateau of Îrân, going from north-west to south-east, and dividing the fertile provinces of the land into two groups; for the Desert is continuous from the southern base of the Alburz mountains, that to the north overlook the Caspian, to the arid ranges of Makrân, which border the Persian Gulf. Thus it measures nearly 800 miles in length, but the breadth varies considerably; for in shape this immense area of drought is somewhat that of an hour-glass, with a narrow neck, measuring only some 100 miles across, dividing Kirmân from Sistân, while both north and south of this the breadth expands and in places reaches to over 200 miles[1].

The medieval Arab geographers refer to the Desert as Al-Mafâzah, 'the Wilderness,' and carefully define its limits. On the west and south-west it was bounded by the Jibâl province, by the

[1] The general outline of the Great Desert is given in Map I (p. 1), details of the northern portion are shown in Map V (p. 185), of the lower part in Maps VI (p. 248), VII (p. 323), and VIII (p. 335). At the present day the Desert, as a whole, is known as the Lûṭ or Dasht-i-Lûṭ (Desert of Lot); the saline swamps and the dry salt area being more particularly known as the Dasht-i-Kavîr, the term Kavîr being also occasionally applied to the Desert as a whole. The etymology of the terms *Lûṭ* (the Arab form of the Biblical Lot) and *Kavîr* is uncertain; see Major Sykes, *Persia*, p. 32.

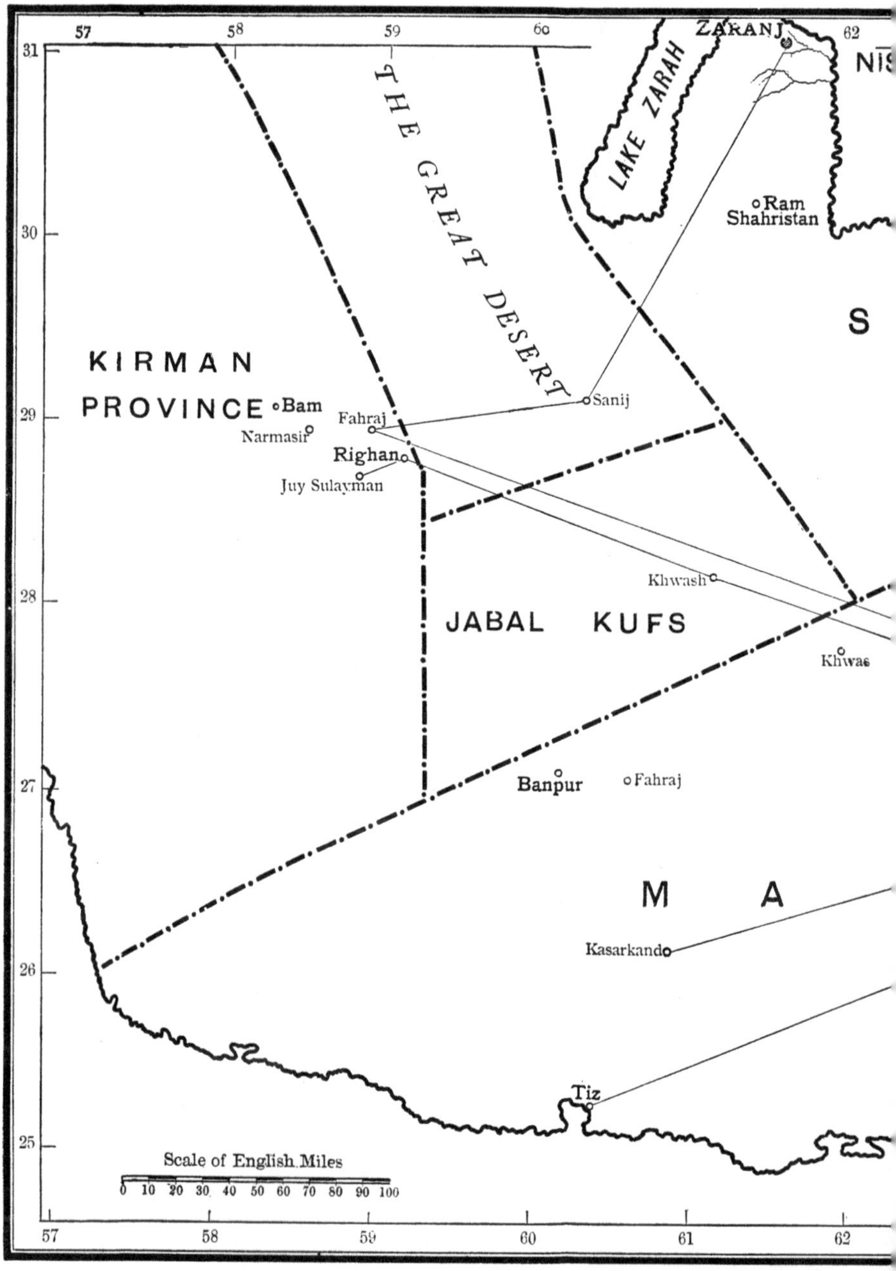

PROVINCE OF MAKRAN,

MAP VII

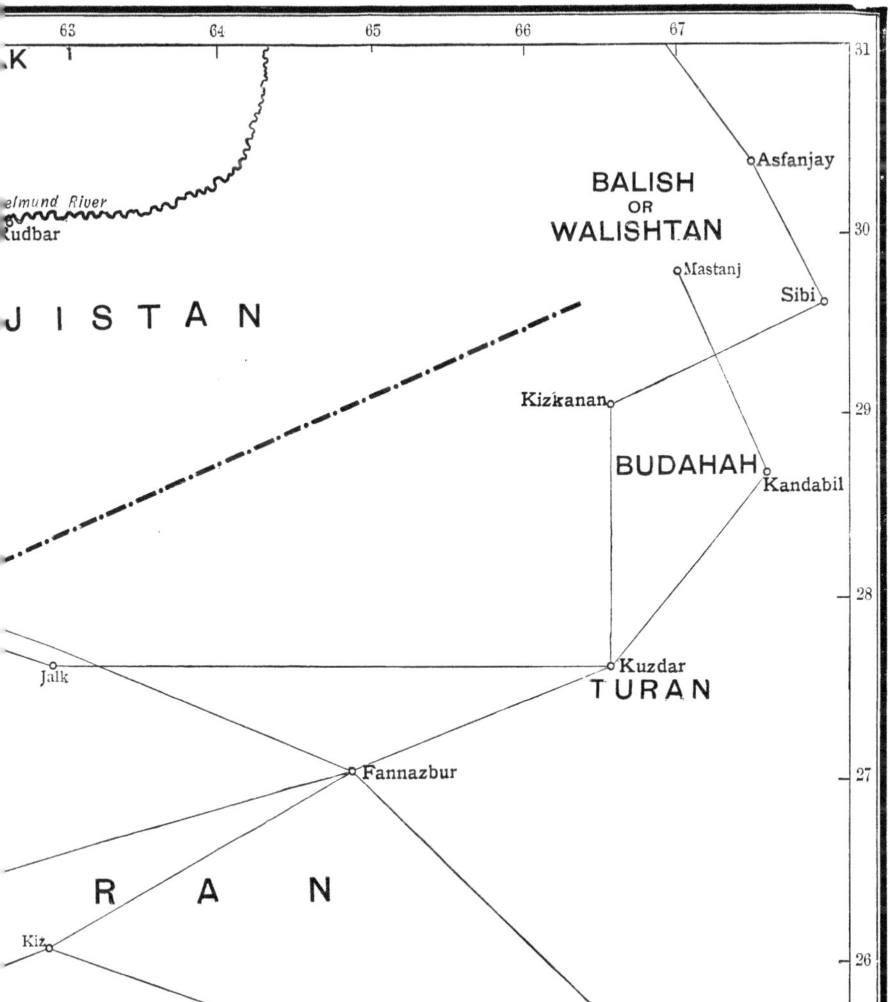

ITH PART OF SIJISTAN

district of Yazd (originally counted as part of Fârs) and by Kirmân, south of which it spread out among the ranges of the Makrân coast. To the east and north-east lay Khurâsân with its dependent and adjacent provinces; namely Ḳûmis to the north of the Desert, and next a corner of Khurâsân proper; then Ḳûhistân, and below this Sijistân at the narrow part opposite Kirmân, Sijistân being coterminous with what is now known as the Balûchistân desert, which in the middle-ages was considered as a part of Makrân.

Both Ibn Ḥawḳal and Muḳaddasî write of the Desert from personal experience, for each had crossed its wastes on more than one occasion. Ibn Ḥawḳal briefly describes it as a No Man's Land, belonging to no province, where robbers from every district found shelter, and where permanent villages, except in three instances, were conspicuously absent. Muḳaddasî enters into the matter in some detail, and of his remarks the following is a summary:—The Desert was, he writes, like the sea, for you could cross it in almost any direction, if you could keep a true line, and pick up the tanks and domes, built above the water-pits, which in the 4th (10th) century were carefully maintained along the main tracks at distances of a day's march. He, Muḳaddasî, had once been 70 days on the passage across, and he speaks from experience of the countless steep passes over the ever-barring ranges of hills, the fearful descents, the dangerous salt swamps (*sabkhah*), the alternate heat and bitter cold. He notices too that there was but little sand, and there were palm-trees and some arable lands hidden away in many of the minor valleys.

At that date the Desert was terrorised by roving bands of the Balûṣ (Balûchî tribesmen), whose fastnesses were in the Ḳufṣ mountains of the Kirmân border, 'a people with savage faces, evil hearts, and neither morals nor manners.' None could escape meeting them, and those they overcame they would stone to death 'as one would a snake, putting a man's head on a boulder, and beating upon it, till it be crushed in'; and when Muḳaddasî enquired why they so barbarously put men to death he was answered that it was in order not needlessly to blunt their swords. 'Aḍud-ad-Dawlah the Buyid, in Muḳaddasî's day, had in part curbed these Balûch brigands, by carrying off a tribe of them to

Fârs as hostages, and caravans were after this tolerably safe, if they had a guide and letters of protection from the Sultan. These Balûṣ, Muḳaddasî adds, went mostly on foot, but possessed a few dromedaries (*jammâz*). Though nominally Moslems, they were more cruel to True Believers than either the Christian Greeks or the heathen Turks, driving their prisoners before them for twenty leagues a day barefoot, and fasting. Their own food was from the nut of the Nabḳ, or Sidr (Lotus) tree, and the men were famous for their power of bearing without complaint both hunger and thirst.

About half-a-century after the time of Muḳaddasî, namely in the year 444 (1052), Nâṣir-i-Khusraw crossed the northern part of the Desert on his return from the pilgrimage to Mecca. He gives no special name to the Great Desert, referring to it merely as the Bîyâbân, 'the waterless land,' but he notes its two chief characteristics and dangers, namely the moving sands (Rîg-ravân) and the salt swamps (Shûristân), the latter often as much as six leagues across. He travelled from Nâyin in the Jibâl province to the central oasis at Jarmaḳ, and thence on to Ṭabas in Ḳûhistân, by the route which will be mentioned presently. His description of the road, however, is vague and adds little to our information. He speaks of the Amîr Gîlakî, of Ṭabas, as in his day keeping such order throughout the Desert that the Ḳufṣ robbers, whom he calls the Kufâj, were powerless to molest travellers; and he mentions that every two leagues along the road he travelled there were cupolas (*gumbad*) over water-tanks, which marked the safe track to be followed, and relieved the wants of the traveller. He remarks that if the tanks were only kept in order, the passage of the Desert could always be effected without much hardship, except for fear of robbers; and his account in this matter is confirmed by the numerous caravan roads, crossing the waste in more than one direction and sufficiently supplied at each stage by water in pits, which are detailed in the itineraries given by Ibn Ḥawḳal and Muḳaddasî[1].

Three far-separated oases were found along the central line of the great waste, and to these naturally the various roads crossing from west to east converged. In the middle-ages these oases

[1] I. H. 287, 288. Muk. 488, 489. N. K. 93, 94. Yak. iv. 147.

were known as Jarmaḳ, Nâband (still so called), and Sanîj; this last according to Muḳaddasî being the only town that the Desert could boast as possessing within its compass.

In the very centre of the upper expansion of the Desert, half-way across from Isfahân to Ṭabas in Ḳûhistân, is the oasis now called Jandak or Biyâbânak, which in the middle-ages was known to the Arabs as Jarmaḳ, and in Persian was written Garmah. It consisted of three hamlets called Jarmaḳ (or Garmah), Biyâdaḳ (or Piyâdah in Persian), and Arâbah. Ibn Ḥawḳal names the whole settlement Sihdih, 'Three Villages'; and Nâṣir-i-Khusraw says there were from ten to twelve hamlets here in the 5th (11th) century. At Piyâdah also there was a small fort, garrisoned by the Amîr Gîlakî, for the safe control of the Desert routes. In this oasis there were palm-trees, and arable fields of some extent where cattle throve; and the three chief settlements, Ibn Ḥawḳal says, all lay within sight of water, the population in the 4th (10th) century numbering over 1000 men. Later authorities add nothing to these details, and in fact down to the time of Mustawfî in the 8th (14th) century the accounts are almost identical, all copying Ibn Ḥawḳal.

Nâband, the second oasis, still bears this name, and it lies at the northern end of the narrow part of the Desert, between Râvar in Kirmân and Khûr in Ḳûhistân. Ibn Ḥawḳal describes Nâband as possessing a Rubâṭ or guard-house, with a score of houses round it, water being plentiful, enough indeed to work a small mill. Palms grew here, and many springs irrigated the fields; and two leagues distant from the place was an outlying spring, surrounded by palms, where there was a domed tank, of evil fame as a noted hiding-place for robbers.

The third oasis lay somewhat further to the south again, and at the very narrowest part of the Desert, at the half-way stage on the road from Narmâsîr in Kirmân to Zaranj, the capital of Sijistân. Here there is a small valley with springs, which is now known to the Persians as Naṣratâbâd, but which the Balûchîs still call Ispî or Isfî. This name is identical with the reading Isbîdh for this oasis, which is otherwise called Sanîj, or Sanîg, by Muḳaddasî. He counts it as a town of Sijistân, while according to Ibn Ḥawḳal it belonged rather to Kirmân. It was, as already said, the only city in the Desert according to the Arab geographers, and Mukaddasî speaks of it as having a considerable population, with much arable land, watered by underground channels; but all around and close up to the houses was the waterless wilderness[1].

The roads across the Desert are given in detail by the geographers of the 4th (10th) century. From the western side, starting from Isfahân and from Nâyin, two roads converged on Jarmak; the first (given by Mukaddasî) is in eight stages, while from Nâyin it was five stages to Jarmak, and there were water-tanks and domes all along the way at distances of a few leagues apart.

From Jarmak, Mukaddasî is our authority for a direct road due north to Dâmghân in Kûmis; the distance was 90 leagues, it being 50 leagues across to a place called Wandah, and thence 40 on to Dâmghân. From Jarmak, going eastward, it was four days' march to a place called Naw Khânî, or Nawjây, with water-domes all along the route at every three or four leagues. At Nawjây the roads bifurcated, going north-east to Turshîz, and south-east to Tabas, both in the Kûhistân province. The distance from Nawjây to Turshîz was four stages, the half-distance being at Bann Afrîdûn (now known as Dih Nâband, a place not to be confused with the oasis of Nâband, just described); and from Jarmak to this Bann Afrîdûn, Mukaddasî also gives a route across the Desert direct, in seven days' march, with a tank (*hawd*) at each stage. From Nawjây, going south-east, Tabas was reached in three marches. The distances between Tabas and Turshîz *viâ* Bann Ibn Khurdâdbih gives in leagues; elsewhere, and as a rule on the Desert routes, only the stages by the day's march (*marhalah*) are given[2].

From Yazd to Tabas, direct, the way went by Anjirah and Khazânah to Sâghand on the Desert border, places already men-

[1] I. H. 289, 293. Muk. 488, 494, 495. N. K. 93, 94. Mst. 183. Yak. iii. 170. The oasis of Biyâbânak (otherwise Jandak or Khur) is mentioned by Tavernier (*Voyages*, i. 769, La Haye, 1718) in the 17th century, and it was visited in 1875 by Col. Macgregor (*Khorasan*, i. 91). Both Nâband and Isfî, or Nasratâbâd, have been visited lately by Major Sykes (*Persia*, pp. 36, 416).

[2] Ist. 231. I. H. 291. I. K. 52. Muk. 491.

tioned as of Fârs (see p. 285). From Sâghand Ibn Khurdâdbih gives the six stages in leagues to Ṭabas, an itinerary which is duplicated by Ibn Ḥawkal and Muḳaddasi, but going by the day's march, and following a not quite identical route. Two stages from Sâghand was the guard-house called Rubâṭ Âb-Shuturân, 'of the Camel-stream,' the water coming from an underground channel, and flowing into a pool. Muḳaddasî describes the guard-house as a fine building of burnt brick, with iron gates, and it was well garrisoned. It had been built by Nâṣir-ad-Dawlah Ibn-Sîmjûr, a famous general of the Buyids, who was governor in these regions during the middle of the 4th (10th) century. Three marches beyond this guard-house the Desert ended; and here the road, as described by Ibn Ḥawkal (repeating Iṣṭakhrî), leaves Ṭabas aside, going in a single march from the stage one march south of this town, to the stage one march north of it, on the road to Bann[1].

The next passage of the Desert starts from the village of Bîrah, of the district called Shûr, meaning 'the Salt-water,' which was on the frontier of Kirmân near Kûhbanân. From here the passage was made in seven or eight stages—each halt at a watering-place—to Kurî, a village on the Desert border of Ḳûhistân, situated a few miles to the south-east of Ṭabas. On this, which was known as the Shûr route, Iṣṭakhrî states that at one point about two leagues to the north of the track there might be seen curious stones, doubtless fossils, in the likeness of various fruits, to wit, almonds, apples, nuts, and pears, while the forms of men and trees were simulated by the rocks here, with likenesses of other created things. In addition to the foregoing route, Muḳaddasî states that there was a road direct from Kûhbanân to Kurî, in 60 leagues, with water in tanks at every second march.

Râvar, as described in Chapter XXI, lies some leagues east of Kûhbanân on the Kirmân frontier, and from this place a road went in five marches to Nâband, the oasis mentioned above, and thence in three marches on to Khûr in Ḳûhistân. There were the usual water-tanks at every three or four leagues along this

[1] I. K. 51. Ist. 236. I. H. 235. Muk. 491, 493.

route also. The town of Khabîṣ, three marches from Mâhân on the Kirmân border, was already almost within the Desert limits (see p. 308); and from here a road is given which reached Khawst (modern Khûsf) in Ḳûhistân in ten marches. The frontier of Ḳûhistân was reached two marches before Khawst, at the village of Kûkûr, where the Desert ended; and on this road, at a place where was the tomb of a certain Al-Khârijî, there were to be found curious white and green pebbles, 'as though of camphor and glass,' while at another place, about four leagues off the road, was a small black boulder of very remarkable appearance[1].

Lastly from Narmâsîr in Kirmân to Zaranj, the capital of Sîstân, the way crossed the narrow part of the Desert, going by the oasis of Sanîj or Ispî, which has been described above. The first stage of this route was to Fahraj on the Desert border, and in four stages it brought the traveller to Sanîj. Ibn Khurdâdbih gives each stage of this route in leagues, Iṣṭakhrî mentioning the day's march only, but the latter gives also a second route to Sanîj by what he calls 'the New Road,' but this was a longer way. From Sanîj it was seven or eight days' march to the city of Zaranj, the frontier of Sîstân being crossed at Gâvnîshak, which was not far from Kundur, a place that is still marked on the map. Between Gâvnîshak and Kundur, and three or four stages south of Zaranj, was a Rubâṭ or guard-house, built by 'Amr the Ṣaffârid in the 3rd (9th) century, which according to Iṣṭakhrî was known as Ḳanṭarah Kirmân, 'the Kirman Bridge'; although, as he is careful to remark, no actual bridge existed here. This place marks an important point, for in the middle-ages the Zarah lake had its borders as far south as this, as will be noticed in the following chapter[2].

[1] Ist. 232, 233, 234. I. H. 292, 293, 294. Muk. 491, 492.

[2] I. K. 49, 50. Ist. 237, 251, 252. I. H. 296, 306, 307. Muk. 492. Sir F. Goldsmid, *Eastern Persia*, i. 256.

The province of Makrân.

The arid ranges of the Makrân coast are, in their general physical features, a prolongation of the Great Desert, and though during the earlier middle-ages the country appears to have been more fertile and populous than it is now, Makrân was never a rich, or, politically, an important province. The chief product of Makrân was the sugar-cane, and the particular kind of white sugar, known to the Arabs as *Al-Fânîdh* (from the Persian *Pânîd*), and made here was largely exported to neighbouring lands[1].

The earlier geographers name many towns as in Makrân, but give scant descriptions of them. The chief commercial centre was the port of Tîz on the Persian Gulf, and the capital of the province was Fannazbûr or Bannajbûr, which lay inland, at the place now known as Panj-gûr. Bannajbûr, according to Muḳaddasî, had in the 4th (10th) century a clay-built fortress, protected by a ditch, and the town was surrounded by palm-groves. There were two city gates, Bâb Tîz opening south-west on the road to the gulf port, and Bâb Ṭûrân opening north-east on the road to the district of that name, of which the capital was Ḳuzdâr. A stream brought water to the city; and the Friday Mosque stood in the market-place, though, according to Muḳaddasî, the people were really only Moslem in name, being savage Balûṣîs (Balûchîs) whose language was a jargon[2].

The ruins of the great port of Tîz lie at the head of what was a fine harbour for the small ships of the middle-ages. Muḳaddasî describes Tîz as surrounded by palm-groves, and there were great warehouses in the town, and a beautiful mosque. The population

[1] I. H. 226, 232, 233. Muk. 475, 476. Yak. iv. 614. The sites of the various medieval towns in Makrân are ably discussed by Sir T. H. Holdich in the *Geographical Journal* for 1896, p. 387, and, in the present state of our information, his conclusions cannot be bettered.

[2] Ḳannazbûr, or Ḳannajbûr, as the name has often been printed, is merely a clerical error for Fannazbûr, by a doubling of the diacritical points over the first letter. Ist. 170, 171, 177. I. H. 226, 232. Muk. 478. Panj-gûr, 'Five Tombs,' is so called after the five martyred warriors of the first Arab conquest. It lies one march west of Ḳal'ah Nâghah, and the surrounding district is also called Panj-gûr. Sykes, *Persia*, p. 234.

was of all nations, as is usual in a great seafaring port; and in the 6th (12th) century the place had, in large measure, acquired the trade of Hurmuz, which had fallen to ruin[1].

Of other towns in Makrân the Arab geographers give only the names, and no descriptions. The names of the well-known town of Bampûr, and Fahraj its neighbour, occur in Muḳaddasî as Barbûr (for Banbûr) and Fahl Fahrah, Yâḳût giving the last under the form Bahrah[2]. The town of Ḳaṣarḳand, north of Tîz, is still a place of some importance; and Kaj, some distance to the east of this, is mentioned as Kîj or Kîz. The names of Jâlk and Dazak also occur; and Khwâsh or Khwâṣ, which is probably the modern Gwasht, lying to the east of Khwâsh in the Sarhad district (already mentioned, p. 317). Râsk was, in the middle-ages, a town of some note on account of its fertile district called Al-Kharûj, but, from the Itineraries, there is doubt whether it can be identical with the present township of this name. Armabîl and Ḳanbalî were two important towns, on or near the coast, about half-way between Tîz and Daybul at the Indus mouth. Iṣṭakhrî describes these as cities of considerable size, lying two days' march apart, and one of them was situated half a league distant from the sea. Their people were rich traders, who had dealings chiefly with India[3].

[1] Muk. 478. Yak. i. 907. For the present ruins of Tîz see Sykes, *Persia*, 101, 110, also Schindler, *J.R.A.S.* 1898, p. 45. See also the history of Afḍal Kirmânî, Houtsma, *Z.D.M.G.* 1881, pp. 394 and 402.

[2] Fahraj a few miles to the east of Bampûr in Makrân, and Fahraj a few miles to the east of Narmâsîr in Kirmân, must not be confused. There was also Fahraj near Yazd.

[3] Ist. 170, 171, 177, 178. I. H. 226, 232. Muk. 475, 476. Yak. i. 769; iv. 332. The spelling *Armayîl* for Armabîl is a frequent clerical error of the MSS. The ruins of Armabîl are probably at Lus Bela, and those of Ḳanbalî at Khayrokot. Sir T. Holdich, *J.R.G.S.*, 1896, p. 400. The earlier Arab geographers in point of fact knew little about Makrân, and the later ones add nothing worth mentioning. Yâḳût only repeats what his predecessors of the 4th (10th) century have said. All that Ḳazvînî (ii. 181) has to tell us of this province is that there was a wonderful bridge there, crossing a river, and formed of one single block of stone. He adds,—'he who crosses it vomits up the contents of his belly, so that naught remains therein, and though thousands should pass over the bridge this always happens to each one. So when any man of that country requires to vomit he has only to cross this bridge.'

The present work does not pretend to deal with medieval India, and indeed the Arab geographers give no systematic account of that country. The Indian port best known to them, beyond the eastern end of the Persian Gulf, was Daybul, then a fine harbour at the principal mouth of the Indus. This was in the Sind province, of which the capital was Al-Manṣûrah, called Brahmanâbâd by the Indians, a great city lying on one of the canals or branches of the lower Indus. The Indus was known to the Arabs as the Nahr Mihrân, and many of the towns along its banks are named, more especially Al-Multân, the great city far up the affluent of the Indus called the Sindarûdh, where there was a famous idol temple. Iṣṭakhrî, who compares the Indus with the Nile for size and importance, notices that the Indian river also had crocodiles like those of Egypt. The sources of the Indus, he says, were in the great mountains to the north, and near the origin of the Oxus. Of the Sind province were the people known to the Arabs under the name of Az-Zuṭṭ, called Jat by the Persians, who are now generally held to be identical with the forefathers of the Gipsies[1].

On the north-eastern frontiers of Makrân, and close to the Indian border, the Arab geographers describe two districts; namely, Ṭûrân, of which the capital was Ḳuṣdâr, and Budahah to the north of this, of which the capital was Ḳandâbîl. Ḳuṣdâr, also spelt Al-Ḳuzdâr, is mentioned among the earlier conquests of Sultan Mahmûd of Ghaznah. Ibn Ḥawḳal describes it as standing on a river (*wâdî*), and having a fortress in its midst. The plain around the town was very fertile, producing vines and pomegranates with other fruits of a cold climate. Muḳaddasî adds that the city lay in two quarters, on either side of the dry river-bed; on one side was the palace of the Sultan and the castle,

[1] Ist. 171, 172, 173, 175, 180. I. H. 226, 227, 228, 230, 234, 235. Muk. 476, 479, 482, 483. The ruins of the port of Daybul, now lying far inland, exist some 20 miles south-west of Thatta, and 45 miles east-south-east of Kurâchî. Manṣûrah is on an old channel of the Indus delta, about 40 miles north-east of Hyderabad. Sind is of course only the old Persian form of the name Hind, but the Arabs used it vaguely to denote the great province to the east of Makrân, which is now in part called Balûchistân and in part is included in modern Sind. Sindarûdh is the River of Sind.

on the other, which was called Bûdîn, dwelt the merchants, whose shops in the market were much frequented by the Khurâsân folk. Muḳaddasî adds that the houses were clay-built, and there were underground channels for the water-supply, but this was bad in quality and scanty.

Ṭûrân, the name given to the Ḳuṣdâr district, was often held to include the lands to the north, known as the Budahah district, of which the chief town, Ḳandâbîl, has been identified with the present Gandava, lying south of Sîbî and east of Kelat. Ḳandâbîl is described by Ibn Ḥawḳal as a large city, standing solitary in a plain, and no date palms grew here. Of its dependencies was the town of Kîzkânân, or Kîkân, which from its position in the Itineraries is to be identified with modern Kelat. Both these towns were often described as of Ṭûrân, some others being also named which it is impossible now to identify, for no sufficient description is given of them, and the readings of the MSS. vary considerably as to orthography[1]. To the north of these districts was Bâlis, or Wâlishtân, with the towns of Sîbî and Mastanj; but these were held by the early geographers to be included in Sijistân, and will therefore be noticed in the next chapter.

The routes across Makrân are in continuation of the roads of the Great Desert already described, and their ultimate point is India. They are unfortunately as a rule only given in a summary way, so many days' march from one town to another, and the distances cannot be considered as reliable. Ibn Khurdâdbih, however, gives the detail of one route in leagues, and stage by stage, though it is impossible now to identify the exact line across the Desert. Starting from Fahraj on the Desert border east of Bam and Narmâsîr in Kirmân, he gives the 14 stages to Fannazbûr, the capital of Makrân; and thence, eastward, the names of three halting-places on the road to Ḳuṣdâr. An almost parallel route, but in the contrary direction, is given by Muḳaddasî, from Ḳuṣdâr to Juy or Nahr Sulaymân, which lay 20 leagues east of Bam, but this road keeps north of Fannazbûr, passing by Jâlk and Khwâṣ[2].

[1] I. K. 56. Ist. 171, 176, 178. I. H. 226, 232, 233. Muk. 476, 478.
[2] I. K. 55. Muk. 486.

From the port of Tîz it was five marches to Kîz, and then two marches on to Fannazbûr, to which city a road also came in from Ḳaṣarḳand, but by an indirect route. From Kîz, and from Ḳaṣarḳand, it is given as six marches to Armabîl, then two to Ḳanbalî, and thence four on to Daybul at the mouth of the Indus[1].

It was reckoned as fourteen marches from Fannazbûr to Daybul. The distances in round numbers are given from Ḳuṣdâr to Ḳandâbîl, and to Kîzkânân (Kelat), also from these places on to Sîbî and Mastanj in Wâlishtân; and the Itineraries close by a summary of the number of days' march that it took to reach Multân and Manṣûrah, the cities on the Indus, from Ḳuṣdâr and from Ḳandâbîl, and from the frontiers of Wâlishtân beyond Sîbî[2].

[1] Ist. 178. I. H. 233. Muk. 485.
[2] Ist. 179. I. H. 233, 234. Muk. 486.

CHAPTER XXIV.

SIJISTÂN.

Sijistân, or Nîmrûz, and Zâbulistân. Zaranj, the capital. The Zarah lake. The Helmund river and its canals. The ancient capital at Râm Shahristân. Nih. Farah, and the Farah river. The Khâsh river and the Nîshak district. Ḳarnîn and other towns. Rûdbâr and Bust. The districts of Zamîn Dâwar. Rukhkhaj and Bâlis, or Wâlishtân. Ḳandahâr, Ghaznah, and Kâbul. The silver mines. The high roads through Sijistân.

Sîstân, which the earlier Arabs called Sijistân from the Persian Sagistân, is the lowland country lying round, and to the eastward of, the Zarah lake, which more especially includes the deltas of the Helmund and other rivers which drain into this inland sea. The highlands of the Ḳandahâr country, along the upper waters of the Helmund, were known as Zâbulistân. Sîstân was also called Nîmrûz in Persian, meaning 'mid-day,' or the Southern Land, a name said to have been applied to the province in regard to its position to the south of Khurâsân. Iṣṭakhrî describes the Sijistân province as famous for its fertility; dates, grapes, and all food-stuffs were grown here abundantly, also assafœtida, which the people were wont to mix with all their dishes[1].

It is to be borne in mind that the Zarah lake was, in the middle-ages, far more extensive than it has come to be at the present day. Besides the Helmund, a great river of many affluents, three other considerable streams drained into the lake, namely, the Khwâsh river, the Farah river, and the river from the neighbourhood of Asfuzâr (Sabzivâr of Herât), which is now known as the Hârûd. In Persian legend, Sîstân and Zâbuliṣtân

[1] Ist. 244. I. H. 301.

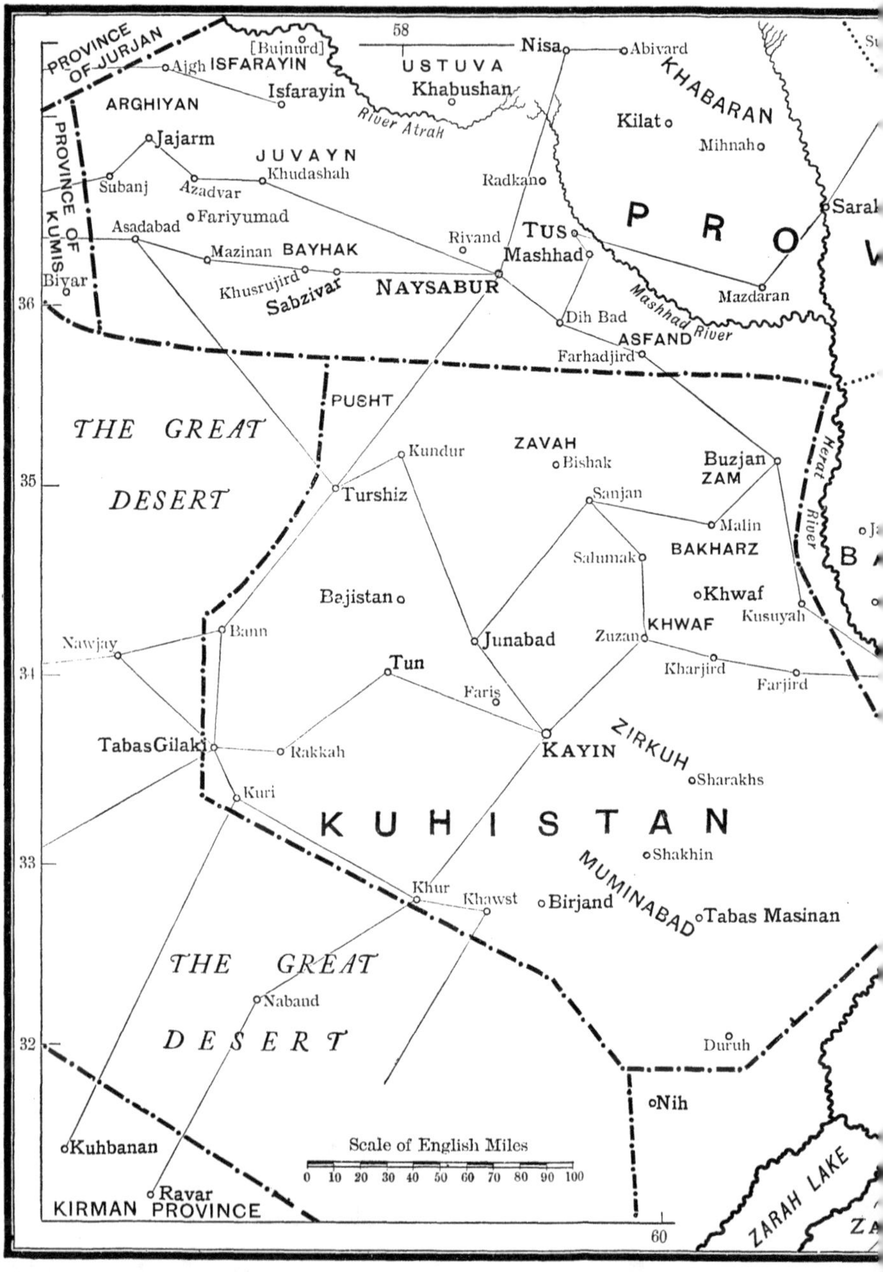

PROVINCES OF KUHIST

MAP VIII

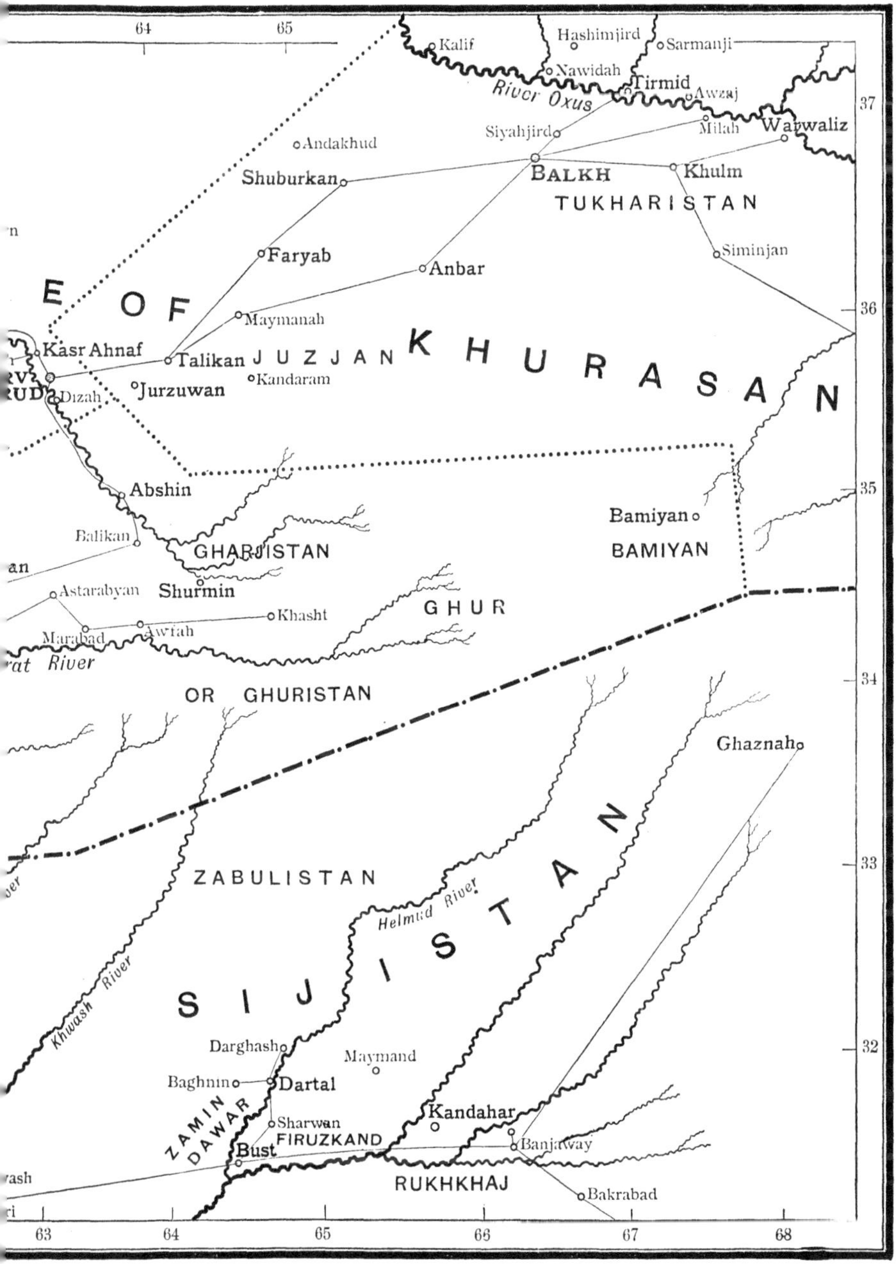

AND PART OF SIJISTAN

were famous as the home of Zâl, the father of the national hero Rustam, whose exploits are still current among the people. In the times of the early Abbasid Caliphate, Sîstân further became known to fame as the place of origin of the Ṣaffârid Amirs, who in the second half of the 3rd (9th) century governed most of southern and eastern Persia, being virtually in the condition of independent princes.

The capital of the province, during the middle-ages, was the great city of Zaranj, destroyed by Tîmûr, of which the ruins still remain, covering a considerable area of ground. The name of Zaranj, however, has now entirely disappeared, and even in the later middle-ages had dropped out of use, the capital of the province being known to the later Arab geographers merely as Madînah Sijistân, 'the City of Sijistân,' the Persian form being the equivalent, Shahr-i-Sîstân, which was in use when Tîmûr finally laid the town in ruins[1]. Under the Sassanian kings Zaranj was already a great city, and at the time of the first Moslem conquest, in the year 20 (641), it is more than once mentioned. It was situated near the Sanârûdh canal, a great branch from the Helmund, which flowed out to the westward, and in flood-time reached the Zarah lake.

Ya'ḳûbî, in the 3rd (9th) century, describes Zaranj as four leagues in circumference, and in the next century we have a detailed notice of the city by Ibn Ḥawḳal. It was then strongly fortified, consisting of an inner town surrounded by a wall having five gates, beyond which lay the suburbs of the outer

[1] The ruins of Zaranj lie round the modern villages of Zâhidân and Shahristân, along the old bed of one of the chief canals from the Helmund, which since the middle-ages has become dry. For the modern condition of these, and other ruined sites, see Sir H. Rawlinson, *J.R.G.S.* for 1873, pp. 280, 283, 284; Sir F. Goldsmid, *Eastern Persia*, i. 301; Sykes, *Persia*, pp. 375, 382, 383. A sketch plan of the chief ruin is given by A. H. Savage Landor in *Across Coveted Lands*, ii. 228. Near Zâhidân is still seen the remains of a tower about 80 feet high, called the Mîl-i-Zâhidân, having a spiral staircase, and two partly legible Kufic inscriptions. This tower, tradition says, was destroyed by Tîmûr; see G. P. Tate, in *J.R.A.S.* 1904, p. 171. Naṣratâbâd, the modern capital of Sîstân, lies a few miles to the south of these ruins; it was known at first under the name of Nâṣirâbâd, which name, however, has now gone out of use. According to Mr Savage Landor it is at the present day also known as Shahr-i-Naṣrîyah.

town, enclosed by the outer wall, which had thirteen gates, these latter opening across a great moat filled with water from springs and from the overflow of the canals. The five gates of the inner town were all of iron. Two, close by one another, opening to the south-east towards Fârs, and known as the Fârs gates, were individually called the Bâb-al-Jadîd and the Bâb-al-'Atîķ, 'the New Gate' and 'the Old Gate.' To the north, towards Khurâsân, was the Bâb Karkûyah, called after the neighbouring town of Karkûyah; the Bâb Nîshak was on the eastern road, toward Bust; while the Bâb-aṭ-Ṭa'âm, 'the Victuals Gate,' which was most in use of all the five, opened on the road leading south through the markets and the gardens lying outside Zaranj.

The Great Mosque, Masjid-al-Jâmi', was in the outer town, standing near the two south-western gates, on the Fârs road, and the prison stood near it, beside the old Government House. Between the Nîshak and the Karkûyah gates, in the north-east part of the town, was the ark or citadel containing the treasury, which had been erected by 'Amr, the second Ṣaffârid prince. His elder brother Ya'ķûb, the founder of the dynasty, had built himself a palace, which subsequently became the new Government House, in that part of the inner town lying between the two south-western gates and the Bâb-aṭ-Ṭa'âm. Near this was also the palace of 'Amr; and these, like all the other houses of the town, were constructed of clay bricks and vaulted, since no beams could be used here for roofing, all woodwork rapidly perishing from the damp climate, and from being bored through by worms. In both the inner and the outer town were many hostels (*fandûķ*), and in the outer town or suburb were the Government offices. The markets of the inner town stood near the Great Mosque. Those of the outer town were extremely populous, and especially famous was that called Sûķ 'Amr, built by the second Ṣaffârid prince, the rents from which, amounting every day to over 1000 dirhams (£40), were divided between the Great Mosque, the town hospital (Bîmaristân), and the Mecca sanctuary.

In the outer town the markets extended for nearly half a league in length, with a continuous line of shops going from the two Fârs gates of the inner wall, to the gate of the outer suburb wall. Throughout Zaranj water was plentiful, being brought from the Sanârûdh by a series of minor canals or watercourses, which entered the inner city at three points—the New Gate, the Old Gate, and the Gate of Victuals. The three together had water-power 'sufficient to turn a mill,' and they flowed into two great reservoir tanks near the mosque, whence the water was distributed throughout the inner town. The houses of the outer town were also well provided by channels with running water, which was an indispensable convenience in this hot climate; and each house had a *Sardâb*, or cellar-room, for living in during the hot season, when the heat of Zaranj was most oppressive. Round the town lay the *sabkhah*, or salt marshes, where date palms grew, environed by the desert sands. Here violent winds blew continually, moving the sands about in a dangerous way and often overwhelming whole villages and devastating the cultivated districts. The ceaseless wind was used by the people to turn their windmills, which were a feature peculiar to this country. The 'moving sands,' however, were a continual source of danger, and Ibn Ḥawḳal gives a long account of how, in the year 360 (970) and odd, the Great Mosque of Zaranj became quite choked up with sand.

Such was Zaranj in the 4th (10th) century, and this description is repeated by Muḳaddasî. He refers also to the riches and the learning of the inhabitants, notes the strongly fortified castle (Ḳal'ah), and the two famous minarets of the Great Mosque, one of which had been built by Ya'ḳûb the Ṣaffârid. The city continued to flourish for many centuries, and even during the Mongol invasion of the year 619 (1222), when Changîz Khân sent his hordes to ravage Sîstân, the capital seems to have escaped devastation, and it was for some time after this date under a Mongol governor. In the early part of the 8th (14th) century, Mustawfî speaks of Zaranj (the name of which the Persians pronounced Zarang) as very flourishing; and the city, he says, was protected from the 'moving sands' of the neighbouring desert by a great dyke (*Band*), stated to have been originally built by the ancient king Gurshâsf, and to have been afterwards restored by King Bahman, son of Isfandiyâr. Mustawfî praises the gardens of Zaranj, which produced excellent and abundant fruit, these gardens being irrigated from the Black Canal (Siyâh Rûd) which was taken from one of the branches of the Helmund river. At the end of the century, however, in 785 (1383), Tîmûr appeared with his armies before the city, which, as already said, was then known as Shahr-i-Sîstân (Sîstân city), and its fate was not long left in doubt. Tîmûr had already taken and destroyed the neighbouring fortress, called the Ḳal'ah or Ḥiṣâr Zarah, which probably stood to the north of Zaranj, near the borders of the lake. The capital of Sîstân closed its gates, and declined to surrender. After a short siege it was taken by storm, all its inhabitants who could be found were massacred, its walls were then razed and its houses destroyed. Since that time Zaranj has come to be a nameless ruin[1].

The Zarah or Zirrah lake (Buḥayrah Zarah), as already said, in medieval times had permanently a far greater extent than is now generally the case; but at all times its area is noted as fluctuating in size, according as the rivers were in flood or drought[2]. It is described by Ibn Ḥawḳal in the 4th (10th) century as having a length of 30 leagues (100 miles), counting from a place called Kurîn in Ḳûhistân to the Sijistân frontier post near Ḳanṭarah Kirmân, at the third stage on the road from Zaranj to Narmâsîr (see above, p. 328). The lake was reckoned as the equivalent of a day's journey (*marḥalah*, about 30 miles) across. It was of sweet water and full of reeds, and was plentifully stocked with fish; its borders, except on the desert side, were dotted with many farmsteads and populous villages, where the fish were caught and dried for export.

The chief water-supply of the Zarah lake came from the great river Helmund, which Yâḳût rightly characterises as 'the

[1] Baladhuri, 392, 394. Ykb. 281. Ist. 239—242. I. H. 297—299, 301. Muk. 305. Mst. 183. A. Y. i. 362.

[2] A number of sketch maps, showing the present condition of the Helmund delta and the lake, are given by Major Sykes, *Persia*, pp. 364, 372. At its southern extremity the great lake basin is in connection with an immense channel—some 50 miles in length, and averaging 350 yards broad, with cliffs 50 feet high—which is called the Shela. This runs in a south-easterly direction into the Gawd-i-Zarah, or 'Hollow of Zarah,' a second lake bed, lying due south of the bend in the lower Helmund, and this Gawd, or hollow, in seasons of flood, receives the overflow of the lake. The Gawd-i-Zarah has an area measuring 100 miles from east to west by about 30 miles across. Sykes, *Persia*, p. 365.

river of the thousand affluents.' He spells the name Hindmand, Hîdmand being a common variant probably due to clerical error, also Hîrmand (or Hîrmîd), and by this name Mustawfî describes the river, which he also calls the Âb-i-Zarah, or Stream of the Zarah (lake). Helmund is the more common modern form. The great river rises in the mountain range lying between Ghaznah and Bâmiyân, which now forms part of Afghanistân, but which, in the middle-ages, was known as the district (or kingdom) of Ghûr. Taking a south-westerly course it passed down through the broad valley known as the Zamîn-Dâwar to the city of Bust, where it was joined on its left bank by the Ḳandahâr river, which watered the country called Rukhkhaj. Bust was the first city the river came to of Sijistân proper, and from here the Helmund began its great semicircular bend, flowing south, then west, and then north to Zaranj, whence turning west again its waters were discharged into the Zarah lake.

When one march, or some 30 miles distant, from Zaranj the Helmund was checked by a series of great dams, which had been built to hold up its waters for irrigation needs, and at this point the greater volume of the main stream was drawn off into five great canals flowing out towards Zaranj and the lake. The first or southernmost of these was the Nahr-aṭ-Ṭa'âm, 'the Victuals Canal,' which irrigated the lands and farms outside the Bâb-aṭ-Ṭa'âm, the gate of Zaranj already mentioned, which lands in part were of the Nîshak district. The next canal was called the Nahr Bâsht Rûdh; and the third was the Sanârûdh, which, starting from the main stream of the Helmund one league from Zaranj, was the waterway to the capital, so that, as Ibn Ḥawḳal remarks, in flood-times a traveller could go by boat all the way from Bust to Zaranj. The fourth canal, which irrigated some thirty villages, was called the Nahr Sha'bah, and the fifth was the Nahr Mîlâ. Beyond this what was left of the main stream of the Helmund entered the channel known as the Nahr Kazak, where its waters were again dammed back for irrigation purposes, except in the flood season, when the overflow escaped direct to the Zarah lake[1].

[1] Ist. 242—244. I. H. 300, 301. Muk. 329. Yak. i. 514; iv. 272, 992, 993. Mst. 216, 226. Muḳaddasî refers to the lake under the name of Buḥayrah-aṣ-Ṣanaṭ, but this possibly is merely a clerical error.

Zaranj, according to the earlier Arab geographers, had not been originally the capital city of Sijistân under the ancient Persian kings. Their capital had stood at Râm Shahristân, otherwise called Abrashahriyâr, a city that had already in the 4th (10th) century been swallowed up by the desert sands, but of which the ruins, with parts of houses, still remained standing, and visible at that date. The situation of this ancient capital is given vaguely as lying three marches from Zaranj, on the left hand of one going from that city towards Kirmân, 'near Darâk and over against Râsak,' two unknown places. It is stated that in older days the main branch canal from the Helmund had brought water to this place, by which all the surrounding lands were fully irrigated. The dam across the great river which fed this canal had, however, suddenly burst, and the waters, pouring down another channel, became permanently diverted. As a result the whole region round the older city lapsed to the state of a desert, and the inhabitants, migrating in a body, founded the city of Zaranj.

At some distance to the west of the Zarah lake, on the Ḳûhistân frontier and close to the border of the Great Desert, is the town of Nih, or Nîh, which is named by earlier Arab geographers as belonging to Sîstân. Muḳaddasî mentions it as a strongly fortified town, the houses of which were built of clay, water being brought down from the hills by underground channels. Nih is also referred to by Yâḳût and Mustawfî, who, however, add no details, except to state that it was founded by King Ardashîr Bâbgân, though at the present day the remains of great fortifications, and the immense ruins found here, would seem to prove that in the middle-ages it had been a place of much importance[1].

Of the rivers flowing into the Zarah lake from the north that which comes down from Asfuzâr (Sabzivâr of Herât), and is now known as the Hârûd, does not appear to be mentioned by the

[1] Ist. 242. I. H. 300. Muk. 306. Yak. iv. 871. Mst. 183. The position of Râm Shahristân is not certain. Sir H. Rawlinson (*J.R.G.S.* 1873, p. 274) would place it at Râmrûd, near the beginning of the Shela, where there are extensive ruins. These ruins, which apparently at the present day are known as Shahr-i-Rustam, Rustam's city, are described, and a sketch plan given, by A. H. Savage Landor in *Across Coveted Lands*, ii. 270. The ruins of Nih are described by Major Sykes, *Persia*, p. 413.

Arab geographers. They notice, however, the Farah river, which takes its rise in the mountains of the Ghûr district. This, the Wâdî Farah, after leaving the hill country, soon entered the province of Sijistân, and came to the city of Farah, which Ibn Ḥawḳal speaks of as lying in a plain, being a large place of clay-built houses, and with sixty dependent villages having many farms where much fruit was grown, more especially dates. Muḳaddasî adds that the city of Farah was in two quarters, occupied respectively by the orthodox Moslems, and by the Khârijite sectaries. One stage south of the city was the bridge over the river called the Ḳanṭarah Farah (Pûl-i-Farah, in Persian), where the high road down to Zaranj crossed from the right bank to the left. This bridge, where there was also a town, was four days' march above Juvayn, and about half-way between the two (according to Ibn Rustah) was a place called Kahan. Near Kahan, one league away to the westward, was a remarkable sand-hill, with strange acoustic properties; for if water, or any small object, were thrown on the sand of this hillock 'a great noise was heard, like a buzzing sound, and very terrible to listen to.' This wonderful sand-hill is also mentioned by Bîrûnî, writing in the 5th (11th) century, and similar acoustic properties of 'the moving sand' have been remarked at the present day in the hillocks of the dunes forming the desert between Sijistân and Ḳûhistân. The modern double town of Lâsh-Juvayn, at the present time a place of much importance, is mentioned by Muḳaddasî, under the form Kuwayn (for Guvayn), as a small city, strongly fortified, in which there was no Friday Mosque, for its inhabitants were all Khârijite sectaries; but except as a stage on the high road, no medieval authority other than Muḳaddasî describes the place, and the name Lâsh is not found.

About half-way between Juvayn and Zaranj the high road crossed the chief overflow canal of the Helmund by a bridge, and a few leagues south of this stood the important town of Karkûyah. This last was one stage north of Zaranj, and gave its name, it will be remembered, to the northern city gate. Karkûyah was peopled by Khârijites, according to Yâḳût, and many ascetics lived here, but it was chiefly remarkable for its great fire-temple, so much venerated by all the Magians of Persia. Ḳazvînî, writing at the close of the 7th (13th) century, gives a long account of this building, which he says was covered by two domes, said to date from the mythical times of the national hero Rustam. Each dome was surmounted by a horn, the two horns curving apart one from the other like the two horns of a bull, and these were relics of the aforesaid hero. Under the twin domes stood the fire-temple, where the sacred fire had never been allowed to become extinguished. A priest, who was at stated times relieved by his fellows, served this temple; and he was wont to stand twenty ells away from the fire, having a veil before his mouth, lest his breath should defile the fire, and he fed the flame continually with span-long logs of tamarisk wood, which he laid on with silver tongs. Kazvînî adds that this was one of the most venerated of the fire-temples of the Magians. Not far from Karkûyah, and three leagues from Zaranj, was the town of Kurunk, which Yâḳût says was commonly pronounced Kurûn, and under this last name it still exists. It was, Yâḳût adds, a pleasant place, full of good things, with a population of Khârijites and weavers[1].

The Khâsh, Khwâsh, or Khuwâsh[2] river flows down to the Zarah lake between the Farah river and the Helmund. It is called by Ibn Ḥawḳal the Nahr Nîshak, Nîshak being the name of the populous district lying due eastward of Zaranj, which gave its name, as already stated, to the eastern gate of the capital. This river also took its rise in the Ghûr mountains, and the town of Khwâsh lies on its banks, being about one day's march from Zaranj. Ibn Ḥawḳal describes Khwâsh as the largest town of this district,

[1] I. R. 174; and with regard to the acoustic sand-hill see Bîrûnî, *Chronology of Ancient Nations*, translated by C. E. Sachau, p. 235 (Arabic text, p. 246). For an example, at the present day, of a sand-hill that gives sounds like 'an Aeolian harp,' see Sir F. Goldsmid (*Eastern Persia*, i. 327), who visited this extraordinary hill, which is at the shrine of Imâm Zâyid, five miles west of Ḳal'ah-i-Kâh. Ist. 244. I. H. 303, 304. Muk. 306, 329. Mst. 215. Kaz. ii. 163. Yak. iii. 42, 888; iv. 263, 269. The site of Karkûyah probably is to be sought among the immense ruins to the south of Pîshâvarân. There is an old bridge here, of two arches, called Takht-i-Pûl; cf. Sir F. Goldsmid, *Eastern Persia*, i. 315. Yate, *Khurasan and Sistan*, 118. The fire-temple was known to the Zoroastrians as the Mainyo Karko.

[2] There were in this region at least three places of this or a similar name; viz. the present river and town of Khâsh, then the town of this name in the Jabal-al-Ḳufṣ (see p. 317), lastly, Khwâṣ of Makrân (see p. 330).

and famous for its date palms. When Yâḳût wrote the name had already come to be more generally pronounced Khâsh, as at the present day. The most famous city of the district, but a smaller place than Khwâsh, was Ḳarnîn or Al-Ḳarnîn, the birthplace of the Ṣaffârid princes Ya'ḳûb and 'Amr, sons of Layth, the famous coppersmith. Ḳarnîn was situated out in the desert plain to the north-west of Khwâsh, and one march from it on the road to Farah. They showed here, Ibn Khurdâdbih remarks, the relics of the stall of Rustam's horse. Muḳaddasî speaks of Ḳarnîn as a small place, but well fortified, having a stream going through the town, which had a Friday Mosque, and possessed suburbs. Mustawfî also refers to it, adding that both corn and fruit were grown in the neighbouring lands, which were very fertile.

Half-way between Ḳarnîn and Farah stood the little town of Jizah, about equal to the former in size, which Ibn Ḥawḳal describes as possessing many villages and farms, for it stood in a very fertile country, amply irrigated by underground water-courses. The buildings of the town were of sun-dried bricks; and Yâḳût adds that in his day the people pronounced the name Gizah. The whole district along the Khwâsh river, known as Nîshak, was, as already said, extremely populous in the 4th (10th) century. Ḥarûrî, 'a populous village belonging to the Sulṭân,' which still exists, lies on the river bank below Khwâsh, where the high road coming in from Bust crossed the Khwâsh river by a bridge of burnt brick. The village of Sarûzan was the next stage on the way to Zaranj, and between these two was situated Zânbûḳ, a strongly fortified hamlet, which Muḳaddasî ranks for size with Juvayn.

One day's journey north of Zaranj, but its exact position is not given in the Itineraries, lay the important town of Aṭ-Ṭâḳ, 'the Arch.' It was very populous, and Muḳaddasî records that grapes in abundance were grown here and in the adjacent farmsteads. Abu-l-Fidâ in the 8th (14th) century, quoting from Ibn Sa'îd, states that this place, which he names Ḥisn-aṭ-Ṭâḳ (the Fortress of the Arch), crowned a high hill at a bend of the Helmund, where, after throwing off the canals to Zaranj, the main stream finally turned westward and flowed out to the Zarah lake; and the town is mentioned, together with the fort of Zarah (Ḳal'ah or Ḥiṣâr-i-Zarah), as having been captured by Tîmûr immediately prior to his attack on Zaranj. In the days of the first Moslem conquest another fortress is mentioned as of this region, namely, Zâliḳ, which is given as lying five leagues from both Karkûyah and from Zaranj. Nothing further, however, is known of it, and in later times the place is not referred to[1].

Bust, approximately, lies in the same latitude as Zaranj, and the direct road from Zaranj thither goes due east by Ḥarûrî as already described, and across the desert. The course of the Helmund, however, doubles the distance by making its semicircular sweep to the south, and half-way along its course stands the town of Rûdbâr. This place is apparently mentioned by Balâdhurî, at the time of the first Moslem conquest, for he speaks of a town called Ar-Rûdhbâr of Sijistân as lying in the direction of Ḳandahâr; and near this Ar-Rûdhbâr was Kishsh (or Kiss), which appears to be the place called Kâj, or Ḳuhîch, at the present day. Rûdhbâr is elsewhere only incidentally mentioned by the Arab geographers; possibly it is identical with the Rûdhbâr described by Iṣṭakhrî as of the Fîrûzḳand district near Bust. This place had many fruitful fields and farms, but the chief export is said to have been salt. Another place of this neighbourhood is Az-Zâliḳân, otherwise spelt Ṣâlaḳân, or Jâliḳân. It is described by Ibn Ḥawḳal as one march from Bust, but in which direction is not stated, and the name does not occur in the Itineraries. It was a town mostly inhabited by weavers, but surrounded by extensive and fruitful lands, well watered by streams, and in the 4th (10th) century it was of about the size of Ḳarnîn.

Bust (or Bast) on the Helmund, at the junction of the river from the Ḳandahâr district, has always been an important place. Iṣṭakhrî mentions that at its gate was the great bridge of boats, 'like those used in 'Irâḳ,' across which the high road came in from Zaranj. Bust was the second largest city of Sijistân in the 4th (10th) century, the people were in easy circumstances, and are described as dressing after the fashion of the men of 'Irâḳ, and as being for the most part merchants who traded with India. The neighbouring lands were extremely fertile, growing dates and

[1] Baladhuri, 393, 395. I. H. 301, 302, 303, 304. I. K. 50. Muk. 306. Yak. ii. 72, 486; iv. 272. Mst. 185. A. F. 343. A. Y. i. 370.

grapes; and Bust was accounted the chief town of all the mountainous country of eastern Sijistân, which included the two great districts of Zamîn-Dâwar and Rukhkhaj. Muḳaddasî states that the city and its fortress, surrounded by great suburbs, stood one league above the junction of the river Khardarûy (the modern Argandâb) with the Hirmand (Helmund). It possessed a fine mosque, and the markets were well stocked. Half-a-league distant, on the Ghaznah road, was Al-'Askar, 'the Camp,' built like a small city, where the Sultan had his residence. In the 7th (13th) century Yâḳût writes that Bust was almost entirely a ruin, and he notices the heat of the climate, while mentioning the abundance of its gardens. At the close of the 8th (14th) century the place and its neighbourhood were devastated by Tîmûr, who marched hither from Zaranj, destroying on his way one of the great dams across the Helmund, known as the Band-i-Rustam, that kept up the head of water which served to irrigate all the western lands of Sîstân[1].

The broad valley, down which the Helmund flows from the mountains of Hindû Kush to Bust, still bears the name, Zamîn-Dâwar, by which the Arab geographers refer to the district. This is the Persian form of which the Arabic equivalent is 'Arḍ-ad-Dâwar or Balad-ad-Dâwar, the meaning being the same, namely, 'the Land of the Gates,' or passes, into the mountains. During the middle-ages this was a fertile and very populous district, with four chief towns, namely, Dartall, Darghash, Baghnîn and Sharwân, with numerous great villages and farmsteads. The chief town of the district was Dartal, Dartall, or Tall as Iṣṭakhrî writes the name, which appears to be identical with the city of Dâwar described by Muḳaddasî. It was a fine large town, with a fortress, garrisoned by horse guards, who in the 4th (10th) century, held this as the frontier post on the road towards the Ghûr mountains. It lay on the bank of the Helmund river, three marches above Bust, and in the account of the first Moslem conquest it is stated that near here was the mountain, Jabal-az-Zûr, where the great idol called Zûr, or Zûn, had been taken as

[1] Baladhuri, 394, 434. Ist. 244, 245, 248. I. H. 302, 304. Muk. 297, 304. Yak. ii. 10, 612; iv. 184. A. Y. i. 370.

booty by the Arabs, this idol being all of gold, with eyes of corundum (*yâkût*).

One march yet higher up the Helmund, and on the same bank as Dartall, was Darghash, while Baghnîn lay one march to the westward of Dartall, in the country held by the Turkish tribes known as the Bishlank, among whom abode the tribe of the Khalaj. These Khalaj Turks afterwards emigrated westward, but Ibn Ḥawḳal in the 4th (10th) century describes them as then living very contentedly in the Zamîn-Dâwar country, 'after the Turk fashion.' A fifth town of the Zamîn-Dâwar was Khwâsh (spelt like the place on the river of that name, just mentioned), which Iṣṭakhrî described as an unwalled city, but protected by a castle. Unfortunately its position is not given, but some authorities count it as belonging to Kâbul.

Between Bust and Dartall, and one march south of the latter town, being apparently not situated on the Helmund river, stood the city of Sarwân or Sharwân, which Ibn Ḥawḳal describes as of the size of Ḳarnîn, but more populous and prosperous. Great quantities of fruit, dates and grapes especially, were exported from its district, and that of Fîrûzḳand, which latter lay south of the Sharwân district and one march to the eastward of Bust[1].

The Rukhkhaj district, occupying the country round about Ḳandahâr, lay to the eastward of Bust along the banks of the streams now known as the Tarnak and the Argandâb. The capital of Rukhkhaj in the middle-ages was Banjaway, the Arabic form of Panj-wây, 'Five Streams,' which is still the name of the district west of Ḳandahâr, near the junction of the two rivers Tarnak and Argandâb. The Rukhkhaj district was immensely fertile during the middle-ages, and wool was exported thence in large quantities, bringing in a good revenue to the treasury. The site of Banjaway city is difficult to fix. It lay on the high road four marches from Bust, at the point where the ways bifurcated, one road going north in 12 marches to Ghaznah, the other east in six marches to Sîbî. It probably was not far from Ḳandahâr,

[1] Baladhuri, 394. Ist. 244, 245, 248. I. H. 302, 304. Muk. 305. Yak. ii. 541; iv. 220. None of these towns of the Zamîn-Dâwar now exist, but Dartall, the capital, must have occupied approximately the site of modern Girishk.

but the distance between the two cities is nowhere given. One league to the west of Banjaway city was the fortress of Kûhak, 'the Hillock,' with a town lying round the fort. Banjaway itself had good fortifications, as well as a fine mosque. It got its water from the neighbouring river.

One stage from here, on the Sîbî road, lay the town of Bakrâwâdh (for Bakrâbâd, which Iṣṭakhrî and Ibn Ḥawḳal give as Takînâbâdh, probably from a clerical error), where there was a Friday Mosque in the town market-place; and this town too stood upon a stream that joined the Ḳandahâr river.

The city of Ḳandahâr (or Al-Ḳunduhâr) is frequently mentioned in the accounts of the first Moslem conquests of the places near the Indian frontier. Balâdhurî says it was reached from Sijistân after crossing the desert, and the Moslems, he adds, attacked the place in boats from the river, destroying the great idol Al-Budd, doubtless a statue of Buddha. After this period only incidental mention of Ḳandahâr occurs—generally as of Hind or the Indian frontier—in Muḳaddasî, Ibn Rustah, and Ya'ḳûbî. Unfortunately no early Itinerary takes us to Ḳandahâr, and in the systematic accounts of the province by Iṣṭakhrî and Ibn Ḥawḳal the name is altogether wanting. Possibly Banjaway replaced it during the earlier middle-ages, for Yâḳût gives no description of the town, and the name only reappears in history when it is spoken of as being devastated first by the Mongols in the early part of the 7th (13th) century, and then again by Tîmûr at the close of the next century[1].

The district round Sîbî was known to the Arab geographers as Bâlis, otherwise Bâlish, or Wâlishtân. The capital city according to Iṣṭakhrî was Sîbî, spelt Sîvî or Sîwah, but the governor generally resided at Al-Ḳaṣr (the Castle), a small town situated one league distant from Asfanjây, or Safanjavî, the second city of the district, the exact site of which has not been identified, but which lay two marches north of Sîbî on the road to Banjaway of Rukhkhaj. The town of Mastang, or Mastanj, is also mentioned by Iṣṭakhrî and Muḳaddasî, who name a number of other villages of this

[1] Baladhuri, 434, 445. Ist. 244, 250. I. H. 301, 302, 305. Muk. 305. Yak. iv. 331. A. Y. i. 376. Dr H. W. Bellew, *From the Indus to the Tigris*, p. 160.

district, which was said to include in all some 2200 hamlets, but no description is given of any of these places[1].

Ghaznah, or Ghaznayn, became famous in history at the close of the 4th (beginning of the 11th) century as the capital of the great Maḥmûd of Ghaznah, who at one time was master both of India on the east and Baghdâd on the west. Unfortunately no adequate description has come down to us of Ghaznah at the time when it was rebuilt and adorned by Maḥmûd with all the plunder of his Indian raids. A generation before this Iṣṭakhrî describes the place as like Bâmiyân, with fine streams but few gardens. He adds that no city of this countryside was richer in merchants and merchandise, for it was as the 'port' of India. Muḳaddasî gives a long list of the names of its districts and towns, most of which, however, it is impossible to identify at the present day. He writes the name *Ghaznayn*, in the dual form, but to what the 'Two Ghaznahs' has reference is not stated, though Ghaznayn in later times is more generally used than the form Ghaznah. Muḳaddasî adds that all the country between this and Kâbul was known as Kâbulistân.

It was about the year 415 (1024) that Maḥmûd had rebuilt Ghaznah, on his return home laden with the spoils of India, and the city then reached its greatest splendour, which lasted for over a century. The Ghûrid Sultan 'Alâ-ad-Dîn, surnamed Jahân-sûz, 'world incendiary,' to revenge his brother's death at the hands of Bahrâm Shâh the Ghaznavid, took Ghaznah by storm in 544 (1149), and afterwards both sacked and burnt the city, which never recovered from this calamity. The tomb of the great Maḥmûd in the mosque nevertheless appears to have been spared, or else it was restored, for Ibn Baṭûṭah saw it here in the 8th (14th) century. He describes Ghaznah as then for the most part in ruins, though formerly, he adds, it had been an immense city. His contemporary Mustawfî speaks of it as a small town, with a very cold climate on account of its great elevation, but he gives no details of any importance[2].

[1] Ist. 179, 244. I. H. 301. Muk. 297.

[2] Ist. 280. I. H. 328. Muk. 296, 297. I. B. iii. 88. Mst. 184. Unfortunately 'Utbî, in his *History of Maḥmûd of Ghaznah*, gives no detailed

As we have seen, the whole of the great mountainous district of the upper waters of the Helmund and the Ḳandahâr rivers was known to the Arabs as Zâbulistân, a term of vague application, but one which more particularly denoted the country round Ghaznah. On the other hand Kâbulistân was the Kâbul country, lying more to the north on the frontiers of Bâmiyân; and this is the division found in the accounts of the conquests of Tîmûr. Already in the 3rd (9th) century Ya'ḳûbî describes Kâbul as much frequented by merchants, who brought back from this country the Kâbulî Ahlîlaj, or myrobalan of the larger sort[1]. Ya'ḳûbî says the chief city was then known as Jurwas, while Iṣṭakhrî in the next century gives the name as Ṭâbân. Kâbul, however, appears also to have been the name in common use, but more especially for the district.

There was here a famous Ḳuhandiz or castle, and the town which was approached by only a single road was well fortified. It was the great emporium of the Indian trade, indigo (*nîl*) being brought here for export to the value of a million gold dînârs yearly (about half-a-million sterling); further, most of the precious stuffs of India and China were warehoused here. As early as the 4th (10th) century the Moslems, the Jews, and the idolaters, had each a separate quarter in Kâbul, where the suburbs, the markets, and the merchants' warehouses were alike famous. Muḳaddasî mentions, too, a wonderful well in the castle; and for him Kâbul is especially the country of the myrobalan. He counts Kâbulistân as an outlying region of Sijistân. Ḳazwînî, in the 7th (13th) century, states that Kâbul was then famous for the breed of she-

description of the capital. See the article on Ghaznah by Sir H. Yule in the *Encyclopaedia Britannica* (9th ed.), x. 560, where a plan is given.

[1] Myrobalan was a name applied during the middle-ages to certain dried fruits and kernels of astringent nature, imported from India, which had a high reputation in the concoction of the medicines of those days. The name is of Greek origin, the Indian fruits used in the manufacture of this condiment are of a variety of species, and one of the best known kinds of myrobalan was that called *Chebulic*, namely, that from Kâbul. The Arabs named the drug (for this it came to be) *Ahlîlaj* or *Halîlaj*, and Ibn Bayṭâr in his *Dictionary of Drugs* (translated by Dr J. Sontheimer, i. 163; ii. 572) has two articles about it; see also Dozy, *Supplément aux Dictionnaires Arabes*, s.v. *Ihlîlaj*, and *Glossary of Anglo-Indian Terms*, by Yule and Burnell, s.v. *Myrobalan*.

camels, known as Bactrian (Bukhtî), the best in all central Asia. Ibn Baṭûṭah, who visited Kâbul in the next century, says that it had then sunk to be a mere village, inhabited by the tribe of Persians known as Afghans (Al-Afghân).

The Kâbul river is an affluent of the Indus, and is formed by the junction of two streams coming down from the Hindû Kush range, the mountains to the north of Kâbul[1]. At the eastern source are the celebrated silver mines, known to the Arabs as Banjahîr (for *Panj-hîr*, or 'Five Hills,' in the dialect of the country), from which large quantities of the precious metal were obtained, and Banjahîr became a mint city under the Ṣaffârid princes in the 3rd (9th) century, the dirhams, of course, bearing the name also of the Abbasid Caliph. Banjahîr city is described by Ibn Ḥawḳal as standing on a hill, and inhabited by 10,000 miners, who were an unruly folk, much given to evil living. Jârbâyah was a neighbouring town, also lying on the Banjahîr, or Kâbul river, which thence flowed out towards the plains of India, past Farwân, a large town with a mosque. Muḳaddasî further mentions the town of Shiyân in the district of Askîmasht, where there was a wondrous spring, and a fine mosque built by the Arab general Ḳutaybah-ibn-Muslim, who had commanded the troops at the time of the first Moslem conquest. Yâḳût gives us a long account of these silver mines with their population of riotous miners. He says that the whole mountain side was hollowed out in caverns, where men worked in the bowels of the earth by torch-light. The people were given over entirely to a species of gambling, men found themselves rich one day and paupers on the morrow; they would recklessly spend 300,000 dirhams (£12,000) in the mere digging of a new shaft. The ruin of the place was due to Changîz Khân; and when Ibn Baṭûṭah, who speaks of the blue waters of the neighbouring stream, came here in the 8th (14th) century, he found no silver mine, but only the disused tunnels of the former workings.

[1] Hindû Kush, in Persian, means (the Mountain that) 'kills the Hindus.' Ibn Baṭûṭah (iii. 84) is one of the first to give this name, which is unknown to the earlier Arab geographers. He explains that the range was so called because many Indian slaves died in crossing it when journeying to Persia.

The products of Sijistân were few in number; and all that Muḳaddasî records is that date-baskets, called *zanabîl*, were made here for export, also ropes of palm-fibre and reed-mats[1].

The high roads in Sijistân all centred in Zaranj, to which in the first place led the desert road from Narmâsîr *viâ* Sanîj, which has been described in the last chapter. From Zaranj northwards, a road went to Herât, passing through Karkûyah, and thence by a bridge over the Helmund overflow to Juvayn on the Farah river. From Juvayn Farah city was reached by a road up the river bank, which crossed the river by the bridge of Farah (mentioned p. 341), beyond which was Farah city. Three marches north of Farah lay Asfuzâr (or Sabzivâr of Herât), the first town in Khurâsân. The distances in leagues along this road unfortunately are not given, only the stages of each day's march, for which Iṣṭakhrî and Ibn Ḥawḳal are the chief authorities[2]. Moreover a good deal of uncertainty exists in the spelling of the names of many of the halting-places.

From Zaranj the road east went to Ḥarûrî on the Khwâsh river, whence taking a straight line across the desert the city of Bust was reached in five marches. At Bust the roads bifurcated, one going to the Zamîn-Dâwar country of the upper Helmund, and another to Banjaway of Rukhkhaj, in the neighbourhood of Ḳandahâr. At Banjaway there was again a bifurcation of the roads, one going north-eastward to Ghaznah, and a second to Sîbî, through the town known as Asfanjây. On these routes too it is to be noted that the distances are again given merely in marches, many of the names of the stages being most uncertain[3].

[1] Ykb. 290, 291. Ist. 278, 280. I. H. 327, 328. Muk. 297, 303, 304, 324. Yak. i. 473; ii. 904, 905; iii. 454. Kaz. ii. 162. A. Y. i. 558. I. B. iii. 85, 89. Mst. 188.

[2] I. R. 174. Ist. 248, 249. I. H. 304, 305. Muk. 350.

[3] Ist. 249—252. I. H. 305—307. Muk. 349, 350.

CHAPTER XXV.

ḲÛHISTÂN.

The province called Tunocain by Marco Polo. Ḳâyin and Tûn. Turshîz and the Pusht district; the Great Cypress of Zoroaster. Zâvah. Bûzjân and the Zam district. Bâkharz district and Mâlin. Khwâf. Zîrkûh. Dasht-i-Biyâḍ. Gunâbâd and Bajistân. Ṭabas of the dates. Khawst, or Khûsf. Birjand and Mûminâbâd. Ṭabas Masînân and Duruh.

The province of Ḳûhistân, like Sijistân, was generally held to be a dependency of Khurâsân by the Arab geographers. Ḳûhistân means 'the Mountain Land,' and the province is thus named in accordance with its distinguishing physical features, the hills here being contrasted with the lowlands of Sijistân, lying to the east of Ḳûhistân on the Helmund delta. Ḳûhistân, as Ibn Ḥawḳal remarks, has for the most part a cold climate from its elevation, and the date palm only grew at Ṭabas Gîlakî on the edge of the Great Desert. In the 4th (10th) century the nomad inhabitants of the country were Kurds, who possessed great flocks of sheep and camels. Without doubt this province is identical with the 'Tunocain kingdom' of Marco Polo, who took the names of its two chief cities (Tûn and Ḳâyin) to be the designation of the whole country[1].

The chief town of Ḳûhistân was Ḳâyin, which Ibn Ḥawḳal describes as protected by a strong fortress, surrounded by a ditch; and the governor's house stood here, also the Friday Mosque.

[1] Ist. 273, 274. I. H. 324, 325. Muk. 301. *Marco Polo*, Yule, i. 87, 131. The name is spelt Ḳûhistân by the Arabs (with dotted Ḳ), and Kûhistân in Persian, where *Kûh* means 'mountain,' but the first vowel in the name is as often as not written short (Kuhistân or Ḳuhistân).

Water was supplied by underground channels, but the gardens were not very fruitful or numerous, for the cold was severe in winter. The city had three gates, and its merchants carried on a considerable trade with Khurâsân. Ibn Ḥawḳal adds that at a place two days' march from Ḳâyin, on the Nîshâpûr road, a kind of edible clay, called *Ṭîn Najâḥî*, was found, and this, he says, was exported to all the neighbouring lands and largely eaten by the people. Ḳâyin was visited in 444 (1052) by Nâṣir-i-Khusraw, who describes the inner town as forming a fortress of great strength. The Great Mosque here had in its sanctuary (*Maḳṣûrah*) the largest arch to be seen in all Khurâsân, and the houses of the town, he says, had all domed roofs. Mustawfî in the 8th (14th) century notes in the first place the central position of Ḳâyin, which was, he says, just 20 leagues distant from every other important place in Ḳûhistân. It was a fine city: all the houses were supplied with water by channels below ground, and had cellar-rooms for the hot weather. The crops matured here very rapidly, and the harvest was early. Corn, fruit, and especially saffron were grown largely in the neighbourhood, and the cattle pastured on these lands quickly put on fat. Mustawfî adds that the population were remarkably dark-skinned.

The city of Tûn lies rather over fifty miles to the westward of Ḳâyin, and a little to the north. Muḳaddasî speaks of it as a populous place, smaller than Ḳâyin, protected by a castle and possessing a fine mosque. Woollen goods were manufactured here, and Nâṣir-i-Khusraw praises its carpets, 400 looms being at work at the time when he passed through the town. Much of the city, however, was in his day in ruin, though the great fort remained. In the eastern suburbs were many fine gardens where pistachios were cultivated. Mustawfî states that Tûn had originally been built 'on the plan of a Chinese town,' but he does not further explain the matter. He speaks of the great castle with its deep dry-ditch; this was surrounded by the streets and bazaars of the outer town. The neighbouring lands were very fertile, for he says that the people had the art of building dykes or dams (*band*) to collect the rain-water and prevent it from flowing away, and on these lands they raised water-melons, noted for their sweet flavour. Much corn and fruit was grown, and silk was produced here abundantly, for the climate of Tûn was temperate, and the underground watercourses very numerous[1].

In the north-west corner of Ḳûhistân is the district of Bûsht, Pûsht, or Busht-al-'Arab, of which the chief towns were Turshîz and Kundur[2]. In the Arab geographers the older form of the name is given as Ṭuraythîth, or Ṭurthîth, later spelt Turshîsh and Turshîs, and it was sometimes counted as of the *Ḥawmah* or domain of Nîshâpûr. Ibn Ḥawḳal speaks of Turshîz as a very populous city, with fertile lands, and in the Pûsht district there were seven other townships with Friday Mosques. Muḳaddasî describes the mosque of Turshîz as in his day rivalling that of Damascus for magnificence; there was also a famous water tank, and the markets were renowned, so that Turshîz was considered the 'store-house of Khurâsân,' where merchandise was exported and imported, to and from Fârs and Isfahân. The neighbouring town of Kundur almost equalled Turshîz in wealth, and in the district immediately round it were 226 large villages.

According to Ibn-al-Athîr in 520 (1126) the Wazîr of Sultan Sanjar the Saljûḳ besieged and plundered Turshîz, which had lately come into the possession of the Ismâîlîs, or Assassins; for the 'Old Man of the Mountain' had recently conquered most of the strong places in the neighbourhood, building many fortresses to overawe all this part of Ḳûhistân. Yâḳût places the advent of the Ismâîlîs as occurring in the year 530 (1136), and relates that the governor of Turshîz had called in the Turkish tribes to aid him against the heterodox Mulâḥids or Ismâîlîans, but they had failed to fight the enemy, and had themselves pillaged the country, thus bringing Turshîz to ruin. In the middle of the

[1] I. H. 324, 325. Muk. 321. N. K. 95. Mst. 184. There is an inscription in the mosque at Ḳâyin dated 796 (1394). Sir F. Goldsmid, *Eastern Persia*, i. 341.

[2] The district of Turshîz exists at the present day, but no town of that name. The small town of Kundur is still marked on the map, and according to Iṣṭakhrî the city of Turshîz lay one march to the westward of it, which would place the site of Turshîz at the Fîrûzâbâd ruins, near the present village of 'Abdulâbâd. In any case the medieval city of Turshîz cannot be identified with Sultânâbâd, the modern capital of the Turshîz district, for this lies east of Kundur.

7th (13th) century Hûlâgû Khân, the Mongol, destroyed the power of the Assassins, and his troops, it is stated, conquered seventy of their castles in the Ḳûhistân province. After this Turshîz quickly recovered its importance; and less than a century later it is described by Mustawfî as one of the chief cities of Ḳuhistân, though still partly in decay. He mentions the four famous castles in the neighbourhood of the place—namely, Ḳal'ah Bardarûd, Ḳal'ah Mîkâl (or Haykâl), Mujâhidâbâd (the Champion's Home), and Âtishgâh (the fire-temple)—which doubtless had been those of the Ismâîlîans. He praises the abundant crops of Turshîz, which he says were exported to all the northern districts round Nîshâpûr. At the close of the 8th (14th) century Turshîz was deemed impregnable from its high walls; but when Tîmûr appeared before it he soon undermined these, and after the sack nothing but ruins remained standing. This was in 783 (1381) and since that time Turshîz has disappeared from the map[1].

Mustawfî states that at the village of Kishmar, near Turshîz, had stood the celebrated cypress-tree, originally planted by Zoroaster as a memorial of the conversion of King Gushtasp to the Magian religion. This tree grew to be larger than any other cypress that had ever been, and according to the *Shâh Nâmah* it sprang from a branch brought by Zoroaster from Paradise. Such too was its power that earthquakes, which frequently devastated all the neighbouring districts, never did any harm in Kishmar. According to Ḳazvînî the Caliph Mutawakkil in 247 (861) caused this mighty cypress to be felled, and then transported it across all Persia, in pieces carried on camels, to be used for beams in his new palace at Sâmarrâ. This was done in spite of the grief and protests of all the Guebres, but when the cypress arrived on the

[1] I. H. 295, 296. Muk. 317, 318. Yak. i. 628; iii. 534; iv. 309. Mst. 183. A. Y. i. 344. Ibn-al-Athir, x. 445. The representative of the Old Man of the Mountain, at the present day (as was proved in the English law courts), is Âḳâ Khân, chief of the Khûjah community at Bombay, and it is curious to find that some of the Ismailian sect still linger in Ḳûhistân, who now pay their tithes to Âḳâ Khân, as their predecessors did to the chief at Alâmût. At the village of Sihdih, to the south of Ḳâyin, Major Sykes (*Persia*, p. 409) found nearly a thousand families of these modern Ismailians, who yearly transmitted a considerable sum to their religious head in India. *Marco Polo*, Yule, i. 145.

banks of the Tigris, Mutawakkil was dead, having been murdered by his son[1].

To the east of the Turshîz district is that of Zâvah. The Zâvah district, or part of it, was also known as Rukhkh, and the chief town was called Bîshak or Zâvah city. The name Rukhkh, when Yâkût wrote, was more commonly pronounced Rîkh. In the 7th (13th) century Zâvah town became celebrated as the abode of a very holy man, Haydar by name, who dressed in felt, in summer was wont to enter the fire, and in winter to stand in the snow, and who founded a sect of Darvîshes known as the Haydarîyah. He was alive at the time of the Mongol invasion of the country in 617 (1220), and was afterwards known as Shaykh Kutb-ad-Dîn (Pole of Religion). When Ibn Batûtah visited Zâvah in the 8th (14th) century, he describes the votaries of the Shaykh as having iron rings fastened for penance in their ears, hands, and other parts of the body, and this the people took to be a proof of their sanctity. Mustawfî describes Zâvah as a fine town, standing in a rich district, with some 50 dependent villages. It had a strong castle built of clay bricks. The irrigation was abundant; corn, cotton, grapes, and much fruit grew here, and silk also was produced. He speaks, too, of the shrine of the Shaykh as greatly venerated in his day. At the present time Zâvah is more commonly the name of the district, the town being generally known as Turbat-i-Haydarî, or 'the Tomb of Haydar,' and the shrine is still a place of pilgrimage[2].

To the east of the Zâvah district, and in the north-east corner of Kûhistân, near the Herât river, was the district of Zâm or Jâm, of which the chief town was in the 4th (10th) century known as Bûzjân. This was a considerable city, and 180 villages were of

[1] Mst. 183. *Shâh Nâmah*, Turner Macan, iv. 1067, eight lines from below. Kaz. ii. 297, where the name is by mistake printed Kishm. The account in Kazvînî (13th century A.D.) of course only represents the tradition. There is nothing about the Kishmar cypress in Tabarî or apparently in any of the earlier Arab chronicles. An amplified version of the story will be found in the *Dabistân*, a work of the 16th century A.D. (transl. by Shea and Troyer, i. 306—309). The cypress of Zoroaster is reckoned to have been 1450 years old. It is possibly the origin of Marco Polo's 'Arbre Sol which we Christians call Arbre Sec.' Yule, *Marco Polo*, i. 131.

[2] Muk. 319. Yak. ii. 770, 910. Kaz. ii. 256. I. B. iii. 79. Mst. 188. Sir F. Goldsmid, *Eastern Persia*, i. 353.

its dependencies. The name Bûzjân was pronounced Bûzkân by the Persians, and in later times it was written Pûchkân. In the 8th (14th) century Mustawfî describes it, under the name of Jâm, as occupying a most fruitful and well-watered district, yielding much silk, for the mulberry-trees grew abundantly. The town was celebrated for the number of its shrines, for many holy men had been buried here, and Ibn Baṭûṭah specially names the saintly Shihâb-ad-Dîn Aḥmad-al-Jâmî, whose descendants had come to own much land in the neighbourhood. The saint indeed was so celebrated that Tîmûr, at the close of the 8th (14th) century, visited his shrine in person, and at the present day the town, which is still a flourishing place, is commonly known as Shaykh Jâm[1].

The district of Bâkharz, or Guwâkharz, lies to the south of Jâm, and to the westward of the Herât river, which here takes its northern course. The chief town of Bâkharz was Mâlin, which from the distances in the Itineraries would appear to have been identical in position with the modern city of Shahr-i-Naw, 'New Town,' and in the 4th (10th) century it was already a populous place. From here both corn and grapes were exported, and cloth-stuffs were also manufactured. Yâḳût explains that the name Bâkharz had originally, in Persian, been *Bâd-Harzah*, 'the place where the wind blows,' and he mentions Jawdhaḳân as among its chief villages, of which 128 might be counted round and about Mâlin. Mustawfî, who gives the name of the chief town as Mâlân, expatiates on its fertility, and especially refers to the 'long melon' of this country, which was famous throughout Khurâsân[2].

South-west of Bâkharz is the district of Khwâf (earlier Khwâb), surrounding the chief town of the same name. Khwâf in the 4th (10th) century was famous for its raisins and pomegranates. Salûmak, later written Salâm, had in early times been the largest town of the district, of which Sanjân (or Sankan) and Kharjird were two other important cities. Under the form Kharkird the

[1] I. K. 24. I. R. 171. Ykb. 278. I. H. 313. Muk. 319, 321. Yak. i. 756; ii. 909; iii. 890. Mst. 188, 197. I. B. iii. 75. A. Y. ii. 211, 229. C. E. Yate, *Khurasan and Sistan*, p. 37.

[2] Muk. 319. Yak. i. 458; ii. 145; iv. 398. Mst. 187.

latter town is mentioned by Ibn Ḥawḳal, who also names Farkird (written Farjird or Faljird by Yâḳût) as lying one march to the east of it, while Kûsûy or Kûsûyah was nearer the Herât river, and to the north of Farkird. Of these three towns Kûsûy was the largest, being a third of the size of the neighbouring city of Bûshanj in Khurâsân, to be described later, to which province many authorities count all three places to belong. The town of Kûsûy possessed many good houses of unburnt brick, and the other two towns, though small, had fine gardens and abundant irrigation. Yâḳût also mentions Sirâwand and Lâz as places of importance in his day in the Khwâf district, but their position is unknown. Mustawfî praises the grapes, melons, pomegranates, and figs of Khwâf, and states that much silk was produced in the district. He names the three towns of Salâm, Sanjân, and Zawzan (or Zûzan) as the chief centres of population in the 8th (14th) century. Zûzan when Muḳaddasî wrote was already famous for its wool-workers, and it was an important point in the road system, for it communicated with Ḳâyin, Salâm (Salûmak), and Farjird. Yâḳût calls Zûzan 'a little Baṣrah' for its trade, and refers to it as a shrine of the Magians. Around it lay 124 important villages[1].

In this central part of Ḳûhistân, Mustawfî, writing in the 8th (14th) century, mentions a number of places which are still found on the map, but which do not occur in the works of the earlier Arab geographers. He refers to the district of Zîrkûh, 'Foot-hills,' as most fertile, producing corn and cotton, which with its silk manufactures were largely exported. This is still the name of the hill country south of Zûzan and east of Ḳâyin, and Mustawfî mentions its three chief towns, Shârakhs, Isfad, and Istind, which exist to this day. To the north-west of Ḳâyin is the district the name of which is written Dasht-Biyâḍ, meaning 'the White Plain,' which the Persians at the present day pronounce Dasht-i-Piyâz. Its chief town was Fâris, and Mustawfî, who praises its nuts and almonds, says it was the Yaylâḳ, or summer quarters, of the people of Tûn and Junâbâd.

[1] Ist. 267. I. H. 313, 319. I. R. 171. Ykb. 278. Muk. 298, 308, 319, 321. Yak. ii. 486, 958; iii. 910; iv. 341. Mst. 188. For the present condition of these places see C. E. Yate, *Khurasan and Sistan*, 128, 129.

This last place, now generally called Gunâbâd, is a considerable town lying to the north-east of Tûn. It is named by Ibn Ḥawḳal Yunâbidh, and by Muḳaddasî Junâwad, and there are some other variants. It was a large place in the 4th (10th) century, with clay-brick houses, and the 70 villages round it were well watered by artificial irrigation. Yâḳût gives the name as commonly pronounced Gunâbidh, for Junâbidh. Mustawfî records that its two strong castles, each on a hill, and on either hand of the town, were called Ḳal'ah Khawâshir and Ḳal'ah Darjân respectively, and from their heights the neighbouring villages, and the desert beyond them, were clearly seen. The sand here, he remarks, did not blow into and invade the garden lands of Gunâbâd, as was the case elsewhere in Ḳûhistân. The water-supply was from underground channels, described as often four leagues in length, coming from springs in the hill-flank, and the terminal shafts or wells at the fountain-head, were, he avers, sometimes as much as 700 ells (*gez*) in depth. Much silk was manufactured here, and corn was exported. Some thirty miles to the north-west of Gunâbâd, and a like distance due north of Tûn, is the small town of Bajistân, which appears to be first mentioned by Yâḳût, who speaks of it as a village in his day; and to this Mustawfî adds that it resembled Tûn, but gives no further details[1].

There were, and still are, two towns called Ṭabas in Ḳûhistân, and for this reason the name often appears in the Arab geographers under the dual form of *Ṭabasayn*. Moreover the name Ṭabasayn, in error, is sometimes applied to one or other of these two towns, the dual form for the single place. The Arab geographers, however, clearly distinguish between the two towns, calling one Date Ṭabas, the other Ṭabas of the Jujube-tree, or Ṭabas-al-'Unnâb.

Ṭabas of the Date—Ṭabas-at-Tamr—was on the border of the Great Desert, where many of the roads crossing it came in, and

[1] Dasht-Biyâḍ, or Dasht-i-Piyâz, is a composite name, Persian and Arabic, very unusual in the nomenclature of Persia. If the last word be really the Arabic Biyâḍ it seems likely that the Persians soon forgot its meaning 'White,' and took it to be a proper name. I. H. 325. Muk. 319, 320, 322. Mst. 183, 184. Yak. i. 497; ii. 120; iv. 206. Fâris at the present time is generally known as Ḳal'ah Kuhnah, 'the Old Castle.' Bellew, *Indus to Tigris*, p. 329.

hence Balâdhurî names it 'the Gate of Khurâsân.' According to Ibn Ḥawḳal, the town was in the 4th (10th) century a somewhat smaller place than Ḳâyin, and it had strong fortifications. The chief feature of the district was the forest of date palms that grew here, for being on the desert border the climate was very hot, and the water-supply from underground channels was abundant. Muḳaddasî speaks of its fine mosque, and of a great tank for storing the drinking-water. There were also excellent hot baths. 'It is (he adds) the only place in Ḳûhistân where there are trees and a running stream; and for the distance of a day's journey thence I passed through villages and palm-groves with running water-courses.'

Nâṣir-i-Khusraw, who passed through Ṭabas in 444 (1052), speaks of it as a fine, populous town, unwalled, but enclosed in its gardens and palm-groves. It was then governed with a strong hand, so that all the neighbourhood was perfectly safe, by a certain Abu-l-Ḥasan ibn Muḥammad Gîlakî—'the native Gîlân'—and to distinguish this from the other Ṭabas, it appears in later days to have been called Ṭabas Gîlakî, after this famous governor, who, from what Nâṣir writes, must have been known far and wide for the vigour of his rule. In the second half of the 5th (11th) century Ṭabas passed into the hands of the Ismâîlîan heretics, and in 494 (1102) the town was besieged and in part destroyed by the army sent against the Assassins by Sultan Sanjar the Saljûḳ. Yâḳût and Mustawfî both refer to Ṭabas of the Date as Ṭabas Gîlakî, and the latter authority notices the place both in his account of the Great Desert, and when describing Ḳûhistân. Besides dates, both lemons and oranges flourished here as they did nowhere else in all Khurâsân, and the water of the neighbouring spring flowed in sufficient abundance to turn two mills. A strong fortress protected the town and the numerous villages lying around the place[1].

On the desert border north of Ṭabas, and about half-way to Turshîz, was the village of Bann, possessing a population of 500 males when Ibn Ḥawḳal wrote, and this place was apparently identical with the stage of Afrîdûn mentioned by Ibn Khur-

[1] Baladhuri, 403. I. H. 324, 325. Muk. 321, 322. N. K. 94. Yak. iii. 513, 514; iv. 333. Mst. 183, 184. Ibn-al-Athir, x. 221.

dâdbih. Ibn Ḥawkal apparently mentions in his itinerary another village called Bann (Bann Ukhrâ), but by the distances given the two stages, if not identical, must have had reference merely to two neighbouring villages of the same name. At the present day Bann is represented by Dih Nâband (not to be confounded with the oasis in the desert of that name described on p. 325). It was an important point where one of the desert roads from Jarmaḳ entered Ḳûhistân[1].

Some three leagues to the south-east of Ṭabas, on the edge of the desert where the Shûr road from Kûhbanân came in, was Kurî or Kurîn, which Balâdhurî mentions as one of the two fortresses of Ṭabas, which it would appear might justify the name of Ṭabasayn being given to Date Ṭabas alone. Ibn Ḥawkal describes Kurî as a meeting point of many roads, where stood a village of a thousand men with many farms. Kurîn, as Muḳaddasî spells the name, was a smaller place than Ṭabas; and of its dependencies—being 12 leagues from Ṭabas and 20 from Tûn—was the village of Ar-Raḳḳah. This last place, when Nâṣir-i-Khusraw visited it in 444 (1052), had grown to be a fine town, with a Friday Mosque surrounded by numerous well-irrigated gardens. About three marches to the south-east of Ṭabas were the two towns of Khûr and Khawst, which respectively were the terminal stages of the two roads across the desert from Râvar and Khabîṣ in Kirmân (see pp. 327, 328). Khûr, according to Ibn Ḥawkal, was smaller than Ṭabas, but had a Friday Mosque; the water-supply was scanty and there were hardly any gardens. The place, too, according to Muḳaddasî, was unfortified.

Khawst on the other hand, though in the 4th (10th) century it had no Friday Mosque, was a place of greater importance. It was well fortified, with a castle to defend it, and the clay-brick houses of the town were surrounded by small gardens, though here too the watercourses gave but a poor supply. Muḳaddasî says the town was larger but less populous than Tûn; there were but few trees, and behind it rose the arid hills of Ḳûhistân. Yâḳût by mistake generally spells the name Jûsf, this being a clerical error for Khûsf or Khûsb, which is the modern form of the name, first given by Mustawfî. Yâḳût, it is true, acknowledges his

[1] I. K. 52. Ist. 231, 236. I. H. 295.

ignorance and uncertainty of the true pronunciation of the name, which he says is sometimes written Jûzf: but in one passage he rightly gives Khawst, when quoting from Muḳaddasî. As just stated the modern spelling first appears in Mustawfî, who describes Khûsf as a small town, with some dependencies, watered by a stream which irrigated its lands, so that excellent crops were produced[1].

About 20 miles east of Khûsf lies Birjand, which at the present day has taken the place of Ḳâyin as the capital town of Ḳûhistân. Birjand is not mentioned, apparently, by any of the Arab geographers before Yâḳût, who in the 7th (13th) century speaks of it as one of the finest villages of this province. Mustawfî in the following century refers to it as an important provincial town, surrounded by many fruitful farms and villages, where, in addition to grapes and other fruits, an abundance of saffron was cultivated. Corn, however, grew badly here. A day's journey to the east of Birjand, is the mountain district still known as Mûminâbâd—'the Believer's Home'—which Mustawfî mentions as dominated by a strong fortress that had formerly been in the hands of the Assassins. This district included many fine villages; and Mustawfî especially mentions Shâkhin, on a stream called the Fashâ Rûd, which still exists some three days' march to the south-east of Ḳâyin[2].

About 50 miles due east of Birjand is the second town of Ṭabas, known to the Arab geographers as Ṭabas-al-'Unnâb, 'of the Jujube-tree,' which the Persians called Ṭabas Masînân. This town Ibn Ḥawḳal describes in the 4th (10th) century as larger than Yunâbidh (Gunâbâd, north-west of Ḳâyin); its houses were built of clay bricks, but the fortifications were then in ruins, and there was no castle. Muḳaddasî speaks of the numerous jujube-trees growing here. Ḳazvînî in the 7th (13th) century states that on the summit of a neighbouring hill was the village called Îrâvah, where there was a fine castle, and gardens with trees, for many

[1] Baladhuri, 403. Ist. 232, 274. I. H. 291, 325. Muk. 321, 322. Yak. ii. 152; iv. 23, 270. Mst. 184. N. K. 94.

[2] Yak. i. 783. Mst. 184. Sykes, *Persia*, pp. 305, 306. Major Sykes, who spells the name *Shahkin*, speaks of an ancient fort near this, possibly that mentioned as formerly held by the Assassins.

streams flowed near the place. Mustawfî remarks of Ṭabas Masînân that the water-supply of the town lands during a drought would hold out for 70 days, while the outlying districts only had sufficient water for seven days. He relates that there was here a pit or well, at the bottom of which the earth was poisonous, so that if anyone by chance swallowed thereof even as much as a grain of millet seed, he forthwith died; hence the water from this well had been carefully closed off. There was another pit or well here which in winter swallowed up all inflowing water, and in summer gave forth continuously enough water to irrigate all the neighbouring lands; and there was also a third well, he says, where, when anyone looked down into it, the image of a fish could be seen. At the present day Ṭabas Masînân, still bearing this distinctive name, is an important place, being also known as Sunnî-khânah (the House of the Sunnîs), for it is now inhabited almost exclusively by Afghân Sunnîs. About 60 miles south of Ṭabas of the Jujube-tree, is the village of Duruh, where there is an ancient fortress on the neighbouring hill-top. Duruh is apparently not mentioned by the earlier geographers. It is first described by Mustawfî, who speaks of Ḳal'ah Duruh as being a very strong place, with a spring of water welling up within the castle precincts. Jujube-trees and corn grew abundantly in the vicinity, with grapes and other fruit in less profusion.

The products of Ḳûhistân were few in number. Muḳaddasî states briefly that these highlands were famous for their carpets and prayer rugs, also for white cloth-stuffs, similar to those that were made in Nîshâpûr[1].

What is known about the high roads crossing Ḳûhistân will be more conveniently dealt with in a later chapter in connection with the roads through Khurâsân. Muḳaddasî and other authorities mention the total distances, by the day's march, between the various towns in Ḳûhistân: but the stages in leagues are not given; and there appear to have been few direct routes crossing this mountainous province.

[1] I. H. 325. Muk. 321, 324. Yak. iii. 513, 514. Kaz. ii. 202. Mst. 184. Sykes, *Persia*, 396, 397.

CHAPTER XXVI.

ḲÛMIS, ṬABARISTÂN, AND JURJÂN.

The province of Ḳûmis. Dâmghân. Bisṭâm. Biyâr. Samnân and Khuvâr. The Khurâsân road through Ḳûmis. The province of Ṭabaristân or Mâzandarân. Âmul. Sâriyah. Mount Damâvand, with the districts of Fâdûsbân, Ḳârin, and Rûbanj. Fîrûzkûh, and other castles. Nâtil, Sâlûs, and the Rûyân district. The fortress of Ṭâḳ and the Rustamdâr district. Mamṭîr and Ṭamîsah. Kabûd Jâmah and the Bay of Nîm Murdân. The province of Gurgân or Jurjân. The river Jurjân and the river Atrak. Jurjân city, and Astarâbâd. The port of Âbaskûn. The Dihistân district, and Âkhur. The high roads through Ṭabaristân and Jurjân.

The small province of Ḳûmis stretches along the foot of the great Alburz chain of mountains which will be described below, and these heights bound it to the north, its fertile lands forming a narrow strip lying between the foot-hills and the Great Desert to the south. The Khurâsân road traverses the province from end to end, going from Ray in the Jibâl province to Nîshâpûr in Khurâsân, and the chief towns of Ḳûmis are, so to speak, strung along this line. At the present day the name Ḳûmis is become obsolete. The province is included for the most part within the limits of modern Khurâsân, while its extreme western end forms an outlying district of Ray or modern Ṭihrân[1].

The capital town of the province was Dâmghân, which the Arabs wrote Ad-Dâmghân, and which in accordance with their usage is often referred to as Ḳûmis (*sc.* Madînah Ḳûmis, 'the

[1] For the map of these provinces see p. 185, Map v. Muk. 353. Yak. iv. 203. Mst. 191. The Arab spelling was Ḳûmis (with dotted ḳ), the Persian form is Kûmis; Mustawfî, however, calls it Diyâr Ḳûmis, 'the Lands of Ḳûmis.'

City of Ḳûmis'), the capital thus taking to itself the name of the province. Dâmghân, according to Ibn Ḥawḳal, had a paucity of water-supply, and hence little cultivation, but the inhabitants manufactured excellent cloth-stuffs which were largely exported. Muḳaddasî reports Dâmghân to have fallen much to ruin at the end of the 4th (10th) century; but it was well fortified, and had three gates, of which he names two, the Bâb-ar-Ray and the Bâb Khurâsân. He says that there were two markets, the upper and the lower; and a fine Friday Mosque stood in the main street, with water tanks 'like those of Marv.' The extraordinary windiness of the town is mentioned by all the later authorities. Yâḳût and others state that there was a ceaseless wind blowing down from a neighbouring valley, so that the trees of Dâmghân were always waving about. Within the city was a great building, dating from the days of the Chosroes, which divided the waters flowing to Dâmghân into 120 channels for irrigation purposes. Excellent pears were grown in the town gardens. The walls of Dâmghân, Mustawfî reports, were 10,000 paces in circuit. Yâḳût adds that one day's journey from Dâmghân (three leagues according to Mustawfî) up in the mountains, and visible from the town, was the great castle of Gird-kûh, which had been a celebrated fortress of the Assassins. This, writes Mustawfî, was called Diz Gunbadân, 'the Domed Fort,' and its district, which was very fertile, was known as Manṣûrâbâd. Mustawfî further speaks of a gold mine in the hills near Dâmghân at Kûh Zar (Gold Mountain), but the situation of the place is not given[1].

The second town of Ḳûmis, for size, was Bisṭâm (or Basṭâm, now Busṭâm), which Ibn Ḥawḳal states to have been situated in the most fertile region of the whole province. Its gardens produced abundant fruit, and Muḳaddasî refers to its magnificent Friday Mosque, which stood 'like a fortress' in the market-place. Nâṣir-i-Khusraw, who visited the town in 438 (1046), appears to regard it as the capital of the province, for he calls it the City of Ḳûmis. He refers to the tomb here, already celebrated, of the great Ṣûfî Shaykh Abu Yazîd, more generally known as Bayazîd Bisṭâmî, who had died and was buried here in 260 (874),

[1] I. K. 23. Ḳud. 201. I. H. 271. Muk. 355, 356. Yak. ii. 539. Kaz. ii. 245. Mst. 191, 204.

and whose shrine is still at the present day greatly venerated. Yâḳût, speaking from personal experience, praises the apples of Bisṭâm, and says that on a neighbouring hill-top stood a great castle with strong walls, said to date from the days of the Chosroes, having been built by Shâpûr Dhû-l-Aktâf (Sapor II). Yâḳût also commends the markets of the city, and its general air of prosperity, and Ibn Baṭûṭah who visited it in the 8th (14th) century confirms this account, referring also to the shrine over the tomb of the Ṣûfî saint[1].

Four leagues from Bisṭâm, on the road towards Astarâbâd, was the town of Khurḳân, a place of some importance in the 7th and 8th (13th and 14th) centuries. Mustawfî refers to it as a village, with a good climate and plentiful water-supply, and it was famous for the tomb of the local saint Abu-l-Ḥasan Kharḳânî. About 50 miles south-east of Bisṭâm, and on the edge of the Great Desert, is the little town of Biyâr, 'the Wells,' which is now called Biyâr-Jumand. Muḳaddasî describes it in the 4th (10th) century as a small town with no Friday Mosque, but possessing a castle, good markets, and fertile fields, where grapes and other fruits were produced. Camels and sheep were also numerous. A small mosque for daily prayers stood in the inner castle, and the town was fortified, having three iron gates in its walls, with a single gate leading to the castle precincts. Mustawfî speaks favourably of the temperate climate and excellent corn crops. Less than half-way between Dâmghân and Ray is the city of Samnân, or Simnân, on the Khurâsân road, of which Muḳaddasî notices the fine Friday Mosque standing in the market-place, with its great water tanks. Mustawfî mentions the pistachios of Samnân as famous, and a varied abundance of fruit was grown. He also mentions Âhûvân, a small town lying between Samnân and Dâmghân, noteworthy for several tombs of holy men, and for the plentiful crops of both corn and fruit that were raised in its neighbourhood[2].

[1] I. H. 271. Muk. 356. N. K. 3. Yak. i. 623. I. B. iii. 82. The city of Shâhrûd, a couple of miles south of Bisṭâm, which is at the present time the centre of trade and population in these parts, is not mentioned by any of the Arab or Persian geographers, so that the Ṣanî'-ad-Dawlah confesses he could not discover when it was built. *Mirât-al-Buldân*, i. 210.

[2] Muk. 356, 357. Kaz. ii. 243. Yak. ii. 424. Mst. 186, 191. Khurḳân

The westernmost town of Ḳûmis, also on the Khurâsân road, and the first important place east of Ray, was Khuvâr, written Al-Khuwâr by the Arabs, which Ibn Ḥawḳal in the 4th (10th) century describes as a pleasant little town, a quarter of a mile in diameter, very populous, with streams that came down from the great Damâvand mountain flowing through its lands. Khuvâr, he adds, was the coldest place of all Ḳûmis, but its fields were very fertile. Ḳazvînî says that much cotton was grown here for export; and Mustawfî records that the place was also famous for its corn and 'Shaltûk,' or rice in the husk. To distinguish this from the town of the like name in Fârs (see p. 279) it was generally spoken of as Khuvâr of Ray, and it is thus mentioned in the campaigns of Tîmûr. Mustawfî, further, says that this Khuvâr was also known as Maḥallah-i-Bâgh—'the Garden Place'—in Persian.

Of the products of Ḳûmis, Muḳaddasî states that a peculiarly valuable kind of cotton napkin was made in this province. These famous napkins (*mandîl*) were woven large and small, plain and ribbed, with a coloured border, and of so fine a texture as to fetch 2000 dirhams (about £80) apiece. Ḳûmis also produced woollen stuffs for robes, and the head-veils called *ṭaylasân*[1].

As we have seen, the province of Ḳûmis was traversed in its length by the great Khurâsân road, and this is given in all the Itineraries, from Ibn Khurdâdbih down to Mustawfî. Leaving Ray the road goes in three marches to Khuvâr, one march beyond which was Ḳaṣr or Ḳarîyat-al-Milḥ (Salt Castle or Village), in Persian called Dih Namak, as given by Mustawfî, which is its present name. The next stage, according to all the Itineraries, was Râs-al-Kalb, 'Dog's Head,' a name not now found on the map, but the situation is that of the strange fortress-town of Lâsgird (a name wanting in all the medieval

is the pronunciation given by Ḳazvînî; the name is identical in form (without vowels) with Kharraḳân in the Jibâl province, with which it must not be confounded.

[1] I. H. 270. Muk. 367. Kaz. ii. 243. Mst. 191, 196. A. Y. ii. 212. The site of Khuvâr is occupied at present by the town of Aradûn, but the surrounding district still preserves the older name, Khuvâr, of its former chief city.

geographers) which now crowns a bluff overlooking the desert plain. Samnân is one long march beyond this, and Dâmghân (which the earlier Itineraries give as Ḳûmis) again one long march to the eastward. One march beyond Dâmghân was Al-Ḥaddâdah (the Forge), which in Mustawfî is given under the alternative name of Mihmân-Dûst ('Guest Friend'). From here it was a day's march up to Bisṭâm; or keeping the lower road the stage was at the post-house, lying two leagues from that city, which was, and is still, known as the village of Badhash, from which you enter the province of Khurâsân, going by the post-road to Nîshâpûr. Further, Muḳaddasî gives the road, in 3 days' march, from Bisṭâm to Biyâr, and from Biyâr it was 25 leagues across the desert back west to Dâmghân[1].

Ṭabaristân or Mâzandarân.

The region of high mountains,—for the most part occupied by what is, at the present day, known as the Alburz chain[2] lying along the south coast of the Caspian Sea, being to the east and

[1] I. K. 22, 23. Kud. 200, 201. I. R. 169, 170 (giving details of the country traversed). Ist. 215, 216. I. H. 274, 275. Muk. 371, 372. Mst. 196. For an illustration representing modern Lâsgird see H. W. Bellew, *From the Indus to the Tigris*, p. 404. In regard to Badhash it is curious that Yâḳût in his Dictionary gives the name once rightly spelt, and then again (but in error) under the letter *n* as *Nadhash*. Yak. i. 530; iv. 773.

[2] Alburz, now generally pronounced Elburz, is the name at the present time given to the great mountain range dividing the high plateau of Persia from the lowlands of the Caspian Sea. This name, however, appears in none of the earlier Arab geographers, who give no single appellation to the range. *Alburz* is Persian, and according to Vullers (*Lexicon Persico-Latinum*, s.v.) is derived from two Zend words signifying 'High Mountain.' Mustawfî (p. 202), who is perhaps the first authority to mention the name, used it in a very vague sense. In his chapter on the mountains of Persia he says that Alburz is a high range that runs continuous with the mountains of Bâb-al-Abwâb (i.e. the Caucasus): 'they are indeed the great mountains which are continuous, and form a chain, extending for over a thousand leagues, from Turkistân (in Central Asia) to the Ḥijâz (in Arabia), so that many consider them to be the (fabled) mountains of Ḳâf (which encircled the earth) and on the west they adjoin the mountains of Gurjistân (Georgia).' For the Alburz peak of the Caucasus see above, p. 181.

to the north of Ḳûmis,—was called Ṭabaristân by the earlier Arab geographers. *Ṭabar* has the signification of 'Mountain' in the local dialect, whence Ṭabaristân would mean 'the Mountain Land.'

In the 7th (13th) century, about the time of the Mongol conquest, the name of Ṭabaristân appears to have fallen into disuse, being replaced by Mâzandarân, which since that date has been the common appellation of this province. Sometimes also Mâzandarân was held to include the neighbouring province of Jurjân. Yâḳût, who is one of the first to mention the name Mâzandarân, writes that he does not know exactly when it came into use; and, though never found in the older books, it was in his day already generally current throughout the country. Practically the terms Ṭabaristân and Mâzandarân were then synonymous, but while the former name was applied primarily to the high mountains, and only included in a secondary use the narrow strip of lowland along the sea-shore running from the delta of the Safîd Rûd to the south-eastern angle of the Caspian, Mâzandarân appears in the first instance to have denoted these lowlands, and then included the mountain region as subsidiary thereto. The name Ṭabaristân is at the present day obsolete.

During the earlier period of the Caliphate this province was politically of little importance, and it was in fact the last portion of the Sassanian kingdom to accept Islâm. For more than a century after the Arab conquest of the rest of Persia the native rulers—called the Ispahbads of Ṭabaristân—were independent in their mountain fastnesses, and until the middle of the 2nd (8th) century their coinage continued to be struck with Pahlavi legends, and the Zoroastrian faith was dominant throughout the forests and fens of the great mountain range. In the 4th (10th) century, according to Muḳaddasî, garlic, rice, and flax, with water-fowl and fish, were the chief products of the country, which, unlike the rest of Persia, had an abundant rainfall. At a later date, according to Ḳazvînî, sericulture flourished, silk being plentifully exported. Wool-stuffs, carpets, veils, napkins, and cloth-stuffs were also largely manufactured, and various woods were cut in the forests, especially box-wood and that called Khalanj, of which arrows, bowls, and other utensils, were made. The houses in Ṭabaristân were built of wood and reeds, for, as Ibn Ḥawḳal remarks, the rains were heavy, both summer and winter. They were built with domed roofs for the like reason[1].

The capital of Ṭabaristân under the later Abbasids was Âmul, though the Ṭâhirid governor, in the 3rd (9th) century, had generally resided at Sâriyah. Âmul, according to Ibn Ḥawḳal, was in his day a larger place than Ḳazvîn and very populous. Muḳaddasî describes the town as possessing a hospital (Bîmaristân) and two Friday Mosques—one, the Old Mosque, standing among trees on the market-place, the New Mosque being near the city wall. Each mosque had a great portico. The merchants of Âmul did much trade. Rice was grown plentifully in the country round, and a large river which ran through the town was used for the irrigation of the fields. To this description Yâḳût adds no new details, but Mustawfî, remarking on the hot, damp climate, says that dates, grapes, nuts, oranges, shaddocks, and lemons grew here abundantly, and the fragrant essences made in the city were celebrated far and wide. The port of Âmul, where its river flowed out into the Caspian, was the small town of 'Ayn-al-Humm, a name which Yâḳût writes Ahlum, and describes as of no great size. Tîmûr ravaged Âmul at the close of the 8th (14th) century, destroying the three castles of Mâhânah Sar, which lay four leagues distant from the city towards the sea-coast.

The second, and the earlier, capital of Ṭabaristân was Sâriyah, now called Sârî, which lies to the eastward of Âmul. Muḳaddasî describes Sâriyah as a populous place where much cloth was manufactured, and its markets were famous. There was a small castle with a ditch, and a Friday Mosque where a fine orange-tree grew, also an immense fig-tree on the town bridge. The bridges of boats here were renowned. Of Sâriyah in later times little is reported; it suffered much in the 7th (13th) century during the Mongol invasion, and when Mustawfî wrote was almost a complete ruin, though its lands produced an abundance of grapes and corn, and silk was still manufactured from the produce of the worms reared here[2].

[1] I. H. 270, 271. Muk. 354. Kaz. ii. 270. Yak. iii. 502. For the word *Ṭabar*, see above, p. 217.

[2] I. H. 271, 272, 275. Muk. 354, 359. Yak. i. 354, 409. Mst. 109. A. Y. i. 391, 571. A. F. 437.

The great mountain of Damâvand dominates the whole of Ṭabaristân, and its snow-capped summit is visible from the plains of Persia a hundred miles and more to the south of Tihrân Mustawfî even says from a hundred leagues distant, and he notes that the peak was always covered with snow. In Persian legend Dunbâvand, as the name is written by the earlier authorities, figures as the home of the Sîmurgh, the fabulous bird which nursed and protected Zâl, the father of Rustam, and Mustawfî relates a number of romantic stories in connection with the national hero. According to Ibn Ḥawḳal the great mountain was visible from Sâvah, 'rising up like a dome in the midst of the other high mountains,' and he was of opinion that no one had ever climbed to the summit, from which, he adds, smoke was always seen to issue. Magicians much frequented it, and many legends were told of it, relating more especially how that ancient tyrant of Persia, Aḍ-Ḍuḥḥâk (Zuhâk), still lived in its recesses.

Damâvand gave its name both to a small town lying on its southern spurs, which Mustawfî writes was also called Pishyân, and to the broad fertile district spreading round its flanks. Of this district, in the 4th (10th) century, the chief town was Wîmah, which with the neighbouring town of Shalanbah, are described by Ibn Ḥawḳal as places famous for their corn lands and vineyards. Yâḳût, who had passed through Wîmah (or Waymah) and found it a ruin, states that the castle of Fîrûzkûh was visible from it. This latter castle he had also visited, and Mustawfî records that it took its water from the head of the stream that flowed out to the plain through Khuvâr of Ray in Ḳûmis. Fîrûzkûh was one of the castles of Mâzandarân which are mentioned as having been besieged and taken by Tîmûr. Another equally famous fortress on the slopes of Damâvand was the castle of Ustûnâvand, or Ustunâbâd, which, according to Ḳazvînî, had never been taken for 3000 years, till in 613 (1216) the Mongols stormed it. Yâḳût, who says it was also called Jarhud and lay 10 leagues distant from Ray, describes it as having been the stronghold of the ancient Magian ruler of the country, the Ispahbad. The last of the line, he adds, was overthrown here by Yaḥyâ the Barmecide, who carried captive the daughters of the Persian chief to Baghdâd, where one of them, called

Bahriyyah, married the Caliph Manṣûr and became the mother of Mahdî, the father of Hârûn-ar-Rashîd. At a later date this great fortress, which had been restored in 350 (961) by Fakhr-ad-Dawlah the Buyid, fell into the hands of the Assassins[1].

The medieval geographers mention the names of many fortresses and towns in Ṭabaristân which are no longer to be found on the map, having been brought to ruin either in the Mongol invasion of the 7th (13th) century, or else stormed and destroyed by Tîmûr, who ravaged Mâzandarân more than once at the close of the 8th (14th) century. Moreover, the names of most of these lost towns and fortresses not occurring in the Itineraries, it is impossible to mark their position, even approximately, on the map. Ibn Ḥawḳal in the 4th (10th) century describes three mountain districts, well wooded and very fertile, which lay south of Sâriyah, about a day's march from this town, and stretching westward towards the frontier of Daylam, in the province of Gîlân. The first of these was the Jabal Fâdûsbân, the Mountains of Bâdûsbân (in the Persian form of the word), this being the name of the ruling family, who as semi-independent chiefs held these districts for nearly 800 years, namely from the time of the Moslem conquest down to the Mongol invasion. The whole of this mountain district was covered with villages, of which the largest was named Ḳariyat Manṣûr, 'Manṣûr's Village,' and another was Uram Khâst or Uram Khâstah with an upper and a lower village, these places all lying about a day's march from Sâriyah, but throughout the mountain side there was no town of sufficient size to have a Friday Mosque.

Adjoining Fâdûsbân was the mountain district called the Jabal Ḳârin after the famous family of this name, which it is said was of Parthian origin; in any case the names of nobles of the Ḳârin occur in the history of the Sassanians, and in Moslem times they still governed this district. The great fortress stronghold of the Ḳârins, which they had held since Sassanian times, was at Firrim, and the chief centre of population was at the town of Sihmâr (or Shihmâr) where there was the only Friday Mosque

[1] Ist. 202. I. H. 265, 270, 271. Muk. 392. Kaz. ii. 195. Yak. i. 243, 244; iii. 930; iv. 944. Mst. 191, 203, 204. A. Y. ii. 577. Fîrûzkûh still exists, but the site of Ustûnâvand appears to be unknown.

of all this region. The position of Firrim, unfortunately, is not exactly given in any of the Itineraries. It is mentioned by Yâḳût, and also in the 8th (14th) century by Mustawfî, who speaks of it as lying on the borders of Ḳûmis. The third mountain region was the Jabal-ar-Rûbanj, lying north of Ray, and therefore nearer to the Daylam frontier. Of this no towns or villages are mentioned, but it is said to have been extremely fertile and well watered, the mountain slopes being covered with trees and thickets[1].

One day's march, or five leagues, to the west of Âmul, in the plain near the coast, was the town of Nâtil or Nâtilah, and a like distance further to the west of this was Sâlûs, or Shâlûs, which Muḳaddasî describes as a city having a castle built of stone, with a Friday Mosque adjoining. The name was also spelt Sâlûsh, and near it lay two other towns, namely Al-Kabîrah and Kajjah. In the accounts of the campaigns of Tîmûr Shâlûs is written Jâlûs, and all this country appears to have been permanently ruined during his wars, together with the mountainous region to the south, namely Rûyân and Rustamdâr[2].

The city of Kalâr, which Yâḳût seems to think was identical with the above-mentioned Kajjah, was one march from Shâlûs, but in the mountains—and from Kalâr it was one march on to the Daylam frontier. There is some confusion in the names, but Kalâr, Kajjah, and Rûyân appear all to refer to neighbouring towns, if not to one and the same town, and Rûyân further was the name of one of the great districts in the mountains on the

[1] Ist. 205, 206. I. H. 268, 269. Yak. i. 212; iii. 324, 890. Mst. 191. For Fâdûsbân the reading *Ḳâdûsiyân* has been wrongly printed in the texts of Iṣṭakhrî and other geographers by a shifting of the diacritical points, and hence these people have often been supposed to represent the ancient Cadusii of Strabo; see Nöldeke, *Geschichte der Perser und Araber zur Zeit der Sassaniden*, p. 151, note 2, who explains that under the Sassanians the Bâdûsbân were the civil governors of the district, as against the Ispahbads, who were the military commanders of this, the frontier province. See also Justi, *Iranisches Namenbuch*, p. 156, s.v. 'Karen,' and p. 245, s.v. 'Patkospan.' For the list of the Bâdûsbân chiefs in Moslem times see G. Melgunof, *Das südliche Ufer des Kaspischen Meeres*, p. 50, and for the Ḳârin chiefs, *idem*, p. 52.

[2] I. H. 275. Muk. 359. I. F. 305. Yak. iii. 13, 237, 504; iv. 726. A. Y. i. 391. Shâlûs is said to be only eight leagues from Ray, but this must be a mistake if it lay on or near the shore of the Caspian.

western border of Ṭabaristân. Abu-l-Fidâ says that the city of Rûyân was also known as Shâristân, and that it crowned the summit of the pass 16 leagues from Ḳazvîn. According to Yâḳût Rûyân was the capital city of the mountain district of Ṭabaristân, just as Âmul was of the lowland plains; it had fine buildings and its gardens were famous for their productiveness. Near Rûyân (or Kalâr) was the little town of Sa'îdâbâd.

The great fortress of Ṭâḳ (the Arch) on the frontier of Daylam, and the last refuge of the Ispahbad prince of Ṭabaristân who was conquered in the time of the Caliph Manṣûr, must have been situated in this district of Rûyân. The place is described at some length both by Yâḳût and Ḳazvînî, who quote older writers. Ṭâḳ was deemed an impregnable stronghold, and had existed since the days of the Sassanian kings of Persia. It was situated high up in the mountains, and was only reached by a tunnel a mile long (it is said) which had been pierced through the encircling cliffs. The tunnel led to an open valley surrounded by precipices in which were many caverns, and from one of these a powerful spring gushed out, and after flowing a short distance disappeared into the depths of a neighbouring cave. Yâḳût adds a long account of the wonders of this place.

At the head-waters of the great Shâh Rûd—the eastern affluent of the Safîd Rûd (see above, p. 170)—lay the district of Rustamdâr, which Mustawfî describes as comprising near 300 villages, and this country, which was watered by the numerous tributaries of the Shâh Rûd, thus lay between Ḳazvîn and Âmul, and to the eastward of the Rûyân district. On the Shâh Rûd, as already described in Chapter XV, p. 221, were the chief castles of the Ismailians or Assassins, and probably in this Rustamdâr district also was Kalâm, described by Yâḳût as an ancient fortress of Ṭabaristân, which had been in the hands of these sectaries, and was destroyed by Sultan Muḥammad, son of Malik Shâh the Saljûḳ[1].

Two leagues to the eastward of Âmul, and on the coast road, lay the town of Mîlah, and three leagues beyond this Barjî, which was one march from Sâriyah. The city of Mamṭîr or Mâmaṭîr,

[1] I. H. 275. Yak. ii. 873; iii. 93, 490, 504; iv. 240, 296, 297. Kaz. ii. 238. A. F. 435. Mst. 190.

one march from either Âmul or Sâriyah, and six leagues from the coast, is identical with the later Bârfarûsh. It had a Friday Mosque, Yâķût says, and much fertile land lay adjacent to the city. Near Sâriyah, and probably to the eastward, were the towns called Nâmiyah (or Nâmishah), with a fine district, 20 leagues from Sâriyah, and Mihrawân, 10 leagues from Sâriyah, where there was a Friday Mosque and a garrison of 1000 men, but unfortunately the exact position of these two places is quite uncertain. On the eastern frontier of Ṭabaristân, and three marches from Sâriyah, on the road to Astarâbâd, from which it was one march distant, lay the town of Ṭamîs, or Ṭamîsah, standing on the great causeway across the marshes which, according to Yâķût, had been built to carry the high road by King Anûshirwân the Just[1].

At the south-east angle of the Caspian is the Bay of Ashurâdah, as it is now named, where a long spit of sand stretches out eastward till it almost reaches the Jurjân coast. This bay with its island or peninsula is described by Mustawfî under the name of Nîm Murdân. The settlement here was very populous in the 8th (14th) century, and was a harbour for ships from all parts of the Caspian. The port was but three leagues distant from Astarâbâd, and the town behind it which carried on a brisk trade was called Shahrâbâd. The neighbouring district, which produced a great deal of silk, and where corn lands and vineyards abounded, was known as Kabûd Jâmah. It had been a very rich country, but was entirely ruined by the wars of Tîmûr at the close of the 8th (14th) century. The city of Rû'ad, or Rûghad, which is also mentioned as passed by Tîmûr on his march into Mâzandarân, was probably of the Kabûd Jâmah district. It was, says Mustawfî, a fair-sized town, being 4000 paces in circuit, and it stood in the midst of many fertile lands, where much corn and cotton, besides various fruits, were grown in abundance.

Of the products of Ṭabaristân, besides the commodities already

[1] I. H. 275. Yak. iii. 503, 504, 547; iv. 398, 642, 699, 733. The earliest mention of Bârfarûsh, under the form Bârah Farûsh Dih ('the Village where Loads are Sold'), occurs in *Haft Iķlîm* of Aḥmad Râzî, a work of the 10th (16th) century; see Dorn, *Muhammedanische Quellen*, iv. p. 99 of the Persian text.

referred to on page 369, Muḳaddasî mentions fine cloth for robes, and stuffs for the *ṭaylasân* veils, also coarse linen cloths that were woven largely for export. Of natural products the Khalanj wood already named was cut and sent away in the rough to be made into bowls and other utensils by the craftsmen in Ray. The Khalanj is described as a tree that produced a variegated and sweet-smelling wood, of which the beads of chaplets were sometimes made, and the best kind grew only on the Ṭabaristân mountains[1].

Jurjân.

The province of Jurjân, or Gurgân, as the Persians pronounced the name, lying at the south-eastern corner of the Caspian, consisted for the most part of the broad plains and valleys watered by the two rivers Jurjân and Atrak. In earlier days it was always held to be a province by itself, though dependent on Khurâsân, but after the changes brought about by the Mongol conquest, it was annexed politically to Mâzandarân. Like other districts near the southern shore of the Caspian it was overrun and devastated by the Mongol hordes in the 7th (13th) century, and then again by Tîmûr at the close of the 8th (14th) century.

Jurjân, as Muḳaddasî writes, being rich in streams, its plains and hills were covered with orchards producing dates, oranges, and grapes in abundance. The most important river of the province was that generally called by its name, the Jurjân river, which Muḳaddasî in the 4th (10th) century states was then known as the river Ṭayfûrî. The river Atrak he does not name. In the 8th (14th) century Mustawfî gives the name as the Âb-i-Jurjân, and says that the Jurjân river rose in the valley of Shahr-i-Naw (New Town), whence, passing through the plain of Sulṭân Darîn, it reached the city of Jurjân, past which it flowed, and thence entered the Caspian, near the island of Abaskûn in the bay of Nîm Murdân. Throughout its course the stream was deep, almost

[1] Muk. 367. Mst. 190, 191. J. N. 339, 341. A. Y. i. 349. The forms of Ashurâdah Bay and of the peninsula have of course changed greatly since the 14th century, when Mustawfî wrote, and the exact sites of the town and port are unknown.

unfordable, so that travellers were often drowned in crossing it; and in flood-time its waters were carried off by channels and used up in irrigation, though much always ran to waste.

The river Atrak is a longer stream than the Jurjân, and rises in the plains of Khurâsân, between Nisâ and Khabûshân, near the sources of the Mashhad river, which latter flows off south-east, and in the opposite direction. The Atrak is very deep and like the Jurjân mostly unfordable, as Mustawfî writes, and flowing along by the Dihistân frontier, on the northern side of the Jurjân province, reaches the Caspian after a course of nearly 120 leagues. The name Atrak is said to be merely a plural form of the word *Turk*, and the River of the Turks was so called from those who once lived on its banks. No name, however, appears to be given to this stream by any of the earlier Arab geographers, and Mustawfî in the 8th (14th) century is one of the first to call it the Atrak, by which appellation it is still known[1].

The capital of Jurjân is the city of the same name, at the present day called Min Gurgân, which Ibn Ḥawḳal in the 4th (10th) century describes as a fine town, built of clay bricks, enjoying a far drier climate than Âmul, for less rain fell in Jurjân than in Ṭabaristân. The city consisted of two parts, one on either side of the Jurjân river, which was here traversed by a bridge of boats, and Jurjân was more properly the name of the eastern half of the town. On the west side lay Bakrâbâd, the suburb, and the two parts of the city together, according to the description of Ibn Ḥawḳal, who had been here, were nearly as large as Ray. The fruit from the gardens round was abundant, and silk was produced in great quantities. The main quarter of Jurjân, that on the east bank, Muḳaddasî calls Shahrastân; it had fine mosques and markets, where the pomegranates, olives, water-melons, and egg-plants, with oranges, lemons, and grapes of the neighbouring gardens were sold cheaply, and were all of superexcellent flavour. The town was intersected by canals, crossed by arched bridges or by planks laid on boats. A Maydân, or public square, faced the

[1] Muk. 354, 367. Mst. 212, 213. J. N. 341. Hfz. 32*a*. The name Atrak is written (and pronounced) with the second vowel short, while the plural of *Turk* is Atrâk; hence the usual explanation of the name is probably erroneous.

governor's palace, and this quarter of the town had nine gates. The defect of Jurjân was the great heat of its climate, and the flies were numerous, as well as other insects, especially bugs of a size so large as commonly to be known as 'the wolves' (Gurgân). Bakrâbâdh, as Muḳaddasî spells the name, was also a populous city with its own mosques, and the buildings extended back for a considerable distance from the river, and for some distance along its western bank.

When Ḳazvînî wrote in the 7th (13th) century Jurjân was famous among the Shî'ahs for the shrine called Gûr-i-Surkh, 'the Red Tomb,' said to be that of one of the descendants of 'Alî, whom Mustawfî identifies as Muḥammad, son of Ja'far-aṣ-Ṣâdiḳ, the sixth Imâm. Mustawfî reports that the city had been rebuilt by the grandson of Malik Shâh the Saljûḳ, and that its walls were 7000 paces in circuit. In the 8th (14th) century, when he wrote, the town lay for the most part in ruins, never having recovered the ravages of the Mongol invasion. He praises, however, the magnificent fruit grown here, and besides those kinds mentioned above names the jujube-tree as bearing freely here, so that trees which were only two or three years old gave good fruit, twice in each season. The population were all Shî'ahs in his time, but they were not numerous. In the year 795 (1393) Tîmûr, who had devastated all Mâzandarân and the neighbouring country, stopped at Jurjân and built for himself here on the banks of the river the great palace of Shâsman, which is especially referred to by Ḥâfiẓ Abrû[1].

The second city of the Jurjân province is Astarâbâd, near the frontier of Mâzandarân. Muḳaddasî describes it as a fine town in the 4th (10th) century, with the best climate of all the region round. Raw silk was its chief product, and in his day the fortress was already in ruin, for the Buyids had ravaged all this country during their wars against the Ziyârids; and Muḳaddasî adds that

[1] I. H. 272, 273. Muk. 357, 358. Kaz. ii. 235. Mst. 190. A. Y. i. 578. Hfz. 32*a*. During the 4th (10th) century Jurjân was governed by a native dynasty, the Ziyârids, whose rule extended over Ṭabaristân and the neighbouring lands. Of these Ziyârids one of the most famous was Ḳâbûs, who died in 403 (1012) and whose tomb, called the Gunbad-i-Ḳâbûs, is still to be seen near the ruins of Jurjân city. C. E. Yate, *Khurasan and Sistan*, pp. 239—241.

there was the Friday Mosque built at the time of the first Moslem conquest still standing in the market-place near the city gate. Yâḳût and Mustawfî merely confirm the above account, praising the climate of Astarâbâd and the abundant supplies, but adding no fresh details. The port on the Caspian of both Jurjân and Astarâbâd was at Âbaskûn, given as one day's march distant from either city, but the site would appear to have been engulfed in the sea during the 7th (13th) century, following on the events of the Mongol invasion. Iṣṭakhrî and Ibn Ḥawḳal, writing in the 4th (10th) century, describe Âbaskûn as a considerable market for the silk trade, being the border station at that time against the Turks and Ghuzz, and the chief port for the coasting trade of the Caspian, sailing towards Gîlân. It was protected by a strong castle built of burnt brick, and the Friday Mosque was in its market-place. Muḳaddasî writes of it as 'the great harbour of Jurjân,' and the Caspian itself, Yâḳût adds, was often called the Sea of Âbaskûn. In history Âbaskûn is celebrated as having been the final refuge of Muḥammad, the last reigning Khwârizm-Shâh, who, fleeing before the Mongol hordes, died here miserably in 617 (1220)[1].

Six days' journey (or 50 leagues) north of Âbaskûn, and four marches from Jurjân city, was the settlement of Dihistân in the district of the same name, the outpost in the 4th (10th) century of the Turk frontier. Ibn Ḥawḳal speaks of Dihistân as lying near the Caspian shore. The only settlements were small villages, with some gardens, but only a sparse population. Adjacent was a shallow bay of the Caspian where boats anchored and much fishing was carried on by the coast people. The chief settlement was called Âkhur, which Muḳaddasî refers to as a city, surrounded by twenty-four villages, 'and these are the most populous of all the Jurjân province.' In Âkhur was a minaret, or tower, which could be seen from a great distance away in the neighbouring desert.

To the eastward of Âkhur was Ar-Rubâṭ, 'the Guard-house, an important settlement at the entrance of the desert route going

[1] Ist. 213, 214. I. H. 273, 274. Muk. 358. Yak. i. 55, 242. Mst. 190, 225. Ibn Serapion (folio 46*b*) states that the town of Âbaskûn lay on the Jurjân river, near where it flowed out into the Caspian. *Mas. Tanbih* 60, 179.

to Khwârizm. Muḳaddasî speaks of it as having three gates, and though in his time it was for the most part in ruin, it was still populous, with good markets and a few well-built houses, and fine mosques. Of these last, the Old Mosque had been built on wooden pillars, Muḳaddasî says, and it was in his day half underground. Another of the mosques had a beautiful minaret. Yâḳût mentions these and some other places in the Dihistân district, namely the villages of Khartîr, Farghûl, and Habrâthân, but he adds no details. Mustawfî, who gives the route from Jurjân to Khwârizm across Dihistân, describes this as the frontier between the Moslems and the heathen Turks and Kurds. The district had a warm climate and a stream watered its fields, but there was little fruit grown here[1].

Four stages from Dihistân on the desert border, where the road started for crossing to Khwârizm, stood the city of Farâvah, which is given by Iṣṭakhrî as a settlement of the Ghuzz Desert. In the 4th (10th) century it was strongly garrisoned by volunteers, and there was a great Rubâṭ, or guard-house, to protect the country lying at the back of it against the Turkish inroads. Its gardens and fields were small in extent and the town or settlement numbered barely a thousand families. Muḳaddasî spells the name Afrâvah, and Yâḳût says that it was a Rubâṭ built by 'Abd Allah, the Ṭâhirid, during the reign of the Caliph Mamûn. From its position there is little doubt that Farâvah is identical with the modern Ḳizil Arvat, a corruption of Ḳizil Rubâṭ, 'the Red Guard-house.' The names only of a number of other places in the Jurjân province are given by Yâḳût, these being the various villages belonging to Jurjân city, or to Astarâbâd. No details, however, are added, their positions are not indicated, and too often the reading of the name is uncertain[2].

Muḳaddasî mentions, among the products for which Jurjân

[1] The ruins of these towns lying on the border of the Khwârizm desert are still to be seen at Misriyân, near the mountains now called the Kören Dâgh, but all cultivation has long ceased in this district, which is now a waterless desert. I. H. 277, 286. Muk. 358, 359. Yak. i. 59, 500; ii. 418, 633; iii. 880; iv. 949. Mst. 190, 197.

[2] Ist. 273. I. H. 324. Muk. 333. Yak. iii. 866. Mst. 197. For these villages see for instance sixteen names given by Yâḳût. Yak. ii. 137, 489, 782; iii. 323, 923, 930; iv. 277, 376, 395, 396, 555, 699, 728, 926, 927.

was famous, a particular kind of face-veil woven of raw silk, which was in his day largely exported to Yaman in southern Arabia. An inferior kind of brocade (*dîbâj*) was also largely manufactured, and of fruits Jurjân was especially famous for its grapes, figs, and olives[1].

The high roads through Ṭabaristân and Jurjân are not numerous, since in the first-named country the mountains are for roads almost impassable. Iṣṭakhrî (duplicated by Ibn Ḥawḳal) and Muḳaddasî give the road from Ray northwards across the great chain to Âmul, passing through Ask and Bulûr (Pulûr), but many of the stages are now difficult or impossible to identify. Travelling westward from Âmul along the coast, Ibn Ḥawḳal and Iṣṭakhrî give the marches through Nâtil and Sâlûs to the frontier of Gîlân (Daylam); also eastward from Âmul to Astarâbâd and Jurjân city. From Jurjân city north to Dihistân the stations are given by Muḳaddasî, as also by Mustawfî in his account of the road from Bisṭâm in Ḳûmis to the capital of Khwârizm. Muḳaddasî also gives the road from Bisṭâm to Jurjân city across the mountain pass, through Juhaynah, which is described by Ibn Ḥawḳal as a fine village on a river. Lastly from Jurjân eastward into Khurâsân Muḳaddasî gives a route in 5 days to Isfarâyin in the Juvayn plain, passing through Ajgh, which is now called Ashk. This district will be described in the following chapter[2].

1 Muk. 367.

2 Ist. 214—217. I. H. 274—276. Muk. 372, 373. Mst. 197.

CHAPTER XXVII.

KHURÂSÂN.

The four Quarters of Khurâsân. The Nîshâpûr quarter. Nîshâpûr city, and Shâdyâkh. The Nîshâpûr district. Ṭûs and Mashhad, with its shrine. Bayhaḳ and Sabzivâr. Juvayn, Jâjarm, and Isfarâyin. Ustuvâ and Kûchân. Râdkân, Nisâ, and Abîvard. Kalât. Khâbarân and Sarakhs.

In old Persian Khurâsân means 'the Eastern Land,' and in the earlier middle-ages the name was applied, generally, so as to include all the Moslem provinces east of the Great Desert, as far as the frontier of the Indian mountains. Khurâsân, therefore, was taken in this larger sense to include all Transoxiana on the north-east, besides Sîjistân with Ḳûhistân on the south, and its outer boundaries were the Chinese desert and the Pamîr towards Central Asia, with the Hindû Kush ranges towards India. Later, however, these limits became more circumscribed, and Khurâsân as a province of medieval Persia may conveniently be held to have extended only as far as the Oxus on the north-east, but it still included all the highlands beyond Herât, in what is now the north-western part of Afghânistân. Further, the country of the upper Oxus, towards the Pamîr, as known to the medieval Arabs, was always counted as one of the outlying districts of Khurâsân.

Arab or medieval Khurâsân is conveniently divided into four Quarters (Rub'), named from the four great cities which at various times were, separately or conjointly, the capitals of the province, to wit Naysâbûr, Marv, Herât, and Balkh. After the first Moslem conquest the capitals of Khurâsân had been at Marv and at Balkh. The princes of the Ṭâhirid dynasty, however, shifted the centre of government westward, and under their sway Naysâbûr became the capital city of the province, being also the chief town of the westernmost of the four Quarters[1].

In modern Persian the name is pronounced Nîshâpûr, the Arab form being Naysâbûr, which is from the old Persian Nîv-Shahpuhr, meaning 'the good (thing, deed, or place) of Shâpûr,' and the city is so called after the Sassanian king Shâpûr II, who had rebuilt it in the 4th century A.D., for Naysâbûr owed its foundation to Shâpûr I, son of Ardashîr Bâbgân. Of the chief towns of the Naysâbûr district, in which was included most of the province of Ḳûhistân already described, long lists are given by the Arab geographers of the 3rd (9th) century, but these are chiefly interesting for the archaic spelling of some of the names, and many places named cannot now be identified[2].

In early Moslem days Naysâbûr was also known as Abrashahr, meaning 'Cloud-city' in Persian, and as such appears as a mint city on the early dirhams of both the Omayyad and Abbasid Caliphs. The name Îrân-shahr—the City of Îrân—is also given to it by Muḳaddasî and others, but probably this was merely used officially and as a title of honour. In the 4th (10th) century Naysâbûr was already a most populous place, measuring from half a league to a league across every way, and consisting of the citadel or fortress, the city proper, and an outer suburb. The chief Friday Mosque stood in the suburb; it had been built by 'Amr the Saffârid, and faced the public square called Al-Mu'askar, 'the Review Ground.' Adjacent thereto was the palace of the governor, which opened on another square called the Maydân-al-Ḥusaynîyîn, and not far from this was the prison—all three buildings standing within a quarter of a league one of the other.

The fortress had two gates, the city four. These last were named Bâb-al-Ḳanṭarah (the Bridge Gate), next the gate of the street of Ma'ḳil, then Bâb-al-Ḳuhandiz (the Fortress Gate), and lastly the gate of the Takîn bridge. The suburbs lying beyond

[1] Ist. 253, 254. I. H. 308, 309, 310. Muk. 295. Mst. 185.

[2] Ist. 258. I. H. 313. I. K. 24. Ykb. 278. I. R. 171. The first syllable of the name Nîshâpûr in old Persian was *Nîv*, or *Nîk*, which in modern Persian exists in *Nîkû*, 'good'; the Arab diphthong Nay(sâbûr) changes in modern Persian to the long vowel, becoming Nîshâpûr, for the Arab *b* is in Persian pronounced *p*. Nöldeke, *Sassaniden*, p. 59.

and round both fortress and city, where the great markets were situated, had many gates. Of these the chief were the gate of the domes (Bâb-al-Ḳubâb), opening west, and on the opposite quarter the war gate (Bâb Jang) towards the Bushtafrûsh district; then to the south was the Bâb Aḥwaṣâbâd, and the names of some others are also given. The most famous market-places were those known as Al-Murabba'ah-al-Kabîrah, and Al-Murabba'ah-aṣ-Ṣaghîrah ('the great quadrangle' and 'the little quadrangle'), of which the great quadrangle was near the Friday Mosque, already mentioned. The little quadrangle was at some distance from the other, in the western part of the suburbs, near the Maydân-al-Ḥusaynîyîn and the governor's palace. A long line of streets flanked by shops went from one quadrangle to the other; and a like street of shops crossed this at right-angles, near the great quadrangle, going south as far as the graveyard known as the Maḳâbir-al-Ḥusaynîyîn, and extending north to the head of the bridge over the river.

In these market streets were many hostels for the merchants, and every sort of merchandise might be found each in its separate mart, while cobblers, clothiers, bootmakers, and men of every trade were abundantly represented. Every house in the city had its own separate underground water channel, the supply coming from the stream of the Wâdî Saghâvar, which flowed down through Naysâbûr from the neighbouring village of Bushtanḳân. These water channels, which were under the inspection of a special officer within the city, often ran as much as a hundred steps below the ground level. Beyond the city the channels reached the surface, and were here used for the irrigation of the garden lands.

No town in all Khurâsân, says Ibn Ḥawḳal, was healthier or more populous than Naysâbûr, being famous for its rich merchants, and the store of merchandise coming in daily by caravan. Cotton and raw silk were its chief exports, and all kinds of stuff goods were manufactured here. Muḳaddasî fully bears out this account, adding some further details. He says that there were forty-two town quarters in Naysâbûr, some of which were of the size of half the city of Shîrâz. The main streets (*darb*) leading to the gates were nearly fifty in number. The great Friday Mosque, which was built in four wards, dated, as already said, from the days of 'Amr the Ṣaffârid. Its roof was supported on columns of burnt brick, and three arcades went round the great court. The main building was ornamented with golden tiles, there were eleven gates to the mosque, each flanked by marble columns, and both the roof and walls were profusely ornamented. The river of Naysâbûr, as noted above, came from the village of Bushtanḳân; it turned seventy mills, and from it the numerous underground watercourses were led off, for the river itself flowed past the place at a distance of a league. Within the city and among the houses there were many wells of sweet water[1].

Yâḳût says that in his day, namely the 7th (13th) century, the name of the city was commonly pronounced Nashâvûr. He declares that in spite of the ruin which had been the result of the great earthquakes in the year 540 (1145), followed by the sack of the place at the hands of the Ghuzz hordes in 548 (1153), he had seen no finer city in all Khurâsân, and its gardens were famous for their white currants (*rîbâs*) and for other fruits. After this Ghuzz inroad, when Sultan Sanjar the Saljûḳ was carried away prisoner, and the city devastated, the inhabitants for the most part removed to the neighbouring suburb of Shâd-yâkh, which was then rebuilt, being surrounded with a wall and enlarged by Al-Mu'ayyad, the governor, who acted in the name of the captive Sultan Sanjar. This suburb of Shâdyâkh, or Ash-Shâdhyâkh, had formerly been a garden, occupied by 'Abd Allah the Ṭâhirid in the early part of the 3rd (9th) century, when he made Naysâbûr the seat of his government. Round his palace, what had been originally the camp of his troops became the chief suburb of Naysâbûr, which, after the Ghuzz invasion, took the place of the capital. Yâḳût, who spent some time at Nîshâpûr about the year 613 (1216), lodged in Shâdyâkh, which he describes. Shortly after this, namely in 618 (1221), the capital was taken and sacked by the Mongols under Changîz Khân, as Yâḳût himself heard and reports, he having by this time sought safety in Mosul. According to his information the Mongols left not one stone standing upon another.

Nîshâpûr, however, must have quickly recovered from the

[1] Ist. 254, 255. I. H. 310—312. Muk. 314—316, 329.

effects of the Mongol invasion, for when Ibn Baṭûṭah was here in the 8th (14th) century it was again a populous city, with a fine mosque encircled by four colleges, while the plain round the city was 'a little Damascus' for fertility, for it was watered by four streams coming from the neighbouring hills. They manufactured here, Ibn Baṭûṭah adds, silk velvets called *kamkhâ* and *nakhkh*, and the markets were much frequented by foreign merchants. Mustawfî, his contemporary, gives a long account of the city of Nîshâpûr and of its district. He says that in the days of the Chosroes, as it was reported, the old town of Naysâbûr had been originally laid out on the plan of a chess-board, with eight squares to each side. Then under the Ṣaffârids Nîshâpûr had increased in size and wealth, becoming the chief city of Khurâsân, till the year 605 (1208), when it was almost completely destroyed by earthquakes. It was after this date, according to Mustawfî, that Shâdyâkh first took its place as the centre of population, this latter city having a wall 6700 paces in circuit. Nîshâpûr, however, was forthwith rebuilt, but again destroyed by the earthquakes in the year 679 (1280), when a third city of Nîshâpûr was refounded on a different site, and this was the place which Mustawfî describes. Its walls then measured 15,000 paces in circuit, and it stood at the foot of the hills, facing south. The water-supply was plentiful, for the Nîshâpûr river, which rose in the mountains two leagues or more to the eastward, had a sufficient current to turn 40 mills before it came to the town. He relates, further, that most of the houses in Nîshâpûr had cisterns for storing water in the dry season.

The present city of Nîshâpûr lies on the eastern side of a semicircular plain, surrounded by mountains, and facing the desert, which is to the south. This plain is watered by many streams coming down from the hills to the north and east, and Mustawfî gives the names of a great number of these, which, after irrigating the lands round Nîshâpûr, become lost in the desert. Five leagues north of the city, at the head-waters of the Nîshâpûr river, was a little lake in the mountains at the top of the pass, called Chashmah Sabz, 'the Green Spring,' from which, according to Mustawfî, two streams running west and east took their rise, the eastern stream flowing down to the valley of Mashhad. This lake appears to have been in the hill called Kûh Gulshân, where there was a wonderful Cavern of the Winds, and from its depths a draught of air and a current of water perpetually issued, the latter sufficiently strong to turn a mill. The lake of Chashmah Sabz is described as a league in circuit, and many wonders were related of it, for it was reported to be unfathomable, and an arrow could not be shot from one bank to the other.

Four districts of the Naysâbûr plain were famous for their fertility, and Muḳaddasî in the 4th (10th) century enumerates these, namely, Ash-Shâmât ('the Beauty Spots'), Rîvand, which still exists to the west of Nîshâpûr, Mâzûl, and Bushtafrûsh. The district of Mâzûl lay to the north, and its chief village was Bushtaḳân (or Bushtanḳân), a league from the city, where 'Amr the Ṣaffârid had planted a famous garden. The currants of this district were especially renowned. The Bushtafrûsh district, now known as Pusht Farûsh, extended for a day's journey eastwards from the Jang Gate of Naysâbûr, according to Muḳaddasî, and from the gardens of its villages, which Yâḳût says numbered 126 in all, apricots were exported in immense quantities. The Shâmât district, Muḳaddasî says, was named Tak-Âb by the Persians, meaning 'whence waters flow,' and its fertility was extraordinary. Rîvand, a small town in the district of the same name, lay one stage west of Naysâbûr; in the 4th (10th) century the town had a Friday Mosque built of burnt brick, and it stood on its own river. Its vineyards were famous and its quinces were in great demand.

One of the main streams of the Nîshâpûr district, according to Mustawfî, was the Shûrah Rûd, 'the Salt River,' which was joined by the waters of the stream from Dizbâd, and after watering many districts ultimately became lost in the desert. A number of other streams are also mentioned by Mustawfî, but many of their names are misspelt and they are now difficult to identify. Some, however, present no difficulty, as for instance the river of Bushtaḳân, rising in the Chashmah Sabz neighbourhood, already mentioned, and the Bushtafrûsh river, both of which in the spring freshets, he says, joined the Shûrah Rûd. Finally, there was the stream named the 'Atshâbâd, or 'Thirst' river, which, though in spring-time it had water enough to turn 20 mills throughout its course of a score of leagues, at other seasons did not give enough to quench a man's thirst, from which cause came its ill-omened name[1].

To the south-east of Nîshâpûr the great Khurâsân high road bifurcates at the stage which the Arabs named Ḳaṣr-ar-Rîḥ, 'Castle of the Wind,' and the Persians Dizbâd or Dih Bâd. Its river has been already mentioned among the streams which flowed to the Shûrah river. From here the road to Marv went due east, that to Herât turning off south-east. On this last, two stages from Dih Bâd, was the village of Farhâdân, which is also called Farhâdhjird by Yâḳût. Its district, which was counted as of Naysâbûr, Muḳaddasî calls Asfand; in Ibn Rustah the spelling given is Ashbandh, and Yâḳût writes Ashfand, adding that this district comprised 83 villages. The old name of the district appears now to be lost, but the village called Farajird (for the older Farhâdhjird) is still marked on the maps at the place indicated by the Itineraries[2].

Due east of Nîshâpûr, but separated from it by the range of mountains in which most of the streams of the Nîshâpûr plain take their rise, lies Mashhad—'the Place of Martyrdom,' or 'Shrine' of the Imâm—now the capital of the Persian province of Khurâsân, and a few miles to the north of it may be seen the ruins of Ṭûs, the older city. Ṭûs, in the 4th (10th) century, was the second city of the Naysâbûr quarter of Khurâsân, and consisted of the twin towns of Aṭ-Ṭâbarân and Nûḳân, while two post-stages distant was the great garden at the village of Sanâbâdh, where lay the graves of the Caliph Hârûn-ar-Rashîd, who died in 193 (809), and of the eighth Imâm 'Alî-ar-Riḍâ, who was poisoned by Mamûn in 202 (817). This village of Sanâbâdh was also known as Barda', meaning 'a pack-saddle,' or as Al-Muthaḳḳab, 'the Pierced[3],' presumably from the windows of the shrine, or for some other fanciful reason.

[1] I. R. 171. Muk. 300, 316, 317. Yak. i. 630; iii. 228—231; iv. 391, 857, 858. I. B. iii. 80, 81. Mst. 185, 206, 219, 220, 226. J. N. 328. For the Chashmah Sabz lake and the Cave of the Winds, see C. E. Yate, *Khurasan and Sistan*, pp. 351, 353. Both places are still famous in Khurâsân.

[2] I. R. 171. Muk. 300, 319. Yak. i. 280; iii. 887. Mst. 196, 197.

[3] Al-Muthaḳḳab was a name given to various fortresses; one near Al-

In the 3rd (9th) century, according to Ya'ḳûbî, Nûḳân was the greater of the two halves of Ṭûs, but in the following century Ṭâbarân had outgrown it, and was the larger city down to the time of Yâḳût, when Ṭûs was ruined by the Mongol hordes. In early days Nûḳân was celebrated for its stone jars made of serpentine (Barâm), which were largely exported; and there were mines for gold and silver, copper and iron, which were profitably worked in the neighbouring hills. Turquoises, and the stone known as 'santalum' (*khumâhan*), also malachite (*dahnaj*), were all found in the neighbourhood of Ṭûs, and brought for sale to the markets of Nûḳân. This part of Ṭûs, however, was rather deficient in its water-supply. The fortress of the adjacent quarter of Ṭâbarân was a huge building, 'visible afar off,' as Muḳaddasî writes, and the markets of this half of the town were well supplied. Its Friday Mosque was beautifully built and finely ornamented. The neighbouring tombs at Sanâbâdh were already in the 4th (10th) century surrounded by a strongly fortified wall, and the shrine, as Ibn Ḥawḳal reports, was constantly thronged by devotees. A mosque had been built near the tomb of the Imâm Riḍâ by the Amîr Fâiḳ 'Amîd-ad-Dawlah, than which, says Muḳaddasî, 'there is none finer in all Khurâsân.' The grave of Hârûn-ar-Rashîd had been made by the side of that of the Imâm, and many houses and a market had been built in the vicinity of the great garden.

The description given by Yâḳût adds little to the above, but he mentions, as one of the most famous tombs at Ṭâbarân, the shrine of the great Sunnî theologian, the Imâm Ghazzâlî, who had died in 505 (1111), after having served some years at Baghdâd as chief of the Niẓâmîyah college. When Yâḳût wrote, in the 7th (13th) century, the name Ṭûs was more generally used to denote the surrounding district, where there were, he says, over a thousand flourishing villages. In 617 (1220), however, all this country,

Maṣṣîṣah (Mopsuestia) has been mentioned in Chapter IX, p. 130. The origin of the name Barda' is not explained. Nûḳân, pronounced Nûgân, is still the name of the north-east quarter and gate of modern Mashhad, leading out doubtless towards Nûḳân of Ṭûs, and the Sanâbâd watercourse at the present day supplies the north-west quarter of Mashhad. I. R. 172. I. K. 24. Yak. iv. 414. C. E. Yate, *Khurasan and Sistan*, 316, 317.

including the two cities of Ṭûs, with the shrines at Sanâbâdh (Mashhad), was devastated and pillaged by the Mongol hordes. From the Mongol sack Ṭûs appears never to have recovered, though the neighbouring shrines under the fostering care of the rich Shî'ahs soon resumed their former splendour; and Mustawfî, in the 8th (14th) century, is one of the first to refer to the Sanâbâdh village as Mashhad, 'the Place of Martyrdom,' a name that it has since always borne.

The Caliph and the Imâm, as Ḳazvînî remarks, lay under one dome, and the latter only was held in honour by the Shî'ahs, who, however, knew not which tomb to revere, for by order of the Caliph Mamûn (son of Hârûn-ar-Rashîd, and the poisoner of 'Alî-ar-Riḍâ), the two graves had been made exactly alike. When Mustawfî wrote, Mashhad had already become a great city, surrounded by immense graveyards with many famous tombs, that of Ghazzâlî, just mentioned, lying to the eastward of the shrines, where also was shown the grave of the poet Firdûsî. Around the city lay the fertile plain known as Marghzâr Takân, 12 leagues long by 5 across, where grapes and figs were more especially grown. The people of the Ṭûs district were, Mustawfî adds, 'a very excellent folk and good to strangers.'

Ibn Baṭûṭah, who visited the Mashhad of Imâm Riḍâ a few years later, gives a careful description of the shrine. Mashhad, was, he says, a large city, plentifully supplied as to its markets, and surrounded by hills. Over the tombs was a mighty dome, covering the oratory, and the mosque with a college (Madrasah) stood adjacent. All these were finely built, their walls being lined with tile-work (*kâshânî*). Above the actual grave of the Imâm was a sort of platform, or casing in wood, overlaid with silver plates, many silver lamps being hung from the beams round about. The threshold of the door into the oratory was overlaid in silver, the aperture being closed by a gold-embroidered silk veil, and the floor under the dome was spread with many fine carpets. The tomb of the Caliph was also covered by a casing of wood, on which candlesticks were set, but it was not held in honour, for, says Ibn Baṭûṭah, 'every Shî'ah on entering kicks with his foot the tomb of Hârûn-ar-Rashîd, while he invokes a blessing on that of Imâm Riḍâ.' The magnificence of the shrine of the Imâm is alluded to by the Spanish envoy Clavijo, who visited the court of Tîmûr in 808 (1405), and on his way passed through Mashhad. In those days it is noteworthy that Christians might enter the shrine, for the Persian Shî'ahs were not then as fanatical in this matter as they are at the present time[1].

Four days' march due west of Nîshâpûr in the district of Bayhaḳ were the two cities of Sabzivâr and Khusrûjird, a league only separating them; Sabzivâr, the chief town, being itself generally known in the middle-ages as Bayhaḳ. The Bayhaḳ district, which extended as far east as Rîvand, measuring 25 leagues across in all directions, comprised according to Yâḳût 321 villages, and he adds that the name Bayhaḳ was from the Persian *Bayhah* or *Bahâyin*, which signified 'most generous.' According to the same authority Sâbzavâr was the more exact name of the town, which the common people had shortened to Sabzvar; and Khusrûjird had originally been the chief town of the district, but the pre-eminence in his day was gone over to Sabzivâr. Mustawfî says that the markets of this town were covered by a wooden roof on arches, very strongly built; grapes and other fruits were grown in the district round, and most of the population in the 8th (14th) century were Shî'ahs[2].

From Bisṭâm in the Ḳûmis province to Nîshâpûr there were two roads. The more direct, the post-road, lies along the edge of the desert, going through Sabzivâr. The longer caravan road is to the north, and curves through the great upland plain of Juvayn, which is separated from the Great Desert by a range of hills. This district of Juvayn, which, according to Muḳaddasî, was also called Gûyân, was very fertile in food-stuffs, and its chief town was Azâdhvâr or Azâdvâr. The Isfarâyin district was in its northern part;

[1] The name of the Imâm is at the present day pronounced Rizâ by the Persians. Ykb. 277. Ist. 257, 258. I. H. 313. Muk. 319, 333, 352. Yak. iii. 154, 486, 560, 561; iv. 824. Kaz. ii. 262. Mst. 186. I. B. iii. 77—79. *Narrative of the Embassy of Ruy Gonzalez de Clavijo*, p. 110 (Hakluyt Society). 'The ambassadors went to see the mosque, and afterwards, when in other lands people heard them say that they had been to this tomb, they kissed their clothes, saying that they had been near the holy [shrine of] Horazan.'

[2] Muk. 317, 318. Yak. i. 804; ii. 441. Mst. 186. For the ruins of Bayhaḳ see C. E. Yate, *Khurasan and Sistan*, p. 398.

while at the western end, on the Ḳûmis border, was the Arghiyân district round Jâjarm. Nearly two hundred villages, according to Yâḳût, were dependencies of Azâdhvâr, which he describes as a populous town with fine mosques, and outside its gate was a great khân for merchants, for its markets were much frequented. The gardens of its villages stretched continuously all down the valley, and the water for their irrigation was brought by underground water-courses from the springs in the southern hills. In the 8th (14th) century, according to Mustawfî, the capital of the Juvayn district had changed to Fariyûmad, some miles to the south of Azâdvâr. Khudâshah, a stage east of Azâdvâr on the caravan road, was also a place of importance, where, at the close of the 8th (14th) century, Ḥâjjî Barlâs, the uncle of Tîmûr, was slain, as is mentioned by 'Alî of Yazd in his history[1].

The town of Jâjarm, also called Arghiyân, which is more particularly the name of its district, had, according to Muḳaddasî, a fine Friday Mosque, and was a well-fortified city, with 70 villages of its dependencies. Yâḳût describes the three towns of Samalḳan or Samanḳan, said to lie east of Jâjarm, Ar-Râwanîr (or Râwansar), and Bân, as being all of the Arghiyân or Jâjarm district, but their exact positions are not given. He also mentions Sabanj or Isfanj, which still exists to the south-west of Jâjarm on the road to Bisṭâm, and this place Mustawfî calls Rubâṭ Savanj. Mustawfî describes Jâjarm as a fair-sized town, which no army could come against, for within the circuit of a day's journey round it the plain was everywhere covered by a grass poisonous to all cattle. On the other hand, at the foot of its castle, there grew two plane-trees (*chinâr*), whose bark if chewed on a Wednesday morning infallibly cured toothache. Mustawfî adds that this bark was largely exported. The district round was very fertile, growing fruit and corn. The Jâjarm river, which ran south and ended in the desert, he names the Jaghân Rûd: it rose by three springs, each of which could have turned a mill, and these after coming

[1] Muk. 318. Yak. i. 230; ii. 165. Mst. 186, 196. A. Y. i. 58. There is some confusion between the names of Khudâshah four leagues to the east of Azâdvâr, and Khurâshah, which is about the same distance to the north of Azâdvar. The two names are written much alike in the Arabic character.

together ran for a course over 12 leagues in length, the water being much used for irrigation[1].

The great plain of Isfarâyin (or Asfarayn), Muḳaddasî says, grew much rice and fine grapes. Its chief town, of the same name, was very populous, and had good markets. Yâḳût states that the town of Isfarâyin was of old called Mihrajân, this, when he wrote in the 7th (13th) century, being still the name of a village near the ruined town, and 51 villages were of its dependencies. The name Isfarâyin, according to Yâḳût, was originally written Asbarâyîn, and meant 'the shield-bearers,' from *asbar*, 'a shield.' Mustawfî relates that in the mosque at Isfarâyin was a great bowl of brass, the largest ever seen, for its outer edge measured a dozen ells in circumference. To the north of the city was the Ḳal'ah-i-Zar, 'Gold Castle,' and the town took its water from a stream that flowed past at the foot of the castle hill. Throughout the surrounding plain nut-trees abounded; the climate was damp, but grapes and corn were grown plentifully[2].

In the marshy plain, where the river Atrak takes its rise to flow westward, while flowing in a contrary direction eastward, the river of Mashhad also has its source, lies the town of Kûchân, which in medieval times was called Khabûshân, or Khûjân. Its district the Arab geographers name Ustuvâ, praising it as a very fertile country; the name is said to mean 'the Highland'; and beyond Ustuvâ, eastwards, was the Nisâ district. Yâḳût, who states that the name of the chief town was in his day pronounced Khûshân, says that 93 villages belonged to it. In the *Jahân Numâ* the name appears as Khûchân, and Mustawfî says that though the name of Ustuvâ for the district was still written in the fiscal registers, it was in his day no longer in common use. The surrounding plain he praises for its fertility, and adds that Hûlâgû Khân, the Mongol, had rebuilt Khabûshân in the 7th (13th) century, his grandson Arghûn, the Îl-Khân of Persia, afterwards greatly enlarging the town. About half-way between Khabûshân

1 Muk. 318. Yak. i. 209, 249, 485; ii. 4, 742; iii. 35, 145. Mst. 186, 196, 220.

2 Muk. 318. Yak. i. 246. Mst. 186. The medieval city of Isfarâyin (the plain is still known by this name) is probably to be identified with the ruins called Shahr-i-Bilḳîs. C. E. Yate, *Khurasan and Sistan*, 378, 379.

and Ṭûs is Râdkân, which is mentioned by Ibn Ḥawḳal, and described by Yâḳût as a small town celebrated as the birth-place of Niẓâm-al-Mulk, the great Wazîr of Malik Shâh, the Saljûḳ[1].

The famous district of Nasâ or Nisâ is the broad valley now known as Darrah Gaz, 'the Vale of Manna.' The city of Nisâ is described by Ibn Ḥawḳal as being a large town, of the size of Sarakhs, having an abundant water-supply from the neighbouring hills. Muḳaddasî praises its fine mosque and excellent markets. Nearly all the houses, he says, had gardens, and rich villages were dotted about the valley all round the town. Yâḳût, however, speaks of Nisâ as most unhealthy, chiefly on account of the guinea-worm (the 'Medina worm,' he calls it), which in summer could hardly be avoided by those living in the place, and the suffering it caused made life unbearable. Ḳazvînî adds that the town was also called Shahr Fîrûz, after the ancient Persian king who was reported to have built it[2].

To the east of Nisâ, beyond the mountain ridge and on the edge of the Marv desert, lies Abîvard, the name being sometimes spelt Bâvard. Muḳaddasî says that its markets, in the midst of which stood the Friday Mosque, were finer even than those of Nisâ, and more frequented by merchants. Mustawfî praises the fruit grown here, and he counts as belonging to Abîvard the great guard-house (*rubâṭ*) at Kûfan, six leagues distant, standing in a village. This guard-house had been built by 'Abd-Allah, the Ṭâhirid, in the 3rd (9th) century; it had four gates, and a mosque was built in its midst. The district in which Abîvard stood was called Khâbarân, or Khâvarân, of which Mihnah, or Mayhanah, was the chief town; further, Yâḳût names Azjah, Bâdhan, Kharv-al-Jabal and Shûkân as among the important places of this district; but Mayhanah, when he wrote, was already in ruins.

[1] I. H. 313. Muk. 318, 319. Yak. i. 243; ii. 400, 487, 730. Mst. 186. J. N. 323. The present town of Bujnurd, lying north of Isfarâyin, and about 60 miles to the north-west of Kûchân, was founded a couple of centuries ago, but near it was an older town called Bizhân, the ruined castle of which, known as the Ḳal'ah, still exists. C. E. Yate, *Khurasan and Sistan*, 195, 196. Sykes, *Persia*, 22.

[2] Ist. 273. I. H. 324. Muk. 320. Yak. iv. 776. Kaz. ii. 311. The city of Nisâ is probably identical with the modern Muḥammadâbâd, the chief town of Darrah Gaz.

In the following, 8th (14th), century Mustawfî speaks of the many fine gardens of the Khâvarân district—he also gives the name as Khavardân—and he says that in its chief town had resided the poet Anvârî, who flourished in the 6th (12th) century, having been the panegyrist of Sultan Sanjar the Saljûḳ[1].

In the mountains, and about half-way between Abîvard and Mihnah, lies the huge natural fortress now known as Kilât-i-Nâdir, after Nâdir Shâh, the celebrated king of the Persia of the 18th century A.D., who stored his treasures here. This stronghold does not appear to be mentioned in any of the Itineraries, or by the Arab geographers of the 3rd and 4th (9th and 10th) centuries, and Yâḳût does not notice it. The earliest mention of Kilât appears to be by 'Utbî, in his *History of Maḥmûd of Ghaznah*, and he merely states incidentally that a certain Amîr went 'from Nîshâpûr to Kilât, which is also in the Arabic fashion written Ḳal'ah.' Mustawfî gives a succinct description of the place, adding that its chief towns were called Jurm and Marinân; further, Kilât had much water, besides arable lands that produced abundantly, and many villages belonged to it of the surrounding districts. In history it first became famous for the siege of the fortress by Tîmûr, at the close of the 8th (14th) century, and after it had fallen into his hands he caused its fortifications to be carefully rebuilt and strengthened[2].

The city of Sarakhs lies on the direct road from Ṭûs to Great Marv, and on the right, or eastern bank of the Mashhad river, which is now known as the Tajand. This river does not appear

[1] Muk. 321, 333. Yak. i. 111, 232, 462; ii. 383, 395, 428; iii. 337; iv. 321, 723. Mst. 189. A. Y. i. 382. J. N. 318. The name of Khâvarân stands for the older form Kharvarân, meaning 'the west country' (the opposite of Khurâsân, 'the east country'), and this small district of the foot-hills on the Marv desert thus preserves at the present day the name applied originally to all western Persia that was formerly not counted as Khurâsân, 'the country of the east.'

[2] 'Utbî, *Kitâb-i-Yamînî*, Arabic text (Cairo, 1286 A.H.), i. 215. Persian text (Tihrân, 1272 A.H.), p. 151. Mst. 187. A. Y. i. 334, 337. J. N. 323. Kilât or Kalât, in Persian, is equivalent to the Armenian Qalaq, signifying 'a city,' and in Arabic appears under the well-known form Ḳal'ah, or Ḳal'at, 'a castle.' Kilât-i-Nâdir was visited by Col. MacGregor (*Journey through Khurasan*, ii. 51) in 1875 and carefully described.

to be named by any of the medieval geographers; it rises, as already described, in the marshes near Kûchân, and at first flows south-east, passing Mashhad. When it has gone about a hundred miles beyond this city it receives from the south, as a great affluent, the Herât river, and thence turning north flows to Sarakhs. At some distance further north, in the latitude of Abîvard, its waters spread out and became lost in the desert sands, at a place called Al-Ajmah, 'the Reed-beds,' where there were many tamarisk trees. Iṣṭakhrî and Ibn Ḥawḳal speak of this river Tajand merely as an affluent of the Herât river. Ibn Rustah, who regards it in the same light, says that two leagues before coming to Sarakhs the Herât river (that is, the lower course of the Tajand) throws off a branch canal that goes direct to this city. Other canals too were taken from it to water the Sarakhs district, more especially one named the Khushk Rûd (Dry River), across which had been built a great masonry bridge, but for a great part of the year even the main stream at Sarakhs carried no water.

Sarakhs in the 4th (10th) century was a great city, being half the size of Marv, with a healthy climate. Camels and sheep were numerous in its pastures, though its arable lands were limited for lack of a constant water-supply. Muḳaddasî praises its Friday Mosque and fine markets, adding that throughout the suburbs there were many gardens. Ḳazvînî, who speaks of it as very populous, says that they made here, for export, scarfs for turbans, and veils that were most beautifully embroidered in gold thread. In the 8th (14th) century Mustawfî describes the walls of Sarakhs as 5000 paces in circuit and protected by a strongly built fortress. Their drinking water, he says, was from the river 'coming from Ṭûs and Herât' (he does not name the Tajand), a fine stream, and of very digestible water, which further served to irrigate the fields round Sarakhs, where melons and grapes grew abundantly[1].

[1] I. R. 173. Ist. 272. I. H. 323, 324. Muk. 312, 313. Kaz. ii. 261. Mst. 189. Modern Sarakhs lies on the west bank of the Tajand.

CHAPTER XXVIII.

KHURÂSÂN (*continued*).

The Marv quarter. The Murghâb river. Great Marv and its villages. Âmul and Zamm, on the Oxus. Marv-ar-Rûd, or Little Marv, and Ḳaṣr Aḥnaf.

The second of the Quarters of Khurâsân, that of Marv, lies along the Murghâb, or Marv river. This river flows down from the mountains of Ghûr to the north-east of Herât, and passing Little Marv turns thence north to Great Marv, where its waters were divided up among a number of canals, after which it became lost in the sands of the Ghuzz Desert, on about the same latitude as the swamps of the Tajand or Herât river, but some 70 miles to the eastward of the latter.

Besides the various towns lying along the Murghâb, the Marv quarter also included the places on the great Khurâsân road, beyond Marv, north-eastward to the Oxus at Âmul, where the crossing for Bukhârâ took place.

The name Murghâb, or Marghâb, is said by Ibn Ḥawḳal to have been originally Marv-âb, 'the Marv-water'; but, says Iṣṭakhrî, Murghâb is the name of the place where its streams rise. Muḳaddasî, who calls the Murghâb the river of the Two Marvs, describes it as flowing past Upper (or Lesser) Marv towards Lower (or Great) Marv. One march south of the latter city its bed was artificially dyked with embankments faced by woodworks which kept the river-bed from changing. This embankment in the 4th (10th) century was under the wardship of a specially appointed Amîr who acted as water-bailiff, with 10,000 workmen under him and horse guards, and saw to the up-keep of the dykes, and the regulation of the water-supply. There was on the embankment a measure which registered the flood-height; in a year of abundance this would rise to 60 barleycorns above the low-level, and the people then rejoiced, while in a year of drought the water would only attain the level of six barleycorns.

At a distance of one league south of Great Marv the waters of the stream were dammed back in a great round pool, whence four canals radiated to the various quarters of the city and suburbs. The height of the pool was regulated by sluices, and it was a great festival when at high flood-time the various dams were cut, and the waters were divided off according to rule. These four main canals were called respectively the Hurmuzfarrah canal, flowing towards the west, next to the eastward that of Mâjân, then the Nahr Zarḳ or Ar-Razîḳ, and finally the Nahr As'adî. Of these four the Nahr-al-Mâjân appears to have carried the main stream of the Murghâb, and after passing through the suburbs of the city, where it was crossed by many bridges of boats, it came out again to the desert plain, and flowed on till the residue of its waters were lost in the swamp. Yâḳût in the 7th (13th) century states that the Murghâb was in his day known as the river Razîḳ (probably identical with the canal already mentioned), a name which he states was often incorrectly spelt Zarîḳ, and the *Jahân Numâ* adds, as a third variant, Zarbaḳ. These names are also mentioned by Mustawfî, who gives Murghâb as the common appellation in his day, and by this name the great river is still known[1].

Great Marv, in the middle-ages, was called Marv-ash-Shâhijân, to distinguish it from Marv-ar-Rûd, Little Marv, and Shâhijân is probably merely the Arab form of the old Persian *Shâhgân*, 'kingly,' or 'belonging to the king,' though Yâḳût and others explain the term as *Shâh-i-Jân* to mean 'of the soul of the king.' Marv, as described by Iṣṭakhrî, Ibn Ḥawḳal, and Muḳaddasî, consisted of an inner citadel (Ḳuhandiz) 'high-built and itself of

[1] Ist. 260, 261. I. H. 315. Muk. 330, 331. Yak. ii. 777. Mst. 214. J. N. 328. The place where the Murghâb ultimately became lost in the sands is called Mâyâb by Ḥâfiẓ Abrû. Hfz. 32*b*. For the places round Marv, see Map x, p. 447. Presumably 60 barleycorns (Sha'îrah) went to the ell.

the size of a town,' surrounded by the inner city with its four gates, beyond which again were extensive suburbs stretching along the banks of the great canals. The four gates of the inner town were the Bâb-al-Madînah, 'the city gate' (S.W.), where the road from Sarakhs came in; the Bâb Sanjân (S.E.) opening on the Banî Mâhân suburb and As'adî canal; the Bâb Dar Mashkân (N.E.) on the road to the Oxus; and lastly the Bâb Bâlin (N.W.). In the 4th (10th) century there were three Friday Mosques in Marv, first the citadel mosque called the Jâmi' of the Banî Mâhân; next the Masjid-al-'Atîḳ, 'the Old Mosque,' which stood at the gate opening on the Sarakhs road, the Bâb-al-Madînah; lastly the New Mosque of the Mâjân suburb, outside this same gate, where the great markets of Marv were found.

The Razîḳ canal flowed into the town, coming to the gate called Bâb-al-Madînah and the Old Mosque, after which its waters were received and stored in various tanks for the use of the inhabitants of the quarter. The Mâjân canal, flowing to the west of it, watered the great Mâjân suburb, which lay round the Maydân, or public square, on which stood the New Mosque, the Government-house, and the prison; all these having been built by Abu Muslim, the great partizan of the Abbasids. To him was principally due their accession to the Caliphate, as history relates, and in a domed house of this quarter, built of burnt brick, the dome being 55 ells in diameter, says Iṣṭakhrî, the place was shown where the first black Abbasid robes had been dyed, that having become the distinguishing colour of the new dynasty.

West of the Nahr Mâjân, as already said, was the canal of Hurmuzfarrah, on the limit of the suburbs of Marv, and along its banks were the houses and quarters built by Ḥusayn the Ṭâhirid, who had transferred many of the markets to this quarter. Yâḳût, at a later date, speaking of the great western suburb of Mâjân, mentions two of its chief streets, namely, the thoroughfare known as Barârjân (for Barâdar-Jân) or 'brother-life' in upper Mâjân, and the street of Tukhârân-bih. The Hurmuzfarrah canal ultimately reached the township of that name, near the swamps of the Murghâb, and the town had its own Friday Mosque. One league distant from Hurmuzfarrah was Bâshân, also a town with its Friday Mosque, while the two hamlets of Kharak (or Kharah) and As-Sûsanḳân, standing a league distant one from the other, lay also on this side of Marv and were likewise of sufficient size for each to have its own Friday Mosque.

One march to the westward of Marv was the town called Sinj (in Muḳaddasî spelt Sink), with a fine Friday Mosque, standing on a canal with many gardens, and beyond it, two marches to the south-west of Marv on the road to Sarakhs, lay the important town of Ad-Dandanḳân. This was small but well fortified, having a single gate, with hot baths (Ḥammâms) outside the wall. Its ruins were seen by Yâḳût in the 7th (13th) century, for it had been pillaged by the Ghuzz in 553 (1158). This was the limit of cultivation of the Marv oasis to the south-west, while Kushmayhan, one march from Marv on the Bukhârâ road, was the limit of cultivation on the north-eastern side. This Kushmayhan, or Kushmâhan, according to Ya'ḳûbî, was famous for the *Zabîb Kushmâhanî*, a kind of raisin. The town also possessed a fine Friday Mosque and good markets; it was watered by a great canal, and there were many hostelries and baths here; much fruit being grown in the surrounding gardens.

Immediately outside the Dar Mashkân gate of Marv, which led to the town of Kushmayhan, had stood the great palace of Mamûn, where he had lived when he held his court at Marv, previous to setting out for Baghdâd to wrest the Caliphate from his brother Amîn. The south-eastern gate of Marv, the Bâb Sanjân, opened on the As'adî canal, along which lay the Banî Mâhân (or Mîr Mâhân) quarter, with the palace of the Marzubân of Marv, the Persian Warden of the Marches. From this gate the road led up the Murghâb river by Al-Ḳarînayn to Marv-ar-Rûd. Six leagues from the city in this direction was the town of Jîranj (or Kîrang, in Muḳaddasî) on the river bank, while one league beyond it lay Zarḳ. Here had stood the mill where Yazdajird III, the last of the Sassanian kings, fled for shelter, and was murdered by the miller for the sake of his jewels. According to Ibn Ḥawḳal, it was at Zarḳ township that the waters of the Murghâb were first canalised, channels being led off to irrigate the gardens round Marv. These gardens had at all times been famous for their melons, also for the assafoetida root (*ushturghâz*) grown here, which was exported to other parts of Khurâsân.

Silkworms, too, were raised here largely, the silk being manufactured into the stuffs for which Marv was celebrated[1].

In the latter half of the 4th (10th) century, when Muḳaddasî knew Marv, a third part of the suburb was already in ruin, and the citadel was in no better state. In the next century, however, the city gained in size and importance under the Saljûḳs, and here Sultan Sanjar, the last of the great Saljûḳs, was buried in 552 (1157), and the remains of his tomb may still be seen at the present day. Yâḳût, who was in Marv in 616 (1219), describes the grave of Sultan Sanjar as lying under a great dome covered with blue tiles, so high as to be visible a day's march away over the plain; and the windows under the dome looked into the adjacent Friday Mosque. It had been built in memory of him, Yâḳût was told, long after the Sultan's death by some of his servants. At the village of Andarâbah, two leagues from Marv, which had been the private property of Sultan Sanjar, the remains of his palace were still standing in the 7th (13th) century, the walls being intact, though all the rest had gone to ruin, as was the case also, Yâḳût adds, with the adjacent village.

Yâḳût describes Marv as in his day possessing two chief Friday Mosques, enclosed by a single wall, one for the Ḥanafites, the other belonging to the Shâfi'ites. He himself lived in Marv for three years, collecting the materials for his great geographical dictionary, for before the Mongol invasion the libraries of Marv were celebrated; 'verily but for the Mongols I would have stayed

[1] Ykb. 280. Ist. 258—263. I. H. 314—316. Muk. 298, 299, 310—312, 331. Yak. i. 534, 827; ii. 610; iv. 507. The town and mill of Zarḳ lay seven leagues from Marv, while the pool where the waters of the Murghâb were divided among the four city canals, of which the Nahr Razîḳ was one, lay at a distance of but one league from Marv. The Razîḳ canal and the Zarḳ mill, therefore, were probably not adjacent, but from the shifting of the diacritical point there is much confusion between Zark or Razḳ, and Zarîḳ or Razîḳ. The name of the mill is sometimes given as pronounced, Zurḳ or Zurraḳ, and the Zarîḳ canal appears as Zarbaḳ, on whose banks, according to some accounts, King Yazdajird came to his death. See Yak. ii. 777, 925; iv. 508. Muḳaddasî (p. 33) records that some two leagues from Marv, but in which direction is not stated, was a small guard-house in which stood a tomb, popularly said to contain the head of Ḥusayn, grandson of the Prophet, but this is a relic that was also shown in divers other localities, and certainly at the time of Ḥusayn's death his head was not sent to Marv.

and lived and died there,' he writes, 'and hardly could I tear myself away.' Thus among others he mentions the two libraries of the Friday Mosque, namely the 'Azîzîyah with 12,000 and odd volumes, and the Kamâlîyah. There was also the library of Sharaf-al-Mulk, in his Madrasah or college, and that of the great Saljûḳ Wazîr the Niẓâm-al-Mulk. Among the older libraries were those founded by the Sâmânids, and one in the college of the 'Umaydîyah; also that in the Khâtûnîyah college, and that which had belonged to Majd-al-Mulk. Finally, and especially, there was the Ḍumayrîyah library in one of the Khânḳâhs, or Darvîsh convents, containing only 200 volumes, but each volume, Yâḳût writes, worth two hundred gold pieces (dînârs), for all the books there were unique and beyond price.

At the approach of the Mongol hordes in 617 (1220) Yâḳût sought safety at Mosul in Mesopotamia, and all the glories of the Marv libraries fell a prey to the flames, which followed in the wake of the Mongol sack of this great city, when nine million corpses are said to have remained unburied among the ruins. The tomb of Sultan Sanjar, Ibn-al-Athîr states, was set on fire by the invaders, together with most of the mosques and other public buildings; and Ḥâfiẓ Abrû adds that they broke down all the great dams and dykes of the Murghâb, which under the early Saljûḳs had been increased in number, and carefully seen to, in order thus to regulate the irrigation of the oasis, which now lapsed into a desert swamp. In the 8th (14th) century, when Ibn Baṭûṭah passed through Marv, it was still one great ruin.

The account which his contemporary, Mustawfî, gives of Marv deals with its past glories in the 2nd (8th) century, when it was under the government of Abu Muslim, who brought the Abbasids to power, and when the Caliph Mamûn resided at this place previous to marching on Baghdâd. Then the Ṣaffârids had removed the capital of Khurâsân to Nîshâpûr, but the Saljûḳs restored the primacy to Marv, and Sultan Malik Shâh built the great wall round the city 12,300 paces in circuit. The crops of the Marv oasis were a marvel of productiveness; Mustawfî reports that seed corn gave a hundred-fold the first year, and from the ungathered overfall some thirty-fold for the second year was obtained, with as much as ten-fold of the original sowing even in the third year. The climate, however, being damp was unhealthy, and the *rishtah*, or guinea-worm, was a terrible scourge. The moving sands of the neighbouring deserts had in his day overwhelmed many of the fruitful districts, but excellent water-melons were still grown, which were dried and largely exported, also grapes and pears.

Mustawfî describes the city of Marv as still almost entirely a ruin, though at the close of the 8th (14th) century it must have regained some of its former splendour, for Tîmûr frequently stopped here in the intervals of his campaigns. He generally lived at a place which 'Alî of Yazd writes Mâkhân, probably a clerical error for Mâjân, which as already said had been in earlier days the name of the great western suburb of Marv, though Yâkût mentions a place also called Mâkhân as a village near the city. Marv was in part restored to its former state of greatness under the reign of Shâh Rukh, the grandson of Tîmûr, who rebuilt much of the city in the year 812 (1409), so that in 821 (1418), when Ḥâfiẓ Abrû wrote, he describes it as once more being in a flourishing condition[1].

On the left bank of the Oxus about 120 miles to the north-east of Marv, where the great Khurâsân road crossed to Bukhârâ and Transoxiana, stood the city of Âmul, and about a hundred miles to the eastward, higher up on the same bank was Zamm, also at a crossing-place. Âmul, which in the later middle-ages was also known as Amûyah, and then came to be called Chahâr Jûy ('Four Canals,' a name the place still bears), is described by Ibn Ḥawḳal as a fertile and pleasant little town, of great importance by reason of the constant passage of caravans going to and coming from the countries beyond the Oxus. All along the road south-west to Marv there were wells at each stage, but otherwise the territory of Âmul was enclosed on all sides by the desert, which here came close up to the river bank. Muḳaddasî praises the excellent markets of Âmul. The town, with its Friday Mosque crowning a small hill, lay a league distant from the Oxus among well-irrigated fields, where there were vineyards. Opposite Âmul,

[1] Ibn-al-Athir, xii. 256. Yak. i. 373; iv. 378, 509, 510. I. B. iii. 63. Mst. 189. A. Y. i. 147, 150, 569. Hfz. 32*b*.

on the right bank of the river, in the Bukhârâ district, was the town of Firabr.

To distinguish this Âmul from the town of the same name which was the capital of Ṭabaristân (see above, p. 370), Yâḳût states that it was known in books as Âmul of Zamm (after the next Oxus passage upstream), or Âmul of the Jayḥûn (Oxus), or Âmul-ash-Shaṭṭ (of the Stream), or further as Âmul-al-Mafâzah (of the Desert). In his day, however, in place of the name Âmul the town had come to be called Amû, or Amûyah, by which denomination it is frequently mentioned in the accounts of the Mongol invasion, and of the campaigns of Tîmûr. It is also known as Ḳal'ah Amûyah, or 'the Amûyah Castle.' In the 11th (17th) century Abu-l-Ghâzî gives the name as Amûyah when dealing with the marches of Changîz Khân, but speaking of the events of his own day writes of Chahâr Jûy, in reference to this Oxus passage, which proves conclusively that the two places are identical. The town of Zamm, also on the Khurâsân bank, as already stated, is the modern Karkhî, and in the middle-ages the town of Akhsîsak faced it on the further side, towards Bukhârâ. Ibn Ḥawḳal speaks of Zamm as a town of the same size as Âmul, but it was only approached on the Khurâsân side by the road up the Oxus bank in four marches from Âmul; for from Zamm direct across to Marv the waterless desert intervened. From Zamm, eastward, Balkh could be reached, and after crossing the Oxus, Tirmidh. Zamm is also briefly mentioned by Muḳaddasî, who speaks of its Friday Mosque standing in the market-place, so that in the 4th (10th) century it must already have been a place of some importance[1].

Coming back now to the Murghâb river, about 160 miles higher up than Great Marv stood Upper, or Little Marv, at that part of the river where, after leaving the Ghûr mountains, it turns north through the desert plains towards Great Marv. Little Marv, or Upper Marv as Muḳaddasî and others call it, is the place known as Bâlâ Murghâb, 'Upper Murghâb,' to the Persians. It is now a complete ruin, and has been so since the invasion of Tîmûr. In the 4th (10th) century, however, Marv-ar-Rûdh, or

[1] Ist. 281, 314. I. H. 329, 363. Muk. 291, 292. Yak. i. 69; ii. 946. A. Y. i. 148, 334, 568. A. G. 124, 329.

'Marv of the River,' as it was then called, was the largest city of this, a most populous district, which had besides four other towns with Friday Mosques. It lay at a bow-shot from the bank of the Murghâb, in the midst of gardens and vineyards, being three leagues distant from the mountains on the west, and two leagues from those on the east. In the market-place was the Friday Mosque, a building according to Muḳaddasî standing on wooden columns, and Ḳudâmah adds that one league from Upper Marv (as he calls it) was the castle of Ḳaṣr-'Amr in the hills, blocking the mouth of a small valley. Yâḳût states that in his day the name Marv-ar-Rûd was pronounced Marrûd by the common folk. It appears to have escaped the utter ruin which was the fate of Great Marv at the hands of the Mongols. At any rate in the 8th (14th) century Mustawfî describes it as still a flourishing place, with a wall 5000 paces in circumference, which had been built by Sultan Malik Shâh the Saljûḳ. The surrounding country was most fertile, grapes and melons were grown abundantly, and living was cheap[1].

One day's march from Marv-ar-Rûd, on the same bank and down the river towards Great Marv, was the castle called Ḳaṣr Aḥnaf, after Al-Aḥnaf ibn Ḳays, the Arab general who in the days of the Caliph 'Othmân, in the year 31 (652), had conquered these lands for Islam. It was a large place, Ibn Ḥawḳal says, with many vineyards round it, and fine gardens, the soil and climate being alike excellent, and Muḳaddasî mentions its Friday Mosque situate in the market-place. At the present day the site of Ḳaṣr Aḥnaf is marked by the village of Marûchak, or Marv-i-Kûchik (Little Marv) as the Persians call the place. In the middle-ages, four leagues above Marv-ar-Rûd, stood Dizah, a town occupying both banks of the Murghâb, the two parts being connected by a stone bridge. This place too had a fine Friday Mosque, and Yâḳût adds that it had originally been called Sinvân.

The hamlets of Panj-dîh (Five Villages) lie below Marûchak on the Murghâb, and the place was visited by Nâṣir-i-Khusraw in 437 (1045) on his way to Mecca; Yâḳût too was there in 616 (1219) and alludes to it as a fine town. The place is also

[1] Kud. 210. Ist. 269. I. H. 320. Muk. 314. Yak. iv. 506. Mst. 190. For the ruins at Bâlâ Murghâb, see C. E. Yate, *Northern Afghanistan*, p. 208.

mentioned in the time of Tîmûr at the close of the 8th (14th) century, when 'Alî of Yazd says it was known as Pandî (but the reading appears uncertain, and some manuscripts give Yandî). During the earlier middle-ages all the country from Little Marv to Great Marv, along the Murghâb, was under cultivation, and studded with villages and towns. Al-Ḳarînayn, already alluded to, was four marches above Great Marv, being two below Marv-ar-Rûd; and half-way between Ḳarînayn and the latter was Lawkar, or Lawkarâ, which Muḳaddasî mentions as a populous place, as big as Ḳaṣr Aḥnaf. Above Marv-ar-Rûd, and all up the Murghâb into the mountains of Gharjistân, there are many flourishing districts, as will be noticed in the next chapter, when speaking of Ghûr in the Herât quarter[1].

[1] Ykb. 291. Ist. 270. I. H. 321. Muk. 299, 314. N. K. 2. Yak. i. 743; iv. 108. A. Y. i. 353. For the ruins at Marûchak, see C. E. Yate, *Afghanistan*, pp. 110, 120, 194.

CHAPTER XXIX.

KHURÂSÂN (*continued*).

The Herât quarter. The Herât river, or Harî Rûd. The city of Herât. Mâlin and towns on the upper Harî Rûd. Bûshanj. The Asfuzâr district. The Bâdghîs district and its towns. Kanj Rustâḳ. Districts of Gharjistân and Ghûr. Bâmiyân.

The Herât quarter of Khurâsân lies entirely in what is now known as Afghânistân, and, for the most part, is watered by the Herât river or Harî Rûd. This river takes its rise in the mountains of Ghûr, and at first flows for some distance westward. In order to irrigate the Herât valley many canals were here led from it, some above and some below Herât city, seven in particular being named by Muḳaddasî as serving to water the fruitful districts round the capital.

The Herât river, flowing from east to west in its earlier course, passes Herât city several miles from its southern gate, near the town of Mâlin. Here there was a bridge over it, unequalled in all Khurâsân for beauty, says Muḳaddasî, it having been built by a certain Magian, and bearing his name on an inscription—'and some say that he afterwards became a Moslem, others that he threw himself into the river, because the Sultan would put his own name upon that bridge.' Mustawfî gives the names of nine of the chief irrigation canals that were taken from the Harî Rûd in the neighbourhood of Herât. Beyond Herât the Harî Rûd passed the town of Fûshanj near its south bank, and turning north flowed on to Sarakhs, before reaching which it took up the waters of the Mashhad river, as has been mentioned in the previous chapter. Beyond, to the north of Sarakhs, its waters were lost in the desert. According to Ḥâfiẓ Abrû the Herât river also bore the name of Khajacharân (the spelling, from the shifting position of the diacritical points, and the true pronunciation are alike uncertain), and he asserts that its source was at a spring not far from the place where the Helmund river took its rise[1].

In the 4th (10th) century, as described by Ibn Ḥawḳal and Muḳaddasî, Herât (written more exactly Harât) was a great city, with a citadel, surrounded by a wall with four gates. These were, the Bâb Sarây or 'Palace Gate' to the north on the Balkh road; then to the west, towards Naysâbûr, the Bâb Ziyâd; the Fîrûzâbâd gate, which Muḳaddasî calls the Bâb Fîrûz, was to the south on the road towards Sijistân; while to the east was the Bâb Khushk towards the Ghûr mountains. These four gates were all of wood, except the Bâb Sarây, which was of iron, says Ibn Ḥawḳal; and the citadel of Herât (called the Ḳuhandiz or Ḳal'ah) had also four gates of the like names, respectively, to the city gates. The city measured half a league square, and the government house was at a place called Khurâsânâbâd, a mile outside the town on the western road towards Fûshanj. At each of the four city gates, within the town, was a market; and outside each gate was an extensive suburb. The great Friday Mosque of Herât stood in the midst of the chief market, and no mosque in all Khurâsân or Sijistân was its equal in beauty. Behind it, on the west side, was the prison.

To the north of Herât the mountains lay two leagues distant from the city, and here the land was desert, not being irrigated. These mountains produced mill-stones and paving-stones, and on the summit of one of the hills was an ancient fire-temple, called Sirishk, which was in the 4th (10th) century much frequented by the Magians. A Christian church also stood at a place lying halfway between this fire-temple and the city. To the south of Herât, down to the Mâlin bridge over the Harî Rûd, the land was like a garden, well cultivated and profusely irrigated by numerous canals, and divided into many districts. Populous villages lay one after the other, for a day's march and more, along the Sijistân road.

[1] Ist. 266. I. H. 318. Muk. 329, 330. Mst. 216. Hfz. 32*a*.

The prosperity of Herât continued unabated till the inroad of the Mongols; and in 614 (1217) when Yâķût was here, some four years before that disastrous event, he considered Herât to be the richest and largest city that he had ever seen, standing in the midst of a most fertile country. His contemporary Ķazvînî, who confirms this account, notes that here might be seen many mills 'turned by wind, not by water,' which was to him an uncommon sight. Herât, however, must have recovered quickly from the effects of the Tartar inroad, and Mustawfî in the following century bears out the statement of Ibn Baṭûṭah that, after Nîshâpûr, it was the most populous city of all Khurâsân. Its walls were then 9000 paces in circuit, and 18 villages lay immediately round the town, watered principally by a canal (Nahrîchah) taken from the Harî Rûd. The grapes of the kind called Fakhrî, and the figs were both superlatively excellent. Already in the 8th (14th) century the people of Herât were Sunnî. It was in the 6th (12th) century, during the supremacy of the Ghûrid dynasty according to Mustawfî, that Herât had reached its greatest splendour. There were then 12,000 shops in its markets, 6000 hot baths, and 659 colleges, the population being reckoned at 444,000.

A strong fortress lay to the north of Herât, when Mustawfî wrote, called the castle of Shamîrân, this having been built on the site of the older fire-temple of Sirishk, mentioned by Ibn Ḥawķal, which was two leagues distant from the city on a hill-top. This fortress also went by the name of the Ķal'ah Amkalchah. At the close of the 8th (14th) century, Tîmûr, after taking possession of Herât, destroyed its walls, and sent most of its artificers to augment the population of his new town of Shahr-i-Sabz in Transoxiana. In the Turkish *Jahân Numâ* it is stated that at that period, in the year 1010 (1600), Herât had five gates; that called Darvâzah-i-Mulk, 'the Government Gate,' to the north, the 'Irâķ gate to the west, that of Fîrûzâbâd to the south, the Khush gate to the east, and the Kipchâķ gate to the north-east—this last being of late origin. The ten Bulûks, or districts, round Herât are also enumerated, but no statement as to the relative positions of these is afforded[1].

[1] Ist. 264—266. I. H. 316—318. Muk. 306, 307. Yak. iv. 958. Kaz. ii. 322. I. B. iii. 63. Mst. 187. J. N. 310—312. A. Y. i. 322, 323. The infor-

Two leagues, or half a day's journey, to the south of Herât, and presumably beyond the great bridge that spanned the Harî Rûd, to which bridge it gave its name, was the town of Mâlin, or Mâlan, with the district of the same name lying a day's journey in extent all round it. This Mâlan was called As-Safalḳât, and Mâlan of Herât, to distinguish it from the place of the same name in the Bâkharz district of Ḳûhistân (mentioned in Chapter XXV, p. 357). It was a small town, surrounded by most fruitful gardens, and the produce of its vineyards was celebrated. Yâḳût who had been there, writes the name Mâlîn, but adds that the people in his day pronounced it Mâlân. Twenty-five villages belonged to its district, and of these he specially mentions four, Murghâb, Bâshînân, Zandân, and 'Absaḳân.

One march to the north-east of Herât lies Karûkh, or Kârûkh, which Ibn Ḥawḳal says was in the 4th (10th) century the largest town of the Herât district after the capital. Apricots and raisins were exported in great quantities from hence to all the neighbouring districts and cities; the Friday Mosque stood in the quarter of the town called Sabîdân, and the houses were built of sun-dried bricks. Karûkh stood in a mountain valley, 20 leagues in length, the whole of which was under cultivation, many villages and broad arable lands lying on its various streams. Its chief river flowed to the Harî Rûd, and appears to be that which Yâḳût names the Nahr Karâgh.

Eastward from Herât, and lying in the broad valley of the Harî Rûd, a succession of towns are mentioned by the geographers of the 4th (10th) century; namely, Bashân, one day's journey from Herât, then Khaysâr, Astarabyân, Marabadh, and Awfah, each situated a day's journey beyond the last, and to the east of it; finally two days' journey beyond Awfah was Khasht, a place that was counted as in the Ghûr district. Of these towns, Awfah was almost as large as Karûkh, and only second to it in importance.

mation given by Ḥâjjî Khalfah, in the *Jahân Numâ*, is in part taken from the monograph on Herât written by Mu'în-ad-Dîn of Asfuzâr in 897 (1492). This monograph has been inserted by Mîrkhwând in the Epilogue (*Khâtimah*) of the *Rawḍat-aṣ-Ṣafâ*, pt vii. 45—51, and it was translated by M. Barbier de Meynard in the *Journal Asiatique*, 1860, ii. p. 461; 1861, i. pp. 438, 473; 1862, ii. p. 269. For the present condition of Herât see C. E. Yate, *Afghanistan*, pp. 25—28.

The other four towns are described in similar terms as being well watered and populous; all were smaller in size than Mâlin, each had gardens and fertile fields, and while Astarabyân grew no grapes, being near the hill country, Marabadh was especially noted for its rice, which was largely exported[1].

One day's march to the west of Herât was the considerable city of Bûshanj or Fûshanj, which apparently occupied the site of the present Ghurian, lying a short distance from the left bank of the Harî Rûd, and to the south of it. Ibn Ḥawḳal describes Bûshanj as about half the size of Herât in the 4th (10th) century, and, like the latter, it lay in a plain two leagues distant from the mountains. The town was well built, and surrounded by trees, among which the juniper throve amazingly, its wood being largely exported. The town was strongly fortified, and was surrounded by a wall and a ditch. There were three gates, the Bâb 'Alî towards Naysâbûr, the Herât gate to the east, and the Ḳûhistân gate to the south-west. Yâḳût, who had seen the town in passing, lying hidden in its wooded valley, gives the name as Bûshanj or Fûshanj. He adds that the Persians pronounced it Bûshang. Mustawfî describes Fûshanj, in the 8th (14th) century, as famous for its water-melons and grapes, of which last there were 105 different varieties. A peculiarity of the place was that it possessed numerous windmills, their origin or invention being popularly attributed to the Pharaoh of Egypt, of the days of Moses, who had once come during a campaign as far east as this city. In 783 (1381) Fûshanj was stormed and sacked by Tîmûr, and this in spite of its high walls and deep water-ditch which are especially mentioned by 'Alî of Yazd. For some unexplained reason the name of Fûshanj after this disappears from history, and at a later date the town of Ghurian, which is now a flourishing place, sprang up on the ruins of the city which Tîmûr had pillaged and destroyed. It is to be added that the three towns of Farjird, Kharjird, and Kûsûy, which have already been described as of the Ḳûhistân province (see p. 358), are often given as belonging to Fûshanj[2].

[1] Ist. 267, 285. I. H. 318, 334. Muk. 50, 298, 307, 349. Yak. i. 470; ii. 950; iii. 605; iv. 247, 397, 499.

[2] Ist. 267, 268. I. H. 319. Muk. 298. Yak. i. 758; iii. 923. Mst. 187.

The Asfuzâr district lies to the south of Herât, on the road towards Zaranj, and in the 4th (10th) century four towns of importance existed here, besides the capital Asfuzâr, namely Adraskar, Kuwârân, Kûshk, and Kuwâshân. Asfuzâr, now the chief town, at the present day goes by the name of Sabzivâr (called Sabzivâr of Herât, to distinguish it from Sabzivâr to the west of Nîshâpûr; see p. 391). In early times, however, Kuwâshân was the largest city of the district, which extended for three days' march from north to south with a breadth across of a day's march. According to Iṣṭakhrî there was here a famous valley, called Kâshkân, with many populous villages, and the river which has its head-waters near Asfuzâr (Sabzivâr) is that now known as the Hârûd of Sîstân, which flows into the head of the Zarah lake to the west of Juwayn. All these towns of Asfuzâr are described as surrounded by fertile lands and gardens. In the Itineraries Asfuzâr bears the second name of Khâstân (or Jâshân, for the reading is uncertain), and it seems not unlikely that Kuwâshân is merely another form of this name, and therefore really identical with Asfuzâr (Sabzivâr). The town of Adraskar, or Ardsakar, as it is also spelt, still exists to the east of Asfuzâr, the name at the present day being written Adraskan. Yâḳût records Asfuzâr as of Sijistân, and Mustawfî speaks of it as a medium-sized town, with many villages and gardens rich in grapes and pomegranates, where already in the 8th (14th) century most of the people were Sunnîs of the Shâfi'ite school. The relative positions of the other towns of the district are, unfortunately, not given in the Itineraries[1].

The high road from Herât northward to Marv-ar-Rûd crosses the great district of Bâdghîs (Bâdhghîs), which occupied the whole stretch of country lying between the Herât river on the west (to the north of Fûshanj) and the upper waters of the Murghâb on the east, where these issue from the mountains of Gharjistân; and Bâdghîs was itself watered by many of the left-

A. Y. i. 312. The Ṣanî'-ad-Dawlah states (*Mirât-al-Buldân*, i. 298) that he passed near and saw the ruins of Bûshanj when travelling down from Nîshâpûr to Herât, near but not at Ghurian.

[1] Ist. 249, 264, 267. I. H. 305, 318, 319. Muk. 298, 308, 350. Yak. i. 248. Mst. 187.

bank affluents of the Murghâb. The eastern part of Bâdghîs, beginning some 13 leagues to the north of Herât, was known as the Kanj Rustâḳ district, and had three chief cities, Baban, Kayf, and Baghshûr, the positions of which can approximately be fixed by the Itineraries. In the remainder of Bâdghîs a list of nine large towns is given by Muḳaddasî, but unfortunately the positions of none of these can be fixed, for they are not mentioned in the Itineraries, and at the present day the whole of this country is an uninhabited waste, having been ruined in the 7th (13th) century by the Mongol invasions. The numerous ruins scattered throughout the district still attest the former state of prosperity of this well-watered country, but the modern names are not those given by the medieval authorities.

The remains of the city of Baghshûr, one of the chief towns of Kanj Rustâḳ, appear to be those now known as Ḳal'ah Mawr. In the 4th (10th) century Ibn Ḥawḳal describes Baghshûr as one of the finest and richest cities of Khurâsân, being of the size of Bûshanj. The governor of the district generally lived at Babnah or Baban, a larger town even than Bûshanj, while Kayf is described as half the size of Baghshûr. All these places had well-built houses of sun-dried bricks, and were surrounded by fertile gardens and farms, for this district was abundantly irrigated by streams, and from wells. Yâḳût, who visited these countries in 616 (1219), confirms the above account of the former riches of Baghshûr and its neighbouring towns, but says that in his day the whole country had gone much to ruin, though this was before the Mongol invasion. Babnah he names Bavan, or Bawn, and he had himself stayed here; having also visited another town called Bâmiyîn, or Bâmanj, which lay at a short distance only from Babnah. The country round he saw to be most fertile, and pistachio trees grew and flourished here abundantly[1].

In regard to the southern part of the Bâdghîs district the

[1] I. R. 173. Ist. 269. I. H. 320. Muk. 298, 308. Yak. i. 461, 481, 487, 694; ii. 764; iv. 333. For the present condition of the Bâdghîs country and its ruins, see C. E. Yate, *Afghanistan*, pp. 67, 68. There are ruined forts and remains at Gulrân, and Sagardân, and Ḳârâ Bâgh (p. 101), also at Ḳal'ah Mawr (pp. 96, 103), and at Ḳârâ Tappah, some of which must be those of the towns named by the Arab geographers.

accounts of its former prosperity are as circumstantial as those describing Kanj Rustâk, but its towns have now completely disappeared from the map, and the medieval names are difficult to locate, or identify with those given to the existing ruins. The capital by all accounts was Dihistân, the position of which may correspond with the present shrine of Khwâjah Dihistân to the north-east of Herât; and Mukaddasî mentions seven other great cities, namely Kûghânâbâdh, Kûfâ, Busht, Jâdhâwâ, Kâbrûn, Kalwûn, and Jabal-al-Fiddah or 'the Silver Hill,' the positions of which can only be very approximately indicated. Dihistân, the second largest city of Bâdghîs, was in the 4th (10th) century a place half the size of Bûshanj, and stood on a hill, its houses built of clay bricks, with good underground chambers for use in the summer heats. It had few gardens, but much arable land. The governor of the province lived at Kûghânâbâdh, a smaller place than Dihistân. Jabal-al-Fiddah, as its name implied, was a town where there was a silver mine in the neighbouring hill, and it lay on the direct road from Herât to Sarakhs, and apparently to the north of Kûghânâbâdh. Fire-wood grew abundantly in its district. The town of Kûfâ was a larger place than Jabal-al-Fiddah, and stood in a plain with excellent gardens; but of the four other towns mentioned by Mukaddasî no details are afforded, except the fact that they all lay near the road running from Herât north to Sarakhs.

Yâkût, who mentions Dihistân as the capital of Bâdghîs, says the name of the district signifies *Bâd-khîz*, 'where the wind rises,' on account of its tempestuous climate. The account which Mustawfî gives of Bâdghîs is difficult to understand, for the names of places have been much corrupted in the MSS. Dihistân was the capital, and the silver mine is referred to under the Persian form of Kûh Nukrah, 'silver mountain'; a third place of importance was Kûh Ghunâbâd (for Kûghânâbâdh), where the governor lived; and a fourth town was apparently called Buzurg-tarîn, but the reading is uncertain. Mustawfî also mentions a town named Kârîz (or Kârîzah), 'the Watercourse,' which he adds was the native place of Hakîm Burkâ'î—'the physician with the face-veil'—commonly known as the Moon-maker of Nakhshab, in other words the Veiled Prophet of Khurâsân, whose revolt in the 2nd (8th) century gave the Caliph Mahdî so much trouble to suppress.

Other places are also mentioned (with many corruptions in the text), reproducing the list given by Muḳaddasî and the earlier Arab geographers, but no details are added. In the 8th (14th) century, according to Mustawfî, Bâdghîs was chiefly remarkable for its pistachio forests; and at the time of harvesting the nuts, great numbers of men assembled here, each gathering what he could carry away, and the nuts being afterwards sold in the neighbouring districts. Such was the abundance of the pistachio trees that Mustawfî adds, 'many make their livelihood for the whole year round by what they can gather here at harvest-time, and it is indeed a wonder to behold.' At the close of the 8th (14th) century the ruin of Bâdghîs appears to have been finally brought about by the passage of the armies of Tîmûr on their devastating march from Herât to Marv-ar-Rûd[1].

To the east of Bâdghîs, at the head-waters of the Murghâb river, is the mountainous region known to the earlier Arab geographers as Gharj-ash-Shâr. The prince of these mountains had the title of the Shâr, and Gharj, according to Muḳaddasî, meant 'mountain' in the local dialect, so that Gharj-ash-Shâr was equivalent to the 'Mountains of the Shâr.' In the later middle-ages this region came to be more generally known as Gharjistân, and as such figures largely in the account of the Mongol invasion. Further, as Yâḳût remarks, Gharjistân, often spelt Gharshistân or Gharistân, was often confounded with Ghûristân, or the Ghûr country, lying to the east of it, which will be more particularly

[1] Ist. 268, 269. I. H. 319, 320. Muk. 298, 308. Yak. i. 461; ii. 633. Mst. 187, 188. J. N. 314, 315. A. Y. i. 308. C. E. Yate, *Afghanistan*, p. 6. The route from Herât to Marv-ar-Rûd, given in the Itineraries of the earlier Arab geographers, goes from city to city through Kanj Rustâḳ, and the southernmost stage (Babnah) is two days' march from Herât. Mustawfî (p. 198) gives a rather different road in seven stages, namely, from Herât in 5 leagues to Hangâmâbâd, thence 5 to Bâdghîs (to be understood doubtless as Dihistân the capital), thence 5 to Bawan (or Babnah), thence 5 to Marghzâr Darrah, 'the Valley of the Meadow-lands,' thence 8 to Baghchî Shûr (Baghshûr), thence 5 to Usrûd, or Lûsrûd, and finally, 4 leagues into Marv-ar-Rûd. For the ruined caravanserais which still apparently mark this route see C. E. Yate, *Afghanistan*, pp. 194, 195, 222.

discussed presently. The Shâr, or prince of Gharjistân, had of old been known to the Arabs as Malik-al-Gharjah (the king of the Gharj people), and in the 4th (10th) century this was a rich district, counting ten Friday Mosques as standing in its various towns.

The two chief cities of Gharjistân were called Abshîn and Shurmîn, the exact sites of which are unknown. Abshîn (Afshîn, or Bashîn) lay a bow-shot distant from the eastern bank of the upper Murghâb, and four marches above Marv-ar-Rûd. Round it were fine gardens, and much rice was sent from thence to Balkh. It had a strong castle, and a Friday Mosque. Shurmîn (or Surmîn) lay in the mountains four marches southward of Abshîn, and likewise four marches from Karûkh to the north-east of Herât. From it they exported currants to all the neighbouring places. The prince of the country, the Shâr aforesaid, resided at neither of these places, but at a great village in the mountains called Balîkân (or Balkiyân). Yâḳût gives the names of two other cities of Gharjistân, namely Sinjah and Baywâr, but except that they lay in the mountains, 'as a man of the country told me,' he cannot indicate their position[1].

The great mountain region to the east and south of Gharjistân was known as Ghûr, or Ghûristân, and it stretched from Herât to Bâmiyân and the borders of Kâbul and Ghaznah, also southward of the Herât river. The medieval geographers refer to it as the country of the head-waters of many great rivers, namely of the Harî Rûd, also of the Helmund, the Khwâsh, and the Farah rivers (which drained to the Zarah lake), while on its Gharjistân frontier rose the Murghâb. The geography of this immense region of mountains is, unfortunately, a complete blank, for the sites of none of the towns and castles mentioned in its history are known. In the 4th (10th) century, according to Ibn Ḥawḳal, Ghûr was infidel land, though many Moslems lived there. Its

[1] Ist. 271, 272. I. H. 323. Muk. 309, 348. Yak. i. 803; iii. 72, 163, 186, 785, 786, 823. Gharjistân of Khurâsân has nothing to do with Gurjistân south of the Caucasus (see Chapter XII, p. 181) now commonly known to us as Georgia, and it is quite a mistake to give the name of Georgia to Gharjistân, as has been done by some writers when describing the Mongol invasion of this region of the upper Murghâb, for there is no Georgia of Afghânistân.

valleys were populous and extremely fertile; it being famous for mines, both of silver and gold, which existed in the mountains towards Bâmiyân and Panj-hîr (see above, p. 350). The richest of these mines was called Kharkhîz. After the fall of the dynasty of Maḥmûd of Ghaznah, the Ghûrid chiefs, at first his lieutenants, became independent, and eventually founded their capital at Fîrûzkûh, an immense fortress in the mountains, the position of which is not known.

The Ghûrid princes ruled independently from the middle of the 6th (12th) century to 612 (1215), when they were defeated by the Khwârizm Shâh, and a few years later the dynasty disappeared at the time of the Mongol invasion. Before this, however, in 588 (1192), the Ghûrids had conquered much of northern India, holding all the country from Dehli to Herât, and after the dynasty had been annihilated by the Mongols the Slave Kings (their Mamlûk generals) continued to rule Dehli in a long line of Sultans, down to 962 (1554).

Ghûr, or Ghûristân, attained its highest point of splendour and riches between 543 and 612 (1148 and 1215) under the Ghûrid princes of the Sâm dynasty. Yâḳût speaks of their great capital at Fîrûzkûh, or Bîrûzkûh (Turquoise Mountain), but gives no details; Mustawfî also briefly refers to this fortress, and says that another of its chief towns was Rûd Hangarân, but the reading is very uncertain. In 619 (1222) the whole country was overrun by Changîz Khân, Fîrûzkûh being stormed and left in ruins. Two other great fortresses are named as having given much trouble to the Mongol troops, namely Kalyûn and Fîvâr, lying ten leagues distant one from the other, but the position of neither is known, and both are said to have been entirely destroyed by Changîz Khân. Ḳazvînî in the 7th (13th) century also names Khûst as one of the great cities of Ghûr, and possibly this is identical with Khasht, the place previously mentioned (p. 410) as near the head-waters of the Harî Rûd. In the time of Tîmûr the only place referred to in Ghûr appears to be the castle called Ḳal'ah Khastâr, but, again, nothing is known of its position[1].

[1] Ist. 272. I. H. 304, 323. Yak. iii. 823; iv. 930. Kaz. ii. 244. Mst. 184, 188. A. Y. i. 150. On Ghûr see the article by Sir H. Yule in the *Encyclopædia Britannica* (9th edition), x. 569.

The city of Bâmiyân was the capital of a great district of the same name which formed the eastern part of Ghûr, and as its very ancient remains show, was a great Buddhist centre long before the days of Islam. Iṣṭakhrî describes Bâmiyân as half the size of Balkh in the 4th (10th) century, and though the town, which stood on a hill, was unfortified, its district was most fertile, being watered by a considerable river. Muḳaddasî names the city Al-Laḥûm, but the reading is uncertain, and he praises it as 'the trade-port of Khurâsân and the treasure-house of Sind.' It was very cold and there was much snow, but in its favour was the fact that bugs and scorpions were conspicuously absent. The city had a Friday Mosque, and rich markets stood in the extensive suburbs, while four gates gave egress from the town. In the 4th (10th) century the Bâmiyân territory included many large cities, the sites of which are now completely lost. The three chief towns are said to have been called Basghûrfand, Sakîwand, and Lakhrâb.

Yâḳût in the beginning of the 7th (13th) century describes in some detail the great sculptured statues of Buddha still to be seen at Bâmiyân. High up in the mountain side, he writes, there was a chamber supported on columns, and on its walls had been sculptured the likenesses of 'every species of bird that Allah had created—most wonderful to see.' Without the chamber-entrance are 'two mighty idols cut in the live rock of the hill-side, from base to summit, and these are known as the Surkh Bud and the Khing Bud [the Red and the Grey Buddha] and nowhere else in the world is there aught to equal these.' Ḳazvînî speaks of a 'Golden House' at Bâmiyân, and likewise describes the two great statues of Buddha; further he mentions a quicksilver (*zîbaḳ*) mine and a sulphur spring as of this neighbourhood. The ruin of Bâmiyân and all its province, even as far east as the Panj-hîr mines, as already mentioned, was due to the wrath of Changîz Khân, whose favourite grandson Mûtûkin, son of Jaghatay, was killed at the siege of Bâmiyân. The Mongol troops were ordered to level with the ground the town walls and all the houses, and Changîz forbade any to build or live here ever again, the name of Bâmiyân being changed to Mav Baliḳ, which in the

Turki dialect means 'the accursed city.' Since that time Bâmiyân has been an uninhabited waste[1].

[1] Ist. 277, 280. I. H. 327, 328. Muk. 296, 303, 304. Yak. i. 481. Kaz. ii. 103. Mst. 188. A. G. 114, 149. For illustrations of the great Buddhist sculptures at Bâmiyân see Talbot and Maitland, in *J. R. A. S.* 1886, p. 323.

CHAPTER XXX.

KHURÂSÂN (*continued*).

The Balkh quarter of Khurâsân. Balkh city and Naw Bahâr. The district of Jûzjân. Ṭalikân and Jurzuwân. Maymanah or Yahûdîyah. Fâryâb, Shaburkân, Anbâr, and Andakhûd. The Ṭukhâristân district. Khulm, Siminjân, and Andarâbah. Warwâlîz and Ṭâyikân. The products of Khurâsân. The high roads through Khurâsân and Ḳûhistân.

Balkh—'Mother of Cities'—gave its name to the fourth Quarter of Khurâsân, which, outside the district of the capital, was divided, west and east, between the two great districts of Jûzjân and Ṭukhâristân.

In the 3rd (9th) century Ya'ḳûbî speaks of Balkh as the greatest city of all Khurâsân. It had had of old three concentric walls, and thirteen gates, and Muḳaddasî adds that it had been called in early days the equivalent, in Persian, of Balkh-al-Bahiyyah, 'Beautiful Balkh.' Outside the town lay the famous suburb of Naw Bahâr, and the houses extended over an area measuring three miles square. There were, says Ya'ḳûbî, two score Friday Mosques in the city. Iṣṭakhrî remarks that Balkh stood in a plain, being four leagues from the nearest mountains, called Jabal Kû. Its houses were built of sun-dried bricks, and the same material was used in the city wall, outside which was a deep ditch. The markets and the chief Friday Mosque stood in the central part of the city. The stream that watered Balkh was called Dahâs, which, says Ibn Ḥawḳal, signifies 'ten mills' (in Persian); the river turns these as it runs past the Naw Bahâr gate, flowing on thence to irrigate the lands and farms of Siyâhjird on the Tirmid road. All round Balkh lay gardens producing oranges, the Nilûfar lily, and the sugar-cane, which, with the produce of its vineyards, were all exported in quantity. Further, its markets were much frequented by merchants.

The city possessed seven gates, namely Bâb Naw Bahâr, Bâb Raḥbah (the Gate of the Square), Bâb-al-Ḥadîd (the Iron Gate) Bâb Hinduwân (the Gate of the Hindus), Bâb-al-Yahûd (the Jews' Gate), Bâb Shast-band (the Gate of the Sixty Dykes), and Bâb Yaḥyâ. Muḳaddasî describes in general terms the beauty, splendour, and riches of Balkh, its many streams, its cheap living, for food-stuffs were abundant, the innumerable broad streets, its walls and its Great Mosque, also its many well-built palaces; and in this state of prosperity Balkh flourished till the middle of the 6th (12th) century, when it was laid in ruins for the first time by the invasion of the Ghuzz Turks in 550 (1155). After their departure the population came back, and rebuilt the city in another but closely adjacent place. In part Balkh before long recovered its former splendour, and thus is described by Yâḳût in the early part of the 7th (13th) century, immediately before its second devastation at the hands of the Mongols.

Of the great suburb of Balkh called Naw Bahâr, where according to Mas'ûdî had stood, in Sassanian days, one of the chief fire-temples of the Guebres, Yâḳût has a long account, which he quotes from the work of 'Omar-ibn-al-Azraḳ of Kirmân, and a similar description is found in Ḳazvînî. Of this fire-temple at Balkh the chief priest had been Barmak, ancestor of the Barmecides, and in Sassanian days his family had been hereditary chief-pontiffs of the Zoroastrian faith in this city. The account given of Naw Bahâr, briefly, is that it was originally built in imitation of, and as a rival to, the Ka'abah of Mecca. Its walls were adorned with precious stones, and brocaded curtains were hung everywhere to cover these, the walls themselves being periodically unguented with perfumes, especially in the springtime, for Naw Bahâr means 'First or Early Spring,' the season when pilgrimage was made to the shrine. The chief building was surmounted by a great cupola, called Al-Ustûn, a hundred ells and more in height, and round this central building were 360 chambers, where the priests who served had their lodgings, one priest being appointed for each day of the year. On the summit of the dome was a great silk flag, which the wind blew out at times to a fabulous distance. This principal building was full of figures or idols, one of which in chief the pilgrims from Kâbul, India, and China prostrated themselves before, afterwards kissing the hand of Barmak, the chief priest. All the lands round Naw Bahâr for seven leagues square were the property of the sanctuary, and these brought in a great revenue. The great Naw Bahâr shrine was destroyed by Aḥnaf ibn Ḳays, when he conquered Khurâsân in the days of the Caliph 'Othmân, and converted the people to Islam[1].

The Mongols in 617 (1220) devastated Balkh, and according to Ibn Baṭûṭah, Changîz Khân ruined the third part of its Great Mosque in his fruitless search for hidden treasure. When Ibn Baṭûṭah visited this district in the earlier half of the 8th (14th) century Balkh was still a complete ruin, and uninhabited, but outside the walls were a number of tombs and shrines that were still visited by the pious pilgrims. In the account of the campaigns of Tîmûr, at the close of the 8th (14th) century, Balkh is often mentioned, and by this date must have recovered part of its former glory. Tîmûr restored the fortress outside the walls called Ḳal'ah Hinduwân, the Castle of the Hindus, which became the residence of his governor, and at a later date he also rebuilt much of the older city.

Balkh at the present day is an important town of modern Afghanistân, and is celebrated for its great shrine, called Mazâr-i-Sharîf (the Noble Tomb), where the Caliph 'Alî—known as Shâh-i-Mardân, 'King of Men'—is popularly supposed to have been buried. According to Khwândamîr this, supposititious, grave of the martyred 'Alî was discovered in the year 885 (1480), when Mîrzâ Bayḳarâ, a descendant of Tîmûr, was governor of Balkh. For in that aforesaid year a book of history, written in the time of

[1] Ykb. 287, 288. Ist. 275, 278, 280. I. H. 325, 326, 329. Muk. 301, 302. Mas. iv. 48. Yak. i. 713; iv. 817, 818. Kaz. ii. 221. The curious passage about Naw Bahâr will be found translated, in full, by M. Barbier de Meynard in his *Dictionnaire Géographique de la Perse*, p. 569. The presence of the idols, great and small, and the (sacred) flags, suggested to Sir H. Rawlinson the idea that Naw Bahâr had been originally a Buddhist shrine, and the name he explained as Naw Vihârah, 'the New Vihârah,' or Buddhist Monastery. See *J. R. G. S.* 1872, p. 510.

Sultan Sanjar the Saljûḳ, was shown to Mîrzâ Bayḳarâ, in which it was stated that 'Alî lay buried at the village of Khwâjah Khayrân, a place lying three leagues distant from Balkh. On the governor forthwith going there and making due search a slab was discovered bearing the inscription in Arabic, 'This is the tomb of the Lion of Allah, and His saint, 'Alî, brother [for cousin] of the Apostle of Allah.' A great shrine was therefore built over this grave, and ever since this has been highly venerated by the people of central Asia, and is still a notable place of pilgrimage[1].

Jûzjân (Al-Jûzajân or Juzjânân) was the western district of the Balkh quarter, through which the road passed from Marv-ar-Rûd to Balkh city. During the middle-ages this was a most populous district, possessing many cities, of which three only now exist under their old names, though the positions of most of the other towns mentioned by the Arab geographers can be fixed from the Itineraries. Though the names are changed, ruins still mark their sites. The whole district was extremely fertile, and much merchandise was exported, especially hides, which were tanned here and carried to all parts of Khurâsân[2].

Three marches distant from Marv-ar-Rûd, towards Balkh, was the city of Ṭâliḳân, the name of which is no longer found on the map, but the ruins and mounds of brick near Châchaktû probably mark its site. Already in the 3rd (9th) century Ṭâliḳân was a town of much importance, and Ya'ḳûbî says that the Ṭâliḳân felts made here were celebrated. The town lay among the mountains, and there was a magnificent Friday Mosque here. Iṣṭakhrî in the following century stated that Ṭâliḳân was as large as Marv-ar-Rûd, and its climate was more healthy. Its houses were built of sun-dried bricks. Near by was the village of Junduwayh, where, according to Yâḳût, in the 2nd (8th) century, the great battle had been fought and won by Abu Muslim at the head of the Abbasid partizans against the Omayyad troops. Shortly after the time when Yâḳût wrote, in 617 (1220), Ṭâliḳân was stormed after a siege of seven months by Changîz Khân, and

[1] I. B. iii. 58, 59. A. Y. i. 176. Khwândamîr, iii. pt 3, p. 238. C. E. Yate, *Afghanistan*, 256, 280.

[2] Ist. 271. I. H. 322. Muk. 298. Yak. ii. 149.

all the population were massacred, its castle being razed to the ground.

In the mountains—with a situation at the foot of hill-spurs and gulleys that, it was said, resembled Mecca—was the town of Jurzuwân, where the governor of the Jûzjân district passed the summer heats. The name of Al-Jurzuwân, as the Arabs called it, the Persians pronounced Kurzuwân or Gurzuvân, and it was also written Jurzubân or Gurzubân. It lay between Ṭâliḳân and Marv-ar-Rûd, in the district towards the Ghûr frontier, and, Yâḳût says, was very populous and full of rich folk. No place of this name now exists on the map, but the ruins at Ḳal'ah Wâlî most probably mark its site[1].

The city of Maymanah, which lay two marches beyond Ṭâliḳân on the Balkh road, still exists as a flourishing town. In the earlier middle-ages it was called Al-Yahûdân, or Al-Yahûdîyah, 'the Jews' Town,' and was often counted as the capital of Jûzjân. Its Friday Mosque, Ibn Ḥawḳal says, had two minarets. Yâḳût, who gives the name also under the form Jahûdân-al-Kubrâ, 'the Great Jewry,' says that it was first settled by the Israelites whom Nebuchadnezzar sent hither from Jerusalem. The name was changed to Maymanah, meaning 'the Auspicious Town,' for the sake of good augury, since 'Jew-town' to the Moslems was a term of reproach, and as Maymanah it exists at the present day. Maymanah is apparently also mentioned by Mustawfî, who speaks of it, in the 8th (14th) century, as a medium-sized town of the hot region, growing corn, fruit, and dates, and taking its water-supply from the neighbouring river. There is, however, possibly some confusion between this Maymanah of Jûzjân, and Maymand for

[1] Ykb. 287. Ist. 270. I. H. 321, 322. Yak. ii. 59, 129; iii. 491; iv. 258. A. G. 114. C. E. Yate, *Afghanistan*, 157, 194, 195, 196, 211. The ruins at Châchaktû (Ṭâliḳân) are 45 miles as the crow flies from Bâlâ Murghâb (Marv-ar-Rûd), which would be an equivalent of the three days' march, in a mountainous country, from this last place to Ṭâliḳân. The name of Châchaktû (written Jîjaktû) is mentioned by 'Alî of Yazd (i. 806; ii. 593) in his accounts of the campaigns of Tîmûr, but Ṭâliḳân is not mentioned by him. The ruins at Ḳal'ah Wâlî (probably Jurzuwân) lie 27 miles from Bâlâ Murghâb. An alternative site might be found at the considerable remains existing near Takht-i-Khâtûn. Either of these places may be Gurzuwân, which it is to be noted was a mint city under the Khwârizm Shâhs.

Maywand in Zâbulistân, half-way between Girishk and Ḳandahâr; and this confusion reappears in the pages of Yâḳût, who writes of Maymand (or Mîmand) of Ghaznah, and says it 'lay between Bâmiyân and Ghûr,' evidently meaning Maymanah or Yahûdîyah. One march from Yahûdîyah or Maymanah was the town of Kandaram, also written Kandadram, the residence, according to Ya'ḳûbî, of the governor of Jûzjân. It was a city of the mountains, Iṣṭakhrî writes, rich in vineyards and nut-trees, and abundantly irrigated by running streams[1].

One of the most important towns of Jûzjân during the middle ages was Al-Fâryâb, the name of which has completely disappeared from the map, but from the position given by the Itineraries the ruins of Fâryâb may be identified as those now known as Khayrâbâd, where there is an ancient fort surrounded by mounds of brick. Al-Fâriyâb, as Ibn Ḥawḳal spells the name, was in the 4th (10th) century a smaller town than Ṭâliḳân, but more fertile and with finer gardens. It was very healthy, and much merchandise was to be found collected here. It had a fine Friday Mosque, which however possessed no minaret. Yâḳût, who also spells the name Fîryâb, gives its position in regard to Ṭâliḳân and Shaburḳân, but adds no details. In 617 (1220), shortly after his time, Fâryâb was completely ruined by the Mongols, and it is only incidentally mentioned by Mustawfî. Between Al-Yahûdîyah and Al-Fâryâb, according to Ibn Ḥawḳal, there stood the city of Marsân, nearly of the size of Al-Yahûdîyah in the 4th (10th) century; and possibly this is identical with the village of Nariyân which Yâḳût mentions as in a like position. Of this mountain region also was the small city of Sân which Ibn Ḥawḳal describes as having many fruitful gardens growing grapes and nuts, for its streams brought water without stint[2].

[1] Ykb. 287. Ist. 270, 271. I. H. 321, 322. Yak. ii. 168; iv. 719, 1045. Mst. 185. C. E. Yate, *Afghanistan*, 339.

[2] Ist. 270. I. H. 321, 322. N. K. 3. Yak. iii. 840, 888; iv. 775. Mst. 188. C. E. Yate, *Afghanistan*, 233. Fâryâb of Jûzjân is called Dih Bâryâb by Nâṣir-i-Khusraw, who passed through it going from Shaburḳân to Ṭâliḳân. It is also given as Bârâb in the *Jahân Numâ* (p. 324), and it is not to be confused with Fârâb, also called Bârâb, which is Otrâr on the Jaxartes, as will be mentioned in Chapter XXXIV.

Shaburḳân, spelt variously Ashbûrḳân or Ushburḳân, also Shubûrḳân or Sabûrghân, which still exists, had in the 3rd (9th) century been once the seat of government of the Jûzjân district, which afterwards was removed to Yahûdîyah (Maymanah), at that time its equal in size. Its gardens and fields were wonderfully fertile, and large quantities of fruits were exported. Yâḳût, who spells the name Shubruḳân or Shufruḳân and Shabûrḳân, says that in 617 (1220), at the time of the Mongol invasion, it was a very populous town, with much merchandise in its markets. A century later Mustawfî speaks of it in similar terms, coupling Shubûrḳân and Fâryâb together, also adding that corn was abundant and cheap here.

One day to the south of Shubûrḳân, and the same distance eastward of Yahûdîyah, was Anbâr, otherwise written Anbîr, which Ibn Ḥawḳal says was larger than Marv-ar-Rûd. Here the governor of the district had his residence in the winter. No town of this name now exists, but by position Anbâr is probably identical in site with Sar-i-pûl, on the upper part of the Shubûrḳân river, still a place of some importance. The town was surrounded by vineyards and its houses were clay-built. It was often counted as the chief city of Jûzjân, and is probably the town which Nâṣir-i-Khusraw visited on his road to Shuburghân, and which he calls the city (or capital) of Jûzjânân. He speaks of its great Friday Mosque, and remarks on the wine-bibbing habits of the people. Out in the plain, to the north-west of Shubûrḳân, lies the town of Andkhuy, the name of which in the earlier geographers is spelt variously Andakhud, Addakhûd, and An-Nakhud. Ibn Ḥawḳal speaks of it as a small town out in the desert, with seven villages lying round it, and, in the 4th (10th) century, for the most part inhabited by Kurds, who possessed many sheep and camels. Yâḳût mentions it, but adds no details; the name also frequently occurs in the accounts of Tîmûr's campaigns[1].

The great district of Ṭukhâristân lay to the eastward of Balkh, stretching along the south side of the Oxus as far as the frontiers

[1] Ykb. 287. Ist. 270, 271. I. H. 321, 322. N. K. 2. Yak. i. 367, 372; iii. 254, 256, 305, 840. Mst. 188, 189, 190. A. Y. i. 805; ii. 593. C. E. Yate, *Afghanistan*, 346.

of Badakhshân, and bounded on the south by the mountain ranges north of Bâmiyân and Panj-hîr. It was divided into Upper Ṭukhâristân, east of Balkh and along the Oxus, and Lower Ṭukhâristân which lay further to the south-east, on the frontiers of Badakhshân. Many towns are mentioned as of Ṭukhâristân by the medieval geographers, but they add few details concerning them, so that excepting in the case of those given in the Itineraries, and those which still exist, it is now impossible to identify the greater number of them.

Two days' march to the east of Balkh is Khulm, described by Muḳaddasî as a small city surrounded by many large villages and districts, with a good climate. Two days' march again from Khulm lay Siminjân and Rûb, two towns near together, which probably are represented at the present day by Haybak, south of Khulm, on the upper course of the Khulm river. Muḳaddasî speaks of Siminjân as a larger town than Khulm; it had a Friday Mosque, and excellent fruit was grown, and Yâḳût describes it as lying in a maze of valleys, which were, or had been, peopled by Arabs of the Tamîm tribe. Mustawfî mentions Siminjân as a large town, already ruined in the 8th (14th) century, but where corn, cotton, and grapes were much cultivated; and under the spelling Saminkân it is mentioned by 'Alî of Yazd in describing the march of Tîmûr from Khulm to the Indian frontier.

Beyond, south-east of Siminjân, was Baghlân, Upper and Lower, and in the latter district, according to Muḳaddasî, was the capital with a Friday Mosque in the 4th (10th) century. Baghlân, or Baḳlân, as the name of the district is spelt by 'Alî of Yazd, apparently lay along the road to Andarâbah, otherwise Andarâb, which is described by Muḳaddasî as having fine markets, being situated among valleys clothed by verdant forests. These valleys, which were on the northern slopes of the Panj-hîr range, had many silver mines in their recesses, according to Ibn Ḥawḳal, who speaks of two rivers, the Nahr Andarâb, and the Nahr Kâsân, as flowing down through this district. Yâḳût, who gives no additional details, spells the name Andarâb or Andarâbah[1].

[1] Ist. 279. I. H. 326. Muk. 296, 303. Yak. i. 372; ii. 827; iii. 142, 518. Mst. 188. A. Y. ii. 19. C. E. Yate, *Afghanistan*, 317. For the relative positions of these places see Map 1, p. 1.

The Khulm river does not flow into the Oxus, but is lost in marshes a few miles to the north of the ruins of the old town. At the nearest bend of the Oxus to Khulm, there was in the 4th (10th) century a strongly fortified guard-house, called Rubâṭ Mîlah, where the road coming in three marches from Balkh crossed the great river into Transoxiana and the Khuttal country. Two marches to the eastward of Khulm was Warwâlîz, or Warwâlîj, which Ibn Ḥawḳal and others describe as a large city in the 4th (10th) century. No town of this name now exists, but by its position in the Itineraries it must have stood very near the site of Ḳunduz. Yâḳût, who apparently by a clerical error gives the name as Wazwâlîn, adds no details, and neither he nor any of the earlier geographers mention Ḳunduz, which is doubtless an abbreviation for Ḳuhandiz, the common name for 'fortress' in Persian, and as such possibly applied to the old castle of Warwâlîz[1].

Two days' march to the east of Warwâlîz lay Ṭâyiḳân, or Ṭâliḳân of Ṭukhâristân, which still exists (not to be confounded with Ṭâliḳân of Jûzjân, described above, p. 423), and which in the 4th (10th) century was one of the most populous towns of the district. Aṭ-Ṭâliḳân, as Muḳaddasî spells the name, though Aṭ-Ṭâyiḳân is the better form, had a large market; it stood in the plain a bow-shot from the hills, and was in the 4th (10th) century about a third the size of Balkh. Its lands were watered by an affluent of the Oxus, called Khuttalâb (sometimes written Khaylâb); and the Watrâb river (or Tarâb, for the readings of these two names are doubtful) appears to have been one of its branch streams, which joined the Khuttalâb above Ḳunduz. The neighbourhood was extremely fertile, and it was a pleasant country; corn and much fruit, according to Mustawfî, were grown, and in the 8th (14th) century, most of the population were weavers. It then possessed a strong fortress, and was surrounded by well-cultivated districts, where grapes, figs, peaches, and pistachios grew abundantly. 'Alî of Yazd frequently mentions Ṭâyiḳân when relating the campaigns of Tîmûr, and according to the older geographers seven days' march east of this was Badakhshân, which will be noticed in the following chapter[2].

[1] Ist. 279. I. H. 326, 332. Muk. 296. Yak. iii. 518; iv. 926.

[2] I. K. 93. Ist. 275, 276, 278, 279. I. H. 326. Muk. 296, 303. A. F.

The most famous exports of Khurâsân, according to Ibn Ḥawḳal, were the silk and cotton stuffs of Naysâbûr and Marv. Both sheep and camels were to be had here cheap, and Turkish slaves—a boy or girl slave, he says, fetching as much as 5000 dînârs (about £2500)—and all food-stuffs were plentiful. Muḳaddasî enters into further details. Naysâbûr was the chief manufacturing centre. Various white cloths were made here; and stuffs for turbans woven in the straight, or across, or diagonally. Veils, thin lining materials, woollens and raw silks, brocades of silk and of silk mixed with cotton, and various linen stuffs and cloths of goat's hair; all these were famous products of Naysâbûr. Here, too, were made cloaks, fine thread, and tabby silks in all varieties. Ironware was forged here, as well as needles and knives. The gardens of Naysâbûr were renowned for their figs, truffles, and rhubarb, and from the mine in the hills of the Rîvand district came the famous turquoises (*fîrûsâj*) of Nîshâpûr.

The towns of Nisâ and Abîvard were noted for their raw silk stuffs, and the cloth that the women wove in these districts. Fox-skin pelisses also were made up here. Nisâ in particular had a special breed of falcons, and produced much sesame seed. From Ṭûs came great cooking pots, a speciality of the town, also mats, and most of the cereals were largely exported. Excellent belts and cloaks were likewise manufactured. Herât produced brocade stuffs of all kinds, preserves made of raisins and pistachios, and divers syrups. Steel, too, was admirably forged in Herât. From the mountainous country of Gharj-ash-Shâr came felts and carpets, saddle cloths and cushions. Gold was found here, and horses and mules were exported largely.

Marv was a great place for all loom work in silk, mixed cotton and silk, and pure cotton, of which veils and all sorts of cloth were woven. The districts round the city produced oil of sesame, condiments and aromatics, and manna. Brass pots were made in Marv, and its bakers produced a variety of excellent cakes. The neighbourhood of Balkh yielded sesame, rice, almonds, nuts, and raisins. Its soap-boilers were famous, and the

472. Yak. iii. 501; v. 24. Mst. 188, 189. A. Y. i. 82, 179. The name is spelt (with or without the article) Ṭâyiḳân, or Ṭâyikân, and, finally, Ṭâliḳân, like the town in Jûzjân.

confectioners here made divers kinds of the so-called 'honey' from grapes and figs, as well as a preserve of pomegranate kernels. Syrups and clarified butter were largely exported; and in the neighbourhood were mines of lead, vitriol, and arsenic. The incense of Balkh too was famous, and its turmeric, unguents, and preserves. From it came hides and cloaks, and from Tirmidh, across the Oxus, soap and assafoetida. As coming from Warwâlîj towards Badakhshân, Muḳaddasî gives a long list of fruits, such as nuts, almonds, pistachios, and pears. Rice and sesame too were largely traded, also various cheeses and clarified butter, and finally horns and furs, more especially fox-skins[1].

The high roads through Khurâsân and Ḳûhistân were as follows. The great Khurâsân road entered Khurâsân beyond Bisṭâm (in Ḳûmis, see p. 365), and from this place to Naysâbûr there were two routes. The northern, or caravan road went from Bisṭâm to Jâjarm, and thence by Âzâdvâr through the plain of Juvayn down to Naysâbûr. This is the road especially given by Mustawfî, and only in sections by Iṣṭakhrî and Ibn Ḥawḳal. The southern, shorter route is the post-road to Naysâbûr, which started from Badhash, already mentioned (p. 368) as two leagues from Bisṭâm. This road keeps along the skirts of the hills with the desert on the right hand, and coming to Asadâbâd, next passes through Bahmanâbâd or Mazînân, where a branch went north to Âzâdvâr. Continuing eastward through Sabzivâr, the post-road finally reaches Naysâbûr, and this is the route described by Ibn Khurdâdbih and in all the earlier Itineraries. From Asadâbâd going south-east, Muḳaddasî says there was a track across this corner of the Great Desert, in 30 leagues, to Ṭurshîz in Ḳûhistân, while from Naysâbûr to Ṭurshîz, the route is given by both Ibn Khurdâdbih and Muḳaddasî. From Naysâbûr north to Nisâ the stages are also given by Muḳaddasî[2].

One stage beyond Naysâbûr at Ḳaṣr-ar-Rîḥ or Dizbâd (Castle of the Wind) the Khurâsân road bifurcated. To the right, south-east, the way went down to Herât, and this will be noticed in the succeeding paragraph. From the Castle of the Wind, turning left

[1] Ist. 281. I. H. 330. Muk. 323—326.

[2] I. K. 23, 52. Kud. 201. I. R. 170 (with descriptive details of the road). Ist. 216, 284. I. H. 275, 333. Muk. 351, 352, 371, 372, 491. Mst. 196.

and north-east, the road went to Mashhad and Ṭûs, and from here by Mazdarân to Sarakhs, at the crossing of the Tajand river. From Sarakhs the desert was crossed to Great Marv, and thence by the desert again the road reached the Oxus bank at Âmul (or Chahâr-Jûy), whence, after leaving Khurâsân, Bukhârâ was the terminus. This stretch of the Khurâsân road from Naysâbûr to Âmul of the Oxus passage is given with but slight variations by nearly all the Itineraries, and most of its stages still exist at the present day under the old names[1].

As already said, the Khurâsân road branched to the right, one stage beyond Naysâbûr, whence Herât was reached. At Sarakhs and Marv there were also bifurcations to the right, these roads both going to Marv-ar-Rûd, and to this city also a road led north from Herât. From Marv-ar-Rûd the main road then led north-east to Balkh, beyond which it crossed the Oxus to Tirmidh. Taking first the Herât road, from the bifurcation at the Castle of the Wind, it was four stages to Bûzjân, and a like distance on to Bûshanj, whence to Herât was a day's march. This road is given by Ibn Rustah and the geographers of the 4th (10th) century, also by Mustawfî. From Bûzjân and from Bûshanj roads respectively went off to the south-west and west, which centred in Ḳâyin, and the distances between the various cities of Ḳûhistân are given by Iṣṭakhrî and others. At Ḳâyin also centred the roads coming from Ṭabas and Khûr on the borders of the Great Desert[2].

From Herât southwards the road went down to Zaranj, passing through Asfuzâr, and crossing the Sijistân frontier between that town and Farah (see above, p. 341). This road is given by Ibn Rustah and the three geographers of the 4th (10th) century. From Herât eastward, up the valley of the Harî Rûd to the Ghûr frontier, the names of the towns one day's march apart are given by the same three authorities. From Herât through Karûkh the distances are also given by the geographers of the 4th (10th) century, in days' marches to Shurmîn and Abshîn in Ghurjistân, whence down the Murghâb Marv-ar-Rûd was reached. And to Marv-ar-Rûd or Ḳaṣr Aḥnaf (Marûchak) the roads are given

[1] I. K. 24, 25. Kud. 201, 202. I. R. 171. Ykb. 279. Muk. 348, 351. Mst. 196 (as far as Sarakhs).

[2] I. R. 172 (with details of road but no distances). Ist. 283, 284, 286. I. H. 332, 333, 335. Muk. 351, 352. Mst. 197.

across Bâdghîs (going by Baghshûr, the capital) in Iṣṭakhrî, Ibn Ḥawḳal and Muḳaddasî, as also by Mustawfî in the 8th (14th) century[1].

From Sarakhs, and from Great Marv, respectively, two roads converged on Marv-ar-Rûd, the first crossing the desert between the two great rivers, the last coming up the Murghâb through the fertile lands and towns on its bank. The desert route, passing by a number of successive Rubâṭs, or guard-houses, is only given by Muḳaddasî, being merely copied by Mustawfî, and in the Turkish *Jahân Numâ*. The road from Great Marv up the Murghâb is given by Ibn Khurdâdbih and Ḳudâmah, also by Muḳaddasî, but by a different route[2].

From Marv-ar-Rûd to Balkh, through the Jûzjân district, Ibn Khurdâdbih and the earlier Itineraries give the road by Ṭâliḳân and thence on, either by Fâryâb and Shaburḳân, or by Yahûdîyah (Maymanah), and Anbâr, to Balkh. Iṣṭakhrî and Muḳaddasî give the distances by the number of marches. Mustawfî has a somewhat different route from Marv-ar-Rûd to Balkh, which passed to the westward of both Ṭâliḳân, which lay six leagues off the road to the right, and Fâryâb, which lay two leagues away likewise on the right hand, reaching Shaburḳân, and eventually, by the Jamûkhiyân bridge Balkh. This route is copied in the *Jahân Numâ*. From Balkh the Oxus was reached opposite Tirmidh in two stages, passing through Siyâhjird[3].

East from Balkh the road went by Khulm and Ṭâyiḳân to the frontiers of Badakhshân, a branch road going south-east from Khulm to Andarâbah and the Panj-hîr mines north of Kâbul. Iṣṭakhrî and Muḳaddasî also give skeleton routes from Balkh across the mountains to Bâmiyân, and thence south by Ghaznah to Ḳuṣdâr, with a branch from Ghaznah eastward to the Indian frontier, but in these routes the stages are uncertain, for the places named are elsewhere unknown[4].

[1] I. R. 173, 174. Ist. 248, 249, 285. I. H. 304, 305, 334. Muk. 348, 349, 350. Mst. 198.

[2] I. K. 32. Kud. 209. Muk. 347, 349. Mst. 196. J. N. 329.

[3] I. K. 32. Kud. 210. Ist. 286. I. H. 322. Muk. 346, 347. Mst. 197. J. N. 329.

[4] Ist. 286. I. H. 334, 335. Muk. 346, 349, 486.

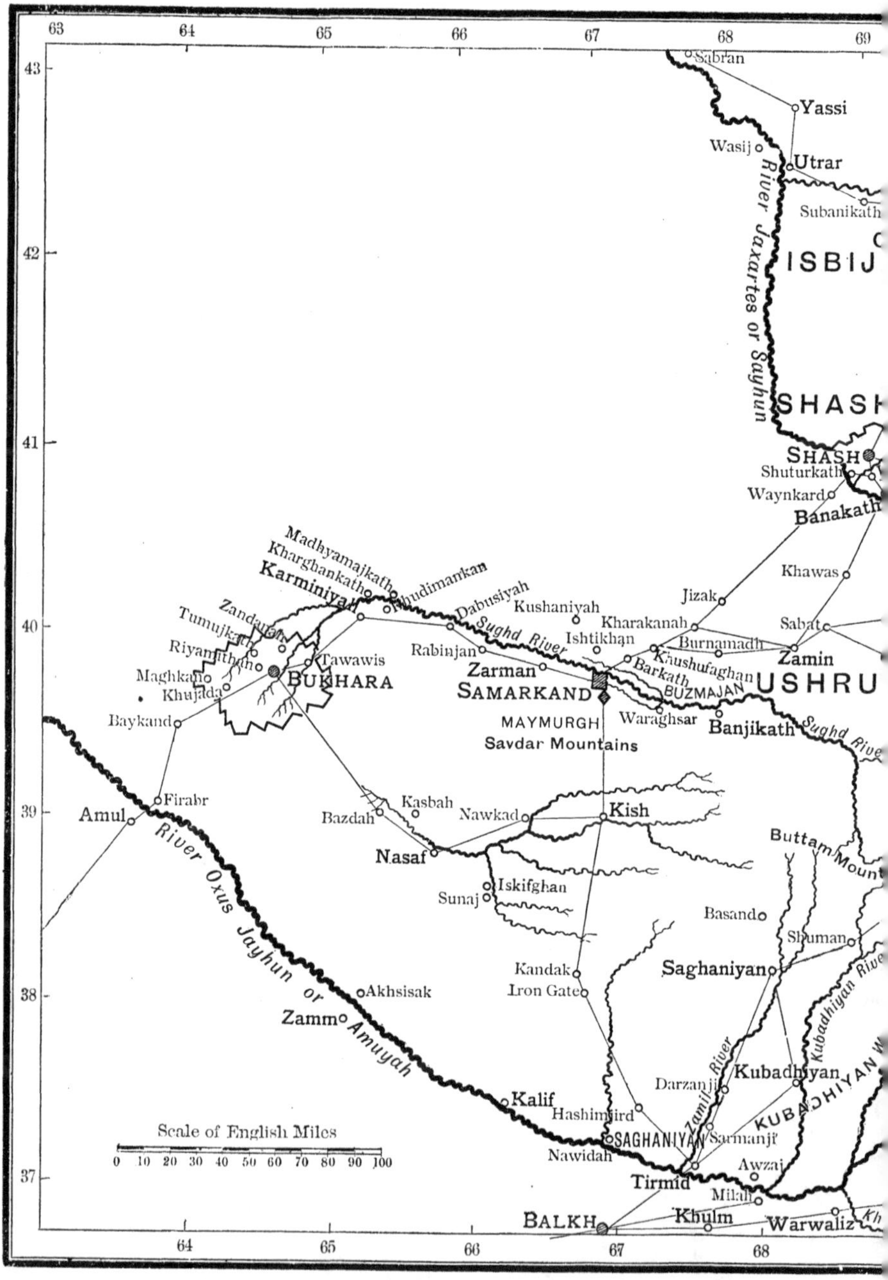

PROVINCES OF TH

MAP IX

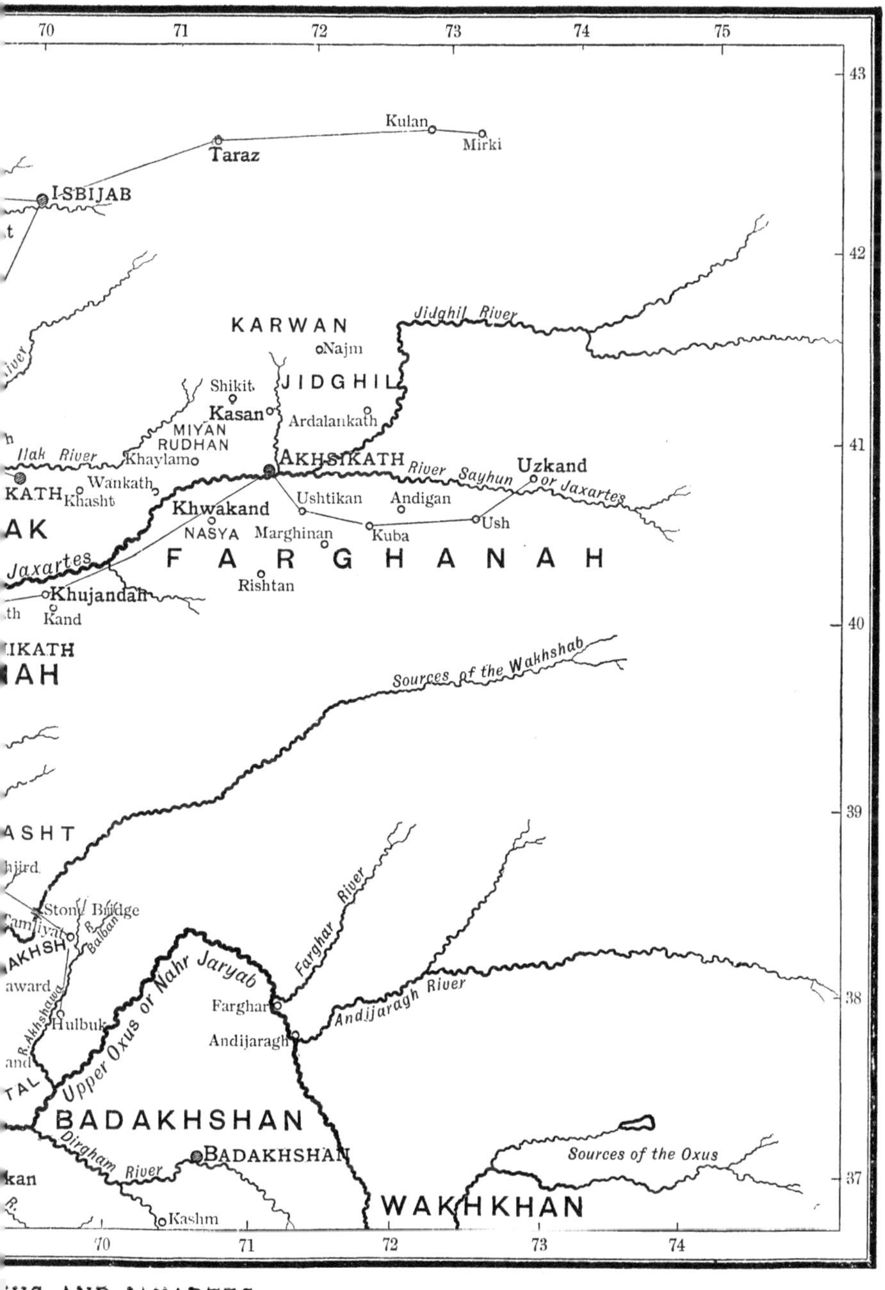

US AND JAXARTES

CHAPTER XXXI.

THE OXUS.

Transoxiana in general. The names Oxus and Jaxartes. The upper affluents of the Oxus. Badakhshân and Wakhkhân. Khuttal and Wakhsh. Ḳubâdhiyân and Ṣaghâniyân, with their towns. The Stone Bridge. Tirmidh. The Iron Gates. Kâlif, Akhsîsak, and Firabr. The Aral Sea or Lake of Khwârizm. Freezing of the Oxus in winter.

The Oxus of old was held to be the boundary line between the Persian-speaking folk and the Turks, between Îrân and Tûrân; and the provinces north, beyond it, were known to the Arabs as Mâ-warâ-n-Nahr, 'Those beyond the River' (Oxus, understood), or otherwise as the Hayṭal. The Hayṭal, in the 5th century A.D., had been the chief enemies of the Sassanian monarchy, and were identical with the Ephthalites of the Byzantine authors, commonly known as the White Huns. To the medieval Arabs, however, the name Hayṭal had come to be employed loosely to mean all the Tûrânian peoples and lands lying beyond the Oxus, and as such it is used by Muḳaddasî.

These lands may conveniently be divided into five provinces. The most important was Sughd, the ancient Sogdiana, with its two capitals Bukhârâ and Samarḳand. To the west of Sughd was Khwârizm, now generally known as Khivah, comprising the Oxus delta; and to the south-east Ṣaghâniyân, with Khuttal and the other great districts on the upper Oxus; to which also belonged Badakhshân, though this lay on the left or south bank, being almost encircled by the great bend of the river beyond Ṭukhâristân. Lastly, the two provinces of the Jaxartes were Farghânah on the upper river, and the province of Shâsh (now Tâshkand) with the districts to the north-west, running down to the outflow of the Jaxartes into the swamps of the Aral Sea.

The medieval Arabs knew the rivers Oxus and Jaxartes under the names, respectively, of Jayḥûn and Sayḥûn, which like the Tigris and Euphrates, the legend said, were the rivers of Paradise. The origin of these names is not quite clear, but apparently the Arabs took them from the Jews, Jayḥûn and Sayḥûn being corrupted forms of two of the rivers mentioned in Genesis ii. 11, 13, to wit the Gihon and the Pison[1].

In the later middle-ages, about the time of the Mongol irruption, the names Jayḥûn and Sayḥûn to a great extent went out of use; the Oxus was then more generally called the Amûyah, or Amû Daryâ, while the Jaxartes, as will be seen in a later chapter, came to be known as the Sîr Daryâ. The origin of the term Amûyah, or Amû, is also not quite clear. According to Ḥâfiẓ Abrû this is explained to be merely the name of the town and district on the Khurâsân bank of the Oxus originally written Âmul (at Chahâr Jûy, see above, p. 403). Possibly, however, the case is inverted, and the true explanation may be that Âmul city came to be called Amûyah or Amû from a local (Persian) name of the great river, which, coming into common use, supplanted the more classical (Arabic) name Jayḥûn. It is further to be observed that, with the Arabs, rivers were very commonly named from the great cities on their banks; hence the Oxus or Amû Daryâ, the River of Amû, also was often called the Balkh river, although that city stands some miles distant from its southern bank. The name Oxus, by which the Greeks knew the great river, is preserved in Wakhsh-âb, the Wakhsh river, which is one of its upper affluents,

[1] Ist. 286, 287, 295. I. H. 335, 347, 348. Muk. 261—268. As has been already mentioned (p. 131), the same names, under the slightly altered forms of Jayḥân and Sayḥân, were given to the Pyramus and Sarus respectively, the two frontier rivers of Cilicia, over against the Greek lands. As to the etymology of these names, it would seem that being taken from a foreign language, and their meaning unknown, the name Sayḥûn was brought into a jingling rhyme with Jayḥûn; and this is the case with many other borrowed names, e.g. in the Ḳurân and Tradition, *Ḳâbîl*, *Hâbîl*, for Cain and ʿAbel; *Ṭâlût*, *Jâlût*, for Saul and Goliath; *Yâjûj*, *Mâjûj*, for Gog and Magog. See Sir H. Yule, in Capt. J. Wood, *The Oxus* (1872), p. xxii.

but by the Arabs the name Wakhsh does not appear ever to have been applied to the main stream.

The sources of the Oxus, as Ibn Rustah and other early geographers rightly state, were from a lake in Little Tibet (At-Tubbat) and on the Pamir (Fâmir). Iṣṭakhrî, who is copied by most subsequent writers, gives the names of four among the many upper affluents of the Oxus. These are not in every case easy to identify, but the following appears to be clearly indicated. The main stream of the upper Oxus was the Nahr Jaryâb, at the present day known as the Panj river, which reached Badakhshân from the east, coming through the country known as Wakhkhân, and the Jaryâb was also known as the Wakhkhâb river. This main stream of the Oxus, coming down from the eastern highlands, makes an immense sweep round Badakhshân, flowing north, then west, and finally south before reaching the neighbourhood of Khulm, and in this course of three-quarters of a circle it receives many great affluents on its right bank. The first of these is the Andîjârâgh, with the town of the same name near its junction with the Oxus; and this appears identical with the present Bartang river. Next there joined the Nahr Fârghar (also written Farghâr, Farghân, or Farghî) flowing down from the Khuttal country, which must be identical with the Wanj river of to-day. Below came in the Nahr Akhshawâ (or Akhsh), almost equalling the main stream of the Oxus, on which stood Hulbuk, the chief town of Khuttal. One of its head-streams was the Nahr Balbân, or Barbân, and these united rivers at the present day are known under the Turkish name of Aḳ-Ṣû or White River. These, therefore, are the four upper affluents of the Oxus as named by Iṣṭakhrî, and he states that their various places of junction were all above the ford, or passage of the main stream at Ârhan.

Also above this ford, but on the left bank, the Badakhshân river, now called the Gukchah, flowed into the Oxus, being known as the Nahr Ḍirghâm. Below the Ârhan ford the Oxus received its great right-bank affluent, namely the Wakhshâb or Wakhsh river, from which the Greeks, as already said, took their name *Oxus*; and this divided the countries of Khuttal and Wakhsh on the east, from the districts of Ḳubâdhiyân and Ṣaghâniyân on the west. The Wakhshâb is the river now known as the Surkhâb, or Red River. Where the Oxus, after curving round three sides of Badakhshân, finally takes its course westward, it receives on its left, or southern bank the rivers of Ṭâyiḳân and Ḳunduz from Ṭukhâristân. These Ibn Rustah calls the Nahr Khuttalâb, and the Nahr Watrâb, respectively, as has been noticed in the previous chapter (p. 428). The two rivers of Ḳubâdhiyân and Ṣaghâniyân—the latter, which flows past Tirmidh, named the Nahr Zâmil by Ibn Rustah—joined the Oxus on its northern or right bank, and had their sources in the Buttam mountains, which here to the north divided the Oxus watershed from that of the Zarafshân in Sughd. These are the last of the affluents of the great river, for west of Balkh the Oxus receives no other stream, and takes its course through the desert, west and north-west, direct to its delta south of the Aral Sea[1].

The country of Badakhshân lay to the eastward of Ṭukhâristân, surrounded on three sides, as we have seen, by the great bend of the upper Oxus. Iṣṭakhrî describes this province as very populous and fertile, with refreshing streams and numberless vineyards. The capital was of the same name, but the Badakhshân (or Gukchah) river was, as already said, known as the Ḍirghâm by the Arabs. For the position of Badakhshân city no Itinerary that has come down to us gives information; but it seems probable, seeing the inaccessible nature of most of the country, that it stood in the valley where the present capital of the country, Fayzâbâd, now stands.

Badakhshân was from the earliest times famed for its precious stones, especially for the balas rubies and for the lapis-lazuli found at the Lâzward mines[2]. Muḳaddasî in the 4th (10th) century states that at the jewel mines was a fort, built by Zubaydah, the wife of Hârûn-ar-Rashîd, and called after her. Besides the ruby, the balas, and lapis-lazuli, the pure rock crystal of Badakhshân was famous, also the bezoar stone. Asbestos was also found here, called by the Arabs *Ḥajar-al-Fatîlah*, 'wick-stone,' for,

[1] I. R. 92, 93. I. K. 33. I. F. 324. Ist. 277, 296. I. H. 348. Muk. 303. I. S. 25*a*, 44*b*. Yak. ii. 171; iii. 469. In Ḳazvînî (i. 177) *Jarbâb* is for Jaryâb and (ii. 353) *Jaryân*; both clerical errors.

[2] Lâzwârd, or Lâzûrd, the name of the mine and mineral, is the origin of the word 'azure.'

being unconsumable, it was used for lamp-wicks. Muḳaddasî adds that of this asbestos fibre they wove mats for table-covers at meals, and when these got soiled by grease, all that was needed was to bake them for a time in an oven, when they became again perfectly clean. In like manner the asbestos lamp-wicks, when clotted with oil, were made as good as new by being put in the fire for an hour, nor, he adds, did they become consumed thereby. Further Muḳaddasî mentions a luminous stone, which in a dark room lighted up all things near it, probably some kind of phosphorescent fluor-spar.

Many of these details are repeated by Ḳazvînî, who, among the other precious stones found in Badakhshân, mentions the garnet, 'a stone like a ruby,' and states that in his day the asbestos stone was supposed by the common people to be formed of the petrified plumes of birds. The chief mines of the Balkhash, or balas ruby, were situated near the city of Yamkân; in the neighbourhood were silver mines, and Abu-l-Fidâ mentions the city of Jirm, which 'Alî of Yazd gives as the name of the Badakhshân river. When Tîmûr invaded Badakhshân in the latter part of the 8th (14th) century the capital was at Kishm, where the king of Badakhshân resided; and one of the chief towns was called Kalâûḳân, but no description is given of these places, and their positions are uncertain.

East of Badakhshân, along the upper Oxus, lay Wakhkhân, described by Ibn Ḥawḳal as on the road into (Little) Tibet, whence came musk. These were infidel lands, and they adjoined the countries called As-Saḳînah and Karrân (or Karrâm); and beyond these again towards Kashmîr was the Bulûr country, 'where for three months you never see the sun for snow and rain.' The silver mines of Wakhkhân were famous in the 4th (10th) century, and gold was found in the beds of its streams. The slave caravans from central Asia came down through this country bringing captured slaves to Khurâsân for the Moslem markets of the West[1].

As already said, the largest affluent of the Oxus was the Wakhshâb, coming in on the right bank from the north, and the

[1] Ist. 278, 279, 297. I. H. 327, 349. Muk. 303. Kaz. ii. 203, 225, 328. A. F. 472. A. Y. i. 179.

great mountainous tract lying in the angle between the Wakhshâb and the Oxus was known as Khuttal, a name that was also vaguely applied to all the infidel lands east and north of Khurâsân[1]. Khuttal included the country called Wakhsh, lying in its northern parts, where the Wakhshâb took its rise. It was, Iṣṭakhrî writes, very fertile, and famous for its fine horses and sumpter beasts; having many great towns on the banks of its numerous streams, where corn lands and fruit orchards gave abundant crops.

In the 4th (10th) century the capital of Khuttal was Hulbuk, where the Sultan lived (probably near the site of modern Khulâb); but the two cities called Munk and Halâward were larger than Hulbuk. Other considerable towns were Andîjâragh (or Andâjâragh) and Farghân (or Fârghar), lying respectively on the rivers of these names; also Tamliyât and Lâwakand, which last was on the Wakhshâb below the Stone Bridge (near modern Kurgân Tappah). Muḳaddasî describes Hulbuk, the capital, as having a Friday Mosque in its midst, and standing on the Akhshawâ river, from which it took its water. The town of Andîjâragh lay near the Oxus bank, where the affluent of the same name came in, and it probably occupied the site of the present Ḳal'ah Wamar. Munk, the largest city of the province, lay to the north of Hulbuk, and east of Tamliyât; while Halâward, on the Wakhshâb, was, according to Muḳaddasî, a finer town than Hulbuk the capital. Tamliyât lay between Munk and the Stone Bridge of the Wakhshâb, and is probably identical with the present Baljuwân; Baljuwân being already mentioned by 'Alî of Yazd when describing the campaigns of Tîmûr[2].

The celebrated Stone Bridge (Ḳanṭarah-al-Ḥijârah) over the Wakhshâb still exists. It is described by Ibn Rustah, Iṣṭakhrî,

[1] There is much confusion in the naming of this country; we have indifferently Khuttal and Khutlân or Khuttalân. According, however, to Ḳazvînî (ii. 352) Khuttalân was the name of a town of the Turks, lying in a gorge between the mountains, the position of which he does not indicate. 'Alî of Yazd (i. 464, and elsewhere), in describing the campaigns of Tîmûr, generally writes Khutlân. The name Khuttal (with its variants) appears in fact to be the same word as Hayṭal, by which name the Arabs knew the Ephthalites or White Huns of Sassanian and Byzantine times.

[2] Ist. 276, 277, 279, 296, 297. I. H. 326, 327, 348, 349. Muk. 290, 291. Yak. ii. 402. A. Y. i. 83.

and many late authorities as crossing the Wakhshâb on the road from Tamliyât to the town of Wâshjird in Ḳubâdhiyân. To the north of this lay the country called Bilâd-al-Kamîdh by Ibn Rustah, beyond which again was the Rasht district at the head-waters of the Wakhshâb. The Stone Bridge, according to Iṣṭakhrî, spanned a deep gorge of the Wakhsh river, at a place where, by reason of the great volume of the stream, more water, it was said, was hemmed in by narrows than at any other known spot on any other river. Ḳazvînî and other writers give a like account, and 'Alî of Yazd also refers to the bridge, giving both the Persian form, Pûl-i-Sangîn, and the Turkish, Tâsh Kûpruk. The place has more than once been described by modern travellers[1].

To the westward of the Wakhsh river, and bounded on the south by the Oxus, lay the district the Arabs named Ṣaghâniyân, which in Persian is written Chaghâniyân. The eastern part of the district was more particularly known as Ḳubâdhiyân, from the city of this name which stands on the first river joining the Oxus to the westward of the Wakhshâb. Ḳubâdhiyân, or Ḳuwâdhiyân, is described by Ibn Ḥawḳal as a smaller town than Tirmidh, and it was known also under the name of Fazz. It was famous for its madder, which was exported to India. The Ḳubâdhiyân river, on which the town lay, is of considerable length, and according to Muḳaddasî there were several important towns in this district, one of which was Awzaj or Ûzaj, probably the present Aywaj, on the northern bank of the Oxus above Tirmidh, and below Rubâṭ Mîlah of the left bank. Yâḳût adds that the fruits of this district were famous.

On the upper waters of the Ḳubâdhiyân river, and west of the Stone Bridge, lay Wâshjird, a town according to Iṣṭakhrî that almost equalled Tirmidh in size; and some distance to

[1] I. R. 92. Ist. 297. I. H. 348. Kaz. ii. 353. A. Y. i. 83, 452. Sir H. Yule in Wood, *The Oxus*, p. lxxxii; Mayef in *Geographical Magazine* for 1875, p. 337, and 1876, p. 328. At the present day the Stone Bridge is described as only ten paces in length, and is abutted on two projecting rocks. The Surkhâb flows below it, hemmed in by lofty and precipitous cliffs, which afford hardly thirty paces' interval for the passage of the stream, which pours down the narrow gorge with a tremendous roar.

the south of it was the great fortress of Shûmân, or Ash-Shûmân. In all this district round Shûmân much saffron was grown for export. Shûmân is referred to by Muḳaddasî as extremely populous, and the town was well built; Yâḳût adding that its population was ever in revolt against their Sultan, and that in his day it was a frontier fortress against the Turks. 'Alî of Yazd, describing the conquests of Tîmûr, frequently mentions it as Ḥiṣâr Shâdmân, and more shortly as Ḥiṣâr, or Ḥiṣârak, and at the present day it is also known as Ḥiṣâr[1].

Ṣaghâniyân city is probably identical with the modern town of Sar-i-Asyâ, on the upper part of the Ṣaghâniyân river, which was also called the Nahr Zâmil. It was, Iṣṭakhrî writes, a larger city than Tirmidh, in the 4th (10th) century, though the latter was more wealthy and populous. Ṣaghâniyân city was defended by a great Ḳuhandiz, or fort, and it stood on both banks of the river. Muḳaddasî likens it to Ramlah in Palestine, and there was a great Friday Mosque in its market-place. Wild-fowl abounded in its neighbourhood, and 6000 villages were counted in its districts, excellent bread being cheap throughout the neighbourhood. The small town of Bâsand, with a great public square and many gardens, lay two marches from Ṣaghâniyân city, among the mountains higher up the river. Lower down the Zâmil, about half-way between Ṣaghâniyân and Tirmidh, lay Dârzanjî, where there was a great guard-house, according to Ibn Ḥawḳal. Excellent wool-stuffs were produced here, and there was a great Friday Mosque in the market-place. South of this again, also near the Zâmil river, was the town of Ṣarmanjî or Ṣarmanjân, likewise with its great guard-house. The place had been famous in the 4th (10th) century for a dole of bread, of the daily value of a dînâr (10 shillings), which was given by its governor, Abu-l-Ḥasan, son of Ḥasan Mâh.

The most important town, however, of the Ṣaghâniyân district was Tirmid (or At-Tirmidh), north of the passage of the Oxus coming from Balkh, and at the place of junction of the Zâmil river. In the 4th (10th) century it was defended by a great fortress, where

[1] Ist. 298. I. H. 350. Muk. 284, 289, 290. Yak. ii. 88; iii. 337; iv. 196. A. Y. i. 49, 52, 450, 452, 464.

the governor lived, and a suburb lay round the town which was enclosed by an inner wall, while a second wall surrounded the suburb. A Friday Mosque of unburnt brick stood in the market-place of the town, but the market buildings were built of kiln-bricks, and the main streets were also paved with the same material. Tirmidh was the great emporium of the trade coming from the north for Khurâsân. The city had three gates, and according to Muḳaddasî was strongly fortified. In the year 617 (1220) it was sacked by the Mongol hordes as they passed south into Khurâsân. After this a new town—as large as the old one, according to Ibn Baṭûṭah, who visited it in the following century—was built two miles above the deserted ruins, and this was soon surrounded by gardens which grew excellent grapes and quinces.

On the right bank of the Oxus, some distance below Tirmidh, was Nawîdah, where those who went from Balkh to Samarḳand direct crossed the river. Nawîdah had a Friday Mosque in the midst of its houses, and was counted as the last town in Ṣaghâniyân on the Oxus. One march north-west of Tirmidh, on the road to Kish and Nakhshab in Sughd, was the town of Hâshimjird, a place of some importance in the 4th (10th) century; and two marches north of this the road passed through the famous Iron Gate.

This defile in the mountains was described by the Chinese traveller, Hwen Thsang, who as a Buddhist pilgrim visited India in 629 A.D.[1] The Arab geographers speak of a town here, and Ya'ḳûbî names it the City of the Iron Gate (Madînah Bâb-al-Ḥadîd), of which he also gives the Persian form, Dar Âhanîn. Iṣṭakhrî, Ibn Ḥawḳal, and Muḳaddasî all name the Bâb-al-Ḥadîd in their itineraries, but add no details. Under the name, in Persian, of the Darband Âhanîn the Iron Gate became famous from the time of Tîmûr, and it is mentioned by 'Alî of Yazd also under the Turkish appellation of Ḳuhlughah. He gives, however,

[1] For a translation of Hwen Thsang's description see Sir H. Yule in Wood, *The Oxus*, p. lxix. The Chinese pilgrim states that in his day the passage was 'closed by folding gates clamped with iron, and to the gates were attached a number of iron bells.' All later accounts omit any mention of gates, which apparently had been removed before the time of Iṣṭakhrî.

no description of the place. This remarkable defile was traversed by Clavijo, the Spanish ambassador to the court of Tîmûr, in August, A.D. 1405. He states that the ravine looked 'as if it had been artificially cut, and the hills rise to a great height on either side, and the pass is smooth, and very deep. In the centre of the pass there is a village, and the mountain rises to a great height behind. This pass is called the Gates of Iron, and in all the mountain range there is no other pass, so that it guards the land of Samarkand in the direction of India. These Gates of Iron produce a large revenue to Tîmûr, for all merchants who come from India pass this way[1].'

The Oxus below the Ṣaghâniyân district took its course through the desert, receiving no important affluents on either bank, and finally reached its delta on the south shore of the Aral Sea, where lay the province of Khwârizm, which will be described in the next chapter. In the stretch of desert several cities lay upon its right and left banks—generally in couples—at the points where the great river was crossed by roads going from Khurâsân to the Turk country, and most of the towns on the Khurâsân side have already been mentioned in the preceding chapter. The town of Kâlif or Kaylif, on the north bank (which still exists), was in the middle-ages faced by its suburb on the Khurâsân side, surrounding the guard-house called Rubâṭ Dhi-l-Kifl; and Kâlif was therefore at that time counted as occupying both banks of the Oxus, Mukaddasî likening it to Baghdâd and Wâsiṭ. On the northern side was the guard-house called after Alexander the Great, Rubât Dhi-l-Karnayn. Yâkût states that Kâlif had a fine castle, it was counted as 18 leagues from Balkh, and was on the road thence to Nakhshab in Sughd. Mustawfî speaks of a great hill near Kâlif, eight leagues in circumference, all of black earth, with water and fine grazing lands on its summit, and he adds that Kâlif in the 8th (14th) century was a large and very strong place.

Below this and opposite Zamm, which has already been

[1] Ykb. 290. Ist. 298, 337. I. H. 349, 350, 400, 401. Muk. 283, 284, 291, 292, 342. I. B. iii. 56. A. Y. i. 49, 59; ii. 593. Clavijo, *Embassy*, p. 122. *Geographical Magazine*, 1875, p. 336; and see 1876, p. 328, for the description of the Iron Gate by Mayef.

described (see p. 404), was the town of Akhsîsak, whence a high road went to Nakhshab. Ibn Ḥawḳal describes Akhsîsak as a small city, its inhabitants going over to Zamm for the Friday prayers, for there was no mosque in their town. The surrounding lands, enclosed on all sides by the desert, were extremely fertile, and the pasture for sheep and camels excellent. Near the right bank of the Oxus, lower down again, and opposite Âmul or Amûyah, stood Firabr, on the road to Bukhârâ, likewise surrounded by a fertile district, and many populous villages. Muḳaddasî writes that Firabr was a league distant from the north bank of the Oxus, and that it was protected by a fortress with guard-houses. The Friday Mosque stood at the town gate towards Bukhârâ, and there was a Muṣallâ, or praying station, with a hostelry outside this where travellers were entertained and a dole given. The grapes of the place were famous. Firabr was also known as Ḳariyat 'Alî, or Rubâṭ Ṭâhir ibn 'Alî, the village or guard-house of these persons[1].

After passing between Firabr and Amûyah, the Oxus held its course for about 140 miles, still through the desert, till it reached Ṭâhirîyah, where the cultivated lands of the delta began. From this point the great river took its course to the Aral Sea, throwing off for nearly 300 miles many irrigation canals which watered the rich province known as Khwârizm during the middle-ages. Since the date of the first Arab conquest the Oxus, in these delta lands, has of course frequently shifted its bed, and the bursting of the great dykes at the time of the Mongol invasion in the 7th (13th) century caused a change in its lower course which will be described later. From the description of the earlier Arab geographers, however, it is still possible roughly to reconstruct the map of Khwârizm in the 4th (10th) century, and it is evident that the Oxus in those days followed a single channel, navigable for boats, down to the swamps on the southern shore of the Aral, which sea the Arabs called the Lake of Khwârizm (Buḥayrah Khwârizm).

The Aral, which was shallow and full of reeds, appears not

[1] Ist. 298, 314. I. H. 349, 350, 363. Kud. 203. Muk. 291. Yak. iii. 862; iv. 229. Mst. 189.

to have been considered navigable; it received on its north-eastern shore the waters of the Jaxartes, but no traffic passed from the Oxus by water to the sister river. The land bordering the eastern coast of the Aral, between the mouths of the Oxus and Jaxartes, was in the 4th (10th) century, and later, known as the Desert of the Ghuzz Turkomans, a name more often given to the Marv desert of eastern Persia. To the earlier Arab geographers the wonder of the Oxus and Jaxartes was the fact that both these rivers froze over in winter, so that caravans of heavily laden beasts could cross on the surface of the river ice, which remained fast frozen, they reported, for from two to five of the winter months, the thickness of the ice reaching five spans and more. Ḳazvînî explains further how in winter the people of Khwârizm had to dig wells through the ice with crowbars till the water below was reached, and the cattle were brought up to drink at these holes, water being carried home to the houses in great jars. Iṣṭakhrî mentions a hill called Jábal Jaghrâghaz, on the Aral Lake shore, below which the water remained frozen all the year through.

The Aral Sea, especially in its southern part and near the creek of Khalîjân where the Oxus flowed in, was famous as fishing ground, but there were no villages or even houses bordering on the lake shore. As already said, all down the course of the Oxus through the delta, great and small canals branched from the right and left bank of the river, and many of these canals were also navigable; their waters finally serving to irrigate the delta lands. On one or other of these canals most of the great towns of Khwârizm had been built, rather than on the Oxus bank, which from the constant shifting of its bed was a source of ever recurring danger. The Oxus was navigable for boats throughout the whole of its lower course, and Ibn Baṭûṭah says that during the summer months the passage down stream from Tirmidh could be accomplished in ten days, cargoes of wheat and barley being thus brought for sale to the Khwârizm markets. The ice in winter made the navigation dangerous or impossible, and Yâḳût relates how in Shawwâl 616 (December 1219), when going from Marv to Jurjânîyah, part of his voyage being by boat on the Jayḥûn, he and his companions came near perishing from the intense cold and the sudden freezing of the river. They were hardly able to land and get up the bank, which was deeply covered with snow, and Yâḳût lost the beast he was riding, he himself barely escaping with his life[1].

[1] Ist. 303, 304. I. H. 353, 354. Kaz. ii. 353. I. B. iii. 5. Yak. i. 191.

CHAPTER XXXII.

KHWÂRIZM.

The province of Khwârizm. The two capitals: Kâth and Jurjânîyah. Old and new Urganj. Khîvah and Hazârasp. The canals of Khwârizm: towns to the right and left of the Oxus. Lower course of the Oxus to the Caspian. Trade and products of Khwârizm.

Khwârizm, in the earlier middle-ages, had two capitals, one on the western or Persian side of the Oxus called Jurjânîyah, or Urganj, the other on the eastern or Turkish side of the stream called Kâth, which in the 4th (10th) century was held to be the capital in chief of the province.

Kâth still exists, but the great medieval city probably stood some miles to the south-east of the modern town. In the earlier part of the 4th (10th) century Kâth came to be in part destroyed by the flood of the Oxus, which at this spot was two leagues in width. The city stood some distance back from the right bank of the main stream, being on a canal called the Jardûr, which ran through the town—the market, for about a mile in length, bordering this canal. Kâth, in these earlier times, had also possessed a fortress (Ḳuhandiz), which the floods had completely destroyed, and here had been the Friday Mosque and the prison, also the palace of the native chief known as the Khwârizm-Shâh. All this quarter of the town, however, had been rendered uninhabitable by the floods when Ibn Ḥawḳal wrote, and a new town had recently been built to the eastward, at a sufficient distance from the Oxus to be safe from the encroachments of the river.

The new city, which Muḳaddasî states was known as the

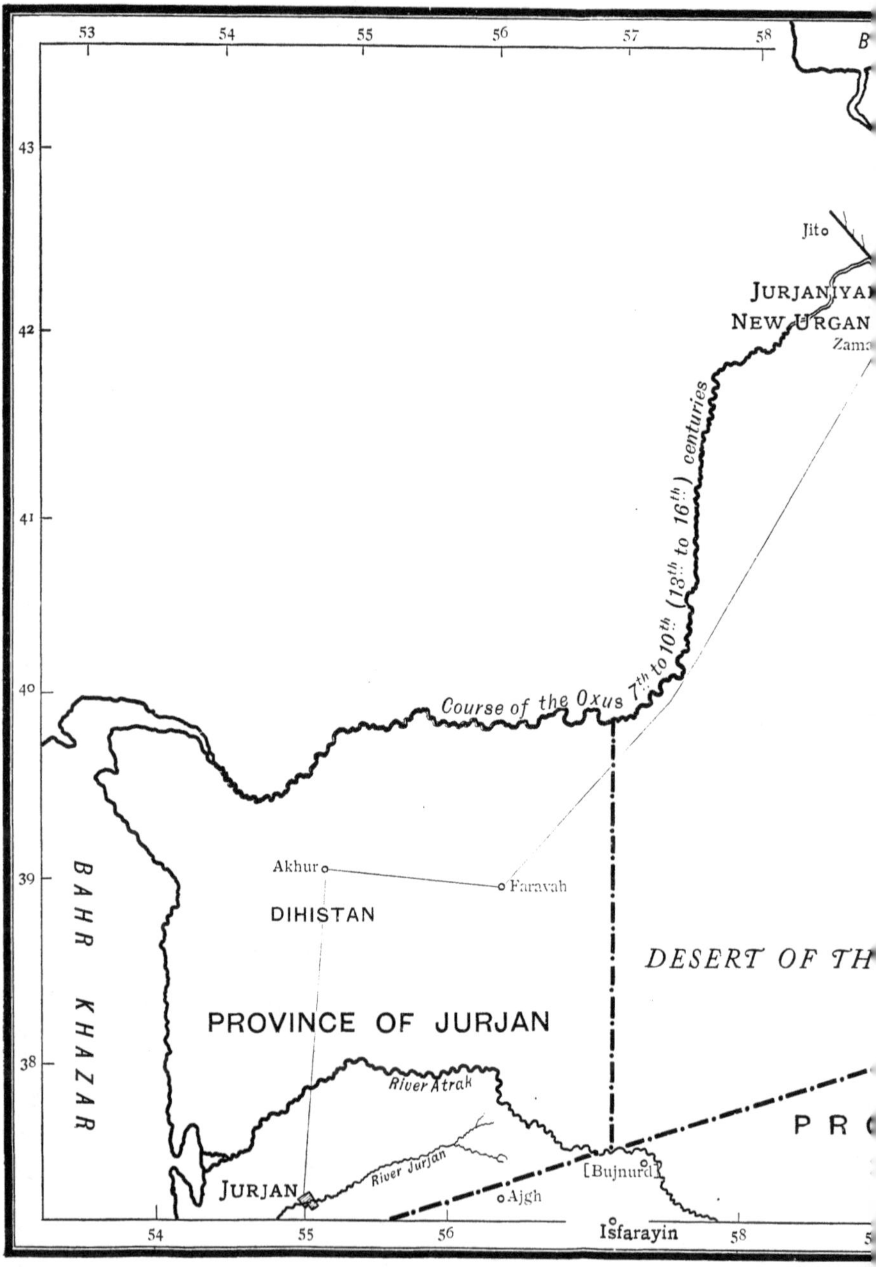

Jit
JURJANIYA
NEW URGAN
Course of the Oxus 7th to 10th (13th to 16th) centuries
Akhur
Faravah
DIHISTAN
BAHR KHAZAR
PROVINCE OF JURJAN
DESERT OF TH
River Atrak
River Jurjan
JURJAN
[Bujnurd]
Ajgh
Isfarayin
PR

PROVINC

MAP X

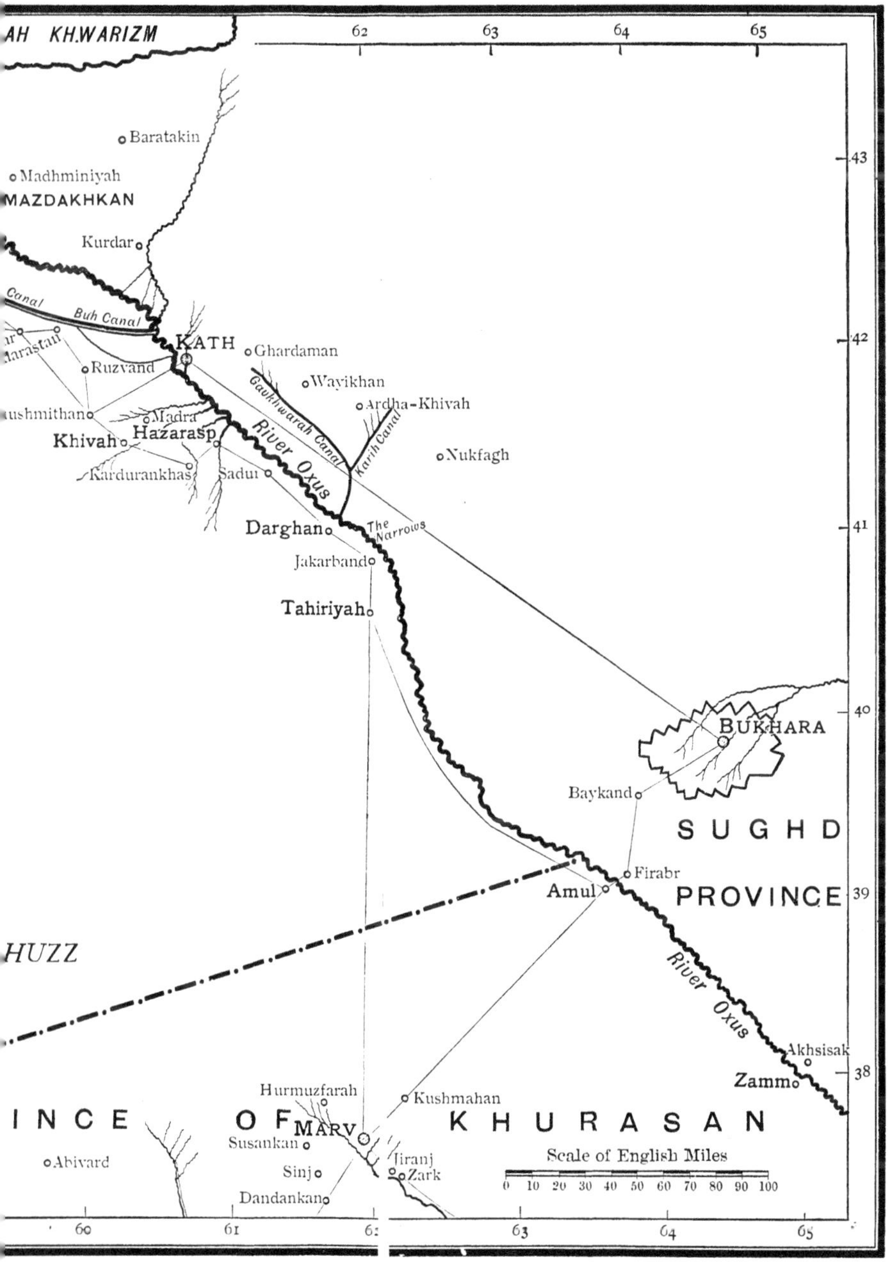

KHWARIZM

Shahristân—'the Capital'—by the Persians, was almost, he says, of the size of Naysâbûr in Khurâsân. In its market-place stood the Friday Mosque, built with columns of black stone, each of a man's height, and above came wooden pillars supporting the beams of the roof. The governor's palace was rebuilt in the new town, the old fortress being left a ruin. Numerous small canals traversed the streets, which says Muḳaddasî were infamously filthy—worse than Ardabîl in Adharbâyjân, for the people used the roadway for their commodity, and even brought the foulness of the gutters into the mosque on their feet when they came to prayers. The markets, however, were rich and well-stored with all kinds of merchandise, and the town architects were very skilful in their buildings, so that Kâth was outwardly a magnificent city. Soon after the close of the 4th (10th) century, however, it appears to have rapidly lost its position as the chief capital of Khwârizm; probably by reason of the recurrently destructive Oxus floods, which ever and anon threw down different quarters of the city; and eventually it sank to be a town of secondary rank.

Coming down to the beginning of the 7th (13th) century, Kâth does not appear to have suffered much during the Mongol invasion, and in the 8th (14th) century Ibn Baṭûṭah, who writes the name Al-Kât, passed through it on his way from Urganj to Bukhârâ, and describes it as a small but flourishing place. There was here a tank, and this at the time of his visit being frozen over, he describes the boys of the town as playing on its surface. At the close of the 8th (14th) century Tîmûr almost destroyed Kâth, but afterwards caused its walls to be rebuilt, and the place is frequently mentioned by 'Alî of Yazd as still in his day an important town[1].

The second capital of Khwârizm which, after the decay of Kâth, became the chief city of the province, was Gurganj, by the Arabs called Al-Jurjânîyah, and at a later date known as Urganj. The chronicles of the Moslem conquest relate that in the year 93 (712), when the Arabs under Ḳutaybah invaded Khwârizm, the capital city which they conquered was called Al-Fîl, 'the Elephant,' a name which was forthwith changed

[1] Ist. 300, 301. I. H. 351, 353. Muk. 287, 288. I. B. iii. 20. A. Y. i. 237, 263, 449.

to Al-Manṣûrah, meaning 'the City of Victory.' This city is said to have stood on the further side of the Oxus, and over against the later Jurjânîyah, but the Oxus flood before long overwhelmed Manṣûrah, and Jurjânîyah succeeded to its place[1].

Jurjânîyah in the 4th (10th) century—though at that time only the second city of the province, Kâth being still the capital—was already the chief centre of trade, and the meeting-place of caravans coming from the Ghuzz country, which exchanged goods with those from Khurâsân. Jurjânîyah lay a bow-shot to the westward of a great navigable canal coming from the Oxus and running a nearly parallel course, and the houses were protected from danger of flood by an immense dyke, with wooden piles to strengthen the embankment. Muḳaddasî in the 4th (10th) century states that the city had four gates, and that it was every day increasing in size. At the Bâb-al-Ḥujjâj, 'the Pilgrims' Gate,' stood a fine palace built by the Caliph Mamûn, with a second palace fronting it, built by Prince 'Alî his son, both overlooking a sandy square, like the famous Rîgistân of Bukhârâ, where the sheep-market was held. With the decay of Kâth, Jurjânîyah soon became the first, and then the sole capital of the Khwârizm province, and in later times it is generally referred to under the name of Khwârizm—'City' being understood.

In the year 616 (1219) Yâḳût was at Jurjânîyah, or Gurganj as he also calls it, shortly before the place was devastated by the Mongols under Changîz Khân; and he writes that he had never seen a mightier city, or one more wealthy or more beautiful. In 617 (1220) all this was changed to ruin. The great canal dykes having been broken down, the waters of the Oxus flowed off by a new course, as will be shown later, and the whole city was laid under water. The Mongol hordes when they marched away left nothing, according to Yâḳût, but corpses and the ruined walls of houses to mark the place of the great city. The capital of Khwârizm, however, in a few years rose from its ruins, rebuilt in a neighbouring spot. This, according to the contemporary chronicle of Ibn-al-Athîr, was in 628 (1231), when New Khwârizm was founded 'in the vicinity of Great Khwârizm.' Before the

[1] The position of Fîl is most uncertain; its name occurs as a mint city on the coins of the Omayyad Caliphs, one example being dated A.H. 79 (698).

Mongol invasion there had existed, according to Yâķût and others, a town known as Little Gurganj, by the Persians called Gurganjak, lying at a distance of three leagues from the capital, Great Gurganj, and it seems probable that Little Gurganj was the spot chosen for New Khwârizm.

New Khwârizm soon took its place as the capital, and is described by both Mustawfî and Ibn Baṭûṭah in the 8th (14th) century. Ķazvînî, who wrote in the latter half of the previous century, states that (New) Gurganj was then famous for its skilful blacksmiths and carpenters, also for its carvers of ivory and ebony bowls and other utensils, like those produced by the people of Ṭarķ near Isfahân. Further, the women here made famous embroideries, and the tailors were renowned. The water-melons of Khwârizm, he relates, were beyond compare, and this latter fact is confirmed by Ibn Baṭûṭah.

Mustawfî, who gives the common name of the city as Urganj, otherwise New Khwârizm, says that it lay ten leagues (probably a mistake for miles) from Old Urganj. Ibn Baṭûṭah, his contemporary, found Khwârizm (as he calls the place) a fine town, well-built, with broad streets, and a teeming population. The market was a magnificent building, like a caravanserai, and near it was the Friday Mosque with its college. Also there was a public hospital, attended, when Ibn Baṭûṭah was here, by a Syrian physician, a native of Saḥiyûn. Near the close of the 8th (14th) century this city of Khwârizm was again almost completely destroyed by Tîmûr, after a siege lasting three months. Tîmûr, however, caused it afterwards to be rebuilt, and the work was completed in 790 (1388). Abu-l-Ghâzî, the prince of Khwârizm, whose account of the lower Oxus course will be given presently, held his court at the beginning of the 11th (17th) century in this city, which he generally names Urganj, and speaks of as a fine place with many gardens; but after this date the town of Khîvah gradually replaced Urganj, becoming the new capital of the province. The ruins of this Urganj, the town built after the Mongol invasion, are those now known as Old Urganj (Kuhnah Urganj)[1].

[1] Anthony Jenkinson was at Urgence (as he spells the name) in 966 (1558), half-a-century before the time of Abu-l-Ghâzî, and describes it as a fine town

Khîvah—which under the Uzbeg chiefs after the time of Tîmûr gradually eclipsed Urganj and became the capital of Khwârizm, giving its name in time also to the whole province—is more than once mentioned as a small town by the Arab geographers of the 4th (10th) century. The older spelling of the name was Khîvaḳ, and this form was in common use down to the time of Yâḳût. Muḳaddasî describes Khîvah as lying on the border of the desert, but watered by a great canal which was brought to it from the left bank of the Oxus. It had a fine public square, also a Friday Mosque, so that in the 4th (10th) century it must already have been a place of some importance. Yâḳût, who says the name was also pronounced Khayvaḳ, speaks of its castle. In the 7th (13th) century its people were all Sunnîs of the Shâfi'ite sect, the rest of the population of Khwârizm following the Ḥanîfite ritual.

At this date Khîvah was already celebrated as the birth-place of the great Shaykh Najm-ad-Dîn, surnamed Al-Kubrâ. He played a foremost part in the defence of Urganj against the Mongols, who put him to death, and his tomb became a place of pious visitation near Urganj, as is described by Ibn Baṭûṭah in the century after his martyrdom. Khîvah is mentioned by 'Alî of Yazd, and he describes an adventure here of Tîmûr, when a young man, who at a later period caused the walls of Khîvaḳ (as the name was then spelt) to be rebuilt. The city in the 11th (17th) century is frequently mentioned by Abu-l-Ghâzî who sometimes lived here, as also at Kât (or Kâth), when not in residence at Urganj; and since his day, and down to the present time, the place has continued to rise in importance, being now the capital of the province called after it [1].

Hazârasp (meaning 'Hundred-horse' in Persian) on the same latitude as Khîvah, but standing nearer to the left bank of the Oxus, is a place of importance that has kept its name unchanged from

with walls 'by estimation four miles about it.' Hakluyt, *Principal Navigations* (Glasgow, 1903), ii. 463. Bal. 421. Ist. 299, 300. I. H. 350, 351. Muk. 288. Yak. ii. 54; iii. 933; iv. 261. A. F. 479. Ibn-al-Athir, xii. 257, 323. Kaz. ii. 349. Mst. 197, 234. I. B. iii. 3—6. A. Y. i. 298, 448. J. N. 345. A. G. 111. *Geographical Magazine* for 1874, p. 78.

[1] Muk. 289. Yak. ii. 512. Kaz. ii. 355. I. B. iii. 6. A. Y. i. 62, 449. A. G. 112, 294.

the Moslem conquest to the present time. Muḳaddasî in the 4th (10th) century describes it as of the same size as Khîvah, the town having wooden gates and being surrounded by a ditch. Yâḳût, who was here in 616 (1219), speaks of it as a strongly fortified and rich town, with excellent markets, where many opulent merchants had their warehouses. Hazârasp was almost surrounded by the waters of its canals, and was only to be reached by a single road, along a causeway coming from the Urganj direction across the level plain which stretched away from the Oxus bank.

About half-way between Ṭâhirîyah, where the cultivation of the delta began, and Hazârasp, the stream of the Oxus passed through a narrow gorge, now known as the Deveh Boyun, 'the Camel's Neck,' where high and precipitous cliffs hemmed in the current to a third of its normal breadth. Iṣṭakhrî calls these narrows by the name of Abûḳshah, or Bûḳshah, adding that the Oxus boatmen feared the spot greatly, on account of the whirlpool and the cataract at the exit of the passage. Mustawfî, who calls this place Tang-i-Dahân-i-Shîr, 'the Narrows of the Lion's Mouth,' says the opposite cliffs were barely 100 *gez* (yards) apart, and there was a guard-house here, on the left bank. Below this, according to him, the Oxus passed by an underground course for a couple of leagues, being completely hidden from sight.

Between Ṭâhirîyah and Hazârasp, on the left bank of the Oxus, there were three towns of some importance during the middle-ages. One march below Ṭâhirîyah, and on the high road, stood Jikarband, surrounded by gardens, with trees growing along its canals. A fine mosque, according to Muḳaddasî, stood in the midst of its market. A march further north, near the narrows of the Oxus, was the city of Darghân, which Muḳaddasî describes as almost of the size of Jurjânîyah. Its Friday Mosque was magnificently ornamented with precious marbles, and the town was two leagues across, being surrounded by nearly five hundred vineyards. Darghân was the first great town in Khwârizm on the road from Marv. Yâḳût, having been here in 616 (1219), describes it as standing on an elevation like the spur of a hill, with its gardens and arable fields stretching between the town and the Oxus bank, which was two miles distant. At the back of the town the desert sands were not far distant. Between Darghân and Hazârasp lay Sadûr on the river bank, a fortified town with a Friday Mosque in its midst, and surrounded by suburbs[1].

The first of the great canals of Khwârizm was taken from the right or eastern bank of the Oxus at a spot opposite Darghân, and was called Gâvkhuwârah, or Gâvkhwârah, the 'Cattle Feeder.' This canal, which was navigable for boats, being two fathoms deep and five fathoms across, went northwards, and irrigated all the lands up to the level of Kâth. Five leagues beyond its point of origin a small canal branched off from it, called the Karîh canal, and this too watered many districts. Four towns of some importance are mentioned by Muḳaddasî as of this eastern bank of the Oxus, each standing about a day's march one from the other, in the fertile districts south of Kâth. The most distant from Kâth was called Nûkfâgh, it stood in the midst of canals, was a fine town, and lay near the desert border. Nearer to Kâth was Ardhakhîvah, which is probably identical with the place called Ḥiṣn Khîvah by Yâḳût, and which he says was 15 leagues distant from Khîvah of the west bank. Ardhakhîvah was a fortress standing under a hill at the beginning of the desert, and having but a single gate. Wâyikhân, also a fortress, surrounded by a ditch and with catapults at its gate, lay one march again to the northward; and then came Ghardamân, one march from Kâth, a well-fortified place with two gates, encompassed by a great water ditch two bow-shots in width.

From the west, or left bank of the Oxus a number of canals were also taken, the first of which was that which was led past Hazârasp to irrigate its district. This also was navigable for boats, though it was only half the size of the Gâvkhuwârah canal; and it led backward, curving round in a direction that, if continued, would have reached the city of Âmul. Two leagues north of Hazârasp the Kardurân-Khwâsh canal branched from the Oxus, flowing past the town so named, which stood half-way between Hazârasp and Khîvah. This canal was larger than that which served Hazârasp, and the town of Kardurân-Khâs (as Muḳaddasî

[1] Ist. 304. I. H. 354. Muk. 288, 289. Yak. ii. 567; iv. 971. Mst. 198, 213.

writes the name) was surrounded by a ditch and had wooden gates. Further north again was the Nahr Khîvah, a still larger canal, by which boats went from the Oxus to that city. A fourth canal, flowing a mile to the northward of the Khîvah canal, was the Nahr Madrâ, which is described as twice as large as the Gâvkhuwârah of the east bank. The town and neighbourhood of Madrâ were watered by it.

Kâth, the eastern capital, as already said, stood back from the Oxus on a canal called the Jardûr, which was taken from the main stream some distance to the south of the city. Two leagues north of Kâth, but from the left or western bank of the Oxus, the great Wadhâk (also Wadâk or Wadân) canal branched off, which was navigable up to the neighbourhood of Jurjânîyah, the western capital of Khwârizm. The point of origin of the Wadâk canal was about a mile to the northward of that of the Madrâ canal, and further north again another canal called the Nahr Buwwah (Bûh or Bûyah) left the Oxus, its waters rejoining those of the Wadâk beyond to the north-west, a bow-shot distant from the village called Andarastân, and about one day's march to the southward of Jurjânîyah. The Wadâk canal was larger than the Bûh, but both were navigable as far as Jurjânîyah, where a dam prevented boats proceeding further northward; and a great dyke, as already said, had originally been built along its bank to keep the city from inundation[1].

The high road north from Khîvah to Jurjânîyah, in the middle-ages, passed through several large towns of which now no trace exists. One march from Khîvah was Ardhakhushmîthân, or Râkhûshmîthân, which Yâķût, who stayed here in 616 (1219), records as being a large city, with fine markets and much merchandise. It was, he says, more populous and more extensive than the city of Naṣîbîn, in Upper Mesopotamia, but it appears to have been ruined by the Mongol invasion. North of this was Rûzvand, a medium-sized town according to Muķaddasî, well fortified and surrounded by a ditch. It had excellent springs of water, and the Friday Mosque stood in its market-place. After passing the village of Andarastân, lying at the

[1] Ist. 301, 302. I. H. 352, 353. Muk. 288, 289, 292, 293. Yak. ii. 512; iv. 230.

junction of the Wadâk and Bûh canals, the town of Nûzvâr was reached, one march south of Jurjânîyah. Mukaddasî describes Nûzvâr as a small well-fortified city, having two iron gates, and surrounded by a ditch crossed by drawbridges, which were taken up at night, being laid on boats. There was a Friday Mosque in its market-place; and without the west gate was a fine bath-house. It is apparently the same town which Yâkût calls Nûzkâth, meaning, he says, 'New Kâth,' or 'New Wall,' and which was utterly destroyed, shortly after he left it, by the Mongol hordes.

Zamakhshar lay between Nûzvâr and Jurjânîyah, and in the 4th (10th) century this town had also drawbridges at its gates. There was a Friday Mosque here, and a strong prison, and it was fortified, having iron gates and a ditch. Yâkût in the 7th (13th) century speaks of this place as a village, and it became famous as the birth-place of one of the great commentators of the Kurân, Az-Zamakhsharî, who was born here in 467 (1075) and died in 538 (1144). Ibn Batûtah, who visited his tomb here in the 8th (14th) century, speaks of Zamakhshar as lying four miles from New Urganj. To the north of Urganj was the shrine of Najm-ad-Dîn Kubrâ already spoken of, and beyond this again, five leagues from Jurjânîyah on the desert border, under the tall cliffs to the west of the Oxus, stood Jîth or Gît, a place often mentioned by the earlier geographers. It was a large town with considerable lands round it, lying at some distance from the left bank of the river, being opposite Madhminîyah at four leagues from the right bank. Jîth appears to be identical in position with the later town called Wazîr (or Shahr-i-Wazîr), which probably replaced it, after the troublous times of the Mongol invasion and the campaigns of Tîmûr. Wazîr is frequently mentioned by Abu-l-Ghâzî, and the name occurs in the *Jahân Numâ*. This Shahr-i-Wazîr, moreover, is probably the town visited and described by Anthony Jenkinson under the somewhat altered form of *Sellizure*, or *Shaysure*, when he was travelling across Khwârizm in the 10th (16th) century[1].

On the right bank of the Oxus, some four leagues north of Kâth, the first of four canals led off, flowing northward, and

[1] Ist. 301. I. H. 352. Muk. 289. Yak. i. 191; ii. 940; iv. 822. I. B. iii. 6. A. G. 195. J. N. 346. Hakluyt, *Principal Navigations*, ii. 461.

after a short distance this was joined by three other small streams, their united waters forming the Kurdar canal. It was said that this, which was of the size of the Wadâk and Bûh canals of the west bank, had originally been an arm of the Oxus, and had flowed out to the north-east into the Aral. The district in the angle between the main stream of the Oxus and the Kurdar canal was called Mazdâkhgân (or Mazdâkhḳân), and it was watered by numerous minor channels taken from the right bank of the Oxus. The district is said to have comprised twelve thousand villages, and Kurdar was its chief town. This is described by Muḳaddasî as a large place and very strong; surrounded by numerous villages, with broad pasture lands for cattle. Two days' march from it, on the north-eastern border of Khwârizm, was the great village called Ḳariyat Barâtakîn (or Farâtagîn), near which were the hill-quarries producing the stone used in the buildings throughout Khwârizm. Barâtakîn in the 4th (10th) century had excellent markets, and a well-built Friday Mosque. To the westward of this place was the city of Madhmînîyah, four leagues from the right bank of the Oxus, opposite Jîth; and from hence down to the shore of the Sea of Aral there were no more cultivated lands, only swamps and reed beds lying at the mouth of the great river[1].

In the 4th century B.C., when Alexander the Great made his conquests in western Asia, the Oxus is described as flowing into the Caspian, and the Greek geographers apparently knew nothing of the Aral Sea. When the change of course from the Caspian to the Aral took place is not known, but though at the present day the Oxus, like the Jaxartes, flows into the Sea of Aral, its old bed to the Caspian still exists, is marked on our maps, and has been recently explored. In the earlier middle-ages the course of the Oxus, as described by the Arab geographers of the 4th (10th) century, is, in the main, that of the present day; but the old bed of the river leading to the Caspian is mentioned by Muḳaddasî, who reports that in former times the main stream had flowed down to a town over against Nisâ in Khurâsân, called Balkhân (or Abu-l-Khân). Later, some two and a half centuries after the time of Muḳaddasî, it seems certain that the Oxus once

[1] Ist. 299, 303. I. H. 350, 353. Muk. 288. Yak. iv. 257.

again resumed its older course. This we learn from the contemporary Persian authors. Hence there appears to be unimpeachable evidence that, from the early part of the 7th (13th) century to near the close of the 10th (16th) century, the Oxus, except for a moiety of its waters which still passed into the Aral by the canals, reached the Caspian along the old bed of the time of Alexander the Great, though at the present day, and since the end of the 10th (16th) century, this channel is once more disused and for the most part dry.

As has been mentioned above the chronicle of Ibn-al-Athîr states that the Mongol hordes in 617 (1220), in order finally to capture Urganj, after a five months' siege broke down the dykes and overwhelmed the city with the waters of the Oxus and its canals, which hitherto had flowed by divers channels to the eastward of the town. The whole country was laid under water, and the overflow after a time began to drain off to the south-west, filling the old bed of the Oxus, and following the line of depression to the Caspian at Manḳishlâgh. The latter Yâḳût, a contemporary of these events, speaks of as a strongly fortified castle standing on the shore of the Sea of Ṭabaristân (the Caspian), into which, he says, the Jayḥûn (the Oxus) flowed. This evidence from incidental notices is further fully corroborated by Mustawfî in the 8th (14th) century, who, in describing the course of the Oxus, states that though a small portion of its waters still drained off through canals from the right bank to the Aral Sea, the main stream after passing Old Urganj turned down the passage called the Steep of Ḥalam, where the noise of its cataract could be heard two leagues away, and thence flowing on for a distance of six days' march, had its exit in the Caspian (Baḥr Khazar) at Khalkhâl, a fishing station.

The position of the 'Aḳabah or Steep of Ḥalam, which the Turks, Mustawfî says, called Kurlâvah (or Kurlâdî), is given by him in his Itinerary, for the town of New Ḥalam stood about half-way between Old Urganj, destroyed by the Mongols in the previous century, and New Urganj which had taken its place. In his article on the Caspian Mustawfî further adds, when speaking of the port on the Island of Âbaskûn (see p. 379), that this island had in his time disappeared beneath the sea 'because the Jayḥûn, which formerly did flow into the Eastern Lake [the Aral] lying over against the lands of Gog and Magog, since the time of the Mongol invasion has changed its course and now flows out to the Sea of Khazar [the Caspian]; and hence, this latter sea having no outlet, the dry land [of the Âbaskûn island] has now become submerged in the rising level of its waters.'

All the above is confirmed by the account of the Oxus written in 820 (1417) by Ḥâfiẓ Abrû, who was a government official of Shâh Rukh, son and successor of Tîmûr, and who must have been well acquainted with the geography of this region from personal knowledge. In two distinct places he writes that, in the year just mentioned, the Oxus, which of old had discharged into the Lake of Khwârizm (the Aral), having taken a new channel, now flowed down by Kurlâvû, otherwise called Aḳranchah, to the Sea of Khazar (the Caspian), adding that the Aral Sea in his time had come almost to disappear. And again, Ruy Gonzalez de Clavijc, the Spanish ambassador who visited these regions in 808 (1405) some years before Ḥâfiẓ Abrû wrote his account, confirms this by his statement that the Oxus 'flows into the Sea of Bâkû,' which can only mean the Caspian. It must be admitted, however, that Clavijo here spoke from hearsay only.

What caused the Oxus once more to discharge into the Aral Sea is unknown, but this great change must have taken place before the close of the 10th (16th) century, for Abu-l-Ghâzî, a native of Urganj, refers to it as though it had become an accomplished fact in 984 (1576), namely some thirty years before he, Abu-l-Ghâzî, was born. The Oxus had, he says, at that date already made itself a new channel, and turning off below Khast Minârahsî (the Tower of Khast), took its way direct to the Aral Sea, thus changing the lands lying between Urganj and the Caspian into a desert for lack of water. And in another passage of his work speaking of former times, among events of the years 928 to 937 (1522 to 1531), he describes how all the way from Urganj to Abu-l-Khân on the Caspian there were arable fields and vineyards along what was still then the course of the lower Oxus. Apparently, however, Abu-l-Ghâzî places the change of bed rather too late, for already in 966 (1558), when Anthony Jenkinson travelled through Russia to Khîvah, he speaks of the Oxus as flowing 'not into the Caspian Sea as it hath done in times past,' for when he saw it the great river already took its course to the Aral Sea, 'the Lake of Kithay,' direct[1].

The chief products of Khwârizm were food-stuffs, cereals, and fruits. The land was extremely fertile and grew large crops of cotton, and the flocks of sheep gave wool. Great herds of cattle pastured on the marshlands near the Aral, and many kinds of cheese and curds were exported. The markets of Jurjânîyah were famous for the various costly furs, brought here from the Bulghâr country of the Volga, and a long list of these is given by Muḳaddasî and others. This list comprises the following skins; marten, sable, fox, and beaver of two kinds, as well as the furs of the squirrel, ermine, stoat and weasel, which were made up into pelisses and short jackets; also artificially dyed hareskins and goatskins, and the hide of the wild ass.

[1] Muk. 285. Yak. iv. 670. Mst. 197, 213, 225. J. N. 360. Hfz. 27 *b*, 32 *b*. A. G. 207, 291. Clavijo, *Embassy*, p. 118. Hakluyt, *Principal Navigations*, ii. 461, 462, 'Voyage of Anthony Jenkinson.' Professor De Goeje in *Das Alte Bett des Oxus* (Leyden, 1875) seeks to discredit the statements of the Persian geographers, and holds that during all the middle-ages the Oxus, as at the present day, flowed into the Aral. The evidence showing that a portion, at any rate, of the Oxus current flowed down the old bed to the Caspian, during more than three centuries, appears to be irrefutable; and it may be added that the late Sir H. C. Rawlinson, who had studied the question as a practical geographer, and knew at first hand the writings of the Arab and Persian authorities, always maintained the opinion that during those several centuries the Oxus did undoubtedly flow into the Caspian. It should be stated that some confusion has arisen from the divers names by which the Moslem geographers denote the Caspian and the Aral. The Caspian is generally referred to as the Sea of Khazar (Baḥr Khazar), from the tribes of the Khazars who inhabited its further shores, but it was also known as the Sea of Ṭabaristân or of Mâzandarân, or of Âbaskûn, or of Jurjân, from the names of the various well-known provinces or districts on its shores. Quite incorrectly the Caspian appears sometimes as the Daryâ Ḳulzum, but Ḳulzum was the name given to the Red Sea. The Aral was generally known as the Buḥayrah Khwârizm, or Lake of Khwârizm, and also as the Lake of Jurjânîyah (the capital of Khwârizm), and this last name being easily misread Jurjân has more especially caused confusion between the Caspian (Baḥr Jurjân) and the Aral (Buḥayrah Jurjânîyah). The Aral was also known to the Persian geographers as the Daryâ-i-Sharḳ, 'the Eastern Sea.' All this, however, does not invalidate the facts clearly recorded by Mustawfî, Ḥâfiẓ Abrû, and Abu-l-Ghâzî.

Among natural and manufactured products were wax, the bark of the white poplar, called Tûz, used for covering shields, fish-glue, fish-bones, amber, khalanj-wood, honcy, and hazel nuts, swords, and cuirasses and bows. Khwârizm also was celebrated for its falcons. Grapes, currants, and sesame were largely grown, and in the looms carpets, coverlets, and brocades of mixed cotton and silk were woven. Cloaks and veils of both cotton and silk stuffs were exported, and various coloured cloths. Locks were of the smith-work of the towns, and they had boats hollowed out of a single tree-trunk, which were used in the navigation of the numerous canals. The chief industry of Khwârizm, however, in the 4th (10th) century, as latterly, was the slave-trade; for Turkish boys and girls were bought or stolen from the nomads of the steppes, and after being educated and made good Moslems, were despatched from here to all the countries of Islam, where, as history relates, they often came to occupy high posts of command in the Government[1].

[1] Ist. 304, 305. I. H. 354. Muk. 325.

CHAPTER XXXIII.

SUGHD.

Bukhârâ, and the five cities within its wall. Baykand. Samarḳand. The Buttam mountains, and the Zarafshân or Sughd river. Karmîniyah, Dabûsiyah, and Rabinjan. Kish and Nasaf, with neighbouring towns. The products of Sughd. Routes beyond the Oxus as far as Samarḳand.

The province of Sughd, the ancient Sogdiana, may be taken as including the fertile lands, lying between the Oxus and Jaxartes, which were watered by two river systems, namely the Zarafshân, or Sughd river, on which Samarḳand and Bukhârâ stood, and the river which flowed by the cities of Kish and Nasaf. Both these rivers ended in marshes or shallow lakes in the western desert towards Khwârizm. More properly, however, Sughd is the name of the district surrounding Samarḳand; for Bukhârâ, Kish, and Nasaf were each counted as separate districts.

Sughd was accounted one of the four earthly paradises, and had attained its greatest splendour in the latter half of the 3rd (9th) century under the Sâmânid Amîrs; in the following century, however, it was still a province fertile and rich beyond compare. Of the two chief cities, Ṣamarḳand and Bukhârâ, it may be said that the former was rather the political centre, while Bukhârâ was considered to be the religious metropolis, but both were equal in rank, and held to be the capitals of Sughd[1].

Bukhârâ was also known under the name of Nûmijkath[2]. In

[1] Ist. 316. I. H. 365. Muk. 261, 262, 266—268. Yak. iii. 394.

[2] This, or Numûjkath, is the true reading of the name which (by an error of the diacritical points) is often wrongly written Bûmijkath. Muk. 267, note *b*. The true pronunciation is fixed by the Chinese pilgrims, who mention Bukhârâ under the name of 'Numi.'

the 4th (10th) century it was a walled city measuring a league across in every direction, which stood in the plain a short distance south of the main arm of the river of Sughd. There were no hills in the neighbourhood, and round it lay many towns, palaces, and gardens, gathered into a compass measuring 12 leagues in length and breadth, and enclosed by a Great Wall that must have been over a hundred miles in circuit. Through this great enclosure passed the Sughd river, with its numerous canals.

The city proper of Bukhârâ, outside the wall and to the north-west, had adjacent to it the fortress, which itself was like a small city. It was the residence of the governor and held the prison and the treasury. Beyond and round the town also were great suburbs, extending as far as the main arm of the river, and along its southern bank. Of the suburbs the chief were those lying to the east, namely the thoroughfares (*darb*) of Naw Bahâr, of Samarḳand, and of Râmîthanah, with others too numerous to mention, whose position cannot now be exactly fixed. The town wall had seven iron gates; Bâb-al-Madînah (the City Gate), Bâb Nûr (or Nûz), Bâb Ḥufrah, the Iron Gate, the Gate of the Fortress, Bâb Mihr or the Banî Asad Gate, and lastly the Gate of the Banî Sa'd. How these were situated is unknown, but the Gate of the Fortress (Bâb-al-Ḳuhandiz) must have been to the north-west, opening on the Rîgistân, the great sandy plain or public square of Bukhârâ which has ever been famous.

The two gates of the fortress were the Bâb-ar-Rîgistân, or Bâb-as-Sahl, 'the Gate of the (Sandy) Plain,' and the Bâb-al-Jâmi', this last opening on the Great Mosque, which also stood on the Rîgistân, at the city Gate of the Fortress above mentioned. The suburbs were traversed by ten main thoroughfares, each of which ended in its gateway, and these are all carefully named by both Iṣṭakhrî and Muḳaddasî. Further there were several gates in the streets shutting off the various quarters of the suburbs one from another, many of these gates being of iron. The Great Mosque was near the fortress, and there were numerous smaller mosques, with markets, baths, and open squares beyond count, and at the close of the 4th (10th) century the Government House stood immediately outside the fortress in the great square called the

Rîgistân. Ibn Ḥawḳal gives a detailed account of the chief canals which, starting from the left bank of the Sughd river, watered Bukhârâ and the gardens in the plain round the city, becoming finally lost in the desert towards the south-west, near Baykand on the Âmul road, for none of them reached the Oxus. The lower course of the river here was known as the Sâm Khâs, or Khwâsh[1].

The ruins of Old Bukhârâ of pre-Islamic days lie some miles to the north-west of the Moslem city, and near the river bank. They were known by the name of Rîyâmîthan, and Muḳaddasî describes them in the 4th (10th) century as still showing immense remains of the ancient city. Within the circuit of the Great Wall round the plain of Bukhârâ there were five flourishing cities. Of these Khujadah, or Khujâdâ, stood one league west of the high road going down from Bukhârâ to Baykand, and three leagues distant from the capital. Muḳaddasî describes it as a large and pleasant town, with a Friday Mosque and a castle. The town of Maghkân was beyond this, being five leagues from Bukhârâ, and three from the high road, close to the western circuit of the Great Wall. Maghkân had a Friday Mosque, was fortified, and had suburbs, besides many villages lying round it, for its lands were amply irrigated.

Tumujkath, or Tumushkath (often by a clerical error written Bumujkath, and Bûmijkaṭh), was a small town to the north-west of Bukhârâ, four leagues distant, and half a league off the high road to the left of one going to Ṭawâwîs. Aṭ-Ṭawâwîs (as the name was often written) means 'Peacocks,' and this was the largest of the five cities within the Great Wall. The town had a flourishing market and was much frequented by merchants from all parts of Khurâsân, its cotton stuffs being exported to 'Irâḳ. It was well fortified and had a castle, and the Friday Mosque stood in the market-place. The last of the five inner towns was Zandanah, which still exists at the present day. It is described as lying four leagues distant from Bukhârâ, to the north. It was well fortified, had a Friday Mosque in the town, and a populous suburb beyond its walls; and Yâḳût adds that the stuffs made here, and called from the town the Zandajî cloths, were widely celebrated.

[1] Ist. 305—309. I. H. 355—358. Muk. 280, 281. Yak. i. 517.

Two leagues outside the Great Wall, and five from Bukhârâ on the road down to the Oxus at Firabr, was the city of Baykand, which still exists. In the 4th (10th) century Baykand possessed but one gate, and was strongly fortified; it had a Friday Mosque in its midst, ornamented with precious marbles and with a finely gilded Mihrâb (or niche showing the Mecca point). There was a market held in the suburbs of Baykand, but no villages surrounded the town; only numerous guard-houses which are reported to have numbered a thousand all told. Beyond the town lay the sandy desert bordering the Oxus.

Throughout the earlier middle-ages Bukhârâ retained its pre-eminence; but in 616 (1219) came the Mongol invasion, and the city was pillaged and utterly ruined. For a century and more it did not recover from this devastation, and in the early part of the 8th (14th) century when Ibn Baṭûṭah visited the place, lodging in the suburb called Fatḥ-âbâd, the mosques, colleges and markets were still for the most part in the state of ruin in which they had been left by Changîz Khân. It was indeed only at the close of the 8th (14th) century, under the rule of Tîmûr, who made Samarḳand his capital, that Bukhârâ, the sister city, regained some of her former splendour[1].

Samarḳand was up stream, and about 150 miles due east of Bukhârâ; being situated at a short distance from the southern bank of the Sughd river, and occupying high ground. The city, which was encircled by a wall with a deep ditch, was protected by a fortress, also on the height, and below, near the river bank, were great suburbs. All round Samarḳand were orchards, and palaces with their gardens, irrigated by canals innumerable, and cypress trees grew here magnificently. Within the fortress had stood the governor's palace, also the prison, but when Ibn Ḥawḳal wrote this stronghold was mostly in ruins; according to Yâḳût it had double gates of iron. The city proper had four gates; namely, Bâb-as-Sîn, 'the China Gate,' to the east, to which steps ascended from the lower level, and from it the river was overlooked; the Bukhârâ Gate to the north; to the west the Bâb-an-Naw Bahâr,

[1] Ist. 313—315. I. H. 362—364. Muk. 281, 282. Yak. i. 737, 874; ii. 952. I. B. iii. 27. E. Schuyler, *Turkistan*, ii. 89.

also on the height; and to the south the Bâb-al-Kabîr, 'the Great Gate,' also known as the Kish Gate.

The city, according to Yâḳût, was 2500 jarîbs in extent (about 750 acres), and within its circuit were many markets and bath-houses. These, with the private houses, had their water brought in by leaden pipes, communicating with a leaden main, which entered the city by the Bâb Kish, the water coming from the canals outside, where it was taken along a great dyke above ground, and in the market-place the leaden channels are described as resting on stone supports. The great market-place of Samarḳand was called the Râs-aṭ-Ṭâḳ, 'the Head of the Arch,' and was a fine square. The Friday Mosque, with the later Government House, stood immediately below the fortress. The houses in the town were built of both wood and clay bricks, and the city population was extremely numerous.

The suburbs of Samarḳand lay along the river bank, on the lower level, and a semicircular wall, two leagues long, surrounded them on the land side, the river to the north, as the chord of the arc, completing the line of defence. This suburb wall was pierced by eight gates, to which the various thoroughfares led, and these were named as follows: first the Bâb Shadâwad, then the Ashbask gate, then those of Sûkhshîn and of Afshînah, next the Bâb Kûhak, or 'Gate of the Mound,' opening on the height where the city and fortress were situated, after which came the Warsanîn gate, the Rîvdad gate, and finally the Bâb Farrukhshîd. The market streets of the suburb all converged on the square of the Râs-aṭ-Ṭâḳ in the city, and all the roadways were paved with stone flags. The markets in the suburb were the centre of trade, being full of merchants and merchandise from all parts, for the city was the great emporium of Transoxiana. Among other goods the paper of Samarḳand was especially famous throughout the East, the art of making it having been introduced from China. The climate of the place was damp, and every house in both city and suburb had its garden, so that viewing Samarḳand from the fortress height it appeared as one mass of trees. To the south rose the hill of Kûhak, a spur from the mountains beyond which lay within a day's march of the city.

The temporary ruin of Samarḳand, as of all Transoxiana, was due to the Mongols, who almost destroyed the city in 616 (1219); so that, when Ibn Baṭûṭah visited it in the following century, he describes it as without walls or gates, with but a few inhabited houses standing in a maze of ruins. The river here (or possibly he refers to the canal from the Sughd river) he names the Nahr-al-Ḳaṣṣârîn, 'the Fullers' River,' and on this stood many water-wheels. The glory of Samarḳand, however, revived shortly after this, at the close of the 8th (14th) century, when Tîmûr made it his capital, rebuilding the town, and founding the great mosques and caravanserais which were seen here by the Spanish ambassador Clavijo in 808 (1405), some of which remain to the present day. The Friday Mosque in particular, according to 'Alî of Yazd, Tîmûr founded on his return from the conquest of India, and its splendour was due to the treasures brought back from this campaign. Clavijo describes Samarḳand at this time as surrounded by an earthen wall; and he states that the city was a little larger than Seville in his native country[1].

The districts round Samarḳand, lying principally to the eastward and south, but also to the north of the Sughd river, were all extremely fertile. Nine leagues to the east of Samarḳand, and likewise on the south side of the river, was the town of Banjîkath (existing at the present day as Penjakant), surrounded by fertile orchards, producing more especially almonds and nuts, with corn lands stretching along its canals. Between this and Samarḳand was the great village of Waraghsar, with its district, where most of the canals watering the lands round Samarḳand had their origin from the river. On the south side of the capital was the Mâymurgh district, with the village of Rîvdad, one league from Samarḳand, and contiguous was the Sanjafaghan district. None of the lands round Samarḳand surpassed Mâymurgh in fertility, it was famous for its splendid trees, and throughout its length and breadth were innumerable villages. To the south of this lay the hill country called the Jibâl-as-Sâvdâr, the healthiest region of the province. Here, according to Ibn Ḥawḳal, at a place called Wazkard was a church belonging to the Christians—probably Nestorians—

[1] Ist. 316—318. I. H. 365—368. Muk. 278, 279. Kaz. ii. 359. Yak. iii. 134. I. B. iii. 52. A. Y. ii. 195. Clavijo, *Embassy*, 169.

which was much visited, and which enjoyed great revenues. The mountain valleys were most fertile, each well watered by its stream, on which stood the farmsteads; and every kind of crop was produced abundantly. The neighbouring district of Ad-Dargham consisted mostly of pasture lands, but grapes grew here abundantly, and on its borders was the Awfar, or Abghar district, with many populous villages, each with pasture lands two leagues across where great herds of cattle were reared. This was the last of the districts to the south of Samarḳand and the river.

On the north bank of the Sughd river, towards the Ushrûsanah province, was the Bûzmâjan, or Bûzmâjaz district, of which the chief town was Bârkath, or Abârkath, four leagues or one day's march distant from Samarḳand to the north-east. Four leagues further to the north lay Khushûfaghan, an important village, in later times known as Râs-al-Ḳanṭarah (Bridge Head). Beyond this again was the Bûrnamadh, or Fûrnamadh district, near the frontier of Ushrûsanah, and next to it the Yârkath district, the furthest to the north; both being famous for their pasture lands.

Seven leagues due north of Samarḳand was the town of Ishtîkhân, with a strong castle and outer suburbs, standing on canals from the Sughd river. Its corn fields were renowned, and Iṣṭakhrî calls it 'the Heart of Sughd' for its fertility. Seven leagues further north, again, was Kushânîyah, or Kushânî, described as a most populous city of Sughd; and its people were all rich or of easy circumstances. Further, as of the north bank, and according to Yâḳût lying only two leagues distant from Samarḳand, was the district of Kabûdhanjakath, with the city called Lanjûghkath, and adjoining it Widhâr, in the hill country, the chief town being of the same name, where celebrated stuffs were made. Lastly the district of the Marzubân —or Warden of the Marches—Ibn Tarkasfî, one of the Sughd Dihḳâns, or provincial nobles, and this lay beyond Widhâr[1].

The Sughd river or Zarafshân ('Gold Spreader'), as it is now called, had its head-streams in the mountain range called the Jabal-al-Buttam, which formed the watershed between the rivers of Sughd on the one hand, and those of Ṣaghâniyân and the

[1] Ist. 321—323. I. H. 371—375. Muk. 279. Yak. i. 277; ii. 447, 890; iv. 234, 276, 944.

Wakhshâb on the other, both, as described in Chapter XXXII, being right-bank affluents of the Oxus. The slopes of the Jabal-al-Buttam, though high and steep, were covered with villages, and there were gold and silver mines here, as well as workings that produced iron, quicksilver, copper, lead, naphtha, and bitumen, while from the district came rosin, turquoises, lignite for burning, and especially sal-ammoniac. This last, which was largely exported, was collected from the deposit of fumes which issued from a cavern. A chamber had been built over the vents, with windows and doors to close at need; and there were here subterranean fires also, according to Iṣṭakhrî, for the sal-ammoniac vapour which appeared as smoke by day was by night visible as a mighty flame. He describes how the fumes were condensed in the chamber, the sal-ammoniac being periodically taken out by men, who, clothed in wet felts, hastily entered and ran out again, by reason of the great heat which otherwise would have burnt them up. The sal-ammoniac fumes, Iṣṭakhrî adds, also issued from many crevasses in the adjacent rocks, and these were enlarged to become new artificial vents. The fumes were only held to be noxious when confined for the purpose of condensation in the chambers, otherwise the vents in the hill sides could be approached with impunity[1].

The source of the Sughd river was at a place called Jan, or Jay, where there was a lake surrounded by villages, the district being known as Wurghar, or Barghar. From the lake the river took its course through mountain valleys, until it reached Banjîkath, after which it came to the village of Waraghsar already mentioned, the meaning of which, in the local dialect, was 'the Dam Head,' for here the waters were divided up and the canals were led off that irrigated both the lands round Samarḳand and the districts on the north bank of the river. Of the canals flowing to Samarḳand two were sufficiently large to carry boats; and Ibn Ḥawḳal gives in a list the names of these various water-courses, and the districts irrigated by each, with their villages.

At Samarḳand the river was crossed by a masonry bridge called Ḳanṭarah Jard, which in flood seasons was sometimes entirely submerged. Below Samarḳand many canals also branched

[1] Ist. 312, 327. I. H. 362, 382.

off to the various districts round Dabûsiyah and Karmîniyah which will be described presently, and then the Sughd river came to the neighbourhood of Bukhârâ. Here the main stream was commonly known as the Bukhârâ river, and already outside the Great Wall of the Bukhârâ district canals began to be led off for the irrigation of the city lands within the wall, and for the district beyond. The names of all these are also enumerated by Ibn Ḥawḳal, with their various villages. Some canals formed a network, flowing back to the main stream, while others were lost in irrigation channels to the south-west. The chief canals leading to Bukhârâ city are described as having been large enough to carry boats[1].

Between Bukhârâ and Samarḳand, on the south side of the Sughd river, there were three important cities in the 4th (10th) century, namely Karmîniyah (which still exists), Dabûsiyah, and Rabinjan. Karmîniyah lay one stage east of Ṭawâwîs, and outside the Great Wall; it was larger than this latter place, very populous and surrounded by villages and fertile lands, which were irrigated by canals from the Sughd river. Yâḳût speaks of its magnificent trees. One stage to the east, again, was the large town of Ad-Dabûsiyah, likewise on a canal from the south bank of the Sughd river, but it had no large villages or dependencies round it.

The small town of Khudîmankan lay one league distant from Karmîniyah, and a bow-shot distant north of the high road. On the north bank of the Sughd river one league above Khudîmankan was the great hamlet of Madhyâmajkath, while Kharghânkath was one league lower down, also on the northern bank and opposite Karmîniyah, from which it was but a league distant. These three hamlets were of sufficient size in the 4th (10th) century for each to have had its Friday Mosque, and Yâḳût reports that Khudîmankan was famous for divers traditionists born here. Arbinjan, or Rabinjan, lay one stage to the east of Dabûsiyah, and was a larger town than this last; to the east again, at the half-way stage between Rabinjan and Samarḳand and seven leagues from this capital, was Zarmân. As of the neighbourhood of Bukhârâ, Muḳad-

[1] Iṣt. 310—312, 319—321. I. H. 359—361, 368—371.

dasî names and describes a large number of other small towns, but unfortunately no distances are given to mark their positions[1].

To the southward, running parallel with the Sughd river and like it ending in marshy lakes, is the shorter stream now known as the Kushkah Daryâ, on which stand Shahr-i-Sabz and Ḳarshî. Shahr-i-Sabz, 'Green City,' was in the earlier middle-ages known as Kish (Kishsh), and is described by Ibn Ḥawḳal as having a castle, the town itself being strongly fortified, with a great suburb lying beyond its gates. Further, beyond the suburb, was a second township, probably that now known as Kitâb, named Al-Muṣallâ, 'the Praying Place,' where stood the hostelries and the palace of the governor. Great markets were found in the suburb, but the Friday Mosque with the prison were in the inner city. This covered a square mile of ground, and its houses were built of wood and unburnt bricks. The neighbouring lands were extremely fertile; all the fruits of the hot region were grown here and exported to Bukhârâ. The inner city of Kish had four gates, called respectively, the Iron Gate, the Gate of 'Ubayd Allah, the Butchers' Gate (Bâb-al-Ḳaṣṣâbîn), and the Inner City Gate. The outer city, or suburb, had two gates, the Bâb Baraknân, so called after a neighbouring village, and the Outer City Gate (Bâb-al-Madînah-al-Khârijah).

The main stream of what is now known as the river Kushkah, was, in the 4th (10th) century, called the Nahr-al-Ḳaṣṣârîn, 'the Fullers' River'; its sources were in the Jabal-Sayâm, and it passed Kish on the south side. To the north ran the Nahr Asrûd, and, one league beyond, the road towards Samarḳand was crossed by the river called the Jây Rûd. To the south, one league from Kish on the road to Balkh, was the Khushk Rûd, 'the Dry River,' and the Khuzâr Rûd lay eight leagues beyond this again. These streams, after irrigating the various districts round Kish, flowed together, and became a single stream, which passed by the city of Nasaf. The Kish territory is described as four days' journey across in every direction, and as famed for its extraordinary fertility. In the neighbouring mountains salt was found, also the manna called Taranjubîn, and various simples which were

[1] Ist. 314, 316, 323. I. H. 363, 365, 375. Muk. 282. Yak. ii. 406, 925; iv. 268.

exported to Khurâsân. In later times Kish attained fame as the birth-place of Tîmûr, who in the latter part of the 8th (14th) century rebuilt the town, where the White Palace—Âk-Saray—became his favourite place of residence. It was at this period that Kish took the name, which it still retains, of Shahr-i-Sabz, 'the Green City[1].'

Rather more than a hundred miles down the river below Kish, and to the westward, is the city now known as Karshî, which the medieval Arabs called Nasaf, and the Persians Nakhshab. In the 4th (10th) century Nasaf had a strong castle, and extensive suburbs lay outside the city, which was surrounded by a wall having four gates, namely the Bâb-an-Najârîyah, the Samarkand Gate, the Kish Gate, and the Bâb Ghûbadhîn. Nasaf stood on the river which, as already said, was the main arm formed by the junction of many streams from the Kish district. On its bank was the palace of the governor, at the place called Râs-al-Kanṭarah, 'the Bridge Head.' The prison lay adjacent to the governor's palace, and the Friday Mosque near the Ghûbadhîn Gate, the great market streets lying in between. Just within the Najârîyah Gate was the oratory, Al-Muṣallâ. Mukaddasî, who praises the excellent grapes of Nakhshab, speaks of its fine markets; the town was surrounded by fertile fields and orchards, but had no great outlying dependencies like those surrounding Kish.

In history Nasaf, or Nakhshab, was famous as the place where in the latter half of the 2nd (8th) century Al-Mukanna'—the celebrated Veiled Prophet of Khurâsân—had first arisen and done miracles. From a well in Nakhshab, night after night, at his command the moon, or its semblance, rose to the wonder of all beholders. To the Persians Mukanna' was generally known as Mâh-sâzandah, or 'Moon Maker,' and, as history relates, the revolt of his followers for many years gave great trouble to the generals of the Caliph Mahdî. As regards Nakhshab city, after the times of the Mongol invasion in the 7th (13th) century, a certain Kapak Khân built himself a palace at a place some two leagues distant from the older town, and 'a palace' in the Mongol language is called *Karshî*, which name was subsequently given to the settle-

[1] Ist. 324. I. H. 375—377. Muk. 282. A. Y. i. 300, 301.

ment that sprang up and replaced the older Nasaf or Nakhshab. Ibn Baṭûṭah sojourned here in the early part of the 8th (14th) century, and describes Ḳarshî as a small town surrounded by gardens. At the close of the century Tîmûr frequently passed his winters at Ḳarshî, and he afterwards built near it the Ḥiṣâr or fortress[1].

There were two towns near Nasaf, in the 4th (10th) century and later, each of which had its Friday Mosque. One of these, the smaller, was Bazdah, or Bazdawah, a strong castle, situate six leagues to the westward of Nasaf on the road to Bukhârâ. The other and larger town was Kasbah, four leagues from Nasaf, also in the Bukhârâ direction, where there were excellent markets according to Yâḳût. Further, between Nasaf and Kish, one stage west of the latter city, was the town or large village of Nawḳad Ḳuraysh; while one stage south-east of Nasaf, on the road to the Iron Gate (see p. 441), was Sûnaj, a large village, with Iskîfghan lying one league from it, both these towns being watered by the Khuzâr river already mentioned[2].

The products, natural and manufactured, of Sughd were numerous. The melons of Bukhârâ were famous all the world over, and its looms produced carpets and prayer rugs, fine cloth for clothes, and coarse carpets such as were spread in great guest-chambers. In the prisons they made saddle-girths; and hides were well tanned, while various sorts of grease and oil were manufactured for export. Samarḳand was above all famous for its paper, and the looms produced red cloth and cloth of silver, with brocades and raw-silk stuffs. Here, too, the copper-smiths made brass pots of a very large size, and other artificers produced stirrups, martingales, and girths, also various sorts of jars and goblets. From the neighbouring districts were exported immense quantities of filberts and walnuts. Karmîniyah, between Bukhârâ and Samarḳand, produced napkins, and from Dabûsiyah came various kinds of cloth and brocade. Rabinjan exported red felts, prayer carpets, and tin cups; also

[1] Ist. 325. I. H. 377, 378. Muk. 282. Kaz. ii. 312. I. B. iii. 28. A. Y. i. 111.

[2] I. H. 376—378. Muk. 283. Yak. i. 604; iii, 197; iv. 273, 825.

hides, hemp cordage, and sulphur. Moreover winter rice was grown in this district[1].

As already said in Chapter XXX (p. 431) the great Khurâsân road crossed the Oxus beyond Amûyah to Firabr, and thence proceeded by Baykand and through the gate in the Great Wall to Bukhârâ. From this capital the road went up the left bank of the Sughd river to Samarḳand, passing through the chief towns of the district, and this part of the highway is given with but little variation by all the earlier authorities, Ibn Ḥawḳal and Muḳaddasî adding the distances between the outlying towns of the Bukhârâ and Samarḳand districts[2].

The high road which passed through Khurâsân to Balkh (see p. 432) crossed the Oxus to Tirmidh, from which branched various roads north through Ṣaghâniyân and Ḳubâdhiyân to Wâshjird, whence by the Stone Bridge the Wakhsh and Khuttal districts were attained. North-west from Tirmidh another road went up to the Iron Gate, and at Kandak, one stage beyond this, bifurcated. Running due north, the road on the right hand went by Kish, and thence on to Samarḳand; while to the north-westward the highway on the left hand led to Nakhshab; whence a branch road turned eastward back to Kish, while the main road crossed the tract of desert to Bukhârâ. These routes, mostly in short distances, are given by Iṣṭakhrî and in part by Muḳaddasî[3].

The delta lands of the Oxus in the Khwârizm province were reached from Âmul on the Khurâsân side by a road going up the left bank to Ṭâhirîyah, where cultivation began, and thence on to Hazârasp. Here one way went to the left by Khîvah to Jurjânîyah (Urganj), while another turned off to Kâth, and the towns on the right bank of the Oxus. These roads are given by Iṣṭakhrî and Muḳaddasî; also the way crossing the desert direct, south-east, from Kâth to Bukhârâ. Further, Mustawfî, in the 8th (14th) century, gives two routes from the south converging on Urganj, one going across the desert north from Farâvah (now Ḳizil Arvâṭ, see p. 380) to Urganj; the other going from Marv, also across the

[1] I. H. 364. Muk. 324, 325.

[2] I. K. 25, 26. Kud. 203. Ist. 334, 342. I. H. 398, 402. Muk. 342, 343.

[3] Ist. 337—341. I. H. 399—403. Muk. 342—344.

desert, and in many places passing the moving sands, and ultimately reaching Ṭâhirîyah on the Oxus. This last road is also given in the *Jahân Numâ*, and from Hazârasp it follows almost identically the road given by the Arab geographers to the capital of Khwârizm at Jurjânîyah[1].

[1] Ist. 338, 341, 342. I. H. 400, 402. Muk. 343, 344. Mst. 197, 198. J. N. 457.

CHAPTER XXXIV.

THE PROVINCES OF THE JAXARTES.

The Ushrûsanah province. Bûnjikath, the capital. Zâmîn and other towns. The Farghânah province. The Jaxartes or Sayḥûn. Akhsîkath and Andîjân. Ûsh, Ûzkand, and other cities. The province of Shâsh. Shâsh city or Binkath. Banâkath or Shâhrukhîyah, and other towns. The Îlâḳ district, Tûnkath city, and the silver mines of Khasht. The Isbîjâb district. Isbîjâb city or Sayrâm. Chimkand, and Fârâb or Utrâr. Yassî and Ṣabrân. Jand and Yanghikant. Ṭarâz and Mîrkî. Outlying towns of the Turks. Products of the Jaxartes countries. Routes to the north of Samarḳand.

The province of Ushrûsanah—also written Usrûshanah, Surûshnah and Sutrûshnah—lay to the east of Samarḳand, between the districts along the right bank of the Sughd river, and those along the left bank of the Jaxartes, for the Ushrûsanah province was of neither river, being a land of plains and hills with no considerable stream running through it. Its eastern frontier was on the Pamir (Fâmir) according to the Arab geographers.

The capital was the city of Ushrûsanah (Madînah Ushrûsanah), otherwise called Bûnjikath, Banjakath, or Bunûjkath, the site of which is identical with the present town of Ura-tepeh[1]. Bûnjikath was in the 4th (10th) century a city of over 10,000 men, built of clay bricks and wood, having an inner part surrounded by a wall, and an outer suburb also walled. The inner city had two

[1] Bûnjikath the capital of Ushrûsanah must not be confounded with Banjîkath (Penjakant) to the east of Samarḳand. The position of the capital of Ushrûsanah is fixed by the Itineraries (see I. K. 29, Ḳud. 207, and Ist. 343), besides present local tradition (Schuyler, *Turkistan*, i. 312). Ist. 325. I. H. 379. Muk. 265. Yak. i. 245, 278, 744.

gates, the upper gate (Bâb-al-A'lâ) and the city gate (Bâb-al-Madînah), and within its precincts were the castle and the prison, the Friday Mosque, and the markets. A great canal passed through the inner city, having many mills upon its bank. The wall that surrounded the suburb, or outer city, included many gardens, and was three leagues in circuit. In this wall were four gates, namely the Zâmîn Gate, the Marasmandah Gate, the Nûjkath Gate, and the Gate of Kahlâbâdh. The town lands were amply irrigated by six small streams, which Ibn Ḥawḳal names. These flowed down from the neighbouring hills, and after a course of about half a league passed through Bûnjikath, having ten mills upon their banks. The town was celebrated for its many charming gardens.

Zâmîn, which still exists, lay to the east of Bûnjikath, and was the point where the great Khurâsân road, coming up through Bukhârâ and Samarḳand, finally bifurcated, one road going north to Shâsh (Tâshkand), the other north-east to Farghânah and beyond. Zâmîn was, in the 4th (10th) century, almost of the size of the capital Bûnjikath; it was a very ancient town, and had been formerly known as Sûsandah, or Sarsandah. It possessed a fine Friday Mosque, and excellent markets, being surrounded by gardens, but it was unwalled. A stream flowed through the town crossed by many bridges of boats. The town of Sâbâṭ likewise exists. It lay between Zâmîn and Bûnjikath, on the road to Farghânah and is described by Muḳaddasî as very populous, embowered by numerous orchards and gardens, lying beside its streams[1].

The names of other towns of Ushrûsanah are given in the lists, but without any description, and the positions of the majority are unknown. Of those still existing, or whose sites can be fixed from the Itineraries, are the following. Dîzak, otherwise Jîzak, lies north-west of Zâmîn; and south of it, on the road from Samarḳand, was the important town of Kharaḳânah. Khâwas or Khâwaṣ is on the road going north from Zâmîn to Shâsh, and Kurkath lies on the frontiers of Farghânah, midway between Sâbâṭ and Khujandah. The position of the two small towns of Mînak and Marasmandah cannot be exactly fixed, for neither is

[1] Ist. 326, 327. I. H. 379, 380. Muk. 277.

given in the Itinerary, but Marasmandah, to judge by the Marasmandah gate of Bûnjikath, must have been in the neighbourhood of the capital. It stood in the hills, had a cold climate, with many streams, but few orchards or gardens, on account of its elevation. Muḳaddasî speaks of its excellent markets; and the Friday Mosque stood in their midst, Marasmandah being a very populous place. Mînak appears to have been in its near neighbourhood, and was celebrated for the great battle fought here by Ḳutaybah, the Arab commander at the time of the first Moslem conquest of Transoxiana. At this place, too, was the castle that had belonged to Afshîn, the general and favourite of the Caliph Mu'taṣim. Near both Marasmandah and Mînak there were iron mines, and tools made here were exported to all parts of Khurâsân, the steel being of excellent quality; so that even in Baghdâd these were much sought after[1].

The great river Jaxartes, as already said (p. 434), was called by the Arabs the Sayḥûn or Sîḥûn. It was, however, more generally known as the Nahr-ash-Shâsh, the river of Shâsh (Old Tâshkand), from the name of the most important city near its banks. In the 8th (14th) century, according to Mustawfî, the Mongol population of these parts knew it under the name of Gil-Zariyân. Since that time, and down to the present day, it has been commonly called the Sîr Daryâ or Sîr Ṣû (River Sîr) by the Turks, this name being mentioned by Abu-l-Ghâzî.

According to Ibn Ḥawḳal the river Jaxartes rose in the Turk country, being formed by the junction of many mountain streams, and it entered the great valley of Farghânah at its eastern end, near the town of Ûzkand; the province of Farghânah lying for a couple of hundred miles and more in length to the north and south of its upper stream[2]. Flowing here due east, the Sayḥûn received numerous affluents during its course through Farghânah, namely the Nahr Kharshân, the rivers of Urast and of Ḳubâ, also the Nahr Jidghil, which is probably the present Naryn river, and some others. Passing on by the walls of Akhsîkath, the capital, the Sayḥûn came to Khujandah, where it finally passed out of the

[1] Ist. 336, 343. I. H. 381, 382, 383. Muk. 278. Yak. ii. 395, 425, 710.

[2] Hence the Naryn, by far the longest of the head rivers of the Sîr, was evidently not considered the main stream by the Arabs.

Farghânah province. Thence, turning northwards, the river next received on its right bank the two streams called the Nahr Îlâḳ and the Nahr Turk, passing to the westward of the districts of Îlâḳ and of Shâsh. Beyond these the Sayḥûn finally came to the Isbîjâb districts, whence, through the deserts of the Ghuzz and the Turks it ultimately flowed out by numerous channels to the Sea of Aral in its north-eastern part. The Arab geographers say that the Sayḥûn was navigable for boats like the Jayḥûn, and for a longer time than was the case with the Oxus the Jaxartes was frozen over in winter, so that caravans could cross it on the ice. Moreover it was counted as only two-thirds of the length of the sister stream[1].

The province of Farghânah, which until within recent years was more generally known as the Khanate of Khûḳand, but which under the Russian government has officially again taken its more ancient name, had for its capital, in the earlier middle-ages, the city of Akhsîkath, which Ibn Khurdâdbih and others call Farghânah city. It lay on the north bank of the Jaxartes. The ruins of this town exist, and in the 10th (16th) century when Bâbar was ruler of Farghânah, under the shortened form of Akhsî it was the second city of the province, Andîjân being then the capital.

Akhsîkath is described by Ibn Ḥawḳal as a large city, with a castle, where stood the Friday Mosque, the governor's palace, and the prison; and outside the inner town was an extensive suburb. The inner city, which measured a mile across in every direction, was intersected by numerous water channels, all connected with a great tank; and there were markets both here and in the suburb, which latter was surrounded by a wall. The inner city had five gates, namely the Ḳâsân Gate, the Mosque Gate (Bab-al-Jâmi'), the Rahânah Gate, next a gate with an uncertain name that may be read as Bakhtar, and finally the Gate of Al-Mardaḳshah. The place was entirely surrounded by gardens, which extended for a distance of a couple of leagues beyond the suburb gates, and on the further, or south side of the Jaxartes were rich pasture grounds. Akhsîkath was apparently

1 I. H. 392, 393. Muk. 22. Yak. iii. 210. Mst. 215. Hfz. 33*a*. J. N. 360. A. G. 13, 181, 290.

ruined, with many of the other cities of Farghânah, in the wars of Muḥammad Khwârizm Shâh at the beginning of the 7th (13th) century, and the Mongol invasion completed the work; after which the capital was removed to Andîjân. In the time of Tîmûr, 'Alî of Yazd gives the name under the form Akhsîkant or Akhsîkat, and as we have seen this was shortened to Akhsî in the days of Bâbar[1].

Andîgân (modern Andîjân), according to Mustawfî, was made the capital of Farghânah by Kaydû Khân, grandson of Ugutay, son of Changîz, in the latter half of the 7th (13th) century. The name of Andigân, or Andukân, occurs in the lists of towns given by Ibn Ḥawḳal in the 4th (10th) century, and is also mentioned by Yâḳût, but nowhere is the town described, though in the account of Tîmûr's campaigns it is frequently referred to by 'Alî of Yazd. From the Itineraries it would seem that the city of Ḳubâ, which in the 4th (10th) century was a place of much importance, must have stood near Andîjân. Ḳubâ, says Iṣṭakhrî, was almost of the size of Akhsîkath, and its gardens were even more extensive. It had a strong castle, where the Friday Mosque stood, in the Maydân or central square; and there was an outer suburb, where was the governor's palace, and the prison. The suburb was surrounded by a great wall, and there were many well-supplied market streets[2].

Half-way between Akhsîkath and Ḳubâ was the town of Ushtiḳân with a Friday Mosque in its market-place; and to the east of Ḳubâ was Ûsh, which already in the 4th (10th) century was a place of great importance. In the castle of Ûsh stood as was usual the governor's palace, and the prison; and the inner town was surrounded by a walled suburb, running up the slope of the neighbouring hill; with three gates, namely the Hill Gate (Bâb-al-Jabal), the Water Gate (Bâb-al-Mâ), and the Mughkadah

[1] I. K. 30. Ist. 333. I. H. 393, 394. Muk. 271. Kaz. ii. 156. A. Y. i. 441; ii. 633. Akhsî is marked on the Russian map given by Schuyler, *Turkistan*, i. 336, a short distance to the south-west of modern Namangan. The termination *Kath*, or *Kât*, is synonymous with *Ḳand*, or *Kant*, and both occur in many names of places in Central Asia, and have the meaning in the Turkish dialects of 'a city' or 'burg,' as Yâḳût (i. 404) very justly remarks. See e.g. Nûzkâth (New Wall) in Khwârizm, mentioned on p. 454.

[2] Ist. 333. I. H. 394, 395. Muk. 272. Mst. 228. Yak. i. 375; iv. 24. A. Y. ii. 633.

Gate. The Friday Mosque stood in a broad Raḥbah, or square, surrounded by markets; and the lands around were plentifully watered by many streams. Near by, on a hill-top, was a guard-house garrisoned by soldiers—volunteers—who watched against the incursions of the Turkish hordes. Beyond Ûsh is Ûzkand, the easternmost city of Farghânah, described as two-thirds of the size of Ûsh. Ûzkand, too, lay in a fertile district, having a castle, a well-fortified inner city, and a suburb with markets that were much frequented by the Turk merchants. A river went by one of the town gates, for the suburb was surrounded by a wall pierced by four gates, and the Friday Mosque stood in the market-place.

That part of Farghânah which lay to the south of the Jaxartes was known as the district of Nasyâ, or Nasâiyah, divided into upper and lower according to its elevation, upper Nasâiyah lying among the hills. Of lower Nasâiyah was the town of Marghînân (modern Marghîlân), a small place in the 4th (10th) century, but with a Friday Mosque in its market. To the west of this lies Rishtân, a large town in early days, also with a fine Friday Mosque. Khûḳand, which in recent times became the capital of Farghânah, and gave its name to the Khanate, is only mentioned incidentally among the cities of upper Nasâiyah, and under the form Khuwâkand or Khuwâḳand.

Khujandah, the first town of Farghânah on the west coming from Samarḳand, lay on the left bank of the Jaxartes, and adjacent to it one league southward was the suburb of Kand. Khujandah was of considerable length along the river strand, but of little breadth; it had a strong castle with a prison. The Friday Mosque was in the city; the governor's palace being in the Maydân, or square, of the suburb. Khujandah is described by Ibn Ḥawḳal as a most pleasant town, and its people possessed boats for going on the Jaxartes river. The outer suburb of Kand was more especially known as Kand-i-Badhâm, 'Kand of the Almonds,' and according to Ḳazvînî it was so called from a particular variety of this fruit, grown here, that was famous for its husk peeling off very easily when the almond was taken in the hand[1].

[1] Ist. 332, 333, 347. I. H. 391, 392, 394, 395. Muk. 262, 272, 345. Yak. i. 404. A. F. 498. Kaz. ii. 372.

Of the cities in the northern part of Farghânah, namely of the lands on the right bank of the Jaxartes, very little is known during early times. Muḳaddasî describes Wânkath as a town with a Friday Mosque and good markets, and from the Itineraries we learn that Wânkath lay seven leagues to the west of Akhsîkath, being one league from the bank of the Jaxartes, and not far from the frontiers of Îlâḳ. North of Wânkath and among the hills was Khayralam, or Khaylâm, a town of the district of Miyân Rûdhân, 'Betwixt the Rivers,' with a fine Friday Mosque and good markets. To the north of this again lay Shikit, or Sikkit, a town where according to Muḳaddasî nuts grew so abundantly that a thousand could be had for a silver dirham; and here too there was a Friday Mosque in the market-place. The town of Ḳâsân still exists, and is described by the earlier geographers as situate in the district of the same name. Yâḳût adds that it had a strong castle, and that past its gate ran the stream which ultimately joined the Jaxartes at Akhsîkath. Further north was the district of Jidghil, of which the chief town was Ardalânkath. To the east of this lay the Karwân district, of which the chief town was called Najm. A number of other towns are also briefly described by Muḳaddasî, but unfortunately there is no indication of their respective positions[1].

To the westward of Farghânah came the district of Shâsh, which, as already said, lay on the right bank or north-east of the Jaxartes. The ruins now known as Old Tâshkand are the site of the city called Shâsh by the Arabs, and Châch by the Persians, which, in the middle-ages, was the greatest of the Arab towns beyond the Jaxartes. The city of Shâsh was also known by the name of Binkath[2], for like many other places in Transoxiana, there was the double nomenclature, Iranian and Turanian.

Shâsh, in the 4th (10th) century, was a city of many walls. There was, in the first place, an inner town, with a castle, or citadel, standing separate, but adjacent, and these two were surrounded by a wall. Outside the inner town was the inner suburb, surrounded by its own wall, and beyond this again lay the outer

[1] Ist. 334, 347. I. H. 396. Muk. 271, 272. Yak. iv. 227.

[2] This is often by an error in the diacritical points written *Bîkath*, e.g. Yak. i. 746.

suburb, with many gardens and orchards surrounded in turn by a third wall. Lastly there was the Great Wall, which, as was the case at Bukhârâ, protected the whole district, making a great semicircular sweep round Shâsh to the northward, from the bank of the Turk river on the east to the Jaxartes on the west.

To return to the inner town and the citadel; this last, within which was the governor's house and the prison, had two gates, one opening on the inner town, the other to the suburbs. The Friday Mosque had been built on the wall of the citadel. The inner town, which was a league across in every direction, contained some important markets and had three gates, first the double gate of Abu-l-'Abbâs, then the Bâb Kish, doubtless to the south where the road from Samarkand came in, and lastly the gate of Al-Junayd. The wall of the inner suburbs had ten gates (Mukaddasî names only eight), and of the outer suburbs seven, which are all carefully enumerated by Ibn Ḥawkal, and in the inner suburbs were found the great markets of Shâsh. The whole city was plentifully supplied by conduits of running water from canals, which afterwards irrigated the numerous orchards and vineyards within the walls.

The Great Wall, at its nearest point, passed at a distance of one league from the gate of the outer suburbs. This wall began on the east at the hill on the Turk river called Jabal Sâblagh, and the extensive plain which it enclosed was known as Al-Ķîlâṣ. The wall was built by 'Abd-Allah ibn Ḥumayd, to protect Shâsh on the north from the incursions of the Turks, and beyond it, at the distance of a league, was dug a deep ditch, going all the way from the hill on the Turk river to the bank of the Jaxartes on the west. The road north from Shâsh to Isbîjâb passed through this wall at the Iron Gate (Bâb-al-Ḥadîd).

In the early years of the 7th (13th) century, Shâsh was in part ruined during the conquests of Muḥammad Khwârizm Shâh, and the Mongol invasion which immediately followed added to the miseries of the people here as elsewhere. The city, however, appears to have recovered rapidly from these misfortunes, and it was again an important place in the 8th (14th) century when Tîmûr halted here. 'Alî of Yazd, who frequently has occasion to mention it in describing the campaigns of Tîmûr, gives the names as Shâsh,

Châch, or Tâshkant; this latter being apparently a popular corruption of the name Shâsh to *Tâsh*, by the Turkish-speaking population, Tâshkant meaning 'the stone city,' under which name it is now become the capital of Russian Turkistân[1].

The Nahr Turk, now known as the river Chirchik, which flows to the south-east of Shâsh, according to Ibn Ḥawḳal rose in the mountains of Jidghil on the north of the Naryn river, and in the district called Baskâm of the Kharlîkh Turks. To the southward of this river and more or less running parallel with it was the Nahr Îlâḳ, now called the river Angran, and immediately below where this joined the Jaxartes stood the city of Banâkath, the second largest town of the Shâsh district. Banâkath, otherwise called Banâkît, or by the Persians Fanâkant, was not fortified in the 4th (10th) century, but it had a Friday Mosque in its market-place. The town stood on the right bank of the Jaxartes where the great Khurâsân road coming up from Samarḳand crossed the river going to Shâsh, and it continued to be a place of great importance till the 7th (13th) century, when it was laid in ruins by Changîz Khân. More than a century later, in 818 (1415), Fanâkant was rebuilt by order of Shâh Rukh, the grandson of Tîmûr, and then received the name of Shâhrukhîyah, under which it is frequently mentioned by 'Alî of Yazd.

The road from Banâkath north to Shâsh passed through the town of Jînânjakath, lying on the south or left bank of the Turk river, some two leagues above its junction with the Jaxartes. This town, though unfortified, was a place of considerable size in the 4th (10th) century, and its houses were built of wood and unburnt brick. Across the Jaxartes to the west, and one march from Jînânjakath on the road to Jîzak, was the small town of Waynkard, which Ibn Ḥawḳal describes as a village of the (Nestorian) Christians. Across the Turk river, and somewhat to the westward in the angle below where it joined the Jaxartes, lay the town of Ushtûrkath, or Shuturkath (Camel City), which was well fortified. This place must have been ruined by the Mongols, for in the latter part of the 8th (14th) century we find it replaced by Chînâs (which still exists), the name of which is frequently

[1] Kud. 27. I. H. 384, 386—388. Muk. 276. Kaz. ii. 362. A. Y. i. 94, 101, 166.

mentioned by 'Alî of Yazd. Ibn Ḥawḳal and Muḳaddasî name more than a score of other cities of the Shâsh districts, but they add no details, and the positions of these places, therefore, cannot now be fixed; though it is evident that in the 4th (10th) century the whole of this country, as also the Îlâḳ district to the south and Isbîjâb to the north, was densely populated, with numerous hamlets that were of the size of towns[1].

The district of Îlâḳ lay to the south of the Îlâḳ river, and north of the great bend of the Jaxartes below Khujandah; and its chief town was called Tûnkath. The district, which was continuous with Shâsh, comprised near a score of important towns, duly enumerated by Ibn Ḥawḳal and others, the sites of which remain undetermined, and it is unfortunately not possible even to discover that of Tûnkath, the capital. According to Ibn Ḥawḳal Tûnkath lay on the river Îlâḳ, and apparently at a distance of eight leagues from Shâsh, of which it is said to have been half the size[2]. There was a strong castle, an inner city, and a suburb surrounded by a wall. Within the castle was the governor's house, the prison and the Friday Mosque both standing at the castle gate. Great markets were found in both city and suburb, and the whole district round was plentifully supplied with running water. All the country lying between Shâsh and Îlâḳ was covered with towns, the names of which are given by Ibn Ḥawḳal, but as already said their positions are unfortunately now lost. One of the most important places mentioned was the populous town of Khâsht (also written Khâsh, Khâs, or Khâṣ), near the silver mines in the Îlâḳ hills on the frontiers of Farghânah. Here, according to Ibn Ḥawḳal, in the 4th (10th) century was a mint, where much gold and silver were coined; and the place was surrounded by numerous villages[3].

To the north of Shâsh, and stretching east from the right bank of the Jaxartes, was the extensive district or province of Isbîjâb or Asbîjâb, with the capital of the same name; and Muḳaddasî

[1] Ist. 328—330, 336, 345. I. H. 384, 385, 388, 405. Muk. 264, 276, 277. A. Y. i. 101; ii. 636.

[2] Tûnkath is sometimes by a clerical error miswritten *Tûkath*, Ist. 331, note *c*. For the distance between Shâsh and Tûnkath see Ist. 344. I. H. 404.

[3] Ist. 331, 332, 345. I. H. 386, 388, 389, 404. Muk. 265, 277, 278.

in the 4th (10th) century mentions nearly fifty towns of this region as well known, of which only a very few can now be identified.

The city of Isbîjâb is identical in site with Sayrâm, lying about eight miles to the eastward of Chimkant on the Arîs or Badam river, which is a right-bank affluent of the Jaxartes[1]. According to Ibn Ḥawḳal, it was a third the size of Shâsh, and consisted of a citadel or castle, with an inner city surrounded by a wall, and the suburb, also walled. It is reported that the whole circuit of the city of Isbîjâb was about a league, and that it stood in a great plain three leagues from the nearest hills, being surrounded by well-watered gardens. The town had four gates, and before each was built a strong Rubâṭ or guard-house. There were markets in both the city and the suburbs, and in the former were situated the governor's house, the prison, and the Friday Mosque. Muḳaddasî mentions the Sûḳ-al-Karâbîs, 'the market of the cotton-merchants,' as especially famous, and the rents for these shops, which were applied to charitable purposes, amounted to 7,000 dirhams (about £300) a month. The city of Isbîjâb appears after the time of the Mongol invasion to have changed its name to Sayrâm, under which it is frequently mentioned by 'Alî of Yazd in his accounts of the campaigns of Tîmûr.

Chimkant, written Chimîkant, is also frequently mentioned by 'Alî of Yazd and appears to be identical with the town which Muḳaddasî writes Jamûkat, and describes as a large, well-fortified city, with a Friday Mosque and suburbs, where there were excellent markets[2].

On the east bank of the Sayḥûn, immediately below where the Chimkant river flows in, is the city at the ford for passing the Jaxartes known originally as Bârâb or Fârâb, and in later times as Utrâr, where in the year 807 (1405) Tîmûr ended his life, when about to set out for the conquest of China. Fârâb, or Bârâb,

[1] Muk. 262—264. Schuyler (*Turkistan*, i. 75) identifies Isbîjâb city with Chimkant, but this is certainly a mistake, for in the *Târîkh-i-Râshidî* (translated by N. Elias and E. D. Ross, p. 171) mention is made of 'Sayrâm which in old books is called Isbîjâb.' The Persian text of this passage will be found on folio 105 *b* of the British Museum MS. Add. 24090.

[2] Ist. 333. I. H. 389. Muk. 263, 272, 275. A. Y. i. 166; ii. 633, 636.

was the name of both district and town, and it was sometimes accounted the capital of the Isbîjâb district; the suburbs of the town in the 4th (10th) century being also known under the name of Kadar. Muḳaddasî speaks of Bârâb as a large city with 70,000 inhabitants; it was strongly fortified and had a citadel or castle, a Friday Mosque, and great markets. In its warehouses much merchandise was stored. Kadar also had its own Friday Mosque, and was the new town. According to Ḳazvînî the city lay among salt marshes, and was celebrated in history as the birth-place of Abu Naṣr-al-Fârâbî, who died in 339 (950), and was accounted the greatest of the Moslem philosophers before Avicenna. According to Ibn Ḥawḳal, however, the actual birth-place of Al-Fârâbî was at Wasîj, a small fortified town lying two leagues distant from Fârâb, where there was a fine Friday Mosque in the market-place. At a subsequent period Fârâb took the name of Utrâr, also spelt Uṭrâr, which was pillaged in the early part of the 7th (13th) century by the Mongol hordes, but was shortly afterwards rebuilt, for it was in its Sarây, or palace, as already said, that Tîmûr died[1].

About half-way between Sayrâm and Uṭrâr was the town of Arsubânîkath, or Subânîkath, which Muḳaddasî speaks of as a fine place, well-fortified, with a Friday Mosque in the inner city, and great suburbs lying without the wall. The district round Subânîkath was called Kanjîdah. One day's march north of Uṭrâr, along the right bank of the Jaxartes, was the town of Shâvaghar, also described by Muḳaddasî as a large place, well-fortified, with a Friday Mosque in its market-place, and surrounded by fertile districts. The name of Shâvaghar does not occur in the later geographers, and from its position it would appear to be identical with Yassî, a place often mentioned by 'Alî of Yazd, and still existing to-day under the name of Ḥaḍrat-i-Turkistân, 'The Presence (of the holy man) of Turkistân,'—he being the patron saint of the Kirghiz, who is buried here. According to 'Alî of Yazd this personage was Shaykh Aḥmad of Yassî, a descendant of

[1] I. H. 390, 391. Muk. 262, 273. Kaz. ii. 405. A. F. 493. I. B. iii. 23. A. Y. i. 166, 275; ii. 646. Ibn Khallikân, No. 716, p. 73. There is often confusion between Fârâb or Bârâb (Uṭrâr) of the Jaxartes, and Fâryâb (see above, p. 425) in Jûzjân, which was also called Bârâb.

Muḥammad ibn Ḥanafîyah, son of the Caliph 'Alî. The Shaykh died here in the early part of the 6th (12th) century, and Tîmûr at the close of the 8th (14th) century built over his tomb the mosque, the magnificent remains of which exist, the shrine being still the object of pilgrimage from all the country round.

One day's journey north of Yassî or Shâvaghar was Sawrân, or Ṣabrân, which stands to the present day, reckoned in the 4th (10th) century as the frontier fortress against the Ghuzz. Here, in peace times, all the neighbouring Turk tribes came to barter with the Moslem merchants. Muḳaddasî depicts Sawrân as a very large town, protected by seven fortifications and walls, one built behind the other. The Friday Mosque was in the inner city, and extensive suburbs lay outside the town. 'Alî of Yazd frequently mentions Ṣabrân when speaking of the campaigns of Tîmûr, and Yâḳût describes its high citadel or castle, which dominated the frontier lands[1].

Among other places on the Jaxartes very frequently mentioned by 'Alî of Yazd, but not noticed by the earlier Arab geographers, is Saghnâḳ, which he gives as the capital of Ḳîpchâḳ and as lying 24 leagues northward from Utrâr. Further to the north again is Jand, mentioned by the earlier geographers, and by Yâḳût, as one of the great Moslem cities of Turkistân beyond the Jaxartes. In the early part of the 7th (13th) century Jand had been devastated by the Mongols. The Aral is often named the Sea of Jand, and here, two marches from the mouth of the Jaxartes, lay the Ghuzz capital, called by the Arabs Al-Ḳariyat-al-Jadîdah (or Al-Ḥadîthah), 'the New Village,' and in later times known as Yanghi-kant or Yangi-Shahr, 'New Town,' in Turkish[2].

About 80 miles to the north-east of Sayrâm (or Isbîjâb) are the ruins of Ṭarâz, near the present town of Aulieh-Ata. Ṭarâz, or Aṭ-Ṭarâz, was an important place as early as the 4th (10th) century, and is described by Ibn Ḥawḳal as the chief commercial

[1] I. H. 390, 391. Muk. 262, 273, 274. Yak. iii. 366. A. Y. i. 466, 557; ii. 9, 636, 642. Schuyler, *Turkistan*, i. 70. The name, which should be written Sawrân or Ṣabrân, is frequently misprinted Ṣîrân in the *Ẓafar-Nâmah* of 'Alî of Yazd.

[2] I. H. 393. A. F. 489. Yak. ii. 127. A. Y. i. 275, 279. For the ruins of Jand see Schuyler, *Turkistan*, i. 62.

town of those Moslems who were engaged in trade with the Kharlakhîyah Turks. Muḳaddasî adds that the city was strongly fortified, with a deep ditch, beyond which lay many gardens, and it was extremely populous. The Ṭarâz river ran by one of the four gates of the city, and there was a Friday Mosque in the market-place. According to Ḳazvînî, Ṭarâz was proverbial for the beauty of its men and women, it had a good climate and its lands were extremely fertile. Also of the Turk country and about one hundred miles due east of Ṭarâz lies Barkî or Mîrkî (modern Merkeh), which Muḳaddasî describes as a medium-sized town, but well fortified, having a castle, and a Friday Mosque that had originally been a (Nestorian) Christian church. There was a great guard-house here in the 4th (10th) century, built by 'Amîd-ad-Dawlah Fâiḳ. one of the Buyid Amîrs. According to the same authority, Kûlân lay one march west of Mîrkî towards Ṭarâz; it was a large and strongly fortified village with a Friday Mosque, and was accounted a place of much importance[1].

In conclusion it is to be observed that Abû-l-Fidâ mentions a number of capital cities of the Turks, the exact positions of which it is difficult now to fix. Of these Balâsâghun was the capital of the Khans of Turkistân during the 4th and 5th (10th and 11th) centuries, and is mentioned by Ibn-al-Athîr in his Chronicle. Its exact site is unknown. Abû-l-Fidâ says, vaguely, that it was near Kâshghâr, but beyond the Jaxartes. The ruins of Almâligh, which was the Mongol capital under Jaghatay, the son of Changîz Khân, have been found near the site of Old Kuljah, on the river Ilih; and its position is indicated by 'Alî of Yazd, who also mentions the Irtish river and the Tulâs. But of all these towns no descriptions are given, and like Kâshghâr, Khutan, Yârkand and other places on the borders of China, the notice in our authorities is merely incidental and unfortunately of no import geographically[2].

The countries of the Jaxartes did not produce any great variety of manufactures, and the slave-trade was the chief industry of the merchants who went thither. Muḳaddasî mentions that

[1] I. H. 390, 391. Muk. 263, 274, 275. Kaz. ii. 365. A. F. 497. Schuyler, *Turkistan*, ii. 120.

[2] A. F. 505. A. Y. i. 485, 494; ii. 218, 219.

at Dîzak (Jîsak) in Ushrûsanah they made excellent felts and cloaks. The natural products of Farghânah were gold and silver from the mines, also turquoises; quicksilver, iron, and copper were likewise obtained, also sal-ammoniac, naphtha, and bitumen. The mill-stones of Farghânah were famous, and stone-coal for burning was common here. From the orchards were exported grapes, apples, and nuts, with perfumes made from roses and violets. Shâsh produced fine white cloth, swords and other weapons, with brass and iron work, such as needles, scissors, and pots. Also saddles of the skin of the wild ass were made, with bows and quivers, dyed hides, and prayer-rugs, as well as a kind of collared cloak. The country round produced rice, flax, and cotton. Finally from Ṭarâz, in the Turk country, came goat-skins; and the Turkistân horses and mules were always and especially famous[1].

In regard to the high roads of these provinces, the continuation of the great Khurâsân road, going north from Samarḳand, crossed the Sughd river, and thence reached Zâmîn in Ushrûsanah, where it bifurcated, the left branch to Shâsh and the lower Jaxartes, the right to the upper Jaxartes and Farghânah. From Zâmîn the direct road to Shâsh crossed the Jaxartes at Banâkath; while a second high road from Samarḳand went by Dîzak, and across the desert to Waynkard, beyond which the Jaxartes was crossed to Shutûrkath, where the road from Banâkath to Shâsh was joined. From Shâsh one road went east to Tûnkath, the capital of the Îlâḳ province, and another north to Isbîjâb, where again there was a bifurcation. Westward from Isbîjâb, one high road went to Fârâb (Utrâr) for the crossing of the Jaxartes, and thence also north along its right bank to Ṣabrân. To the right, eastward from Isbîjâb, the other road went to Ṭarâz, and thence to Barkî or Mîrkî, the last Moslem town of Turk lands in the 4th (10th) century, and from this place Ibn Khurdâdbih and Ḳudâmah give the stages across the desert to Upper Nûshanjân on the frontiers of China, which place is probably to be identified with Khutan[2].

The road to Farghânah which, as already said, bifurcated from

[1] I. H. 397, 398. Muk. 325. Kaz. ii. 405.

[2] I. K. 26—29. Kud. 203—206. Ist. 335—337, 343—346. I. H. 398, 399, 403—405. Muk. 341—343. Mustawfî unfortunately gives no routes

the continuation of the Khurâsân road at Zâmîn, went by Sâbâṭ (where the road to Bûnjikath, the capital of Ushrûsanah, turned off) to Khujandah on the Jaxartes. From here, keeping along the south bank of the river, and up stream, Akhsîkath, the capital of Farghânah, at the crossing of the Jaxartes, was reached. The distances from Akhsîkath to the various towns lying to the north of the upper Jaxartes are given by Iṣṭakhrî and Ibn Ḥawḳal; while from the capital of Farghânah eastward the continuation of the high road by Ûsh to Ûzkand is found in Ibn Khurdâdbih and Ḳudâmah. Further Muḳaddasî gives notes of the way from Ûzkand into the Turk country, and ultimately to the frontiers of China. The account is difficult to follow, but, as with Ibn Khurdâdbih and Ḳudâmah, the last stage is Upper Nûshajân, or Barsakhân, the conjectural Khutan[1].

beyond the Oxus. For the route to Khutan and China see the article on the Wall of Gog and Magog by Professor M. J. De Goeje in *Mededeeling der Koninklijke Academie Amsterdam*, for 1888, p. 123. For the route followed by 'Abd-ar-Razzâḳ, the Ambassador of Shâh Rukh, who travelled to China and back between 822 and 825 (1419 and 1422), see the Persian text and translation by E. Quatremère in *Notices et Extraits*, vol. XIV. pt. i. p. 387, also the notes by Sir H. Yule in *Cathay and the Way thither*, pp. cxcix—ccix.

[1] I. K. 29, 30. Kud. 207—209. Ist. 335, 343—347. I. H. 398, 399, 403—406. Muk. 341, 342.

INDEX.

A. Arabic. P. Persian. T. Turkish.

www.ingramcontent.com/pod-product-compliance
Lightning Source LLC
LaVergne TN
LVHW051043220926
848485LV00001BA/1